W0050865

Degenerative Diseases of the Retina

Degenerative Diseases of the Retina

Edited by

Robert E. Anderson

University of Oklahoma Health Sciences Center
Oklahoma City, Oklahoma

Matthew M. LaVail

University of California, San Francisco
San Francisco, California

and

Joe G. Hollyfield

The Cleveland Clinic Foundation
Cleveland, Ohio

Springer Science+Business Media, LLC

Library of Congress Cataloging-in-Publication Data

Degenerative diseases of the retina / edited by Robert E. Anderson, Matthew M. LaVail, and Joe G. Hollyfield.
p. cm.
"Proceedings of the Sixth International Symposium on Retinal Degenerations, held November 4-9, 1994, in Jerusalem, Israel"--T.p. verso.
Includes bibliographical references and index.
ISBN 978-1-4613-5774-2 ISBN 978-1-4615-1897-6 (eBook)
DOI 10.1007/978-1-4615-1897-6
1. Retinal degeneration--Congresses. I. Anderson, Robert E. (Robert Eugene) II. LaVail, Matthew M. III. Hollyfield, Joe G. IV. International Symposium on Retinal Degenerations (6th : 1994 : Jerusalem)
[DNLM: 1. Retinal Degeneration--congresses. 2. Retinitis Pigmentosa--congresses. 3. Photoreceptors--congresses. 4. Apoptosis--congresses. 5. Disease Models, Animal--congresses. WW 270 D317 1995]
RE661.D3D445 1995
617.7'3--dc20
DNLM/DLC
for Library of Congress 95-36351
CIP

Proceedings of the Sixth International Symposium on Retinal Degenerations, held November 4–9, 1994, in Jerusalem, Israel

ISBN 978-1-4613-5774-2

Originally published by Plenum Press, New York in 1995
Softcover reprint of the hardcover 1st edition 1995

10 9 8 7 6 5 4 3 2 1

Elaine Ruth Berman, Ph.D.

Professor of Experimental Ophthalmology,
Hadassah–Hebrew University Medical School, Jerusalem, Israel

This book is dedicated to Elaine Ruth Berman for her pioneering studies on the biochemistry of the retinal pigment epithelium and characterization of the interphotoreceptor matrix, for her lifetime of contributions to the knowledge of the biochemistry of the eye, and for her efforts in organizing the Sixth International Symposium on Retinal Degenerations, held in Jerusalem in 1994.

PREFACE

In 1984, we organized a two-day symposium on retinal degenerations as part of the biennial meeting of the VI International Society for Eye Research, held in Alicante, Spain. The success of this first meeting led to the second held, two years later in Sendai, Japan, organized as a satellite of the VII ISER. We were fortunate that these meetings began at a time of vigorous research activity in the area of retinal degenerations, thanks to the financial support of the Retinitis Pigmentosa Foundation and the strong encouragement of its scientific director, Dr. Alan Laties. Significant advances were made so that every two years scientists were eager to meet to share their findings. The programs included presentations by both basic and clinical researchers with ample time for informal discussions in a relaxed atmosphere. Many investigators met for the first time at these symposia and a number of fruitful collaborations were established.

This book contains the proceedings of the VI International Symposium on Retinal Degenerations held November 6-10, 1994, in Jerusalem. As with the other meetings, some new areas were covered. One session was devoted to apoptosis, an important process involved in cell death in inherited retinal degenerations. Another session was on invertebrate photoreceptors, where numerous mutations have now been identified that lead to altered function or degeneration of the retina. All participants were invited to submit chapters and most complied. We thank them for their contributions.

The meeting received financial support from a number of organizations. We are happy to thank and acknowledge important contributions from The Foundation Fighting Blindness (formerly the RP Foundation Fighting Blindness, Baltimore, Maryland; The National Institutes of Health, Bethesda, Maryland; R.P. Ireland — Fighting Blindness, Dublin; Swiss RP Association, Zurich; and the Louisiana State University Medical Center Foundation, New Orleans, Louisiana.

The untiring efforts of Dr. Elaine R. Berman, the local organizer, guaranteed the success of the meeting. Her choice of the historic Kibbutz Ramat Rachel provided an ideal setting. We thank her and the other organizers, Drs. Saul Merin and Baruch Minke, for helping to make our trip to Jerusalem an unforgettable experience.

Thanks also go to Mrs. Ann Koval for her help with all of the correspondence related to the meeting, to Ms. Pamela Eisenhour for her help in organizing the manuscripts for the publisher and to Ms. Maureen B. Maude for help in editing the text. Finally, we thank Plenum Press for publishing this volume.

Robert E. Anderson
Matthew M. LaVail
Joe G. Hollyfield

THE EDITORS

Robert E. Anderson, M.D.,Ph.D., is Professor of Ophthalmology and Biochemistry and Molecular Biology and Director of the Oklahoma Center for Neurosciences at The University of Oklahoma Health Sciences Center in Oklahoma City, Oklahoma. He received his Ph.D. in Biochemistry (1968) from Texas A&M University and his M.D. from Baylor College of Medicine in 1975. In 1968, he was a postdoctoral fellow at Oak Ridge Associated Universities. At Baylor, he was appointed Assistant Professor in 1969, Associate Professor in 1976, and Professor in 1981. He joined the faculty of the University of Oklahoma in January of 1995.

Dr. Anderson has published extensively in the areas of lipid metabolism in the retina and biochemistry of retinal degenerations. He has edited six books, five on retinal degenerations and one on the biochemistry of the eye.

Dr. Anderson has received the Sam and Bertha Brochstein Award for Outstanding Achievement in Retina Research from the Retina Research Foundation (1980), the Dolly Green Award (1982), a Senior Scientific Investigator Award (1990) from Research to Prevent Blindness, Inc., an Award for Outstanding Contributions to Vision Research from the Alcon Research Institute (1985), and the Marjorie Margolin Prize. He has served on the editorial boards of *Investigative Ophthalmology and Visual Science, Journal of Neuroscience Research,* and *Current Eye Research* and is currently on the editorial boards of *Neurochemistry International* and *Experimental Eye Research*. Dr. Anderson has received grants from the National Institutes of Health, the Retina Research Foundation, the Foundation Fighting Blindness, and Research to Prevent Blindness, Inc. He has been an active participant in the program committees of the Association for Research in Vision and Ophthalmology (ARVO) and is currently a trustee representing the Biochemistry and Molecular Biology section. He has served on the Vision Research Program Committee and Board of Scientific Counselors of the National Eye Institute and the Board of the Basic and Clinical Science Series of The American Academy of Ophthalmology. Dr. Anderson is a past Councilor and current Treasurer for the International Society for Eye Research.

Matthew M. LaVail, Ph.D., is a Professor of Anatomy and Ophthalmology and Vice-Chairman of Anatomy at the University of California at San Francisco School of Medicine. He received his Ph.D. in Anatomy (1969) from the University of Texas Medical Branch in Galveston and was subsequently a postdoctoral fellow at Harvard Medical School. Dr. LaVail was appointed Assistant Professor of Neurology-Neuropathology at Harvard Medical School in 1973. In 1976, he moved to UCSF, where he was appointed Associate Professor of Anatomy. He was appointed to his current position in 1982, and in 1988 he also became director of the Retinitis Pigmentosa Research Center at UCSF.

Dr. LaVail has published extensively in the research areas of photoreceptor-retinal pigment epithelial cell interactions, retinal development, circadian events in the retina, genetics of pigmentation and ocular abnormalities, inherited retinal degenerations, and light-induced retinal degeneration. He is the author of more than 90 research publications and has edited six books on inherited and environmentally induced retinal degenerations.

Dr. LaVail has received the Fight for Sight Citation (1976), the Sundial Award from the Retina Foundation (1976), the Friedenwald Award from the Association for Research in Vision and Ophthalmology (ARVO; 1981), a MERIT Award from the National Eye Institute (1989), and an Award for Outstanding Contributions to Vision Research from the Alcon Research Institute (1990), and the Award of Merit from the Retina Research Foundation (1990). He has served on the editorial board of *Investigative Ophthalmology and Visual Science* and is currently on the editorial board of *Experimental Eye Research*. Dr. LaVail has been an active participant in the program committees of ARVO and the International Society for Eye Research (ISER), and he is currently a Trustee of ARVO.

Joe G. Hollyfield, Ph.D., is a Professor of Ophthalmology and Neuroscience at Baylor College of Medicine, Houston, Texas. He received his Ph.D. in Zoology in 1966 from the University of Texas at Austin and was a postdoctoral fellow at the Hubrecht Laboratory in Utrecht, The Netherlands. He was appointed Assistant Professor of Anatomy assigned to Ophthalmology at Columbia University College of Physicians and Surgeons in New York City in 1969, and was promoted to Associate Professor in 1975. From 1977 to 1995 he was a member of the Cullen Eye Institute, Baylor College of Medicine, where he was promoted to Professor in 1979. He was Director of the Retinitis Pigmentosa Research Center in the Cullen Eye Institute from 1978 to 1995. He is currently the Director of Opthalmic Research at the Cleveland Clinic Foundation, Cleveland, Ohio.

Dr. Hollyfield has published over 100 papers in the area of cell and developmental biology of the retina and retinal pigment epithelium in both normal and retinal degenerative tissue. He has edited six books, five on retinal degenerations and one on the structure of the eye.

Dr. Hollyfield has received the Marjorie W. Margolin Prize (1981, 1994) and the Sam and Bertha Brochstein Award (1985) from the Retina Research Foundation, the Olga Keith Wiess Distinguished Scholars' Award (1981), a Senior Scientific Investigator Award (1988, 1994) from Research to Prevent Blindness, Inc., an award for Outstanding Contributions to Vision Research from the Alcon Research Institute (1987), the Distinguished Alumnus Award (1991) from Hendrix College, Conway, Arkansas, and the Endre A. Balazs Prize (1994) from the International Society for Eye Research (ISER). He has previously served on the editorial boards of *Vision Research and Survey of Ophthalmology*. He is currently Editor-in-Chief of *Experimental Eye Research*. He has received grants from the National Institutes of Health, The Retina Research Foundation, Fight for Sight, Inc., The Foundation Fighting Blindness, and Research to Prevent Blindness, Inc. Dr. Hollyfield has been active in the Association for Research in Vision and Ophthalmology (ARVO) serving as a member of the Program Committee, Trustee and President. He is also the Immediate Past President and former Secretary of ISER.

CONTENTS

I. Apoptosis

1. Apoptosis in Retinitis Pigmentosa 1
   Zong-Yi Li and Ann H. Milam

2. Oxidative Damage and Responses in Retinal Nuclei Arising from Intense Light Exposure 9
   D. T. Organisciak, R. K. Kutty, M. Leffak, P. Wong, S. Messing, B.Wiggert, R. M. Darrow, and G. J. Chader

3. Light-Induced Apoptosis in the Rat Retina *in Vivo*: Morphological Features, Threshold and Time Course 19
   Charlotte E. Remé, Michael Weller Piotr Szczesny, Kurt Munz, Farhad Hafezi, Jörg-Joachim Reinboth, and Matthias Clausen

4. Inhibitory Effects of Cycloheximide and Flunarizine on Light-Induced Apoptosis of Photoreceptor Cells 27
   Suhui Li, Cheng-Jong Chang, Andrew S. Abler, and Mark O. M. Tso

5. Apoptosis of Photoreceptors and Lens Fiber Cells with Cataract and Multiple Tumor Formation in the Eyes of Transgenic Mice Lacking the P53 Gene and Expressing the HPV 16 E7 Gene under the Control of the Irbp Promoter 39
   David S. Papermaster, Kim Howes, Nancy Ransom, and Jolene J. Windle

II. Role of Neurotrophic Factors in Retinal Degenerations

6. Function, Age-Related Expression and Molecular Characterization of PEDF, a Neurotrophic Serpin Secreted by Human RPE Cells 51
   Joyce Tombran-Tink

7. Nitric Oxide in the Retina: Potential Involvement in Retinal Degeneration and Its Control by Growth Factors and Cytokines 61
   O. Goureau, F. Becquet, and Y. Courtois

8. *In Vitro* Expression of Epidermal Growth Factor Receptor by Human Retinal Pigment Epithelial Cells 69
Yusuf K. Durlu and Makoto Tamai

III. Vertebrate Models of Retinal Degenerations

9. New Retinal Degenerations in the Mouse 77
Thomas H. Roderick, Bo Chang, Norman L. Hawes, and John R. Heckenlively

10. Chorioretinal Interface: Age-Related Changes in Rodent Retina 87
P. M. Leuenberger, Y. Gambazzi, and E. Rungger-Brändle

11. Fractionation of Interphotoreceptor Matrix Metalloproteinases 95
James J. Plantner, Timothy A. Quinn and George J. Dwyer

12. Peptides from Rhodopsin Induce Experimental Autoimmune Uveoretinitis in Lewis Rats 105
Grazyna Adamus, Henry Ortega, Lundy Campbell, Anatol Arendt, and Paul A. Hargrave

13. Retinal Degeneration in Rats Induced by Vitamin E Deficiency 111
E. El-Hifnawi, H. -J. Hettlich, and C. Falk

14. Retinal Pigment Epithelial Cells Cultured from RCS Rats Express an Increased Membrane Conductance for Calcium Compared to Normal Rats 119
O. Strauß and M. Wienrich

15. Effect of Sugars on Photoreceptor Outer Segment Assembly 129
Monica M. Stiemke and Joe G. Hollyfield

16. Regulatory Influences on the Glycosylation of Rhodopsin by Human and Bovine Retinas 139
Edward L. Kean, Jermin Ju, and Naiqian Niu

17. The Dolichol Pathway in the Retinal Pigment Epithelium of the Embryonic Chick 149
Edward L. Kean and Naiqian Niu

18. Morphological and Biochemical Studies of the Retinal Degeneration in the Vitiligo Mouse: A Model with Perturbed Retinoid Metabolism 155
Sylvia B. Smith and Barbara N. Wiggert

19. Immunological Aspects of Retinal Transplantation in Retinal Degeneration Rodents 163
Luke Qi Jiang, Duco Hamasaki, Jessica Zuletta, and Marianela Jorquera

20. Iron and Hereditary Degeneration of the Retina 177
R. N. Etingof, N. D. Shushakova, and M. G. Yefimova

21. Receptor Degeneration Is a Normal Part of Retinal Development 187
Juliani Maslim, Rupert Egensperger, Horstmar Holländer, Martin Humphrey, and Jonathan Stone

22. Development of Opsin and Synapses in Monkey Photoreceptors 195
A. Hendrickson, E. Dorn, K. Bumsted, and A. Szel

23. The Nature of Newly Formed Capillaries in Experimental Naphthalene-Induced Retinal Degeneration in Rabbit 203
Nicola Orzalesi, Luca Migliavacca, and Stefano Miglior

24. Retinal Cell Responses to Argon Laser Photocoagulation 209
Martin F. Humphrey, Yi Chu, Claudia Sharp, Krishna Mann, and Piroska Rakoczy

IV. Invertebrate Models of Retinal Degenerations

25. *Drosophila* as a Model for Photoreceptor Dystrophies and Cell Death 217
William S. Stark, David Hunnius, Jennifer Mertz, and De-Mao Chen

26. Abnormal Ca^{2+} Mobilization and Excessive Photopigment Phosphorylation Lead to Photoreceptor Degeneration in *Drosophila* Mutants 227
Baruch Minke and Zvi Selinger

27. The Role of Dominant Rhodopsin Mutations in *Drosophila* Retinal Degeneration 235
Phani Kurada, Timothy D. Tonini, and Joseph E. O'Tousa

28. The Role of the *Retinal Degeneration B* Protein in the *Drosophila* Visual System: Function of *Drosophila* rdgB Protein in Photoreceptors 243
D. R. Hyde, S. Milligan, and T. S. Vihtelic

29. *Drosophila* Visual Transduction, a Model System for Human EYE Disease? Retinal Degenerations 255
Jude Fitzgibbon and David Hunt

30. Characterization of Vertebrate Homologs of *Drosophila* Photoreceptor Proteins . . 263
Paulo A. Ferreira and William L. Pak

31. Retinal Pathology in Retinitis Pigmentosa: Considerations for Therapy 275
Ann H. Milam and Zong-Yi Li

V. Molecular, Cellular and Clinical Studies

32. Genotype-Phenotype Correlation in Autosomal Dominant Retinal Degeneration with Mutations in the Peripherin/*RDS* Gene 285
M. Nakazawa, E. Kikawa, Y. Chida, K. Kamio, Y. Wada, T. Shiono, and M. Tamai

33. Genetic Studies in Autosomal Recessive Forms of Retinitis Pigmentosa 293
A. F. Wright, D. C. Mansfield, E. A. Bruford, P. W. Teague, K. L. Thomson, R. Riise, Jay. M, M. A. Patton, S. Jeffery, A. Schinzel, N. Tommerup, and M. Fossarello

34. Clinical Features of Autosomal Dominant Retinitis Pigmentosa Associated with the Ser 186TRP Mutation of Rhodopsin 303
K. Rüther, C. L. v. Ballestrem, A. Müller, S. Kremmer, A. Eckstein, E. Apfelstedt-Sylla, A. Gal, and E. Zrenner

35. Mutations in the Gene for the *B*-Subunit of Rod Photoreceptor CGMP-Specific Phosphodiesterase (PDEB) in Patients with Retinal Dystrophies and Dysfunctions 313
A. Veske, U. Orth, K. Rüther, E. Zrenner, T. Rosenberg, W. Baehr, and A. Gal

36. Molecular Genetic Studies of Retinal Dystrophies Principally Affecting the Macula 323
Kevin Evans, Cheryl Y. Gregory, Sujeewa Wijesuriya, Marcelle Jay, Amresh Chopdar, and Shomi S. Bhattacharya

37. Molecular Analysis of the Human GAR1 Gene: A Candidate Gene for Retinal Degeneration 331
Michelle D. Ardell, Ajay Makhija, and Steven J. Pittler

38. Guanylate Cyclase-Activating Protein (GCAP): A Novel Ca^{2+}-Binding Protein in Vertebrate Photoreceptors 339
Wolfgang Baehr, Iswari Subbaraya, Wojciech A. Gorczyca, and Krzysztof Palczewski

39. Abnormal Cone Receptor Activity in Patients with Hereditary Degeneration 349
Donald C. Hood and David G. Birch

40. Abnormal Rod Photoreceptor Function in Retinitis Pigmentosa 359
David G. Birch and Donald C. Hood

41. Scotopic Threshold Responses and Rod Intensity-Response Functions as Sensitive Indicators of the Carrier Status in X-Linked Recessive Retinitis Pigmentosa 371
A. Iannaccone, E. M. Vingolo, R. Forte, P. Tanzilli, B. Grammatico, C. De Bernardo, E. Rispoli, G. Del Porto, and M. R. Pannarale

42. ERG Findings in Two Patients with Autosomal Dominant Congenital Stationary Night Blindness and HIS258ASN Mutation of the β-Subunit of Rod Photoreceptor CGMP-Specific Phosphodiesterase 377
T. Rosenberg, A. Gal, and S. E. Simonsen

43. Docosahexaenoic Acid Abnormalities in Red Blood Cells of Patients with Retinitis Pigmentosa 385
Dennis R. Hoffman, Ricardo Uauy, and David G. Birch

44. Ligand-Binding Properties of Recombinant Human IRBP 395
J. M. Nickerson, V. Mody, C. DeGuzman, K. A. Heron, K. Marciante, J. Boatright, J. S. Si, and Z. Y. Lin

45. Cataractogenesis in Retinitis Pigmentosa: Spectroscopic Fluorescence Analysis of Aqueous Humor Composition 403
Enzo M. Vingolo, Andrea Bellelli, Monica Santori, Luigi Pannarale, Renato Forte, Alessandro Iannaccone, and Roberto Grenga

Index 409

1

APOPTOSIS IN RETINITIS PIGMENTOSA

Zong-Yi Li and Ann H. Milam

Department of Ophthalmology RJ-10
University of Washington
Seattle, Washington 98195

SUMMARY

Retinitis pigmentosa (RP) is a group of inherited retinal diseases characterized by progressive death of photoreceptors. Photoreceptor death in several rodent models of human RP was recently demonstrated to involve apoptosis. We sought to determine if photoreceptor death occurs by apoptosis in retinas from patients with RP. Donor retinas were screened using two labeling methods for *in situ* demonstration of fragmented nuclear DNA, a hallmark of apoptosis. To improve the sampling, portions of the retinas were processed as flat mounts. We examined 15 retinas from patients (age 24 to 87 yrs) who had different genetic forms of RP and retained photoreceptors at the time of death, as well as seven retinas from normal donors (age 19 to 68 yrs). Scattered labeled nuclei were present in the retinas from two RP donors who were 24 and 29 yrs old at the time of death. Labeled nuclei were not found in the normal retinas or in the retinas from the older RP donors. This study provides initial evidence that photoreceptor death in human RP occurs by apoptosis, as found previously in the rodent inherited retinal dystrophies. Apoptosis is a relatively rapid process in other types of cells, and it may be difficult to demonstrate in human RP retinas, where photoreceptor death is protracted over several decades. In addition, the likelihood of documenting this rather brief event in the retinas of older RP donors is small because they contain greatly reduced numbers of photoreceptors. Photoreceptor dysfunction results from several different gene defects in animal models and RP patients, but photoreceptor death in each case appears to occur by the final common pathway of apoptosis. Agents to inhibit apoptosis warrant evaluation as new therapy for inherited photoreceptor diseases.

INTRODUCTION

Elucidation of the mechanisms of cell death is an important and rapidly progressing field of research [1]. In necrosis, as caused by a number of conditions (ischemia, complement attack, metabolic poisons, trauma and others), cells lose surface membrane ion pumping activities, swell and lyse. Necrosis usually involves groups of cells and the resultant cellular

Degenerative Diseases of the Retina, Edited by Robert E. Anderson et al.
Plenum Press, New York, 1995

debris is ingested by inflammatory cells. A cardinal feature of apoptosis, a form of programmed or physiological cell death, is the requirement for synthesis of new RNA and protein. Like necrosis, apoptosis can be triggered by a variety of causes, but typically involves single or small groups of cells within a population of unaffected cells of the same type. The characteristic cytologic features of apoptosis are loss of intercellular junctions, condensation and fragmentation of the nucleus and cytoplasm, and formation of surface blebs (apoptotic bodies), which are ingested by neighboring cells with little or no inflammation. One hallmark of apoptosis is internucleosomal fragmentation of DNA, which can be demonstrated by agarose gel electrophoresis and the terminal dUTP nick end labeling (TUNEL) technique [2]. Both methods were used recently to demonstrate apoptotic photoreceptors in retinas of Royal College of Surgeons rats [3; 4], mice with mutations in the *retinal degeneration* and *retinal degeneration slow/peripherin* genes [3; 5-7], and transgenic mice with a mutant *rhodopsin* gene that causes retinitis pigmentosa (RP) in humans [5; 6]. The various mutant genes may lead to photoreceptor dysfunction by different disease mechanisms but in each case, death of photoreceptor cells appears to occur by the final common pathway of apoptosis.

We sought to determine if photoreceptor death occurs by apoptosis in human RP, and report the results of screening donor RP retinas by two techniques to demonstrate apoptotic cells *in situ*.

MATERIALS AND METHODS

The study was conducted on donor retinas from 15 patients with RP (Table 1). Control retinas with similar *post mortem* times (2.5 to 25.5 hrs) were obtained from normal donors of similar ages (19, 25, 30 [two], 38, 66, and 68 yrs). The eyes were fixed with 4% paraformaldehyde and 0.5 to 2.5% glutaraldehyde in 0.13 M phosphate buffer. Five of the RP donors had the simplex or isolate form; three had multiplex RP, two had X-linked RP, two had autosomal dominant RP of unknown genetic type, one had autosomal dominant RP caused by the threonine-17-methionine rhodopsin mutation [8], one was a symptomatic carrier of X-linked RP, and one had Usher Syndrome type II.

Two to four samples per retina, each approximately 5 mm in diameter, were dissected from central and peripheral regions that retained at least one layer of photoreceptors. The retinal samples were processed as 12 μm cryosections and as flat mounts using the TUNEL [2] and *in situ* nick translation labeling methods [9; 10]. Our data on cell death in transgenic mouse retinas indicate that the *in situ* nick translation method is more sensitive than the TUNEL method on flat mounted retinas [Al-Ubaidi, Naash, Li and Milam, unpublished]. To improve penetration of reagents, the retinal samples to be flat mounted were frozen/thawed, treated with ethanol (10, 25, 40, 25 and 10%; 10 min each) and 1% Triton X-100 in 0.01M phosphate buffered saline (30 min at room temperature [RT]). All steps were performed in a humidity chamber with agitaton. The incubation steps were modified for the retinal flat mounts from the methods published for tissue sections [2; 5; 9; 10]. For the TUNEL method, the retinal specimens were immersed in TdT buffer (140 mM sodium cacodylate, 1mM cobalt chloride in 30 mM Tris-base, pH 7.2) for 10 min (2X), and incubated in a solution of 0.3 enzyme units (eu)/μl terminal deoxynucleotidyl transferase (TdT) (Gibco BRL Technologies, Inc., Gaithersburg, MD) and 0.01 mM biotinylated-11-dUTP (Sigma, St. Louis, MO) in TdT buffer for 2-3 hrs at 37°C. The reaction was stopped by transferring the retinas to TB buffer (300 mM sodium chloride, 30 mM sodium citrate) for 15 min at RT. For *in situ* nick translation, the retinas were immersed in NT buffer (5mM magnesium chloride, 10 μg/ml BSA, 10 mM 2-mercaptoethanol in 50 mM Tris-HCl, pH 7.5) for 10 min (2X) and incubated

Table 1. Characteristics of donor RP retinas and results of *in situ* DNA end labeling

RPF #	Age/Sex	*Post mortem* time (hrs)	Diagnosis	Nuclear label
310	24 M	11.5	Usher Syndrome II	+
342	29 M	16.5	Simplex	+
320	31 F	10.0	Simplex	–
114	39 M	5.5	X-Linked (XL)	–
311	44 M	12.0	Multiplex	–
215	46 M	1.0	XL	–
266	54 F	3.2	Simplex	–
231	56 M	3.3	Simplex	–
207	65 M	11.0	Autosomal Dominant	–
303	68 M	2.2	Autosomal Dominant	–
316	68 M	8.5	Autosomal Dominant*	–
271	72 F	4.0	Symptomatic XL Carrier	–
340	73 F	9.0	Multiplex	–
184	76M	3.1	Multiplex	–
371	87M	5.0	Simplex	–

*Threonine-17-methionine rhodopsin mutation [8].

in 0.2 eu/μl DNA polymerase I (Sigma) and 0.01 mM biotinylated-11-dUTP and 0.01 mM dATP, dGTP, dCTP (Sigma) in NT buffer for 2-3 hrs at RT or 37°C. The reaction was terminated by transferring the retinas to PBS at 4°C. For both methods, the specimens were rinsed in PBS for 10 min (2X). The endogenous peroxidase was quenched with 0.3% H_2O_2 in 10% methanol for 30 min. The retinas were rinsed in PBS and reacted with ExtrAvidin-peroxidase (Sigma) diluted 1:200 in distilled water for 2-3 hrs at 37°C. The retinas were rinsed with PBS and immersed in diaminobenzidine (DAB, Sigma) solution (0.5mg/ml in 0.05M Tris-HCl, pH 7.6) for 15 min at RT, and incubated in the DAB solution with 0.05 μl/ml H_2O_2 for 20 min at RT. To avoid confusion with the brown color of melanin granules, some samples were processed using the red substrate, VIP (Vector Labs, Inc., Burlingame, CA). The samples were rinsed in PBS and mounted photoreceptor side up in glycerol/fix. As positive controls, samples of normal and RP retinas were pretreated with 1μg /ml DNase I (Sigma) in TM buffer (10 mM magnesium sulphate in 50 mM Tris-HCl, pH 7.4) for 15 min at 37°C before the labeling procedure [2; 5]. As negative controls, samples were processed by the same procedure with omission of the TdT for the TUNEL or DNA polymerase I for the nick translation method.

Small retinal samples which contained DAB labeled nuclei were re-embedded in LR-White resin for electron microscopic immunocytochemistry [11], which was performed using antibodies against rhodopsin (monoclonal antibodies 4D2 (1:20) and 1D4 (1:20) from Dr. R. Molday); cone opsins (polyclonal antibodies against red/green (1:100) and blue (1:10) cone opsins from Drs. C. and K. Lerea); arrestin (1:200; polyclonal antibody from Dr. H. Shichi); glial fibrillary acidic protein (1:200; GFAP; polyclonal antibody from DAKO Corp., Carpinteria, Calif.); cellular retinaldehyde-binding protein (1:200; CRALBP; polyclonal antibody from Dr. J. Saari); and HAM-56 (1:200; monoclonal antibody from DAKO Corp.), which is specific for human monocytic macrophages [12].

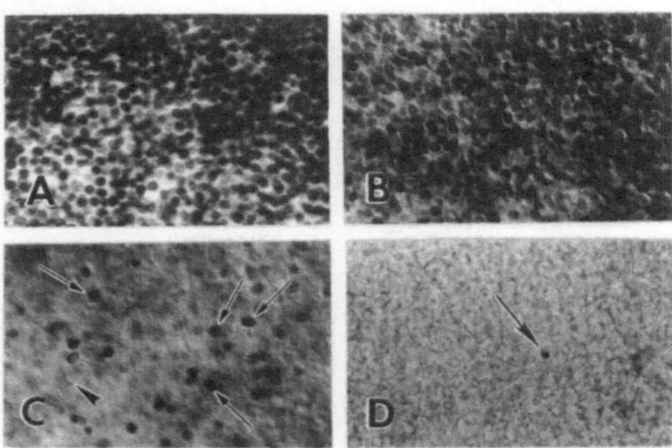

Figure 1. (A - C). Positive control. A normal retina was treated with 1μg/ml DNase prior to *in situ* nick translation. The labeled nuclei were photographed by focussing in different retinal layers. (A) Photoreceptor layer. (B) Inner nuclear layer. (C) Ganglion cell layer, arrows indicate ganglion cell nuclei. The red blood cells are not labeled (arrowheads). (D). A labeled photoreceptor nucleus (arrow) in retina from a 29 yr old RP patient (RPF# 342) is surrounded by unlabeled cells. The labeled nucleus is shrunken and condensed in comparison with the normal photoreceptor nuclei, as shown in Figure 1A. Nomarski phase interference, X 210.

RESULTS

Controls

Normal and RP retinas pretreated with DNase had intense labeling of all nuclei with both methods. The labeled nuclei in the photoreceptor, inner nuclear and ganglion cell layers could be identified by focussing at different levels (Fig. 1). The negative controls, in which TdT or DNA polymerase was omitted, showed no nuclear labeling.

Experimental Samples

Rods and cones in the RP retinas were reduced in number, usually to 1-2 rows of nuclei, and had short outer and inner segments. No labeled nuclei could be identified in the sections of the RP and normal retinas using either method. The flat mounted retinal preparations provided much better sampling and two of the RP retinas (donor ages 24 and 29 yr) contained scattered labeled nuclei in the photoreceptor layer (Fig. 1D, 2A, B, C; Table 1). The labeled nuclei were rare, on the order of 1 to 5 per 10 mm^2 of retina. They were scattered in the outermost layer of the retinas and occurred singly, in pairs or in small clusters. Fig. 2A illustrates the largest number of labeled nuclei found in a single field. Nuclei of cells of the retinal pigment epithelium (RPE) and inner retinal neurons in the same areas were unlabeled (Fig. 2A). No labeled nuclei could be identified in the flat mounted samples from the normal retinas or from the other RP retinas.

Electron microscopy revealed that the labeled nuclei were condensed and shrunken, and belonged to small, dense cells in the photoreceptor layer that lacked inner and outer segments, cytoplasmic organelles, and intercellular junctions (Fig. 2B). The dense cells were unreactive with anti-rhodopsin, -cone opsins or -arrestin, although neighboring photoreceptors with inner and outer segments were well labeled with these antibodies. We looked for but were unable to find labeled nuclei in rods that showed delocalization of rhodopsin labeling to the inner segment surface membrane, as this abnormality has been considered to

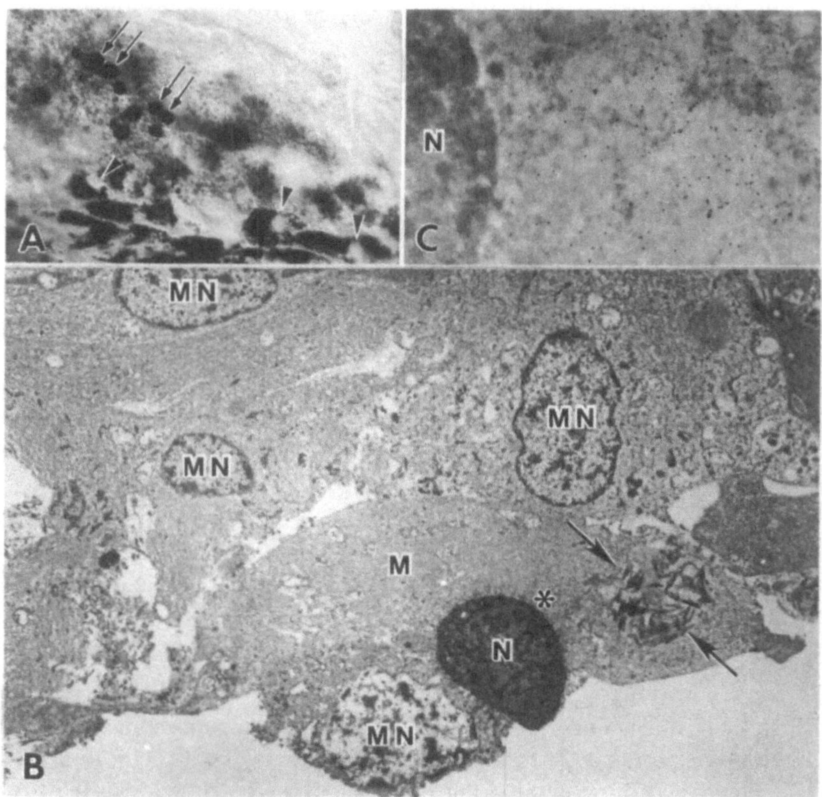

Figure 2. (A) DNA end-labeled nuclei (arrows) in flat mounted donor retina from 29 yr old RP patient (RPF#342). Note unlabeled nuclei (arrowheads) in adjacent RPE cells which contain dark melanin granules and have migrated into the retina. X330. (B) Electron microscopy of labeled nucleus (N) in small, dense cell in the photoreceptor layer of this RP retina. Note densification of the cytoplasm and absence of cytoplasmic organelles. The adjacent Müller cell processes (M) are hypertrophic and contain membranous cytoplasmic inclusions (arrows). Area (*) is shown at higher magnification in Figure 1C. MN, Müller cell nucleus. X4,140. (C) Higher magnification view of area (*) in Figure 2B processed by the immunogold procedure for demonstration of GFAP, which is localized to the process of a reactive Müller cell. N, labeled nucleus. X17,130.

represent a step in rod cell degeneration [3; 8; 13; 14]. The Müller cells adjacent to the labeled cells had hypertrophic processes with CRALBP (+) cytoplasm and prominent GFAP (+) filaments (Fig. 2C). The expression of GFAP in reactive Müller cells is consistent with ongoing photoreceptor death in the RP retinas [8; 11]. Some of the Müller cells contained unusual membranous inclusions in the cytoplasm (Fig. 2B), which were unreactive with all of the immunomarkers tested. We have observed similar Müller cell inclusions in other RP retinas [Milam, unpublished], and their occurrence adjacent to TUNEL-positive cells suggests that they may represent phagocytosed apoptotic debris.

Rods and cones in one retina (RPF#342) were organized in rosettes with phagocytes in the lumina, as found in a previously reported RP retina (RPF#184; [11]). Some of the phagocytes contained melanin granules and TUNEL-positive material, as well as membranous inclusions resembling those in the Müller cell processes. The phagocytes were weakly immunoreactive for CRALBP but were negative for GFAP (a marker for reactive Müller cells) and with HAM-56 (a marker for monocytic macrophages), suggesting that they were RPE cells.

DISCUSSION

The presence of scattered apoptotic nuclei in cells of the photoreceptor layer is consistent with the progressive death of rods and cones that occurs in RP. The finding of apoptotic cells in retinas from patients with different genetic forms of RP agrees with reports that photoreceptor death due to different gene mutations in rodents occurs by the final common pathway of apoptosis. The labeled nuclei were found only in the retinas from the two RP patients who had died in the third decade of life, and were not found in the normal retinas or those from the older RP patients. A higher than normal rate of photoreceptor death in the retinas from the two youngest donors may correlate with the loss of vision experienced by many RP patients in the first three decades of life. Unfortunately, information was not available on the visual histories of our two youngest RP donors.

However, the number of labeled nuclei was surprisingly low, in spite of use of flat mounted retinas to increase sampling and the *in situ* nick translation method to increase sensitivity. Like previous studies [3-6], we have found numerous labeled nuclei in retinas from young mice with genetically-caused photoreceptor degeneration [Al-Ubaidi, Naash, Li and Milam, unpublished]. Why were so few nuclei labeled in the human RP retinas? There are several possibilities.

- First, TUNEL labeling is observed in apoptotic cells in non-retinal tissues for only a brief time period (3 hr in liver cells [15]) and may be difficult to document in RP retinas because photoreceptor death extends over decades in the life of the patient. We found only a few labeled nuclei (1 to 5 per 10 mm^2) in the retinas from the two youngest RP patients. Assuming an apoptosis period of 3 hr, the calculated rate of apoptosis in the RP retinas would be on the order of 8 to 40 photoreceptors per 10 mm^2 per day. No measurements are available on the time period in which an apoptotic photoreceptor is positive by TUNEL labeling, but if this value were shorter than 3 hr, this calculated rate of apoptosis would represent an underestimate. It is also possible that greater numbers of apoptotic cells would be found in still younger RP patients, were such retinas available for comparable analysis. A recent study [16] addressed the question of changes in the rod cell population in normal human retinas with aging. It was calculated that a slow, steady loss of a small number of rods at a given time could result in a major loss in the cells over decades. For example, death of only 20 rods per 10 mm^2 per day would account for the observed 30% loss of rods between age 34 and 90 yrs [16]. It is possible that death of relatively small numbers of photoreceptors at a given time, but in excess of that found with normal aging, leads to the marked loss of photoreceptors in RP.
- Second, the likelihood of documenting this rather brief event is poor in the retinas from older RP patients because they retain many fewer rods and cones. We could not find labeled nuclei in the retinas of the next youngest (31 and 39 yr old) RP donors in our series, probably because they each had more advanced disease, retaining only a monolayer of cones in the macula and only scattered photoreceptors elsewhere in the retina. It will be important to examine additional human RP retinas for apoptosis, although well preserved retinas from young RP donors that retain significant numbers of photoreceptors are difficult to obtain.

- Third, *post mortem* autolysis is a problem with the human retinas not encountered with retinas of experimental animals. We used donor retinas with the shortest post *mortem times* available, but this problem, which we have not found addressed in the apoptosis literature, may affect the labeling results. However, we feel that the importance of documenting the mechanism of cell death in the human retinas outweighs the inherent technical problems.

Several features of apoptosis in the RP retinas resemble this process in other tissues. The labeled nuclei were condensed, although we did not find examples of peripheral chromatin margination or nuclear fragmentation. The dying cells, which were shrunken and dense, lacked cytoplasmic organelles and intercellular junctions, and were negative for certain proteins (opsins, arrestin) characteristic of differentiated rods and cones. They were also negative for macrophage and Müller glia markers, suggesting but not proving that the dying cells in the photoreceptor layer were photoreceptors. We were unable to find labeled nuclei in rods that showed delocalization of immunoreactive rhodopsin to the inner segment surface membrane, suggesting that these cells, while degenerate, had not yet entered the irreversible apoptotic pathway that includes DNA fragmentation.

No inflammatory infiltrate was observed around the apoptotic cells but there was evidence of phagocytosis of cellular debris by adjacent retinal cells, including Müller cells and phagocytes probably derived from the RPE. Recent reports indicate that a similar phagocytic function in dystrophic rodent retinas is performed by cone photoreceptors [3] and by wandering phagocytes of unknown origin [4; 5]. Our finding of membranous debris in the Müller cell processes corroborates early observations of phagocytosis by Müller cells in RP retinas [17]. Phagocytosis of apoptotic residue by Müller cells has also been demonstrated in developing human retina [18]. The presence of apoptotic cells in RP would be consistent with the up regulation in RP retinas of clusterin [19], which also occurs in brain neurodegenerations involving apoptosis. DNA laddering by agarose gel electrophoresis has been demonstrated in dystrophic rodent retinas [4-7], but the small number of apoptotic nuclei identified in the RP retinas in our study probably would not yield sufficient amounts of fragmented DNA for detection by this method.

The cascade of events in apoptosis and mechanisms of physiologic control of this process are topics of active investigation. Efforts are underway to develop methods or agents to regulate apoptosis in various neoplastic and inflammatory diseases. It is also desirable to prevent premature neuronal cell death in inherited diseases of the central nervous system and retina. Our results provide initial evidence that photoreceptor death in human RP occurs by apoptosis and suggest that agents to inhibit apoptosis warrant evaluation as new therapy for this disease.

ACKNOWLEDGMENTS

The authors thank Drs. F. Wong, D. S. Papermaster, M. O. M. Tso, V. Sarthy, C. Curcio, J. Saari, and K. Palczewski for helpful advice; Mr. D. Possin, Ms. J. Chang, and Ms. I. Klock for technical assistance; Mr. C. Stephens and Mr. R. Jones for photographic help; and Drs. R. Molday, C. and K. Lerea, H. Schichi, and J. Saari for gifts of antibodies.

REFERENCES

1. Schwartzman, R. A. and Cidlowski, J. A., 1993, Apoptosis: the biochemistry and molecular biology of programmed cell death, *Endocr. Rev.* 14:133-151.

2. Gavrieli, Y., Sherman, Y. and Ben-Sasson, S., 1992, Identification of programmed cell death *in situ* via specific labeling of nuclear DNA fragmentation, *J. Cell Biol.* 119:493-501.
3. Papermaster, D. S. and Nir, I., 1994, Apoptosis in inherited retinal degenerations, In: Apoptosis, Mihich, E. and Schimke, R. T., eds., New York: Plenum Publishing Corp., pp 15-30.
4. Tso, M. O. M., Zhang, C., Abler, A. S., Chang, C.-J., Wong, F., Chang, G.-Q. and Lam, T. T., 1994, Apoptosis leads to photoreceptor degeneration in inherited retinal dystrophy of RCS rats, *Invest. Ophthalmol. Vis. Sci.* 35:2693-2699.
5. Chang, G.-Q., Hao, Y. and Wong, F., 1993, Apoptosis: final common pathway of photoreceptor death in *rd, rds*, and rhodopsin mutant mice, *Neuron*. 11:595-605.
6. Portera-Cailliau, C., Sung, C.-H., Nathans, J. and Adler, R., 1994, Apoptotic photoreceptor cell death in mouse models of retinitis pigmentosa, *Proc. Natl. Acad. Sci., USA*. 91:974-978.
7. Lolley, R. N., Rong, H. and Craft, C. M., 1994, Linkage of photoreceptor degeneration by apoptosis with inherited defect in phototransduction, *Invest. Ophthalmol. Vis. Sci.* 35:358-362.
8. Li, Z.-Y., Jacobson, S. G. and Milam, A. H., 1994, Autosomal dominant retinitis pigmentosa caused by the threonine-17-methionine rhodopsin mutation: retinal histopathology and immunocytochemistry, *Exp. Eye Res.* 58:397-408.
9. Dawson, B. S. and Lough, J., 1988, Immunocytochemical localization of transient DNA strand breaks in differentiation myotubes using *in situ* nick translation, *Devel. Biol.* 127:362-376.
10. Wijsman, J. H., Jonker, R. R., Keijzer, R., van de Velde, C. J. H., Cornelisse, C. J. and van Dierendonck, J. H., 1993, A new method to detect apoptosis in paraffin sections. *In situ* end-labeling of fragmented DNA, *J. Histochem. Cytochem.* 41:7-12.
11. Milam, A. H. and Jacobson, S. G., 1990, Photoreceptor rosettes with blue cone opsin immunoreactivity in retinitis pigmentosa, *Ophthalmol.* 97:1620-1631.
12. Gown, A. M., Tsukada, T. and Ross, R., 1986, Human atherosclerosis II. Immunocytochemical analysis of the cellular composition of human atherosclerotic lesions, *Am. J. Path.* 125:191-207.
13. Roof, D. J., Adamian, M. and Hayes, A., 1994, Rhodopsin accumulation at abnormal sites in retinas of mice with a human P23H rhodopsin transgene, *Invest. Ophthalmol. Vis. Sci.* 35:4049-4062.
14. Sung, C.-H., Makino, C., Baylor, D. and Nathans, J., 1994, A rhodopsin gene mutation responsible for autosomal dominant retinitis pigmentosa results in a protein that is defective in localization to the photoreceptor outer segment, *J. Neurosci.* 14:5818-5833.
15. Bursch, W., Paffe, S., Putz, B., Barthel, G. and Schulte-Hermann, R., 1990, Determination of the length of the histological stages of apoptosis in normal liver and in altered hepatic foci of rats, *Carcinogenesis*. 11:847-853.
16. Curcio, C. A., Millican, C. L., Allen, K. A. and Kalina, R. E., 1993, Aging of the human photoreceptor mosaic: evidence for selective vulnerability of rods in central retina, *Invest. Ophthalmol. Vis. Sci.* 34:3278-3296.
17. Friedenwald, J. S. and Chan, E., 1932, Pathogenesis of retinitis pigmentosa with a note on the phagocytic activity of Müller's fibers, *Arch. Ophthalmol.* 8:173-181.
18. Penfold, P. L. and Provis, J. M., 1986, Cell death in the development of the human retina: phagocytosis of pyknotic and apoptotic bodies by retinal cells, *Graefe's Arch. Clin. Exp. Ophthalmol.* 224:549-553.
19. Jones, S. E., Meerabux, J. M. A., Yeats, D. A. and Neal, M. J., 1992, Analysis of differentially expressed genes in retinitis pigmentosa retinas. Altered expression of clusterin mRNA, *FEBS Lett. 300:279-282.*

2

OXIDATIVE DAMAGE AND RESPONSES IN RETINAL NUCLEI ARISING FROM INTENSE LIGHT EXPOSURE

D. T. Organisciak,[1] R. K. Kutty,[2] M. Leffak,[1] P. Wong,[2] S. Messing,[1] B.Wiggert,[2] R. M. Darrow,[1] and G. J. Chader[2]

[1] Department of Biochemistry and Molecular Biology
Wright State University
Dayton, Ohio 45435
[2] Laboratory of Retinal Cell and Molecular Biology
National Eye Institute, NIH
Bethesda, Maryland 20892

INTRODUCTION

Prolonged or high intensity visible light exposure leads to photoreceptor cell damage and loss by incompletely understood mechanisms. In the rat retina, the sequence of events associated with intense light damage is triggered by the bleaching of rhodopsin (1), and the extent of damage is modulated by other photoreceptor cell proteins involved in visual transduction (2). Thus, environmental light-rearing conditions that alter the steady state levels of rhodopsin, α-transducin and s-antigen (arrestin) can affect the ultimate fate of visual cells (2,3). However, irrespective of their prior light-rearing environment, when rats are pretreated with natural or synthetic antioxidants, retinal light damage is less than in unsupplemented animals (4-8). This indicates that intense light exposure also results in oxidative reactions within the photoreceptor cell.

At this time, there is little information about the mechanisms by which retinal light damage leads to visual cell death. It is known that the destructive reactions leading to photoreceptor cell loss occur to a large extent during the dark period following acute-intense light exposure. It is also known that the entire visual cell is affected, including the rod outer- and inner-segments, the nucleus and even the synaptic end plate (1,9), at an early time during light exposure. Recent studies implicating nuclear damage in photoreceptor cell death suggest that both oxidative DNA damage and apoptosis can occur (10,11) and that transcriptional changes in retinal mRNA are found during and after intense light exposure (12,13).

In this study, we describe some of the nuclear changes associated with retinal light damage in rats. We have found that the loss of visual cell rhodopsin, retinal DNA strand breaks and damage to the opsin gene are reduced in rats treated with the antioxidant dimethylthiourea (DMTU) before light exposure. More recently, we have found that DMTU

Degenerative Diseases of the Retina, Edited by Robert E. Anderson et al.
Plenum Press, New York, 1995

treatment is associated with a decrease in the expression of mRNA for retinal heme oxygenase 1 (HO-1), a 32 kDa heat shock protein (12) that is a marker for oxidative stress and a delay in the expression of mRNA for testosterone-repressed prostatic message (TRPM-2), a marker of cellular apoptosis (13). Oxidative DNA damage also occurs after intense light exposure, as measured by a seven-fold elevation of 8-hydroxydeoxyguanosine (8-OHdG). Our findings support the hypothesis that intense light exposure initiates oxidative damage of retinal DNA and enhances the transcription of retinal genes associated with cellular protection and programmed cell death.

MATERIALS AND METHODS

Animal Maintenance and Light Exposure

Male albino Sprague-Dawley rats were purchased from Harlan Industries (Indianapolis, IN) as weanlings and maintained in a weak cyclic light environment for 40-50 days before use. The cyclic light environment consisted of 12 hrs white light per day (on 08:00, off 20:00) at an illuminance of 20-40 lux. All rats were fed rat chow ad libitum (Teklad, Inc., Madison, WI) and had free access to water. Animals between the ages of 60-70 days were dark adapted overnight and then exposed to intense light for up to 24 hrs, beginning at 09:00. Intense light exposures were in green Plexiglas chambers transmitting light of 490-580 n meters (1), at an illuminance of 1500-1750 lux (~800 $\mu W/cm^2$). Twenty four hrs before, and just prior to light exposure, some rats were given IP injections of DMTU, 500 mg/kg dissolved in saline. 1,3-dimethylthiourea was obtained from Aldrich Chemical, Inc. (Milwaukee, WI). The DMTU-treated and -untreated rats were simultaneously exposed to intense light. Food and water were available to rats during the light exposure period. At appropriate times during or after light exposure, rats were sacrificed in a CO_2-saturated chamber and used for rhodopsin determinations. For DNA and RNA analyses, retinas were excised from experimental animals and frozen in liquid N_2. The overall paradigm of light treatment and tissue sampling times is shown in Figure 1. All procedures employed in this study conformed to the ARVO Resolution on the use of Experimental Animals in Research.

Rhodopsin, DNA and RNA Analyses

Rhodopsin. Whole eye rhodopsin levels were determined, to estimate the loss of visual cells from intense light exposure. These measurements were performed 2 wks after light treatment. The level of rhodopsin in experimental animals was compared to the level

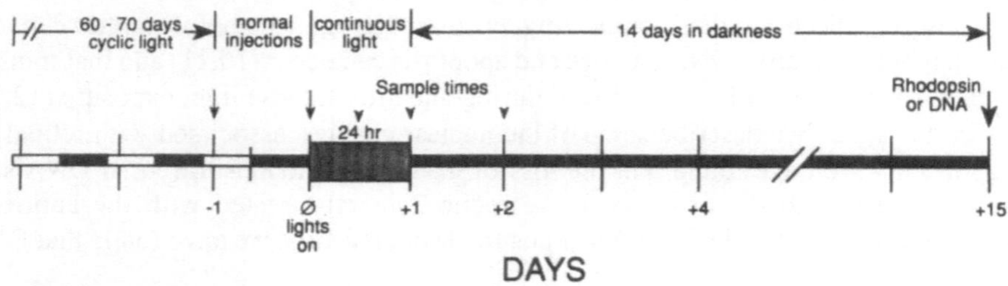

Figure 1. Experimental time course for intense light exposure.

in unexposed control rats maintained in darkness for the same 2 wk period. Procedures for tissue dissection, rhodopsin extraction, and measurement have been described (14,15).

8-OHdG. Individual retinas were thawed in Applied Biosystems 1X Lysis Buffer (urea/ CDTA/sarcosyl/Tris-Cl, pH 7.9), 2 mg/ml proteinase K, and incubated overnight at 37°C. Proteinase was inactivated by heating at 68°C for 15 min. Samples containing ~300 μg of retinal DNA were digested with RNase A (200 μg/ml) and purified manually by organic solvent extraction using phenol:chloroform:isoamyl alcohol (24:24:1), and chloroform (16). DNA was precipitated with ethanol and redissolved in 40 mM Tris-Cl, pH 8.0, 15 mM $MgCl_2$, 1 mM $CaCl_2$, 20 mM NaCl, and digested overnight with phosphodiesterase I (0.025 u), DNase I (200 u), and alkaline phosphatase (0.05 u). Samples were clarified for HPLC by centrifugation through a 3,000 MW cutoff ultrafilter.

Denaturing Gel Electrophoresis and Hybridization. Retinas, previously frozen in liquid N_2, were thawed in 1X Lysis Buffer and digested with proteinase K and RNase as above (16). DNA was purified by automated extraction with phenol:chloroform, and chloroform on an Applied Biosystems Nucleic Acid Extractor 340A, and precipitated with ethanol. DNA was denatured by heating and electrophoresed overnight on alkaline 1.5% agarose gels (30 mM NaOH, 1 mM EDTA). The gels were neutralized (40 mM Tris-Cl, pH 8, 1 mM EDTA) and stained with ethidium bromide.

Opsin Analysis. DNA was dissolved in 10 mM Tris-Cl buffer (pH 7.4) containing 1 mM EDTA, and digested with restriction enzyme BamH1 using the reaction buffer and conditions recommended by the supplier (Promega). DNA was repurified by phenol:chloroform extraction, precipitated, and electrophoresed on 0.5% denaturing agarose gels (30 mM NaOH, 1 mM EDTA). DNA was blotted to nylon membranes by standard capillary transfer (16), and visualized by hybridization to a 6.7 kb opsin genomic fragment probe (generously provided by Dr. C. Barnstable) labeled with α-^{32}P-dCTP using random primer extension. Final stringency washes were in 0.1X SCC, 0.1% SDS, at 65°C. Filters were exposed to Kodak XAR film at -80°C for autoradiography, and hybridization signals were quantitated from the filters using a Molecular Dynamics PhosphorImager.

RNA Analysis Total RNA was extracted from tissue samples using RNAzol B (Tel-Test, Inc., Friendswood, TX) and subjected to 1.2% agarose gel electrophoresis in the presence of 1.2% formaldehyde. The RNA was then capillary blotted onto an Immobilon N membrane (Millipore Corporation, Bedford, MA), UV cross-linked and hybridized using probes for rat HO-1, TRPM-2 or β-actin, as previously described (12,13).

The HO-1 probe was generated by RT-PCR from spleen poly (A)+ RNA preparations, using 5'- AAG GAG GTG CAC ATC CGT GCA and 5'- ATG TTG AGC AGG AAG GCG GTC primers (12), which correspond to the 64-632 bp region of the cDNA sequence reported for rat HO-1 (17). The TRPM-2 cDNA probe was obtained directly from available plasmids (18). Human β-actin probe was obtained from Clontech, Palo Alto, CA. Probes were radio-labeled with [α-^{32}P]dCTP by random priming to a specific activity of not less than 10^6 cpm/μg (19). Hybridization was carried out at 42°C in the presence of 50% formamide. The blots were washed under stringent conditions and exposed to Kodak X-OMAT AR film. The intensities of individual bands on the film were estimated using an LKB Ultrascan XL Laser Densitometer. Membranes were then stripped and reprobed with the TRPM-2 probe and subsequently with the β-actin probe. The HO-1 signals were normalized using the corresponding actin signals as controls for RNA loading.

Table 1. Rhodopsin levels in rats exposed to intense light (n mol/eye)*

	Light						
	0 hrs	8 hrs		16 hrs		24 hrs	
DMTU	– or +	–	+	–	+	–	+
Rhodopsin	2.1 ± 0.1	2.0 ± 0.2	2.1 ±0.2	1.6 ± 0.3	2.0 ± 0.1	0.8 ± 0.3	1.8 ± 0.1
% of unexposed control	100	95	100	76	95	38	86

*Rhodopsin was measured 2 wks after intense light exposure. Results are the mean ± SD for 4-16 pairs of eyes.

Ed: C layou table OK? —Co

RESULTS

Retinal Damage as Assessed by Rhodopsin Loss

Whole eye rhodopsin levels were measured 2 wks after light exposure to estimate the relative losses of photoreceptor cells in DMTU-treated and -untreated rats. As shown in Table 1, visual cell loss from intense light depends on the duration of exposure. In rats not given DMTU, 24 hrs of intense light resulted in rhodopsin levels that were an average of 62% lower than found in controls. In comparison, only about 24% of the rhodopsin was lost in –DMTU rats after 16 hrs of light exposure. For rats pretreated with DMTU, the average loss of rhodopsin was only 14% after 24 hrs of light exposure; there was no significant rhodopsin loss following 16 hrs of light exposure. Table 1 also shows that 8 hrs of light exposure does not lead to photoreceptor cell loss in either the DMTU-treated or -untreated rats.

Retinal DNA Damage as Measured by 8-OHdG

Table 2 indicates that oxidative changes in retinal DNA occur both during and after intense light treatment. As measured by the relative levels of 8-OHdG, a marker of DNA damage (20), there was extensive damage after light exposure. For rats not treated with DMTU, 8-OHdG was about 1.5 fold higher than control after 24 hrs of light. Six hrs after the 24 hr period of intense light exposure, 8-OHdG was 7 fold higher than in control retinas and 4 fold higher than in retinas of 24 hr light-treated rats.

Table 2. 8-OHdG in retinal DNA (residues/10^5 deoxyguanosine residues)*

Light exposure	0	24	24 hrs
Dark period	0	0	6 hrs
8-OHdG	3.5	5.8	25.1
% of control	100	165	717

*Retinal DNA was extracted under conditions which minimize artifactual oxidation and digested extensively to mononucleosides (20). The digest was filtered and analyzed by HPLC by Dr. G. Claycamp (University of Pittsburgh). Results are the average of 2 separate determinations.

DNA Damage Detectable by Alkaline Gel Electrophoresis

Irregular chromatin condensation and nuclear pycnosis are among the earliest observed changes in retinal histology after exposure of rats to intense light. To determine whether these morphological alterations are associated with DNA damage at the molecular level, rats were exposed to light for up to 24 hrs, and the retinal DNA was examined on denaturing agarose gels (Figure 2). Under the alkaline conditions of electrophoresis, retinal DNA from rats exposed to longer light treatments contained progressively shorter DNA strands (lanes 1, 3, 5). Thus, with increasing duration of light treatment, there was a readily detectable increase in the number of lesions in retinal DNA.

In rats exposed to intense light for 8- or 24-hrs, subsequent dark treatment (24 hr) further increased the amount of shortened DNA strands (compare lanes 3 & 7; 5 & 9), indicating that the initial light treatment triggered a pathway of DNA damage which continued in the dark. Both the initial DNA damage and the subsequent dark reaction were strongly inhibited by the antioxidant DMTU (compare odd and even # lanes), suggesting an essential role for free radicals in the DNA damage process.

Effects of Visible Light on the Opsin Gene

In order to quantitate the incidence of light-induced DNA damage, a region of retinal DNA containing the opsin gene was examined. Rats were exposed to light for 0, 4 or 8 hrs, without (– lanes) or with (+ lanes) DMTU pretreatment. DNA was isolated, digested with BamH1 to yield a 14 kb fragment containing the opsin gene, electrophoresed on denaturing agarose gels and transferred to nylon membranes for visualization by hybridization to a radiolabeled opsin genomic probe (Figure 3, upper panel). A single DNA strand break in the 14 kb opsin fragment would produce smaller fragments of greater electrophoretic mobility, decreasing the 14 kb signal detectable by the opsin probe.

Quantitation of the hybridization signals (normalized per μg DNA) showed a progressive decrease in signal from the opsin DNA region with increasing light exposure (Figure 3, lower panel). The loss of signal (~50% by 8 hr) suggests that, on average, 50% of the copies in the opsin gene region had been damaged. Examination of the entire retinal genome (Figure 2) suggests that the extent of damage in the opsin gene is representative of damage

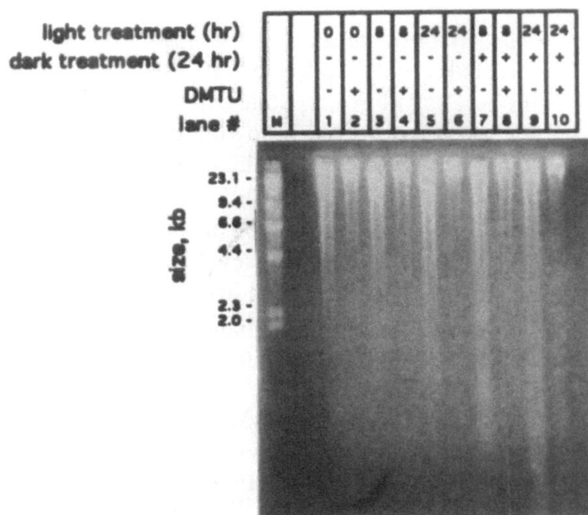

Figure 2. DNA strand break analysis by alkaline gel electrophoresis. Rats were exposed to light for the indicated times, without (lanes 1, 3, 5, 7, 9) or with (lanes 2, 4, 6, 8, 10) DMTU pretreatment. Some rats were then moved to a dark environment for 24 hr following light treatment (lanes 7-10). DNA was prepared as described in materials and methods. M; size marker (in kb), lambda phage DNA digested with restriction enzyme Hind III.

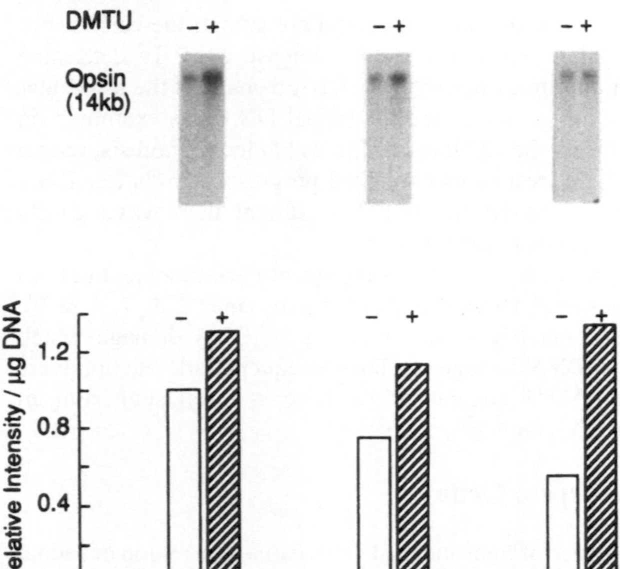

Figure 3. Light induced damage in the opsin gene of retinal DNA. Rats were exposed to light for the indicated times, without (– lanes) or with (+ lanes) DMTU pretreatment. DNA was prepared as described (16), and quantitated by ethidium bromide fluorescence. Following electrophoresis on an alkaline agarose gel, the gel was neutralized and stained with ethidium bromide for requantitation of the DNA. Southern transfer to nylon membranes, and hybridization followed standard procedures. Relative intensity; hybridization signal per mg DNA relative to the -DMTU, unexposed rat retinal DNA in lane 1.

to the overall genome, confirming that visible light treatment can cause a substantial increase in retinal DNA damage.

The results of DMTU treatment were striking. Even in rats not exposed to intense light, DMTU increased the amount of signal from the undamaged 14 kb opsin gene DNA (Figure 3 "0 hrs light"). This implies that, even under cyclic light rearing conditions, there is a substantial load of steady-state DNA damage which reflects the equilibrium between DNA damage and repair. With 4 hrs and 8 hrs of light treatment, DMTU significantly protected the opsin DNA from attack, corroborating the view that free radicals play an important role in light-induced DNA damage.

Changes in Retinal Gene Expression Determined by Northern Analysis

The effect of intense light exposure on the retinal expression of HO-1 mRNA in DMTU-treated rats and -untreated rats was investigated by Northern blot analysis. Figure 4 shows the results of three separate experiments (panels A-C) in which a marked increase in HO-1 mRNA was observed following light exposure (control). In unexposed animals, the HO-1 signal was barely detectable. Relative to the actin signal, increases in HO-1 message were 27- and 70-fold following 12- and 24-hrs of light, respectively. Keeping the animals in darkness for 24 hrs following 24 hrs of light exposure caused the HO-1 signal to decrease considerably from the elevated level, i.e., from 70- to 37-fold. Animals pretreated with DMTU (treated) showed a much lower induction of HO-1 than those not treated with the antioxidant (control), 2- vs 27-fold at 12 hrs; and 12- vs 70-fold at 24 hrs. HO-1 expression, however, continued to increase in the DMTU-treated animals during the 24 hr dark period after light exposure.

In contrast to HO-1 mRNA, TRPM-2 mRNA was readily detectable in control retinas not exposed to intense light (Figure 5). As with HO-1 expression, however, the signal was markedly increased upon light exposure. Moreover, pretreatment of rats with DMTU delayed the light- induced increase in TRPM-2 mRNA levels seen in rats not given DMTU for up to

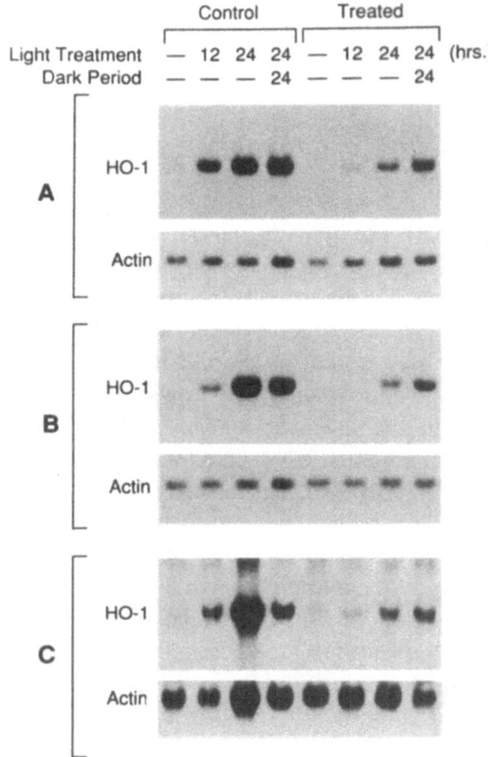

Figure 4. Light induced changes in HO-1 mRNA. Retinal RNA preparations (15 mg) from 2 eyes of individual animals were analyzed for HO-1 and β-actin by the Northern blot technique; control = untreated rats, treated = DMTU treated rats. Panels A, B and C represent 3 different experiments (12) [with permission].

12 hrs. The levels of TRPM-2 mRNA in DMTU-treated animals after 12 hrs of light exposure were similar to those found in rats not exposed to light. TRPM-2 expression, therefore, also appears to be affected by oxidative stress. However, after 24 hrs of light exposure, TRPM-2 levels in the DMTU-treated animals were comparable to those observed in untreated animals exposed to 12 hrs of light. This increase, as well as the increase in HO-1 expression, may be

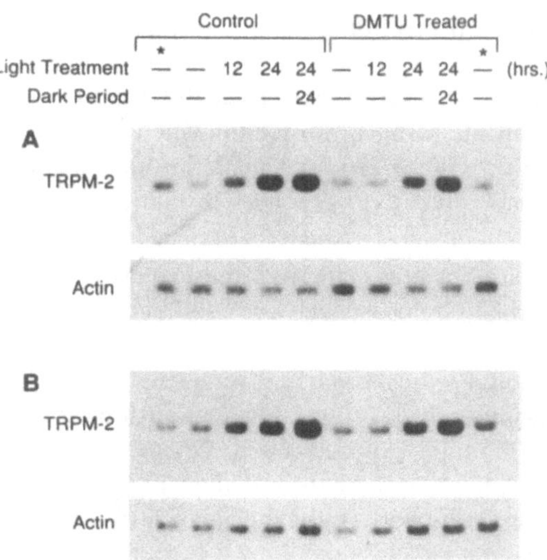

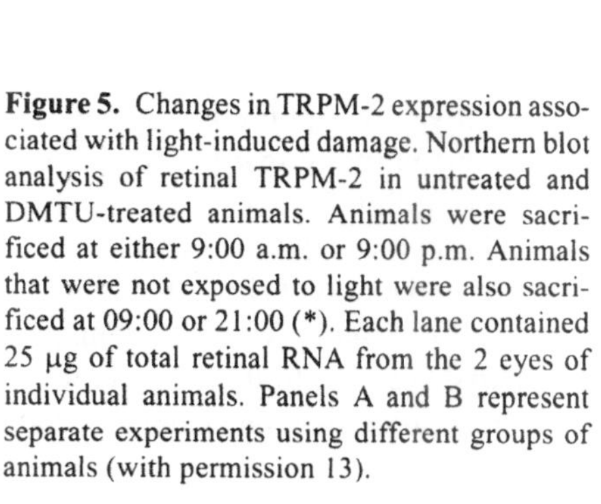

Figure 5. Changes in TRPM-2 expression associated with light-induced damage. Northern blot analysis of retinal TRPM-2 in untreated and DMTU-treated animals. Animals were sacrificed at either 9:00 a.m. or 9:00 p.m. Animals that were not exposed to light were also sacrificed at 09:00 or 21:00 (*). Each lane contained 25 μg of total retinal RNA from the 2 eyes of individual animals. Panels A and B represent separate experiments using different groups of animals (with permission 13).

due to the depletion of DMTU from the retina over the subsequent 12-24 hr period of light exposure or darkness.

DISCUSSION

This study demonstrates remarkable changes in retinal nuclei from rats exposed to intense visible light. As measured by ethidium bromide staining of total retinal DNA or by specific damage to the opsin gene, significant DNA damage occurs as a result of light exposure. The light-induced effect on retinal DNA appears to be oxidative in nature because 8-OHdG levels were elevated and DMTU treatment of rats decreased both the extent of DNA strand breaks and the overall loss of visual cell rhodopsin. Although both natural and synthetic antioxidants have previously been shown to decrease retinal light damage in rats (4-8), this report is, to our knowledge, the first study demonstrating a protective effect by DMTU on retinal DNA. DMTU treatment also reduced the expression of HO-1 mRNA and delayed the over-expression of TRPM-2 mRNA in rat retinas. These mRNAs have previously been shown to be increased by intense light exposure (12,13). Taken together, our findings support the hypothesis that visible light leads to an increase in oxidative stress in the retina which may lead to visual cell death.

Whereas the mechanism of oxidative damage in the retina remains unknown, the time course of DNA damage indicates that it begins early in the pathological process and that it progresses rapidly in the dark period after exposure. This was apparent from the loss of integrity of the opsin gene after only 4-8 hrs of light exposure, and by the duration-dependent increase in DNA damage over the entire retinal genome. As determined by the additional increase in DNA damage after light exposure and the 7-fold increase in 8-OHdG, oxidative DNA damage appears to be a major manifestation of visual cell loss after light treatment.

This relationship, however, is complicated by the process of DNA repair which presumably occurs almost simultaneously with damage. Whereas DNA damage was apparent after only 8 hrs of light exposure, rhodopsin loss measured as an end point of damage 2 wks later did not occur until at least 16 hrs of light treatment. The retina also increased its expression of HO-1 mRNA following 12 hrs of light exposure. HO-1 is a 32 kDa heat shock protein, inducible by oxidative stress in cells where it is thought to afford protection against oxidative damage (12). At this time, it is unknown whether HO-1 is expressed exclusively in photoreceptors, in the inner retinal layers, or in all retinal cell types. Increased HO-1 expression, however, suggests that the retina exhibits multiple forms of protection when subjected to light-induced oxidative stress. Additional work will be required to define the relationship between DNA damage and repair and other protective responses in the retina.

The mechanism(s) of visual cell death from light also remains an open question. Tso and associates (10) have presented evidence for both oxidative and apoptotic changes in retinal photoreceptors upon light exposure. Our findings of increased retinal DNA damage and increases in both HO-1 and TRPM-2 mRNA expression in light-exposed retinas (13) strongly suggest that both processes occur. The fact that DMTU delayed the onset of TRPM-2 expression also suggests a relationship between this apoptotic cell marker (18) and oxidation in the light-exposed retina. Whether visual cell death arises from oxidative DNA damage, apoptosis, or both, the rat model of *in vivo* light-induced retinal degeneration offers a unique opportunity to study the mechanism of cellular death and protective mechanisms that maintain visual function.

REFERENCES

1. Noell WK, Walker VS, Kang BS, Berman S. Retinal damage by light in rats. *Invest Ophthalmol.* 1966;5:450-473.
2. Organisciak DT, Xie A, Wang HM, Jiang YL, Darrow RM, Donoso LA. Adaptive changes in visual cell transduction protein levels: Effect of light. *Exp Eye Res.* 1991;53:773-779.
3. Penn JS, Thum LA. A comparison of the retinal effects of light damage and high illuminance light history. In: Hollyfield JG, Anderson RE, LaVail MM, eds. *Degenerative Retinal Disorders: Clinical and Laboratory Investigations.* New York: Alan R Liss; 1987:425-438.
4. Organisciak DT, Wang HM, Li ZY, Tso MOM. The protective effect of ascorbate in retinal light damage of rats. *Invest Ophthalmol Vis Sci.* 1985;26:1580-1588.
5. Li ZY, Tso MOM, Wang HM, Organisciak DT. Amelioration of photic injury in rat retina by ascorbic acid: A histopathologic study. *Invest Ophthalmol Vis Sci.* 1985;26:1589-1598.
6. Lam S, Tso MOM, Gurne DH. Amelioration of retinal photic injury in albino rats by dimethylthiourea. *Arch Ophthalmol.* 1990;108:1751-1757.
7. Organisciak DT, Jiang YL, Wang HM, Bicknell I. The protective effect of ascorbic acid in retinal light damage of rats exposed to intermittent light. *Invest Ophthalmol Vis Sci.* 1990;31:1195-1202.
8. Organisciak DT, Darrow RM, Jiang YL, Marak GE, Blanks JC. Protection by dimethylthiourea against retinal light damage. *Invest Ophthalmol Vis Sci.* 1992;33:1599-1609.
9. Kuwabara T, Funahashi M. Light effect on the synaptic organ of the rat. *Invest. Ophthalmol. Vis. Sci.* 1976; 15:407-411.
10. Shahinfar S, Edward DP, Tso MOM. A pathologic study of photoreceptor cell death in retinal photic injury. *Curr Eye Res.* 1991;10:47-59.
11. Messing SL, Darrow R, Leffak M, Fleischman D, Organisciak DT. Visible light-induced damage to retinal DNA in vivo. *Invest. Ophthalmol. Vis. Sci.* 1994; 35:2138.
12. Kutty RK, Kutty G, Wiggert B, Chader GJ, Darrow RM, Organisciak DT. Induction of heme oxygenase-1 in the retina by intense light: Suppression by the antioxidant dimethylthiourea. *Proc Nat Acad Sci. USA* 1995; 92:1177-1181.
13. Wong P, Kutty RK, Darrow RM, Shivaram S, Kutty G, Fletcher RT, Wiggert B, Chader G, Organisciak DT. Changes in TRPM-2/Clusterin expression associated with light induced retinal damage in rats. *Biochem Cell Biology* (in press).
14. Delmelle M, Noell WK, Organisciak DT. Hereditary retinal dystrophy in the rat: Rhodopsin, retinol and vitamin A deficiency. *Exp Eye Res.* 1975; 21:369-380.
15. Organisciak DT, Wang HM, Kou AL. Rod outer segment lipid opsin ratios in the developing normal and retinal dystrophic rat. *Exp Eye Res.* 1982; 34:401-412.
16. Maniatis T, Fritsch EF, and Sambrook J. (1982). Molecular Cloning: A Laboratory Manual (Cold Spring Harbor, New York: Cold Spring Harbor Laboratory).
17. Shibahara S, Muller R, Taguchi H, and Yoshida T. (1985) Cloning and expression of cDNA for rat heme oxygenase. *Proc Natl Acad Sci USA*, 82, 7865-7869.
18. Wong P, Pineault J, Lakins V, Taillefer D, Leger JG, Wang C, Tenniswood MP. Genomic organization and expression of the rat TRPM-2 (clusterin) gene, a gene implicated in apoptosis. *J Biol Chem.* 1993; 268:5021-5031.
19. Feinberg A.P. and Vogelstein B. A technique for radiolabeling DNA restriction endonuclease fragments to high specific activity. *Anal Biochem.* 1983; 132, 6-13.
20. Claycamp G. Phenol sensitization of DNA to subsequent oxidative damage in 8-hydroxyguanine assays. *Carcinogenesis* 1991; 13:1289-1292.

3

LIGHT-INDUCED APOPTOSIS IN THE RAT RETINA IN VIVO

Morphological Features, Threshold and Time Course

Charlotte E. Remé,[1] Michael Weller,[2] Piotr Szczesny,[1] Kurt Munz,[1] Farhad Hafezi,[1] Jörg-Joachim Reinboth,[1] and Matthias Clausen[1]

[1] Department of Ophthalmology

[2] Section of Clinical Immunology, Department of Internal Medicine University Hospital, Zürich, Switzerland

SUMMARY

Apoptotic cell death is observed in acute retinal lesions in the albino rat induced by relatively low light levels and short exposure durations (1000 lux and 3000 lux diffuse, white fluorescent light for 2 h). At higher illuminances and extended post - exposure intervals, necrotic cell death is prominent. In our model, the threshold for apoptosis is at 1000 lux for 2 hrs. Distinct morphological signs of apoptosis with chromatin - and cytoplasmic condensation in rod photoreceptors appear within 60 minutes after the onset of light exposure and increase thereafter. 24 hours after exposure, presumed apoptotic bodies, cellular debris and necroses prevail. In situ DNA end labeling reveals DNA breaks in rod nuclei. Agarose gel electrophoresis after DNA extraction from light exposed retinae shows the "ladder" formed by DNA fragments of nucleosomal size (180 to 200 bp or multiples). In the pigment epithelium, apoptosis is seen 24 h after the light exposure. Potential mediators of apoptosis may include arachidonic acid and some of its metabolites which are evoked by light exposure and induce apoptosis in other cells.

INTRODUCTION

Apoptosis is known as a specific type of cell death first described in tissue- and organ development as histogenetic cell death. This physiological cell death is "programmed" and regulated by gene expression. Apoptosis is also observed in numerous mature tissues such as the immune system, neurons, various cell- and tissue cultures and cancer cells and is induced by a wide variety of physiological and non-physiological stimuli. Similarly, various agents can suppress apoptotic cell death (24,30). Multiple genes are directly involved in regulating cell death and a large number of genes is expressed coinciding with the occurrence of apoptosis (23). Furthermore, there are multiple known or suspected signalling molecules

Degenerative Diseases of the Retina, Edited by Robert E. Anderson et al.
Plenum Press, New York, 1995

contributing to the regulation of apoptosis. Essential biochemical steps in apoptotic cell death, however, are not fully elucidated.

In the retina, apoptosis has been described during development (13,31). In donor eyes of patients who had suffered from the inherited dystrophy retinitis pigmentosa (RP), an altered gene expression was found that possibly leads to apoptotic cell death (8). In several animal models of RP, apoptotic cell death is observed and possibly linked to the known genetic defects (5,10,14,27). An induced retinal degeneration occurs after bright light exposure in rodent models (12). Apoptotic cell death was observed after relatively long exposure to high levels of green light (1,25) and after short term exposure to moderate levels of diffuse white light (26). So far, no other light damage studies reported on apoptotic cell death in the retina.

The present study shows that light exposure induces apoptotic cell death in retinal photoreceptors as defined by light- and electron microscopy and confirmed by in situ DNA end labeling, and nucleosomal size DNA fragmentation.

MATERIALS AND METHODS

All experiments were performed in triplicate and conducted in accordance with the ARVO Resolution of the Use of Animals in Research. Male albino rats of about 300 gr (8 - 10 weeks of age) were dark adapted for 36 h and exposed to 1000 lx or 3000 lx of diffuse, white, overhead fluorescent light for 2 h, followed by 2 h of darkness and transferred to their regular light - dark cycle. Dark adapted unexposed rats served as controls. Rats were retrieved from the respective light exposure or from darkness following light exposure every 15 minutes. After 24 h, another group of light exposed rats was retrieved. After decapitation of the animals, the eyes were rapidly enucleated and processed for light- and electron microscopy and histochemistry or the retinae were gently extruded through a slit in the cornea (29) and processed for gel electrophoresis. For light- and electron microscopy, the upper and lower central retinae were trimmed under a dissecting microscope equipped with a red filter and processed for microscopy (20). For histochemical in situ end labeling of DNA breaks, the TUNEL method (TdT-mediated dUTP-biotin nick end labeling) was used (6). For gel electrophoresis, fragmented DNA was isolated by phenol dichloromethane extraction and visualized on ethidium bromide stained agarose gels (28).

RESULTS

Our model of acute, light-induced photoreceptor lesions reveals initial alterations in rod outer segment (ROS) tips consisting of dilations and vesiculations of disks that increase with increasing illuminance levels and extend towards the base of ROS but do not affect the remainder of the cell (20). At an illuminance of 1000 lx for 2 h, a turning point is observed with changes seen in both photoreceptor nuclei and inner segments. The latter changes are consistent with the characteristics of apoptotic cell death. In specimens from rats exposed to 1000 lx for 2 h, apoptosis in photoreceptors is confined to the lower central retina with few scattered pycnotic nuclei and condensed cytoplasm in the inner nuclear- and ganglion cell layer. 1 h after the onset of light exposure, several photoreceptor nuclei show condensed chromatin and scattered inner segment (IS) densifications. 2 h of light exposure result in chromatin condensations and distinct IS densifications in numerous photoreceptors (Fig. 1a). 24 h after the light exposure, a large amount of photoreceptors shows condensed chromatin but IS densifications are no longer discernible. Some cells show cytolytic changes consisting of swollen and lightened IS. The prominent feature is cellular debris and presumed apoptotic bodies with scattered macrophages in the area of the pigment epithelium (PE) and photore-

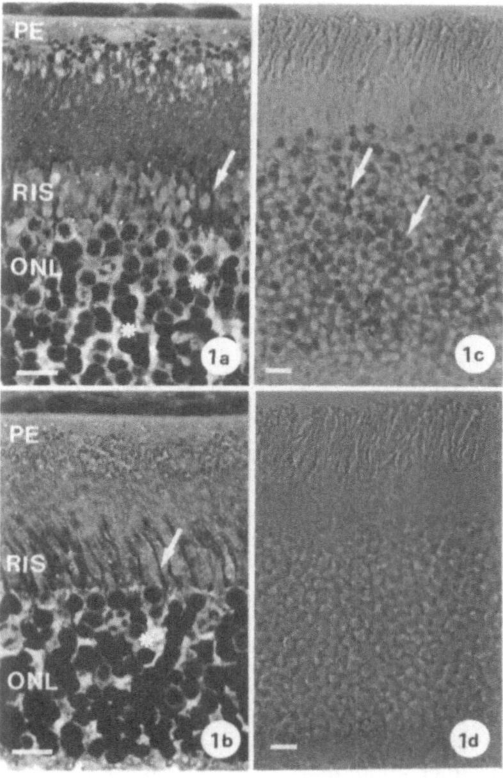

Figure 1. Light micrographs depicting retinae exposed to 1000 lx (a, c, d) or to 3000 lx (b) of white light for 2 hours. a) The outer nuclear layer (ONL) shows numerous densely stained nuclei (*) and multiple normal appearing chromatin patterns, the cytoplasm of several rod inner segments (RIS) is condensed (↑). Bar = 10 μm. b) The ONL shows abundant densely staining nuclei (*) with only few normal appearing chromatin patterns. Numerous RIS show condensed cytoplasm (↑). Note the burst of phagosomes in the PE. Bar = 10 μm. c) TUNEL staining shows numerous stained nuclei in the ONL indicating DNA - breaks typical of apoptosis. Bar = 10 μm. d) Control section for TUNEL staining with no positive cells after the omission of cobalt chloride, the cofactor for terminal transferase. Bar = 10 μm.

ceptor layer. Notably, apoptotic changes in the PE first appear 24 h after light exposure. These consist of peripheral chromatin clumping and condensation of the cytoplasm (Fig. 2 a-c).

At an illuminance of 3000 lx for 2 hrs, apoptosis in photoreceptors occurs in the upper and lower central retina with few scattered pycnotic nuclei and condensend cytoplasm in the

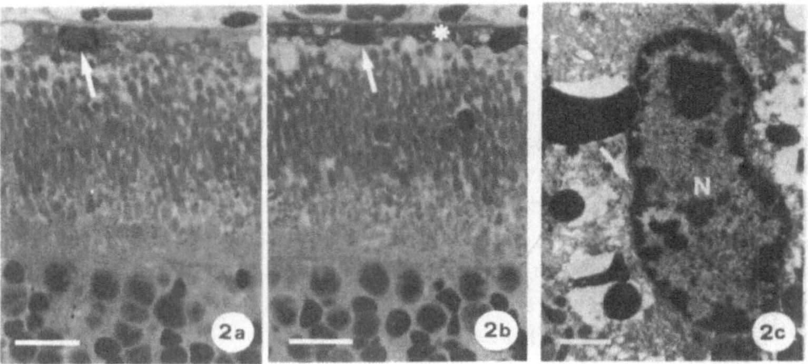

Figure 2. Micrographs depicting pigment epithelial cells from rats exposed to 1000 lx of white light for 2 h, followed by 2 h of darkness and returned to the regular LD - cycle for 24 h. a) Peripheral chromatin clumping (↑) is distinct. Bar = 10 μm b) Nuclei show highly condensed chromatin (↑), the cytoplasm is shrunken and condensed (*). Bar = 10 μm c) Electron micrograph showing peripheral chromatin clumping (↑) of a PE cell nucleus (N). Bar = 1 μm.

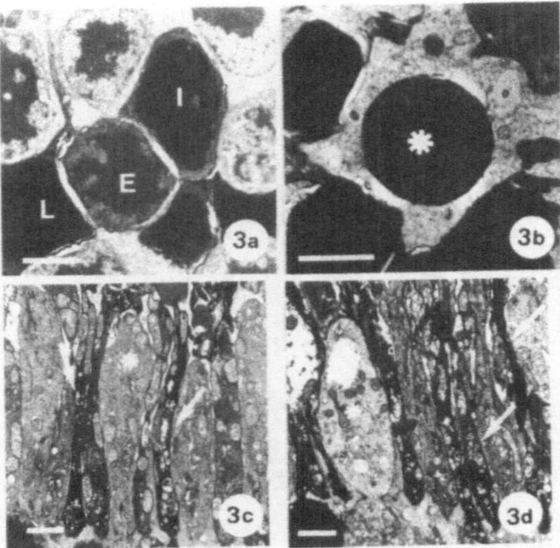

Figure 3. Electron micrographs showing features seen in photoreceptors after exposure to 1000 lx or 3000 lx white light. a) 60 minutes after the onset of light exposure, various stages of chromatin changes are observed indicating various stages of apoptosis. E: early stage, I: intermediate stage, L: late stage. Bar = 2 µm. b) End - stage nucleus (*) showing extremely condensed chromatin and rounded shape visible after two hours of light exposure. Bar = 2 µ. c) 60 minutes after the onset of light exposure, distinct condensation of RIS (↑) is apparent next to normal looking cells (*). Bar = 2 µm. d) RIS showing the typical signs of necrotic cytolysis (*) with swollen cell body, translucent cytoplasm and mostly destroyed cellular organelles. Adjacent cells show apoptotic condensation of their cytoplasm (↑). Bar = 2 µm.

inner nuclear layer and the ganglion cell layer. 1 h after the onset of light exposure, photoreceptors show numerous nuclei with condensed chromatin and dense IS with a marked increase after 2 h of light exposure (Fig. 1b). At this time, cytolysis in IS is observed. These changes distinctly increase during the dark period following the light exposure. 24 h after the light exposure, there are abundant pycnotic nuclei but no IS densifications. Numerous presumed apoptotic bodies and cellular debris with macrophages in the outer retina are seen. Cytolysis in IS is distinct. The PE reveals apoptosis 24 h after light exposure with peripheral chromatin clumping and condensed cytoplasm.

Electron microscopy clearly demonstrates different stages of photoreceptor chromatin condensations from initial weak densifications to extremely electron dense, shrunken endstage nuclei. Furthermore, IS alterations consisting of cytoplasmic densification and shrinkage or cytolytic changes revealing swollen and electron lucent cytoplasm with fragmentation of organelles is confirmed (Fig. 3 a - d).

Histochemical in situ end labeling confirmed the presence of DNA breaks in photoreceptor nuclei at 1000 lx and 3000 lx exposure (Fig. 1 c,d). In specimens exposed to 1000 lx for 2 h, staining is confined to the lower, central retina corresponding to the morphological appearance of apoptosis. In specimens exposed to 3000 lx for 2 h, numerous dark staining cells appear in the ONL and scattered stains in the inner nuclear- and ganglion cell layer. In addition, background staining is fairly uniformly increased in all cellular layers.

Gel electrophoresis demonstrates nucleosome - size DNA breaks of 180 - 200 bp or multiples which results in the "ladder" appearance characteristic of apoptosis in many but not all tissues (2,11). After 1000 lx exposure, only minor fragmentation is seen whereas after 3000 lx, distinct bands of fragmented DNA are recognized. 24 h following exposure to 3000 lx, the ladder is obscured by a smear of degraded DNA which is consistent with the development of cellular necrosis (Fig. 4a, b).

DISCUSSION

Our model of acute light-induced photoreceptor lesions clearly reveals the turning point from changes seen only in ROS to alterations affecting the entire photoreceptor cell

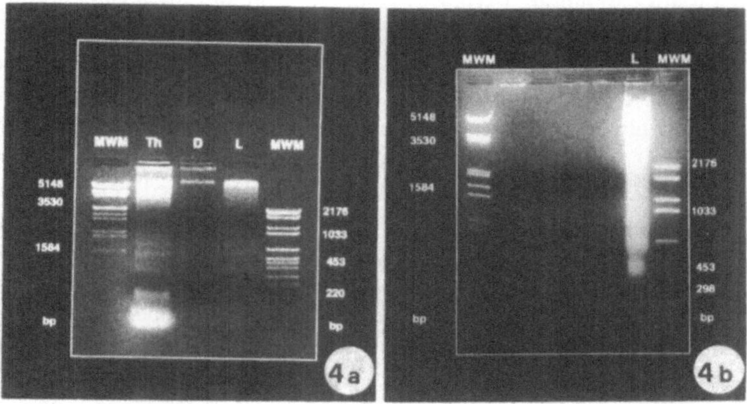

Figure 4. After DNA extraction from light exposed retinae, nucleosomal size DNA fragments of 180 - 200 bp subunits or multiples appear in 1,8% agarose gel electrophoresis as a "ladder". D: dark adapted retinae without visible DNA fragmentation, L: light exposed retinae showing DNA fragmentation, T: thymus extract from young rats showing DNA fragmentation as positive controls, M: Molecular weight markers. a) After exposure to 3000 lx for 2 hrs, a distinct ladder is apparent. b) 24 hours following an exposure to 3000 lx, a ladder can hardly be recognized, whereas distinct irregular DNA fragments indicating cellular necrosis are visible.

(21). The relatively synchronized onset of the apoptotic response may permit us to modify this response with various agents and to study regulative mechanisms and gene expression.

The different stages of apoptotic cell death in the light damaged retina resemble those found in other tissues (3). The morphology of photoreceptor chromatin and the time of appearance of apoptotic bodies, however, are different from that described in many other cell types. No peripheral chromatin clumping occurs after light exposure but rather a progressive "in situ" condensation of the central parts leaving a highly condensed endstage nucleus. The occurrence of presumed apoptotic bodies is seen only 24 h after light exposure. At this time point, cellular fragments resulting from necrotic cell death are also found and cannot consistently be distinguished from late stage apoptotic bodies. This is supported by gel electrophoresis where at 24 h the ladder is obscured by a smear of degraded DNA resulting from cellular necrosis. Perhaps the removal of apoptotic bodies requires more time in the retina than is usually observed in other tissues resulting in a coincidence with necrotic debris. Alternatively, the removal is very fast and was missed in our study.

Autophagic vacuoles are abundant in apoptotic and normal rod inner segments confirming our earlier work on light - induced autophagy (18). Autophagic degradation of cytoplasmic organelles such as mitochondria, endoplasmic reticulum or parts of the Golgi zone may precede apoptotic death of the entire cell.

In general, no clear distinctions are available between pycnotic and apoptotic nuclei. Pycnosis is collectively used for a highly condensed and shrunken nucleus. This description, however, does not imply the information whether or not the process is apoptotic. Only additional investigations such as in situ end labeling, gel electrophoresis and different stages of chromatin and cytoplasmic changes observed by electron microscopy help to elucidate underlying mechanisms.

Apoptosis in the PE was found to lag behind apoptosis in photoreceptors, occuring 24 h after light exposure. The reason for this difference in timing is unkown. Conceivably, apoptosis in the PE is induced by secondary tissue changes following the light exposure.

The apoptotic response in our model occurs rapidly after the onset of light exposure compared to induced apoptotic responses in several other tissues. Which potential mediators may be available for activation? Earlier work in our laboratory had shown a light evoked release of arachidonic acid (AA) as a function of illuminance time and intensity (9). Similarly, a light dependent synthesis of AA metabolites was shown (17), including leukotriene B_4 (LTB $_4$), 5-hydroxyperoxy- eicosatetraenoic acid (5-HPETE) and thromboxane B_2 (TXB $_2$). It is noteworthy that AA and oxygenated metabolites are potent inducers of apoptosis (4) via gene expression in other tissues. Similar mechanisms may occur after light exposure in the retina. Reactive oxygen intermediates (ROI) induced by several mechanisms have been postulated as strong effectors of apoptosis (4). Intracellular ROI may activate proapoptotic genes through oxidative stress - responsive nuclear transcription factors (4). In retinal light damage, oxidative stress may occur through several photochemical mechanisms and may constitute an important cause of acute cellular lesions. Depending on the light dose and the experimental paradigms, apoptotic cell death may thus be a distinct process during early stages of light damage. Exposure to high illuminance levels for long durations (days to weeks) may evoke predominantly necrosis, whereas relatively short exposures (few hours) to moderate light levels may induce mainly apoptosis. The occurrence of oxidative stress even at relatively low light levels is supported by our earlier work, where a radioprotective agent and a platelet - activating factor antagonist significantly reduced acute ROS lesions in the retina (19,22). The prevention of apoptosis of cultured neurons by cycloheximide was attributed to the shunting of cysteine from protein synthesis to glutathione (15). The protooncogene bcl-2 blocks apoptosis in multiple experimental situations (16) and acts in an antioxidant pathway (7). Similar mechanisms may be able to protect the retina from apoptotic cell death.

The concept of dose dependence of apoptotis may be important in retinal pathology and ageing. Gradual cell loss during ageing and in the course of induced retinal degeneration may occur through apoptosis. Repeated exposures to bright light or other noxious agents may lead to augmented apoptotic cell death over a lifetime of an individual. The susceptibility to such factors may be genetically determined.

ACKNOWLEDGMENTS

Supported by the Swiss National Science Foundation grant No. 3100-040791.94/1, Bruppacher Foundation, Zuerich and Wilhelm Sander Foundation, Muenchen, Germany (Ch.E. R), M. Weller is a postdoctoral fellow of the German Research Foundation (DFG).

REFERENCES

1. Abler, A. S., Chang, C. J., Fu, J. and Tso, M. O. M., 1994, Photic injury triggers apoptosis of photoreceptor cells, *Invest. Ophthalmol. Vis. Sci.*, Suppl. 35:1517.
2. Arends, M. J., Morris, R. G. and Wyllie, A. H., 1990, Apoptosis. The role of the endonuclease, *Am J Pathol*, 136:593-608.
3. Bursch, W., Paffe, S., Putz, B., Barthel, G. and Schulte-Hermann, R., 1990, Determination of the length of the histological stages of apoptosis in normal liver and in altered hepatic foci of rats, *Carcinogenesis*, 11:847-53.
4. Buttke, T. M. and Sandstrom, P. A., 1994, Oxidative stress as a mediator of apoptosis, *Immunol Today*, 15:7-10.
5. Chang, G. Q., Hao, Y. and Wong, F., 1993, Apoptosis: final common pathway of photoreceptor death in rd, rds, and rhodopsin mutant mice, *Neuron*, 11:595-605.
6. Gavrieli, Y., Sherman, Y. and Ben-Sasson, S. A., 1992, Identification of programmed cell death in situ via specific labeling of nuclear DNA fragmentation, *J Cell Biol*, 119:493-501.
7. Hockenbery, D. M., Oltvai, Z. N., Yin, X. M., Milliman, C. L. and Korsmeyer, S. J., 1993, Bcl-2 functions in an antioxidant pathway to prevent apoptosis, *Cell*, 75:241-51.

8. Jones, S. E., Meerabux, J. M., Yeats, D. A. and Neal, M. J., 1992, Analysis of differentially expressed genes in retinitis pigmentosa retinas. Altered expression of clusterin mRNA, *Febs Lett*, 300:279-82.
9. Jung, H. and Remé, C.E., 1994, Light-evoked arachidonic acid release in the retina: illuminance/duration dependence and the effects of quinacrine, mellitin and lithium. Light-evoked arachidonic acid release, *Graefes Arch Clin Exp Ophthalmol*, 232:167-75.
10. Lolley, R. N., Rong, H. and Craft, C. M., 1994, Linkage of photoreceptor degeneration by apoptosis with inherited defect in phototransduction, *Invest Ophthalmol Vis Sci*, 35:358-62.
11. Oberhammer, F., Fritsch, G., Schmied, M., Pavelka, M., Printz, D., Purchio, T., Lassmann, H. and Schulte-Hermann, R., 1993, Condensation of the chromatin at the membrane of an apoptotic nucleus is not associated with activation of an endonuclease, *J Cell Sci*, 104:317-26.
12. Organisciak, D. T. and Winkler, B. S., 1994, Retinal Light Damage: Practical and theoretical considerations, *Progress in Retinal and Eye Research (Pergamon Press Ltd., Oxford, England)*, 13:1-29.
13. Penfold, P. L. and Provis, J. M., 1986, Cell death in the development of the human retina: phagocytosis of pyknotic and apoptotic bodies by retinal cells, *Graefes Arch Clin Exp Ophthalmol*, 224:549-53.
14. Portera-Cailliau, C., Sung, C. H., Nathans, J. and Adler, R., 1994, Apoptotic photoreceptor cell death in mouse models of retinitis pigmentosa, *Proc Natl Acad Sci U S A*, 91:974-8.
15. Ratan, R. R., Murphy, T. H. and Baraban, J. M., 1994, Macromolecular synthesis inhibitors prevent oxidative stress-induced apoptosis in embryonic cortical neurons by shunting cysteine from protein synthesis to glutathione, *J Neurosci*, 14:4385-92.
16. Reed, J. C., 1994, BCL-2 and the regulation of programmed cell death, *J Cell Biol*, 124:1-6.
17. Reinboth, J. J., Gautschi, K. and Remé, C. E., 1994, Light damage in the rat retina: arachidonic acid metabolites may mediate inflammatory response, *Invest. Opthalmol. Vis. Sci.*, Suppl. 35:2138.
18. Remé, C. E., Aeberhard, B. and Schoch, M., 1985, Circadian rhythm of autophagy and light responses of autophagy and disk-shedding in the rat retina, *J Comp Physiol A*, 156:669-677.
19. Remé, C. E., Braschler, U. F., Roberts, J. and Dillon, J., 1991, Light damage in the rat retina: effect of a radioprotective agent (WR-77913) on acute rod outer segment disk disruptions, *Photochem Photobiol*, 54:137-42.
20. Remé, C. E., Malnoè, A., Jung, H. H., Wei, Q. and Munz, K., 1994, Effect of dietary fish oil on acute light-induced photoreceptor damage in the rat retina, *Invest Ophthalmol Vis Sci*, 35:78-90.
21. Remé, C. E., Szczesny, P. J. and Munz, K., 1994, Light damage in the rat retina: from treshold lesion to inflammatory response and recovery or scar formation, *Invest. Ophthalmol. Vis. Sci.*, Suppl. 35:2138.
22. Remé, C. E., Wei, Q., Munz, K., Jung, H., Doly, M. and Droy-Lefaix, M. T., 1992, Light and lithium effects in the rat retina: modification by the PAF antagonist BN 52021, *Graefes Arch Clin Exp Ophthalmol*, 230:580-8.
23. Schwartz, L. M. and Osborne, B. A., 1993, Programmed cell death, apoptosis and killer genes, *Immunol Today*, 14:582-90.
24. Schwartzman, R. A. and Cidlowski, J. A., 1993, Apoptosis: the biochemistry and molecular biology of programmed cell death, *Endocr Rev*, 14:133-51.
25. Shahinfar, S., Edward, D. P. and Tso, M. O., 1991, A pathologic study of photoreceptor cell death in retinal photic injury, *Curr Eye Res*, 10:47-59.
26. Szczesny, P. J., Remé, C. E. and Munz, K., 1994, Is apoptosis of photoreceptors and pigment epithelium an early response in light damage of the rat retina?, *Invest. Ophthalmol. Vis. Sci*, Suppl. 35:2137.
27. Tso, M. O., Zhang, C., Abler, A. S., Chang, C. J., Wong, F., Chang, G. Q. and Lam, T. T., 1994, Apoptosis leads to photoreceptor degeneration in inherited retinal dystrophy of RCS rats, *Invest Ophthalmol Vis Sci*, 35:2693-9.
28. Weller, M., Constam, D. B., Malipiero, U. and Fontana, A., 1994, Transforming growth factor-beta 2 induces apoptosis of murine T cell clones without down-regulating bcl-2 mRNA expression, *Eur J Immunol*, 24:1293-300.
29. Winkler, B. S. and Giblin, F. J., 1983, Glutathione oxidation in retina: effects on biochemical and electrical activities, *Exp Eye Res*, 36:287-97.
30. Wyllie, A. H., Kerr, J. F. and Currie, A. R., 1980, Cell death: the significance of apoptosis, *Int Rev Cytol*, 68:251-306.
31. Young, R. W., 1984, Cell death during differentiation of the retina in the mouse, *J Comp Neurol*, 229:362-73.

4

INHIBITORY EFFECTS OF CYCLOHEXIMIDE AND FLUNARIZINE ON LIGHT-INDUCED APOPTOSIS OF PHOTORECEPTOR CELLS

Suhui Li, Cheng-Jong Chang, Andrew S. Abler, and Mark O. M. Tso

Georgiana Dvorak Theobald Ophthalmic Pathology Laboratory
Department of Ophthalmology and Visual Sciences
University of Illinois at Chicago
1855 W. Taylor Street, Room L217, Chicago, Illinois

INTRODUCTION

The pathogenesis of photic retinopathy has been actively investigated for many years. Although the exact pathogenic mechanism involved in light-induced photoreceptor degeneration remains unknown, certain hypotheses were made based on previous animal studies [1-8]. Free radical formation and lipid peroxidation are among the most widely accepted hypotheses regarding the pathogenesis of photic retinopathy [1-5]. In addition, possible roles for protein synthesis and alteration of intracellular Ca^{2+} concentration in light-induced photoreceptor cell death have been suggested [6-8]. Protein synthesis inhibitors, such as cycloheximide and Ca^{2+} channel overload blockers, such as flunarizine, were both demonstrated to have ameliorative effects on retinal photic injury [6-8]. These findings provided supportive evidence of the possible involvement of protein synthesis and alteration of intracellular Ca^{2+} concentration in retinal photic injury. However, the mechanism whereby these two factors ameliorated light-induced photoreceptor cell death was not determined.

Using morphological criteria, Shahinfar et al observed the presence of apoptotic photoreceptor cells in the early phase of photic retinopathy [6]. This intriguing observation opened a new approach to the understanding of light-induced photoreceptor cell death.

The regulation and mechanisms of apoptosis are complex. A number of intracellular events were described in apoptotic cells [9]. The activation of calcium/magnesium-dependent endonuclease and the resultant cleavage of the double-stranded DNA into monomers and multimers of 180-200 base pair DNA subunits is believed to be an important step involved in apoptosis [10,11]. Protein synthesis [12-19] and alteration of intracellular calcium concentration [17,20-25] are among the other intracellular events associated with apoptosis. However, these two intracellular events are not required for all examples of apoptosis [26,27]

Degenerative Diseases of the Retina, Edited by Robert E. Anderson et al.
Plenum Press, New York, 1995

and have not been shown to occur in apoptosis of photoreceptor cells secondary to photic injury.

In the present study, we attempted to explore the regulatory mechanisms of light-induced photoreceptor cell death. We studied the effect on photic retinopathy of intravitreally injected cycloheximide, a protein synthesis inhibitor, and flunarizine, a Ca^{2+} channel overload blocker. The criteria used to define apoptosis included morphology of the light-injured photoreceptor cells, terminal deoxynucleotidyl transferase (TdT)-mediated biotin-dUTP nick end labeling (TUNEL), and agarose gel electrophoresis of retinal DNA. These morphological, immunohistochemical and biochemical criteria for apoptosis were successfully applied to the apoptotic photoreceptor cells by Chang et al in rd, rds and rhodopsin mutant mice and by Tso et al in RCS rats [28,29].

MATERIALS AND METHODS

Preparation of Reagents

Cycloheximide and flunarizine (Sigma Chemical Co., St. Louis, MO) were dissolved in 0.9% sterile saline and 40% dimethyl sulfoxide (DMSO) in phosphate-buffered saline (PBS), respectively. The concentrations used for intravitreal injections were 0.25 mg/ml of cycloheximide and 1.5 mg/ml of flunarizine. The doses of cycloheximide and flunarizine were designed based on those used in previous studies [6,7,13,30].

Animal Preparation

Forty-eight 35-day-old male albino Lewis rats (Harlan Sprague Dawley, Inc., Indianapolis, IN), weighing 180-220 grams, were divided into six groups of eight rats each. All rats were kept in 12-hour cycles of light (5 foot candles) and darkness for 14 days and dark-adapted for 24 hours. They were then exposed to fluorescent light (490-580 nm, 300-320 foot candles) for 12 hours at 26°C. Immediately after light exposure, intravitreal injections of 2 ul of cycloheximide (0.25 mg/ml) (group 1), normal saline (vehicle control for cycloheximide) (group 2), flunarizine (1.5 mg/ml) (group 3), or 40% DMSO in PBS (vehicle control for flunarizine) (group 4) were given to the first four groups. The injection site was 1 mm behind the limbus at the junction between the peripheral retina and ciliary body. Rats in group 5 were similarly punctured with a needle but did not receive an intravitreal injection. They served as light-exposed, punctured controls. Group 6 did not receive any intravitreal injection or puncture and served as light-exposed, non-treated controls. All rats were allowed to recover in total darkness until they were killed 24 hours after light exposure.

From each group the left eye of each animal was enucleated for DNA agarose gel electrophoresis. Four of the eight right eyes from each group were prepared for TUNEL, and the remaining four right eyes were prepared for morphologic and morphometric studies. All procedures involving animals were in accordance with the Resolution on the Use of Animals In Research established by the Association for Research in Vision and Ophthalmology.

Morphologic and Morphometric Study

Eyes enucleated for morphologic study were fixed in 4% buffered formaldehyde and 1% glutaraldehyde. The anterior segment was removed, and the posterior segment was divided into superior, inferior, nasal, and temporal quadrants, and the tissue was osmicated, dehydrated in graded alcohol, and embedded in epoxy resin. Sections from all quadrants were then evaluated morphologically for light damage. Morphometry of the outer nuclear layer (ONL) was performed by measurement of the ONL thickness along the entire length of the section of each quadrant using a customized image processing system described previously [7]. Measurements obtained from four quadrants of each eye were averaged to obtain a representative ONL thickness for the entire eye. Ultrathin sections from the representative areas of the superior quadrants were examined by electron microscopy.

TUNEL

Eyes enucleated for TUNEL were kept in Davison's fixative overnight at 4°C. The eyes were opened superiorly at the ora serrata to remove the lens. They were then bisected vertically through the optic nerve head, processed, and embedded for paraffin sectioning. TUNEL was performed *in situ* on 5-um-thick paraffin sections essentially as described by Gavrieli et al [31] except that the proteinase K step for tissue digestion was omitted. In this technique, the tissue sections were deparaffinized, rehydrated, and incubated in methanol containing 0.3% H_2O_2 to block endogenous peroxidase. The sections were washed in distilled water. After a rinse in TdT buffer (30 mM Trizma base, pH 7.2, 140 mM sodium cacodylate, and 1 mM cobalt chloride), the sections were incubated with TdT buffer containing TdT and biotinylated dUTP for 60 minutes. Then, TB buffer (300 mM sodium chloride and 30 mM sodium citrate) was used to terminate the reaction. The sections were washed with double distilled water and incubated with peroxidase-conjugated streptavidin. Staining was developed in diaminobenzidine (DAB).

Quantification of TUNEL-Positive Photoreceptor Nuclei

TUNEL-positive nuclei in the ONL were quantified from two selected segments of each retina, each of which measured 0.4 mm in length. The first segment was chosen to be located 0.4 mm superior to the optic nerve head, and the second segment was 0.4 mm superior to the first segment. Two sections from each retina were counted. The average number of TUNEL-positive nuclei in each 0.4-mm retinal segment was used to represent the density of TUNEL-positive photoreceptor nuclei of each retina.

Retinal DNA Agarose Gel Electrophoresis

Eyes enucleated for DNA agarose gel electrophoresis were bisected at the ora serrata and the lens was removed. The retinas were dissected from the underlying tissues and were transferred to liquid nitrogen. They were then kept frozen in -70°C until extraction and analysis of DNA with a previously described method [32]. With this technique, the frozen retinas were thawed and vortexed in a buffer containing 50 mM of ethylenediaminetetraacetic acid (EDTA), 0.5% SDS and 20 mM of Tris-hydrochloride at pH 8.0. RNase A was added to a final concentration of 100 ug/ml. After incubation for 30 minutes at 42°C, proteinase K was added to a final concentration of 400 ug/ml. The samples were then incubated at 55°C until clear lysates were produced. The lysates were extracted with phenol/chloroform/isoamyl alcohol (25:24:1) and chloroform/isoamyl alcohol (24:1) before precipitation of DNA with 3M of sodium acetate and ice cold ethanol. The DNA was analyzed for internucleosomal cleavage by electrophoresis through a 2% agarose gel. DNA in the gel was visualized by ultraviolet light after staining with ethidium bromide and was photographed using a Polaroid MP-4 system.

Statistical differences in the ONL thickness and the counts of TUNEL-positive photoreceptor nuclei between the drug-treated and non-treated groups were analyzed by one-way analysis of variance.

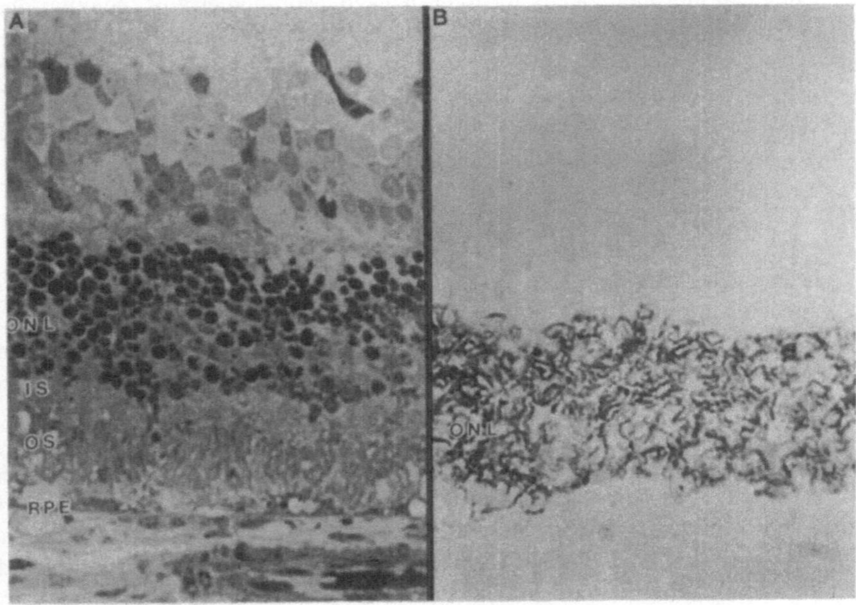

Figure 1. Morphological changes **(A)** and TUNEL labeling **(B)** of light-exposed, non-treated control retinas at 24 hours of dark recovery after 12 hours of light exposure. IS: inner segments. OS: outer segments.

RESULTS

Morphological Evaluation by Light Microscopy

The light-exposed control retinas that received no treatment exhibited thinning and disorganization of the ONL with marked loss of photoreceptor cells and presence of numerous densely stained and shrunken nuclei. Shrunken nuclei were occasionally noted in the inner segments. The inner and outer segments were distorted and shortened. The retinal pigment epithelium (RPE) was mildly vacuolated (Figure 1A). The morphological changes of the light-exposed, punctured control retinas were comparable to those of the light-exposed, non-treated control retinas (data not shown).

Retinas of the vehicle (0.9% normal saline) controls for cycloheximide exhibited mild thinning and disorganization of the ONL with mild loss of photoreceptor cells and presence of scattered, densely stained, and shrunken nuclei. The inner and outer segments were markedly distorted. The RPE showed mild vacuolation (Figure 2A). The cycloheximide-treated retinas appeared similar to the vehicle control retinas; however, the inner and outer segments were better aligned (Figure 2B).

Retinas of vehicle (40% DMSO) controls for flunarizine showed mild thinning and disorganization of the ONL with scattered densely stained and shrunken nuclei. The inner and outer segments were well aligned. Macrophages were noted in the subretinal space. Focal loss of RPE was seen (Figure 2C). The ONL of the flunarizine-treated retinas appeared similar to the vehicle control retinas but with less densely stained, and shrunken nuclei. The inner and outer segments were distorted. The RPE was only mildly vacuolated. No focal loss of the RPE was observed (Figure 2D).

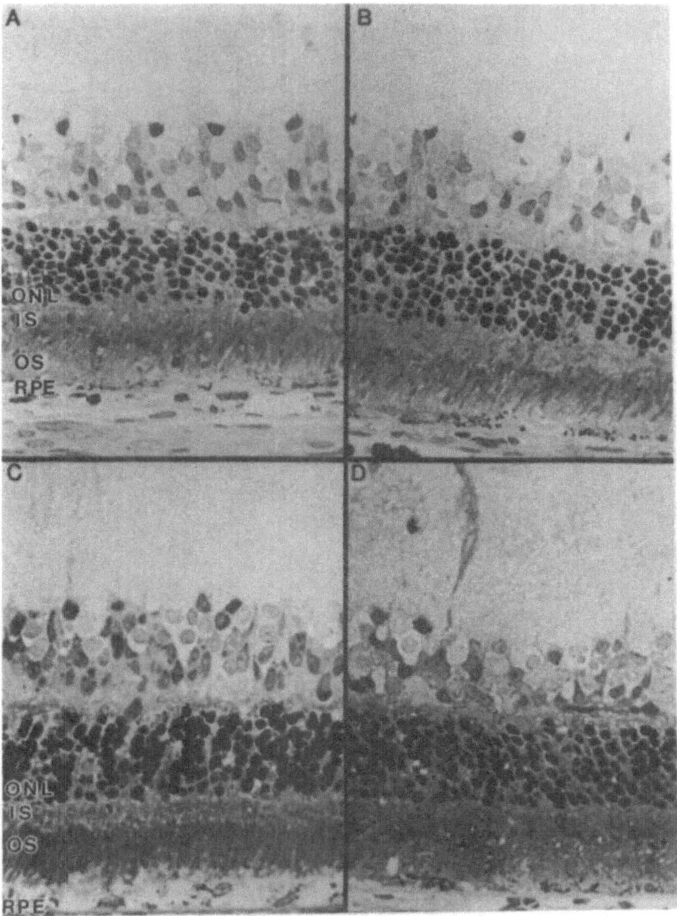

Figure 2. Effects of cycloheximide and flunarizine on retinal morphological changes at 24 hours of dark recovery after 12 hours of light exposure. **(A)** Retina of vehicle (0.9% normal saline) control for cycloheximide. There was mild thinning and disorganization of the ONL and scattered densely stained, and shrunken nuclei. Inner segments (IS) and outer segments (OS) were markedly distorted. RPE showed mild vacuolation. **(B)** Cycloheximide-treated retina. Compared with retina of vehicle control, the ONL appeared similar but the IS and OS were better aligned. RPE was mildly vacuolated. **(C)** Retina of vehicle (40% DMSO) control for flunarizine. There was mild thinning and disorganization of the ONL and scattered densely stained, and shrunken nuclei. IS and OS were well aligned. Macrophages were noted in the subretinal space. Focal loss of RPE was seen and the remaining RPE showed marked vacuolation. **(D)** Flunarizine-treated retina. ONL appeared similar to that of the vehicle control retina but with less densely stained and shrunken nuclei. However, the IS and OS appeared more distorted than those of the vehicle control retina. Mild vacuolation of RPE was seen. No focal loss of RPE was observed.

Morphologic Evaluation by Electron Microscopy

Morphologically, the light-injured photoreceptor cells in the treated and control retinas exhibited typical ultrastructural features of apoptosis by electron microscopy. The light-exposed, non-treated control retinas are shown in Figure 3. The light-injured photoreceptor cells were scattered throughout the ONL. There was early central condensation of nuclear chromatin with islands of densification. The densely stained, and shrunken nuclear

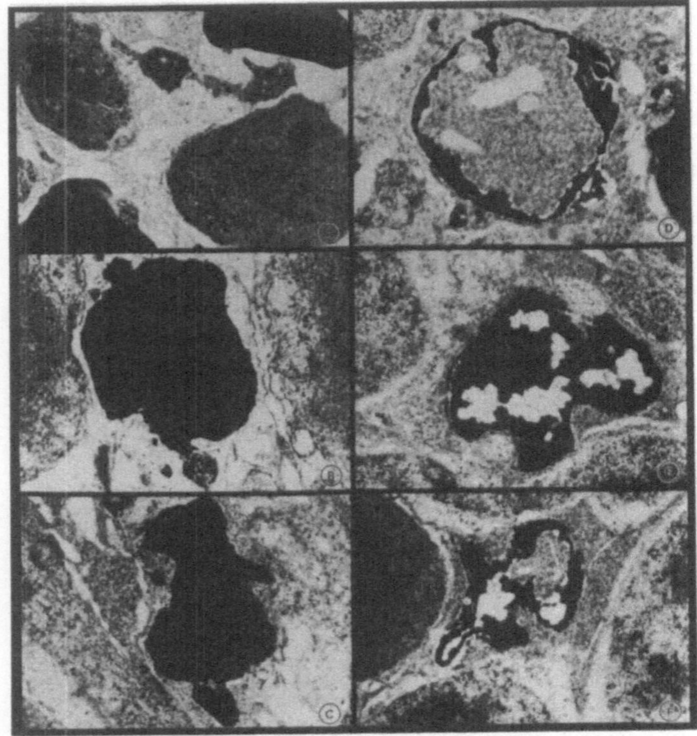

Figure 3. Representative nuclear changes in the light-injured photoreceptor nuclei at 24 hours of dark recovery after 12 hours of light exposure. **(A)** Early nuclear changes included central condensation of chromatin with islands of densification. Margination of chromatin to the periphery of the nucleus was observed with a perinuclear halo. **(B)** Chromatin became uniformly dense. **(C)** Chromatin aggregated into circumscribed dense masses that gave an electron-lucent appearance centrally. **(D, E)** Further chromatin margination gave an electron-lucent appearance centrally. **(F)** End stage of photoreceptor cell degeneration showed dissolution of chromatin, breaks in nuclear membrane, and disintegration of cytoplasmic contents (X 15,000).

chromatin underwent margination followed by disintegration with subsequent formation of apoptotic bodies. The cellular membrane and the intracellular organelles remained intact during this process.

Morphometry of the ONL

Morphometrically, no statistically significant difference in ONL thickness was noted between the saline-treated and cycloheximide-treated retinas ($p = 0.97$) or between 40% DMSO-treated and flunarizine-treated retinas ($p = 0.83$) (data not shown).

TUNEL

The light-exposed, non-treated retinas revealed extensive and intense labeling of the photoreceptor nuclei throughout the ONL (Figure 1B).

Retinas of the vehicle (0.9% normal saline) controls for cycloheximide showed intense, ring-like but less extensive, labeling of the photoreceptor nuclei throughout the ONL

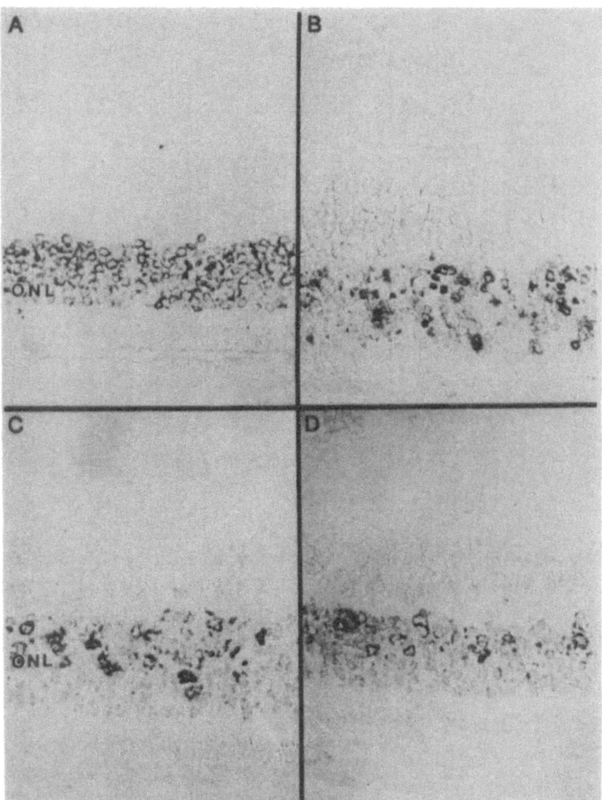

Figure 4. Effects of cycloheximide and flunarizine on TUNEL of photoreceptor cells at 24 hours of dark recovery after 12 hours of light exposure. **(A)** Retina of vehicle (0.9% normal saline) control for cycloheximide. Intense, ring-like but less extensive labeling of the photoreceptor nuclei was seen throughout the entire ONL. **(B)** Cycloheximide-treated retina. Only scattered photoreceptor nuclei were labeled. **(C)** Retina of vehicle (40% DMSO) control for flunarizine. Scattered but intensely labeled photoreceptor nuclei were seen throughout the entire ONL. **(D)** Flunarizine-treated retina. Scattered and fewer labeled photoreceptor nuclei were seen mainly along the inner aspect of the ONL.

(Figure 4A). In contrast, only scattered photoreceptor nuclei were labeled in the cycloheximide-treated retinas (Figure 4B).

In flunarizine control retinas, scattered, intensely labeled photoreceptor nuclei were seen. In contrast, flunarizine-treated retinas showed fewer labeled photoreceptor nuclei. The labeled photoreceptor nuclei were distributed mainly along the inner aspect of the ONL (Figure 4D).

Quantification of TUNEL-Positive Photoreceptor Nuclei

By cell counting, cycloheximide- and flunarizine-treated retinas had less TUNEL-positive photoreceptor nuclei than did their corresponding vehicle control retinas ($p = 0.046$ and $p = 0.025$, respectively) (Figure 5).

Retinal DNA Agarose Gel Electrophoresis

The cycloheximide- and flunarizine-treated groups showed significant inhibition of DNA fragmentation, whereas the three control groups showed the typical ladder pattern of DNA fragmentation seen in apoptosis (Figure 6).

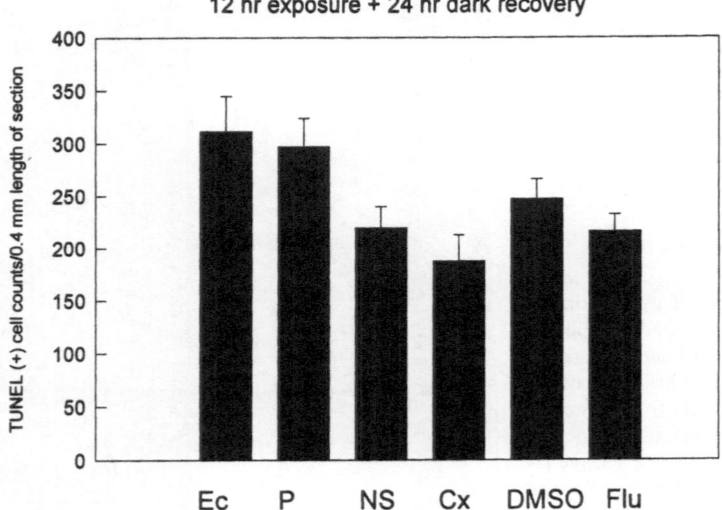

Figure 5. Effects of cycloheximide and flunarizine on the average number of TUNEL-positive photoreceptor nuclei (mean ± S.D.) in the retina at 24 hours of dark recovery after 12 hours of light exposure. There were statistically significant differences between the saline-treated (NS) and cycloheximide (0.25 mg/ml)-treated (Cx) retinas ($p = 0.046$). A significant reduction of TUNEL-positive photoreceptor cells also was noted in the 40% DMSO-treated and flunarizine (1.5 mg/ml)-treated (Flu) retinas ($p = 0.025$). Ec: light-exposed, non-treated controls. P: exposed, punctured controls.

DISCUSSION

Our previous research suggested that photoreceptor cells died via apoptosis in the early phase of photic injury on morphologic ground [6]. Apoptosis appears to be an energy-dependent, multi-step process of cell death that typically involves the loss of scattered individual cells without a marked inflammatory response [9]. Numerous studies have implicated ongoing or *de novo* protein synthesis and elevated intracellular calcium in the process of apoptosis [12-14,17-19], although occasional exceptions have also been reported [26,27]. We recently showed that cycloheximide, a protein synthesis inhibitor, and flunarizine, a calcium channel overload blocker, ameliorated photic retinopathy [6-8]. We speculated that these agents might act by blocking apoptosis. Therefore, we performed this study to determine if intravitreal injection of cycloheximide or flunarizine inhibited light-induced apoptosis of photoreceptor cells, as judged by morphology, morphometry, TUNEL, and electrophoretic analysis of retinal DNA.

The light-injured photoreceptor cells (Figures 1-3) in our study exhibited typical light and electron microscopic features of apoptosis [6,9]. Biochemically [11], apoptosis is characterized by internucleosomal double-stranded DNA cleavage, producing DNA fragments that are multiples of 180-200 bp, which appear in agarose gel electrophoresis as a ladder pattern. In contrast, the random breakdown of DNA in necrosis yields a smear after gel electrophoresis. The double-stranded DNA nicks in apoptosis could also be demonstrated by the positive staining with TUNEL in paraffin sections [31]. The light-injured photoreceptor cells in our study were shown to undergo apoptosis by agarose gel electrophoresis of retinal DNA and by TUNEL.

Our results indicated that both cycloheximide and flunarizine inhibited light-induced apoptosis of photoreceptor cells. The cycloheximide- and flunarizine-treated groups showed

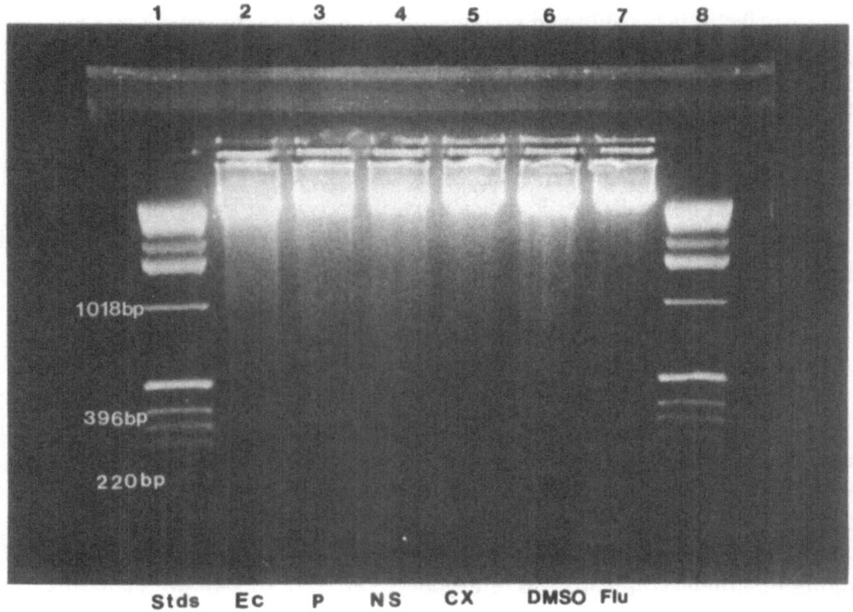

Figure 6. Retinal DNA agarose gel electrophoresis. Lanes 1 and 8 contained molecular weight standards (Stds). Lanes 2 through 6 contained the retinal DNA from light-exposed control (Ec) group, exposed, punctured control (P) group, normal saline (NS)-treated group, cycloheximide (Cx)-treated group, 40% DMSO-treated group, and flunarizine (Flu)-treated group, respectively. The three control groups showed the typical ladder pattern of apoptosis, whereas the cycloheximide- and flunarizine-treated groups showed significant inhibition of DNA fragmentation.

a better preserved outer retina, fewer TUNEL-positive nuclei, and less internucleosomal DNA cleavage, compared with their respective control groups. During this study, we also noted that dry needle puncture or injection of saline into the subretinal space prevented, to a mild degree, light-induced apoptosis of photoreceptor cells, suggesting that intraocular puncture may cause the release or production of factors that rescue photoreceptor cells from an apoptotic fate. To minimize the confounding influence of intraocular puncture on our analysis, we performed all injections 1 mm behind the limbus at the junction between the peripheral retina and ciliary body.

Inhibition of protein synthesis blocks apoptosis in many tissue systems. Wyllie et al [17] showed that administration of protein synthesis inhibitor cycloheximide resulted in suppression or delay of apoptosis in glucocorticoid-treated rat thymocytes. Cohen and Duke [18] noted decreased apoptosis in glucocorticoid-treated lymphocytes when the cells also were treated by cycloheximide. Martin et al [12] studied apoptosis of dissociated sympathetic neurons following deprivation of nerve growth factor. and the death process was prevented by cycloheximide, puromycin, actinomycin D, and anisomycin. Oppenheim et al [13] used an *in vivo* chick embryo model to show that apoptosis of developing neurons induced by target deprivation (limb removal) and axonotomy was prevented by cycloheximide, puromycin and actinomycin. Ghibelli et al [14] demonstrated that cycloheximide could rescue mouse L cells from apoptosis induced by heat shock. Johnson and Deckworth [19] demonstrated RNA and protein synthesis inhibitors prevented the cell death of sympathetic sensory and motor neurons induced by trophic factor deprivation both *in vitro* and *in vivo*. However, not all apoptotic processes involved protein synthesis. Cycloheximide had an inhibitory effect on apoptosis induced by cytotoxic T lymphocytes [26]. Furthermore, cell death caused by tumor necrosis factor was facilitated by inhibition of protein synthesis [27]. The seemingly contradictory roles of protein synthesis in apoptotic processes in different tissue systems need to be further explored. Umansky [33] proposed that there are two groups of proteins in the cell. Activation of one group initiates apoptosis, whereas, activation of the second group inhibits apoptosis.

Calcium overload has been noted in many apoptotic processes. Duke et al [21] described a calcium-dependent endonuclease, which produced DNA fragmentation in glucocorticoid-induced apoptosis in thymocytes. Wyllie and coworkers [17] showed that calcium ionophore induced DNA fragmentation in thymocytes and that this effect could be blocked by cycloheximide or actinomycin D. McConkey et al [22,23] showed that increased intracellular level of Ca^{2+} in immature thymocytes could activate apoptotic DNA fragmentation. They further demonstrated that apoptosis in thymocytes induced by anti-CD3 antibodies resulted in a sustained calcium increase which preceded endonuclease activity and cell death [24,25]. Furthermore, the presence of intracellular or extracellular calcium chelators blocked both DNA fragmentation and apoptosis. The exact regulatory mechanism of intracellular Ca^{2+} on apoptosis is not clear. It might regulate apoptosis through its effect on a Ca^{2+}- and Mg^{2+}-dependent endonuclease that mediates DNA fragmentation [27].

Although both cycloheximide and flunarizine were previously proved to be effective in ameliorating light-induced photoreceptor degeneration, little was known about their mechanism of action [6-8]. Our study demonstrated that these two drugs may ameliorate light-induced photoreceptor degeneration by inhibiting apoptosis of photoreceptor cells. These observations also implicate protein synthesis and increase of intracellular Ca^{2+} concentration in the regulation and/or execution of light-induced apoptosis of photoreceptor cells.

CONCLUSIONS

Apoptosis is involved in light-induced photoreceptor cell death. Light-induced apoptosis of photoreceptor cells could be inhibited by intravitreal injection of cycloheximide and flunarizine, suggesting that protein synthesis and elevated intracellular Ca^{2+} are involved in light-induced apoptosis of photoreceptor cells.

ACKNOWLEDGMENTS

We gratefully acknowledge Cynthia Haynes and Shihong Li for their assistance in this study. This work was supported in part by research grants EY06761, Amelioration of Retinal Photic Injury; EY01903, Pathology of Retinal Dysfunction; core grant EY01792 from the National Eye Institute, Bethesda, Maryland; gifts from the estate of Harry J. Willwerth, Palatine, Illinois; the Lions of Illinois Foundation, Hillside, Illinois; and an unrestricted research grant from Research to Prevent Blindness, Inc., New York, New York. Dr. Mark O. M. Tso is a Research to Prevent Blindness Senior Scientific Investigator and is on leave from the University of Illinois at Chicago as Professor of Ophthalmology, Department of Ophthalmology and Visual Sciences, Chinese University of Hong Kong.

REFERENCES

1. Feeney L, Berman ER. Oxygen toxicity: membrane damage by free radicals. Invest Ophthalmol Vis Sci 1976;15:789-792.

2. Noell WK. Possible mechanisms of photoreceptor damage by light in mammalian eyes. Vision Res 1980;20:1163-1171.
3. Anderson RE, Rapp LM, Wiegand RD. Lipid peroxidation and retinal degeneration. Current Eye Res 1984;3(1):223-227.
4. Organisciak DT, Wang HM, Xie A, Reeves DS and Donoso LA. Intense-light mediated changes in rat rod outer segment lipids and proteins. Prog Clin Biol Res 1989;314:493-512.
5. Tso MOM. Experiments on visual cells by nature and man: in search of treatment for photoreceptor degeneration. Invest Ophthalmol Vis Sci 1989;30:2430-2454.
6. Shahinfar S, Edward DP and Tso MOM. A pathologic study of photoreceptor cell death in retinal photic injury. Current Eye Res 1991;10:47-59.
7. Edward DP, Lam TT, Shahinfar S, Li J and Tso MOM. Amelioration of light-induced retinal degeneration by a calcium overload blocker flunarizine. Arch Ophthalmol 1991;109:554-562.
8. Li J, Edward DP, Lam TT, Tso MOM. Amelioration of retinal photic injury by a combination of flunarizine and dimethylthiourea. Exp Eye Res 1993;56:71-78.
9. Tomei LD, Cope FO. Apoptosis: the molecular basis of cell death. Cold Spring Harbor, NY: Cold Spring Harbor Laboratory Press;1991.
10. Compton MM. A biochemical hallmark of apoptosis: internucleosomal degeneration of the genome. Cancer Metastasis Rev 1992;11:105-119.
11. Deckwerth TL, Johnson EM Jr. Temporal analysis of events associated with programmed cell death (apoptosis) of neurons deprived of nerve growth factor. J Cell Biol 1993;123:1207-1222.
12. Martin DP, Schmidt RE, DiStefano PS, Lowry OH, Carter JG, Johnson EMJ. Inhibitors of protein synthesis and RNA synthesis prevent neuronal death caused by nerve growth factor deprivation. J Cell Biol 1988;106:829-844.
13. Oppenheim RW, Prevette D, Tytell M, Homma S. Naturally occurring and induced neuronal death in the chick embryo in vivo requires protein and RNA synthesis: evidence for the role of cell death genes. Dev Biol 1990;138:104-113.
14. Ghibelli L, Nosseri C, Oliverio S, Piacentini M, Autuori F. Cycloheximide can rescue heat-shocked L cells from death by blocking stress-induced apoptosis. Exp Cell Res 1992;201:436-443.
15. Martin SJ. Protein or RNA synthesis inhibition induces apoptosis of mature human CD4+ T cell blasts. Immunol Lett 1993;35:125-134.
16. Inouye M, Tamaru M, Kameyama Y. Effects of cycloheximide and actinomycin D on radiation-induced apoptotic cell death in the developing mouse cerebellum. Int J Rad Biol 1992;61:669-674.
17. Wyllie, AH, Morris RG, Smith AL, Dunlop D. Chromatin cleavage in apoptosis: association with condensed chromatin morphology and dependence on macromolecular synthesis. J Pathol 1984;142:67-77.
18. Cohen JJ, Duke RC. Glucocorticoid activation of a calcium-dependent endonuclease in thymocyte nuclei leads to cell death. J Immunol 1984;132:38-42.
19. Johnson EM Jr, Deckwerth T. Molecular mechanisms of developmental neuronal death. Annu Rev Neurosci 1993;16:31-46.
20. Jones DP, McCondey DJ, Nicotera P, Orrenius S. Calcium-activated DNA fragmentation in rat liver nuclei. J Biol Chem 1989;264:6398-6403.
21. Duke RC, Chervenak R, Cohen JJ. Endogenous endonuclease-induced DNA fragmentation: an early event in cell-mediated cytolysis. Proc Natl Acad Sci USA 1983;80:6361-6365.
22. McConkey DJ, Hartzell P, Duddy SK, Hakansson H, Orrenius S. 2,3,7,8 tetrachlorodibenzo-p-dioxin kills immature thymocytes by Ca^{2+}-mediated endonuclease activation. Science 1988;242:256-259.
23. McConkey DJ, Nicotera P, Hartzell P, Bellomo G, Wyllie AH, Orrenius S. Glucocorticoid activate a suicide process in thymocytes through an elevation of cytosolic Ca^{2+} concentration. Arch Biochem Biophys 1989;269:365-370.
24. McConkey DJ, Orrenius S, Fondal M. Cellular signaling in programmed cell death (apoptosis). Immunol Today 1990;11:120-121.
25. McConkey DJ, Jondal M, Orrenius S. Cellular signaling in thymocyte apoptosis. Semin Immunol 1992;4:371-377.
26. Collins RJ, Harmon BV, Souvlis T, Pope JH, Kerr JF. Effects of cycloheximide on B-chronic lymphocytic leukaemic and normal lymphocytes in vitro: induction of apoptosis. Br J Cancer 1991;64:518-522.
27. Belloma G, Perotti M, Taddei F et al. Tumor necrosis factor alpha induces apoptosis in mammary adenocarcinoma cells by an increase in intranuclear free Ca^{2+} concentration and DNA fragmentation. Cancer Res 1992;52:1342-1346.
28. Chang GQ, Hao Y, Wong F. Apoptosis: Final common pathway of photoreceptor death in rd, rds and rhodopsin mutant mice. Neuron. 1993;11:595-605

29. Tso MOM, Zhang C, Abler AS, et al. Apoptosis leads to photoreceptor degeneration in inherited retinal dystrophy of RCS rats. Invest Ophthalmol Vis Sci 1994;2693-2699.
30. Rich KM, Hollowell JP. Flunarizine protects neurons from death after axotomy or NGF deprivation. Science 1990;248:1419-1421.
31. Gavriela Y, Sherman Y, Ben-Sasson SA. Identification of programmed cell death in situ via specific labeling of nuclear DNA. J Cell Biol 1992;119:493-501.
32. Tilly JL, Hsueh AJ. Microscale autoradiographic method for the qualitative and quantitative analysis of apoptotic DNA fragmentation. J Cell Physiol 1993;153(3):519-526.
33. Umansky SR. Apoptotic process in the radiation-induced death of lymphocytes, in Tomei LD, Cope FO, eds. Apoptosis: The molecular basis of cell death. Cold Spring Harbor, NY: Cold Spring Harbor Laboratory Press 1991:193-208.

5

APOPTOSIS OF PHOTORECEPTORS AND LENS FIBER CELLS WITH CATARACT AND MULTIPLE TUMOR FORMATION IN THE EYES OF TRANSGENIC MICE LACKING THE P53 GENE AND EXPRESSING THE HPV 16 E7 GENE UNDER THE CONTROL OF THE IRBP PROMOTER

David S. Papermaster,[1,2] Kim Howes,[2] Nancy Ransom,[1] and Jolene J. Windle[1,3]

[1] Department of Pathology
[2] Department of Cellular and Structural Biology
[3] The Cancer Research Center
University of Texas Health Science Center at San Antonio
San Antonio, Texas 78284-7750

SUMMARY

Because of the clinical heterogeneity of human inherited retinal degenerations, we have initiated a study to seek common themes in the pathogenesis of cell death of photoreceptors by apoptosis. The interstitial retinol binding protein (IRBP) promoter was used to drive expression of the human papilloma virus 16 (HPV16) E7 gene in the retina and other ocular tissues in mice. The result is the death of photoreceptors as they undergo terminal differentiation. Lens fiber cells also die after a period of inappropriate proliferation and abnormal differentiation to form cataracts. Cross-breeding these transgenic mice to mice lacking the p53 gene leads to formation of several ocular tumors by one month of age if both copies of the p53 gene are missing. With one copy of the gene, the mice develop retinal tumors after a much longer latency and at a lower incidence; the tumors that do arise have lost their normal copy of the gene. The lack of the p53 gene does not eliminate apoptosis of either the retina or the lens in these transgenic mice although the rate of destruction of photoreceptors is slightly delayed. The retinal tumors apparently arise from precursors that survive amid a dying cell layer.

Degenerative Diseases of the Retina, Edited by Robert E. Anderson et al.
Plenum Press, New York, 1995

INTRODUCTION

The death of photoreceptors in inherited retinal degenerations presents dilemmas at a clinical, cellular and molecular level. Unlike the inherited degenerations of mice and rats, human families have a highly varied degree of penetrance in many forms of retinitis pigmentosa, for example, so that some family members are significantly affected while others are virtually spared yet they carry the same mutation. Moreover, in a single family, some members may have one clinical form of the disease, e.g. death of the peripheral retina, while others have greater loss in the macula. The sources of this heterogeneity are not explained by the mutations that have so far been described since some of these problems of heterogeneity are characteristic of families carrying mutations of the rhodopsin gene or the rds/peripherin gene which should affect all rods and all photoreceptors, respectively. At the cellular level, an increasing body of evidence favors a mechanism of programmed cell death called apoptosis in both experimental retinal degenerations of mice and rats and now in humans [1-6] .

Several years ago a group of laboratories independently and simultaneously discovered evidence of apoptosis in *rds* and *rd* mice and RCS *(rdy)* rats [2-6] and in albino rats exposed to constant bright light [7]. The cells demonstrated the classical morphologic features of apoptosis including condensation of the cell's contents, pyknosis and fragmentation of the nucleus, formation of apoptotic bodies and phagocytosis of dying photoreceptors by adjacent photoreceptor cells in addition to the RPE. The nuclear DNA became fragmented into nucleosome-sized multiples of 180 base pairs [3] and the nuclei could be labeled by various end-labeling techniques that permit the apoptotic cells to be detected *in situ*.[2, 5-7].

The stereotypical response of photoreceptors not only to inherited lesions but also to those induced by excess light, retinal detachment and transgenes of various sorts suggests that these cells live on the edge of survival and are triggered by diverse stimuli to initiate an internal death program. Little is yet understood about the genes involved in cell death or the degree to which the pathway(s) can be abrogated but the recognition of the apoptotic mechanism in some if not all inherited retinal degenerations has altered the direction of research in many laboratories including ours. One fascination of this field of inquiry is that the outcomes of experiments do not always parallel the theories that are derived from the study of invertebrates or cells in tissue culture. Thus the investment of energy into studies of the impact of genes on photoreceptor survival in the intact eye becomes a valuable counterpoint to *in vitro* research.

The obvious next questions revolve around two issues. a) How does any mutation stimulate the cell to respond by activating apoptosis? and b) If the apoptotic pathway(s) could be interrupted, could the photoreceptor cell survive in a useful state, even if the mutation were uncorrected?

An attempt to evaluate the first question in one inherited retinal degeneration by the addition of exogenous growth factors to the eyes of RCS rats induced partial rescue of their retinas [8]. This result suggested that in this mutation, at least, growth factors or sensitivity to them may be a limiting component in maintaining the health of the photoreceptors. Similar rescue was generated in constant light-exposed albino rats [9]. Unfortunately, subsequent studies with various mouse strains indicate that the results might not be universal and might be species and strain specific [10].

Clinical studies already provide a hint about the answer to the second question, since a 65 year old patient with sector ADRP who lost the photoreceptors of the bottom half of his eye and of his fovea had a Thr17Met mutation of his rhodopsin gene [11] . Clearly half of his photoreceptors tolerated the existence of the mutation sufficiently to preserve vision for a long time. Since his son is similarly affected, the pattern as well as the initiating mutation are heritable. This family's experience suggests that gene(s) ameliorating the impact of the rhodopsin mutation already exist and need to be discovered since they have potential therapeutic value

To evaluate apoptotic pathways, we have developed an approach that exploits insights into apoptosis arising from studies of tumor supressor genes. Tumors gain a survival advantage, in part, because mutation of these genes contributes to tumor cell survival by abrogating apoptosis. The p53'gene is the most commonly mutated tumor suppressor gene in human cancer [12]. Loss of function of p53 permits cells to escape G1/S arrest or apoptosis in response to DNA damage. pRb, the protein product of the *Rb* tumor suppressor gene, binds to E2F-1, a transcription factor that controls the expression of many genes which initiate DNA replication. The function of E2F-1 is regulated by binding of pRb so that loss-of-function mutations of the *Rb* gene lead to inappropriate entry of cells into the mitotic cycle in addition to other effects [13]. Several oncogenic viruses gain their oncogenic properties because of the capacity of some of the proteins coded by the viral genome to bind to p53 and to pRb thereby inactivating them as if they were mutated. The SV40 T antigen (TAg) binds both proteins and can induce retinoblastoma in transgenic mice [14-16]. Human papilloma virus (HPV) 16 which is implicated in the pathogenesis of human cervical cancer, encodes two oncoproteins: E7 which binds and inactivates pRb and its homologues, and E6 which inactivates p53 by accelerating its destruction. When introduced into transgenic mice, constructs containing TAg [15, 16] or E7 [17] under the control of the IRBP promoter generate important and unexpectedly complex results.

MATERIALS AND METHODS

Transgenic mice bearing the SV40 TAg or the HPV16 E7 gene under the control of the IRBP promoter (IRBP-TAg mice and IRBP-E7 mice, respectively) were generated as previously described [16, 17]. IRBP-E7 mice were crossed to mice lacking the p53 gene (p53-/- mice) as previously described [17]. Eyes were obtained immediately after sacrifice and were either processed for light or electron microscopic study in paraffin or Epon, respectively. Paraffin sections were stained with hematoxylin and eosin and apoptotic cells were localized by TUNEL (Terminal-deoxyribonucleotidyl transferase-mediated dUTP-biotin Nick End Labeling) as described by Gavrieli et al [18] on sections that were counterstained by methyl-green. DNA fragments were separated by agarose electrophoresis as previously described [17].

RESULTS

IRBP-SV40 TAg Transgenic Mice

These mice invariably develop tumors of their entire developing photoreceptor layer (Figure 1) [16]. Since SV40 TAg binds both p53 and pRb, it was logical to test the effects of the action of each of these intrinsic proteins individually. The effect of loss of p53 was tested in vivo by several laboratories which generated mice lacking the *p53* gene by the technique of gene replacement (so-called gene "knock-out" mice). These mice develop normally except for their tendency to generate tumors, especially lymphomas, as they age [19]. Thus, the p53 gene is apparently not needed for normal development of many tissues, including the retina, despite the major role of apoptosis in remodeling the developing retina [20, 21]. The role of pRb in retinal development is not so easily assessed by the gene

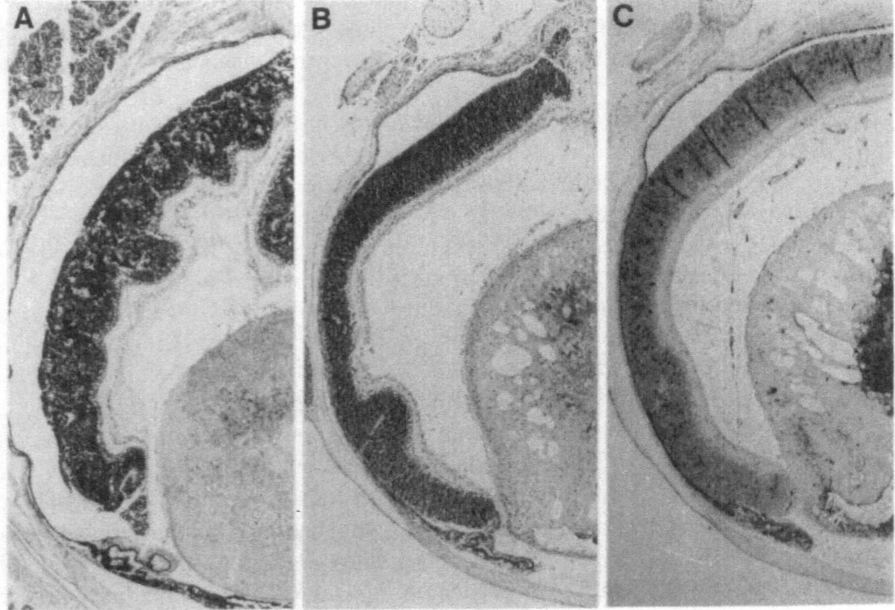

Figure 1. Retinas of IRBP-SV40 TAg or IRBP-E7 transgenic mice. A. IRBP-SV40 TAg mouse at P10.5. A tumor of small oval cells, some forming rosettes, obliterates the entire retinal photoreceptor layer. Note the formation of vacuoles and abnormally proliferating fiber cells in the lens. (H & E stain, X41). B. IRBP-E7 mouse at P4. The ventricular layer has separated from the ganglion cell layer of the retina. The lens is vacuolated and dying cells fill the center. (H & E stain, X60). C. Serial section of IRBP-E7 retina and lens at P4. Apoptotic cells are widely scattered throughout the retinal ventricular layer. Apoptotic cells are also detected in the fiber cells of the lens. (TUNEL assay, counterstained by methyl-green, X60)

knock-out approach since mice lacking both copies of the Rb gene die in utero about E14 apparently as a consequence of profound disturbance of their hematopoetic and central nervous systems. Heterozygotes, moreover, who are a genetic model of the children at risk for hereditary retinoblastoma, do not develop ocular tumors but rather form aggressive pituitary tumors and their eyes are normal [22, 23]. Thus, an alternative approach was needed to interfere with the function of the Rb protein selectively in ocular tissues. This could be accomplished, in part, by expressing the HPV 16 E7 gene under the control of the IRBP promoter since IRBP expression appeared to be confined to photoreceptor cells and pinealocytes [24].

IRBP-E7 Transgenic Mice Lose Their Photoreceptors by Apoptosis

Because E7 binds to and inactivates pRb, an anticipated result was that IRBP-E7 mice might also develop retinoblastomas since the IRBP-E7 construct should be expressed by E13 and thereafter as the photoreceptors begin to terminally differentiate and complete their last divisions shortly after birth. The result was, however, that photoreceptor cells began to die rapidly toward the end of embryonic life. Between P1 and P10 they nearly disappeared completely. This is shown not only by histologic comparison of the normal littermates and their IRBP-E7 siblings, but also by staining the dying cells with the TUNEL assay (Figures 1B and 1C and Figure 2). Agarose gel electrophoresis of the retinal DNA revealed the typical DNA "ladder" that is characteristic of internucleosomal cleavage in apoptotic cells [17].

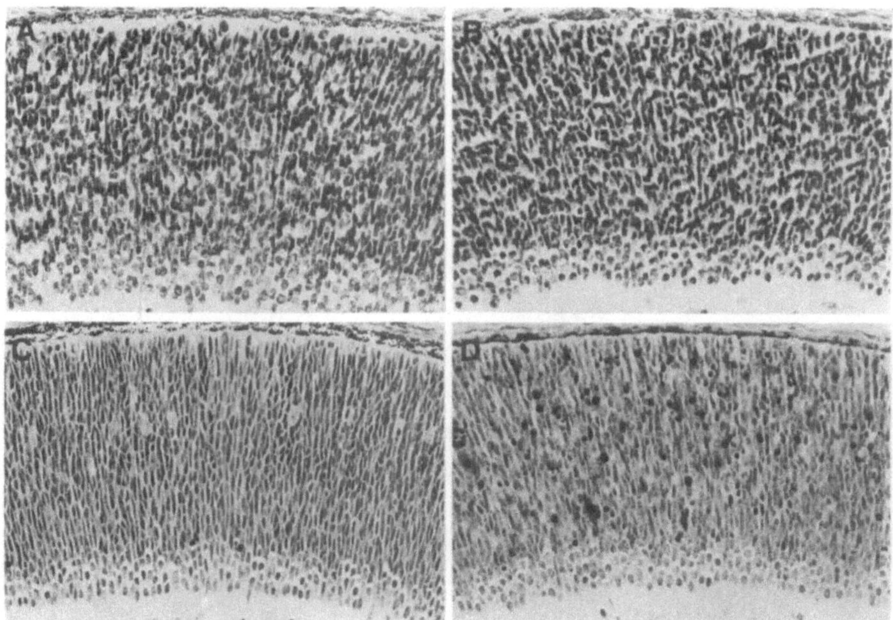

Figure 2. Comparison of the extent of apoptosis in normal and IRBP-E7 mice at P4. (A) Normal littermate control. Few pyknotic nuclei are observed. (H&E, X296) (B) IRBP-E7 transgenic mouse at P4 . (H&E, X296) The photoreceptor layer is filled with pyknotic nuclei. (C) Normal littermate control. The few apoptotic cells in the normal mouse retinas reflect the last stages of photoreceptor post-natal development [20]. (TUNEL, X296). (D) IRBP-E7 transgenic mouse at P4 stained for the presence of cleaved DNA by the TUNEL assay. Abundant stained nuclei are scattered throughout the ventricular layer (TUNEL, X296).

Effects of the p53 Gene on Apoptosis and Tumorigenesis in IRBP-E7 Transgenic Mice

Apoptosis of photoreceptors in the IRBP-E7 transgenic mice was initially unexpected. This result raises some significant questions about human retinoblastoma. If children that are heterozygous for *Rb* mutations lack functional pRb in their retinoblastomas, why do they get tumors and not apoptotic foci in their retinas? Could the difference arise only because of the species difference and the smaller target population of the mouse eye? Could an additional genetic lesion be necessary for the generation of retinoblastoma? Did the formation of retinoblastomas in the IRBP-TAg transgenic mice indicate a role for p53 in the pathogenesis of human retinoblastoma in addition to the loss of function of pRb?

To address some aspects of these questions, we bred IRBP-E7 mice to a p53-/- background. The results were both gratifying and perplexing. Indeed, tumors developed rapidly in the photoreceptor layer. By one month, all E7/p53-/- mice had retinal tumors. The E7/p53+/- littermates had apoptotic photoreceptor layers. This led us to raise the possibility that apoptosis and retinoblastoma represent "alternative states" in response to specific manipulations of these genes that were so important in control of the cell cycle [17]. As we generated more mice over the next few months and gained the opportunity to look at larger numbers of mice at earlier and later stages of development, a far more complex picture emerged.

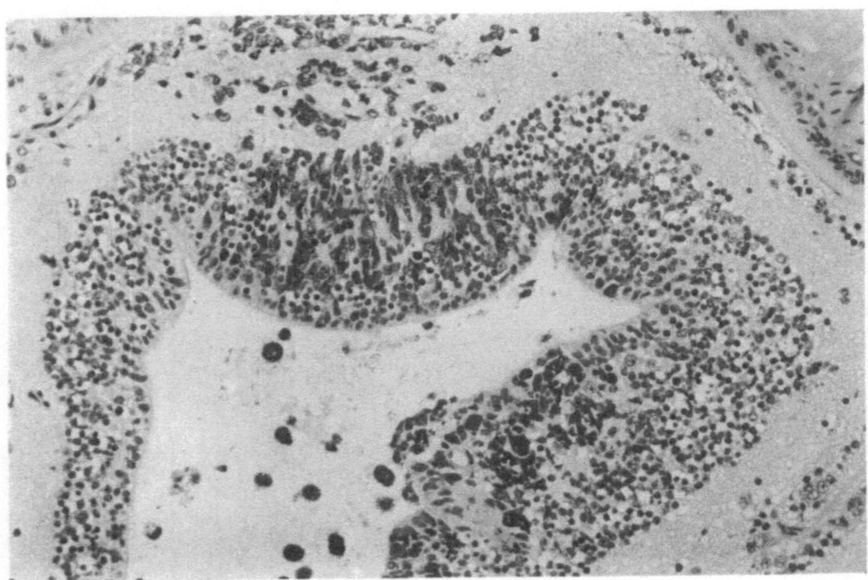

Figure 3. Retina of an IRBP-E7/p53-/- mouse at 9 weeks. Nearly all photoreceptor nuclei are missing as a consequence of apoptosis despite the absence of the p53 gene in this transgenic mouse. Amid the cells of the inner nuclear layer are several independent small nests of tumor cells. The ganglion cell layer is well separated but some of its cells are pyknotic. (H&E, X250)

First of all, as we examined E7/p53-/- mice from two weeks of age to one month of age, we discovered that the retinas actually contained two processes occurring simultaneously. *The photoreceptors continued to die apoptotically nearly as fast as they did in the E7/p53+/- and E7/p53+/+ mice.* The detailed comparison of apoptotic rates in these various combinations is an experiment that is currently in progress. So far, our results indicate that the absence of p53 leads to a slight delay in the onset of apoptosis in the retina but does not protect the retina from eventual destruction, i.e. the E7 gene can induce a p53 independent apoptosis of the retina in addition to contributing to oncogenesis.

Secondly, the eyes of E7/p53-/- mice developed several different types of tumors. Tumorigenesis in these young mice, however was highly dependent on the state of the p53 gene. None of the E7/p53+/- mice developed retinal tumors by one month of age. With one copy of the gene, the mice develop retinal tumors after a much longer latency and at a lower incidence. Genotyping of these tumors demonstrated that they had lost the normal copy of the p53 gene in the process (Howes et al, unpublished data). This means that a precursor cell had survived in the apoptotic retina that was capable of forming a tumorigenic clone several months later. The search for these precursors is now in progress using labels to detect cells synthesizing DNA.

In addition to their relatively undifferentiated multicentric retinoblastomas, the IRBP-E7/p53-/- mice also formed undifferentiated non-melanotic tumors of the posterior retinal pigment epithelium and posterior papillary adenocarcinomas, apparently of retinal origin, although the exact layer from which these tumors arose is obscured by their invasion of (or origin from) the RPE. The ciliary body doubled of its size and became a papilloma. The anterior pigmented epithelium of the iris and ciliary body was transformed into a non-adherent tumor of pigmented cells growing virtually as a suspension culture in the anterior chamber. Some of these cells settled down on the anterior lens capsule as a

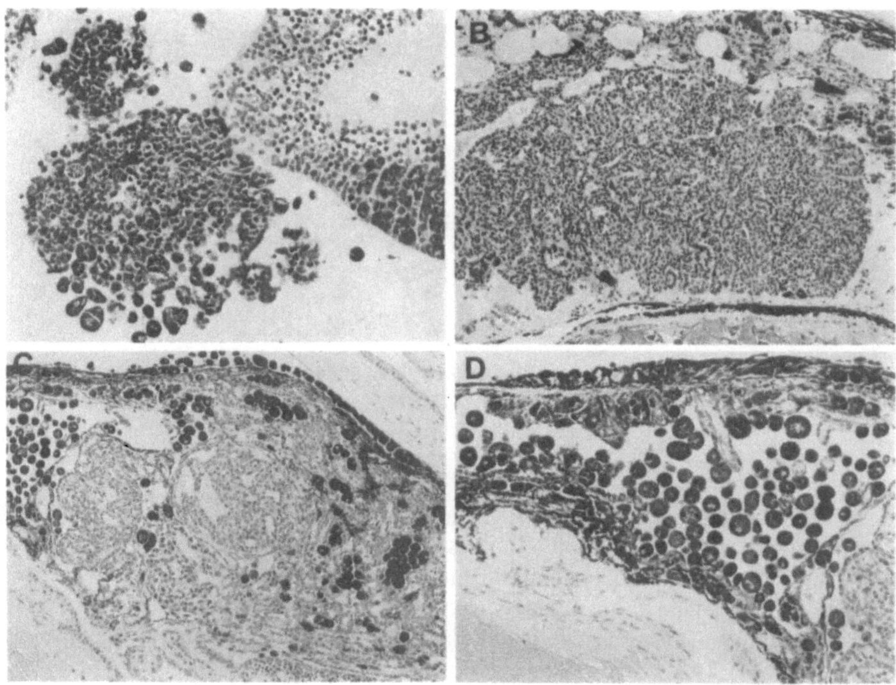

Figure 4. Tumors arising in the retinas, pigment epithelium, ciliary body and anterior uveal pigmented epithelium in IRBP-E7/p53 -/- mice at two months of age. (all H&E) (A) Poorly differentiated retinoblastoma. (X185) (B) Retinal (?RPE) adenocarcinoma invading the posterior retina. (X88) (C) Ciliary body papilloma (X160) (D). Pigmented epithelial cells in free suspension in the posterior and anterior chambers (X172).

monolayer (Figure 4). Similar alteration of some of the posterior RPE cells led to infiltration of giant pigmented cells into the retina.

This result was, therefore, an embarrassment of riches. Prior studies of the expression of IRBP have not yet described its expression in the RPE and uveal tissue. We wonder, therefore, if the extra-retinal tumors in these ocular tissues indicate that IRBP is expressed elsewhere in the eye in addition to photoreceptor cells. Regardless, the formation of the poorly differentiated retinoblastoma clearly indicated that our experimental protocol of mating IRBP-E7 mice to p53 knock-out mice had unlocked the oncogenic potential of the E7 gene—and the loss of Rb gene function—in the mouse retina.

Effects of IRBP-E7 on the Lens in the Presence and Absence of p53 Function

A disorderly population of nucleated fiber cells collects in the posterior lens by P1 in IRBP-E7 transgenic mice. Eventually, over the next ten days, these cells begin to die, initially by apoptosis and finally by necrosis as well. This is readily discerned in cells stained by the TUNEL assay: apoptotic cells have a dark brown nucleus while necrotic cells have pale staining nuclei and brown staining cytoplasm and intercellular spaces. The cleavage of the DNA into small fragments, the breakdown of the nuclear envelope and the escape of the cleaved DNA into the cytoplasm and the intercellular spaces as the plasma membrane disintegrates explains the pattern of TUNEL staining in necrotic areas. As a consequence of

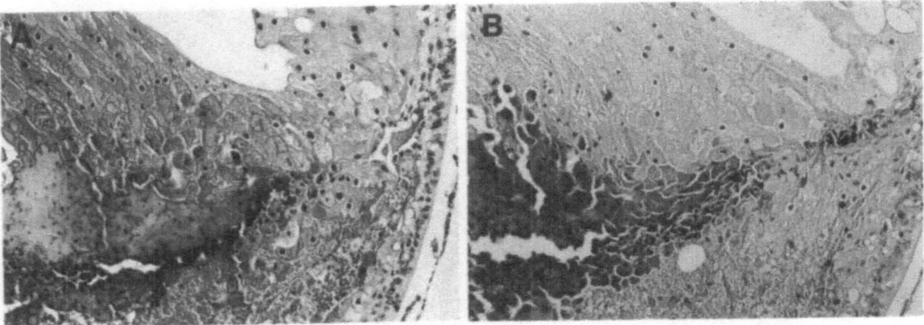

Figure 5. Apoptotic and necrotic cells accumulate in the lens of IRBP-E7/p53-/- mice. Vacuoles form and necrotic cells accumulate centrally. (A) H&E X170. (B) TUNEL assay counterstained by methyl green X170.

this catastrophe, the lens becomes an opaque and shrunken mass. Despite this assault on the posterior fiber cells, the anterior epithelium is relatively unscathed, although preliminary analysis of numbers of apoptotic cells in the anterior epithelium indicates that they are not entirely spared and are also dying at an accelerated rate. This fate of lens epithelium is not affected by the status of the p53 gene in a major way although there may be a slight delay in onset of apoptosis. The process progresses inexorably in all combinations of E7 x p53-/- transgenic offspring. Cataracts were also observed previously in IRBP-TAg mice [16].

DISCUSSION

These experiments were initiated to develop insights into the relationships of the Rb gene to the formation of retinoblastoma. We now are offered, as a consequence of the induction of photoreceptor apoptosis by introduction of the E7 gene into the retina, an opportunity to look at the impact of perturbation of the cell cycle and differentiation of photoreceptors as tools to understand the vulnerability of rods and cones to die by apoptosis. The insights available from this new model of inherited retinal degeneration, while dependent on an entirely different mechanism of injury than all other known mutations that cause retinal degeneration, are highly relevant to the examination of the effects of perturbation of pRb and p53 function *in vitro* and in other cell systems *in vivo*. The advantage of the study of these perturbations in an *in vivo* setting is that we are not altering the natural environment or the intercellular relationships of the developing cells except by the introduction of the transgene. It remains to be determined if these genes are also involved in other inherited forms of retinal apoptosis.

Although abundant evidence closely links p53 to the induction of apoptosis of many cell types lacking functional pRb, not all cells demonstrate a mutual interaction in all conditions. For example, lymphocytes of p53-/- mice are resistant to irradiation induced apoptosis but retain their sensitivity to glucocorticoid induced cell death by apoptosis[25, 26]. Furthermore, since the retina, and indeed many other tissues of developing mice employ apoptosis extensively during differentiation and yet have no discernible abnormality in p53-/- mice, the gene is clearly an important but not uniquely required element in the control of orderly cell proliferation or of the control of apoptosis during development. What is interesting about our experimental results is that they also demonstrate a p53 independent pathway to photoreceptor apoptosis after pRb inactivation. It is too soon to know if inactivating pRb by E7 expression invokes the same p53 independent pathway employed by the brain and retina during normal development or another alternative pathway.

Finally, these transgenic mice have introduced us to new complexities in the differentiation of lens fiber cells. In normal development, the lens vesicle completes the obliteration of the vesicle lumen by elongation of the posterior fiber cells until they contact the anterior epithelium and form tight junctions. These cells begin the synthesis of their characteristic crystallin molecules at an early stage [27] and then lose their nuclei and mitochondria. The fiber cells must last the entire lifetime of the animal in a clear state for normal transmission of light to the retina. Throughout post-natal life, a population of cells at the lateral equator of the lens continually differentiate from cuboidal epithelial cells into long fiber cells that join at suture planes and enlarge the lens diameter. By contrast to this orderly pattern, the expression of E7 in the lens leads to its destruction by both apoptosis and necrosis, not only in the embryo after E16 but also in the adult as new cells form.

Why should this gene be expressed in the lens? At first this was puzzling since prior reports of the expression of IRBP-CAT (chloramphenicol-acetyl transferase) constructs which were used to reveal the tissue-specific expression of the IRBP promoter did not reveal any expression of the construct outside the retinas of mice at E16 [24]. Recently, however, transient lenticular expression of IRBP from E11 to E15 was described by Farr *et al.*[28] which may indicate how our IRBP-E7 construct became a problem for the lens as well as the retina.

Regardless of the mechanism leading to expression of E7 in the lens of these IRBP-E7 mice, they consistently develop cataracts as a consequence of apoptosis and necrosis of fiber cells. Although there appears to be a slight but significant slowing of the apoptosis of the retina in IRBP- E7 x p53-/- hybrids, compared to those with at least one normal copy of the p53 gene, the effect of loss of p53 function in the level and rate of lens fiber cell apoptosis is slight at best.

In this respect, our results differ somewhat from nearly parallel studies of the effects of expression of E7 and E6 in the lenses of transgenic mice bearing these viral genes under the control of the α-crystallin promoter. Pan and Griep [29] discovered that α-crystallin-E7 transgenic mice also accumulate nucleated proliferating cells in the posterior lens which then die by apoptosis and lead to cataract formation. By contrast, however, they could reduce the impact of this E7 construct by mating these mice to transgenic mice expressing the HPV 16 E6 gene in the lens using the same promoter. E6 inactivates p53 function. Greater than 30% of the apoptosis observed in α-crystallin-E7/p53+/+ lens fiber cells was blocked in α-crystallin-E7xE6 offspring. They concluded, therefore, that the apoptosis of fiber cells as a consequence of perturbation of pRb by the expression of E7 in lens was altered by destruction of p53 function by E6, i.e. that the loss of functional pRb induced lenticular apoptosis by a pathway that was at least partially dependent on p53 function. Similar conclusions were reached by Morgenbesser *et al* [30] who ablated pRb function by generation of Rb knockout mice and tested the impact of the loss of this gene on lens cell differentiation in the presence or absence of the p53 gene by crossbreeding Rb-/+ heterozygous mice to p53 knockout mice. The homozygous Rb-/- offspring die in utero. This mating generated embryos with an Rb-/-/p53-/- and Rb+/-/p53-/- genotype that could be compared to related Rb-/-/p53+/- and Rb-/-/p53+/+ embryos. At E14, the lenses of the mice with at least one intact copy of both the Rb gene and the p53 gene were normal. Those lacking the Rb gene developed apoptosis of their fiber cells if at least one copy of the p53 gene was intact, while those lacking the p53 gene exhibited a profound decrease in the amount of apoptosis of their lens fiber cells.

Thus, in our experiments, the expression of E7 under the control of the IRBP promoter in the presence or absence of p53 function or IRBP-TAg leads to apoptosis of the fiber cells. By contrast, E7 expression controlled by the α-crystallin promoter or ablation of Rb in knock-out mice induces apoptosis of the proliferating fiber cells in a manner that is somewhat more dependent on p53 function. We are actively pursuing the reasons for these differences because they indicate that the outcome of alteration of the function of the Rb and p53 genes in the apoptosis of the lens and the retina is not simply a question of the presence or absence of their activity in the cell. The outcome may also depend on the exact timing or extent of the perturbation of their function.

It is now apparent that even in these relatively "controlled" conditions where we understand which genes we have perturbed, at what developmental stage they should be affected, and which tissues should express the transgene, that differing results can arise under slightly different conditions *in vivo*. At this early stage in the study of experimentally induced apoptosis and retinoblastoma, we see some hints of the complexity of outcome found in patients with inherited retinal degeneration. Hopefully, we will somehow learn ways to ablate retinal apoptosis without setting the stage for development of retinal tumors and will develop candidate genes for eventual consideration in clinical trials of gene therapy of human retinal degeneration and conversely, for retinoblastoma.

ACKNOWLEDGMENTS

The authors wish to thank Ann Murphree for preparation of histologic sections, Shelly Friedkin for assistance with some of the TUNEL assays and the National Institutes of Health for support of this project (grants EY-6891 to DSP and EY 9213 to JJW).

REFERENCES

1. Li, Z-Y. and Milam, A. H., 1995, Apoptosis in Retinitis Pigmentosa. *This volume*, p. 1-8.
2. Papermaster, D. S. and Nir, I., 1994 Apoptosis in inherited retinal degenerations. Mihich, E. and Schimke, R.H., eds. In "*Apoptosis*", NY: Plenum Press; pp. 15-30.
3. Chang, G. Q., Hao, Y., and Wong, F., 1993, Apoptosis: Final common pathway of photoreceptor death in *rd*, *rds*, and rhodopsin mutant mice. *Neuron* 11: 595-605.
4. Lolley, R.N., Rong, H. and Craft, C.M., 1994, Linkage of photoreceptor degeneration by apoptosis with the inherited defect in phototransduction. *Invest Ophthalmol Vis Sci.* 1994;35:358-362.
5. Portera-Cailliau, C., Sung, C.-H., Nathans, J. and Adler, R., 1994, Apoptotic photoreceptor cell death in mouse models of retinitis pigmentosa. *Proc Nat Acad Sci USA* 91:974-978.
6. Tso M.O.M., Zhang C., Abler A.S., Chang, C.-J., Wong, F., Chang, G.-Q., and Lam, T. T., 1994, Apoptosis leads to photoreceptor degeneration in inherited retinal dystrophy of RCS rats. *Invest Ophthalmol Vis Sci.* 35:2693-2699.
7. Shahinfar S., Edward D.P. and Tso M.O.M., 1991, A pathologic study of photoreceptor cell death in retinal photic injury. *Curr Eye Res.* 10:47-59.
8. Faktorovich, E.G. , Steinberg, R.H., Yasumura, D., Matthes, M.T. and LaVail, M.M., 1990, Photoreceptor degeneration in inherited retinal dystrophy delayed by basic fibroblast growth factor, *Nature* 347:83—86
9. LaVail, M.M., Unoki, K., Yasumura, D., Matthes, M.T., Yancopoulos, G.D. and Steinberg, R.H., 1992, Multiple growth factors, cytokines, and neurotrophins rescue photoreceptors from the damaging effects of constant light, *Proc. Natl. Acad. Sci. USA.* 89:11249-11253.
10. Yasumura, D., Matthes, M.T., Lau, C., Unoki, K., Steinberg, R.H. and LaVail, M.M., 1995, Attempts to rescue photoreceptors with survival factors in mice with inherited retinal degenerations or constant light damage. *Invest. Ophthalmol. Vis. Sci.* 36: Suppl. 1. S252.
11. Li Z.-Y., Jacobson S.G. and Milam A.H., 1994, Autosomal dominant retinitis pigmentosa caused by the threonine-17-methionine rhodopsin mutation: Retinal histopathology and immunocytochemistry. *Exp. Eye Res.* 58:397-408.
12. Levine, A., Momand, J. and Finlay, C.A., 1991, The p53 tumour suppressor gene. *Nature* 351:453-456.
13. Nevins, J.R., 1992, The E2F transcription factor-a link between the Rb tumor suppressor protein and viral oncoproteins. *Science* 258:424-429.
14. Windle, J.J., Albert, D.M., O'Brien, J.M., Marcus, J.M., Disteche, C.M., Bernards, R. and Mellon, P.L., 1990, Retinoblastoma in transgenic mice. *Nature* 343:665-669.

15. Al-Ubaidi, M.R., Font, R.L., Quiambao, A.B., Keener, M.J., Liou, G.I., Overbeek, P.A. and Baehr, W., (1992), Bilateral retinal and brain tumors in transgenic mice expressing simian virus 40 large T antigen under control of the human interphotoreceptor retinoid-binding protein promoter. *J. Cell Biol.* 119:1681-1687.
16. Howes,. K.A., Lasudry, J.G.H., Albert, D.M. and Windle, J.J., 1994, Photoreceptor cell tumors in transgenic mice. *Invest. Ophthamol Vis Sci* 35:342-351.
17. Howes, K.A., Ransom, N., Papermaster, D.S., Lasudry, J.G.H., Albert, D.M. and Windle, J.J., 1994, Apoptosis or Retinoblastoma: Alternative fates of photoreceptors expressing the HPV-16 E7 gene in the presence or absence of p53. *Genes Dev* 8:1300-1310.
18. Gavrieli, Y., Sherman, Y. and Ben-Sasson, S., 1992, Identification of programmed cell death *in situ* via specific labeling of nuclear DNA fragmentation, *J. Cell Biol.* 119:493-501.
19. Donehower, L.A, Harvey, M., Slagle, B.L., McArthur, M.J., Montgomery, C.A., Jr, Butel, J.S., and Bradley, A., (1992), Mice deficient for p53 are developmentally normal but susceptible to spontaneous tumours. *Nature* 356:215-220.
20. Young, R.W., (1984), Cell death during differentiation of the retina in the mouse. *J. Comp. Neurol.* 229:362-373.
21. Penfold, P.L. and Provis, J.M., (1986), Cell death in the development of the human retina: Phagocytosis of pyknotic and apoptotic bodies by retinal cells. *Graefe's Arch. Clin. Exp. Opthalmol.* 224:549.
22. Lee, E.Y., Chang, C.Y., Hu, N., Wang, Y.C., Lai, C.C., Herrup, K., Lee, W.H. and Bradley, A., 1992, Mice deficient for Rb are nonviable and show defects in neurogenesis and haematopoiesis *Nature.* 359:288-294.
23. Jacks, T., Fazeli, A., Schmitt ,E.M., Bronson, R.T. Goodell, M.A. and Weinberg, RA., 1992, Effects of an Rb mutation in the mouse *Nature.*359:295-300.
24. Liou, G.I., Geng, L., Al-Ubaidi, M.R., Matragoon, S., Hanten, G., Baehr, W. and Overbeek, P.A., (1990), Tissue-specific expression in transgenic mice directed by the 5'-flanking sequences of the human gene encoding interphotoreceptor retinoid-binding protein. *J. Biol. Chem.* 265:8373-8376.
25. Clarke, A.R., Purdie, C.A., Harrison, D.J., Morris, R.G., Bird, C.C., Hooper, M.L. and Wyllie, A.H., 1993, Thymocyte apoptosis induced by p53-dependent and independent pathways. *Nature.* 362:849-852.
26. Lowe, S.W., Schmitt, E.M., Smith, S.W., Osborne, B.A. and Jacks, T., 1993, p53 is required for radiation-induced apoptosis in mouse thymocytes *Nature.* 362:847-849.
Van de Kamp, M. and Zwann, J., 1973, Intracellular localization of lens antigens in the developing eye of the mouse embryo. *J. Exp. Zool.* 186:23-32.
28. Farr, E.A., Borst, D.E., Chader, G.J., and Bradley, D.J., (1994), Developmental regulation of the mouse interphotoreceptor retinol binding protein gene in the fetal and post-natal eye and pineal gland. *Invest. Ophthamol. Vis. Sci.* Suppl. 35: abstract #2199.
29. Pan H. and Griep, A.E.,1994, Altered cell cycle regulation in the lens of HPB-16 E6 or E7 transgenic mice: Implications for tumor suppressor gene function in development. *Genes & Dev* ;8:1285-1299.
30 Morgenbesser, S.D., Williams, B.O., Jacks ,T. and DePinho, R.A., 1994, p53-dependent apoptosis produced by Rb-deficiency in the developing mouse lens. *Nature.* 371:72-4.

6

FUNCTION, AGE-RELATED EXPRESSION AND MOLECULAR CHARACTERIZATION OF PEDF, A NEUROTROPHIC SERPIN SECRETED BY HUMAN RPE CELLS

Joyce Tombran-Tink

National Institutes of Health
National Eye Institute
Laboratory of Retinal, Cell and Molecular Biology
Bethesda, Maryland 20892

INTRODUCTION

The retinal pigment epithelium (RPE) is a highly specialized neuroepithelium that develops in advance of and lies adjacent to the neural retina where it plays a critical role in retinal homeostasis. RPE cells are multifunctional in nature and have been compared to macrophages (Elner et al., 1981, Young and Bok, 1969), oligodendrocytes (Steinberg and Wood, 1974), astrocytes (Immel and Steinberg, 1986), melanocytes (Feeney-Burns, 1980) and hepatocytes (Bok 1985). The unique geography of this single-celled epithelial layer allows it to establish a definitive blood-retinal barrier (Cunha-Vaz 1979) and to function as a transporting and absorbing epithelium (Miller and Steinberg 1982; Misfeldt et al., 1976). In the vertebrate retina, a closed, extracellular microenvironment exists that is bounded by the apical membrane of RPE cells distally and photoreceptor inner segments and Muller cell processes proximally. Tight junctions effectively isolate it from most larger blood components and it is thought that most of the IPM components are synthesized by surrounding cells. This highly specialized matrix constitutes a novel conduit for the transport of nutrients, metabolites or trophic factors between the two cell layers and also facilitates intercellular communication. Thus, analysis of the IPM is of importance because it contributes to our understanding of essential aspects of retinal development, homeostasis and visual function. RPE cells form a functional complex with photoreceptor neurons of the retina and interacts with them through the IPM. Cultured RPE cells synthesize and secrete several trophic factors including a photoreceptor-survival promoting factor (PSPA) (Hewitt et al., 1990), PDGF (Campochairo 1988), FGF (Plouet 1988), TGF-α (Fassio et al., 1988) and TGF-β (Connor et al., 1988). Evidence of its potential to synthesize and secrete these factors *in vivo*, into the IPM (Hageman and Johnson 1991) makes the role of the RPE in relationship to the developing retina, photoreceptor cell pathology and the visual process of even greater interest.

Degenerative Diseases of the Retina, Edited by Robert E. Anderson et al.
Plenum Press, New York, 1995

NEUROTROPHIC ACTIVITY OF RPE-CM AND PEDF

In an attempt to isolate other trophic factors secreted by RPE cells, conditioned-medium from a well-defined human fetal RPE cell culture (Pfeffer 1990) was analyzed for neurotrophic activity using the human Y79 retinoblastoma cell line as a target cell in our model system. Y79 cells are of neural retinal origin and thought to be the tumorous counterpart of early undifferentiated retinoblasts. Because they have been shown to retain several neural retinal characteristics they serve as a good model for examining some aspects of retinal cell differentiation (Kyritsis et al., 1984, Bogenman et al., 1988). We found that human fetal RPE-conditioned medium (hfRPE-CM) induced a high level (>90 %) of "neuronal" differentiation in morphologically undifferentiated Y79 cells (Tombran-Tink and Johnson 1989, Tombran-Tink et al., 1991). Specifically, an elaborate outgrowth of neurites were seen to project from hfRPE-CM treated Y79 cells. A 50 kD protein, pigment epithelium-derived factor (PEDF), was subsequently isolated from hfRPE-CM and found to contain this biological inductive potential in our defined assay system using the Y79 cells (Fig. 1) (Tombran-Tink et al., 1991). Biochemical changes were seen as well as morphological differentiation when these cells were treated with PEDF. The differentiated Y79 cells exhibit enhanced expression of neuron-specific enolase and synthesize the 200 kD subunit neurofilament protein, both considered to be molecular markers for neuronal cell maturation (Tombran-Tink et al., 1991).

PEDF MAINTAINS NEURONAL CELL SURVIVAL

In addition to its neurite-promoting activity, PEDF maintains neuronal survival of the differentiated Y79 cells (Fig. 2) (Tombran-Tink and Johnson 1989) as well as cerebellar granule cells in culture (Taniwaki et al., 1994). Y79 cells maintain a differentiated, non-proliferative phenotype for more than 30 days (35-40 days) after attachment if maintained in serum-free, defined medium supplemented with either 50% hfRPE-CM or 50 ng/ml purified PEDF. Well-differentiated, attached Y79 cells however, retract their processes, detach from the substratum and subsequently die within 11-15 days if not supplemented with conditioned medium or PEDF after the initial treatment in suspension.

PEDF ISOLATION FROM IPM AND VITREOUS

We have also found that bovine IPM contains neuronal inductive activity on Y79 retinoblastoma cells similar to that seen by hfRPE-CM (Tombran-Tink et al., 1992). A comparison of the percentage of neuronal cell differention between hfRPE-CM, human IPM and purified PEDF from hfRPE-CM shows that, by day 10 post-attachment, all treated Y79 cultures exhibited over 85% neuronal differentiation (Fig. 3) even though the percentage varied within the first few days of attachment. From these data, we hypothesized that PEDF may also be secreted *in vivo*, into the IPM, by adult RPE cells (Tombran-Tink et al., 1992). This hypothesis was later confirmed when a PEDF polyclonal antibody was obtained and used in western blot analysis to show the presence of a similar 50 kD protein in both fetal and adult IPM of a number of other vertebrate species including human, monkey and bovine (Tombran-Tink et al., 1995). The PEDF antibody recognizes at least four isoforms of secreted human and bovine PEDF as assessed by two dimensional gel analysis. It is thus apparent that PEDF not only is synthesized and secreted by fetal RPE cells *in vitro* and *in vivo* but also by adult RPE cells. In this regard, the presence of PEDF in both fetal and adult IPM

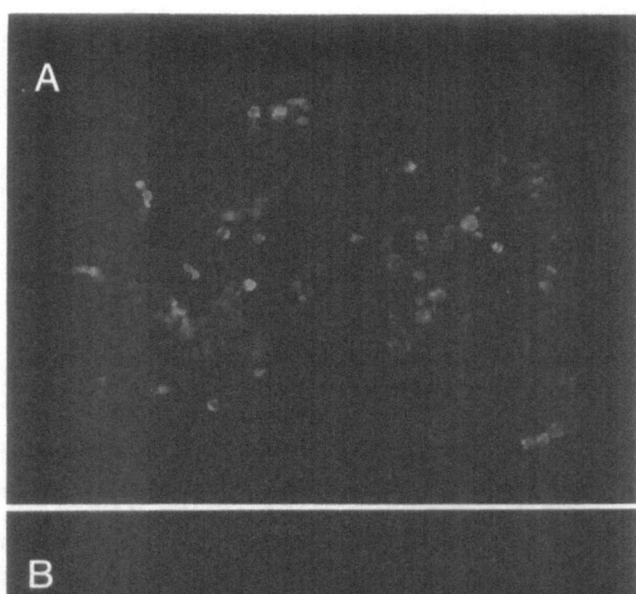

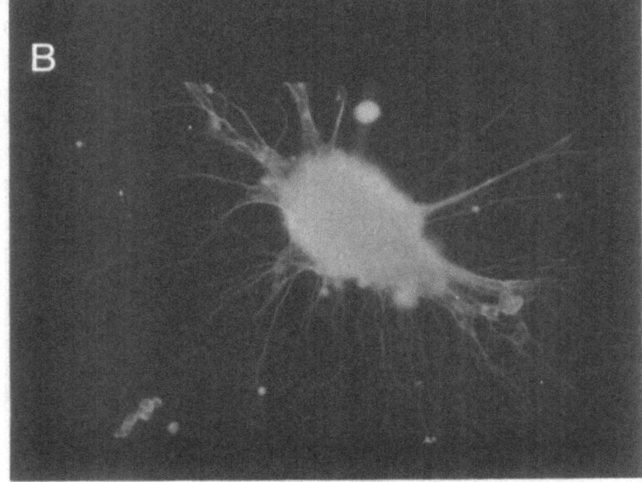

Figure 1. A: Morphologically-undifferentiated Y79 retinoblastoma cells grown in serum-free medium and attached to a poly-D-lysine substratum (Tombran-Tink et al., 1991). The cells are GFAP positive as detected by a polyclonal GFAP antibody. B: Y79 cells pre-treated for 7 days in suspension with 50 ng/ml of purified PEDF and attached to a poly-D-lysine substratum. Within 3 days of attachment, extensive neurite outgrowths are seen in PEDF-treated cells. The micrograph shows immunofluorescence for neuron-specific enolase (NSE) in well differentiated cells as detected by a polyclonal NSE antibody.

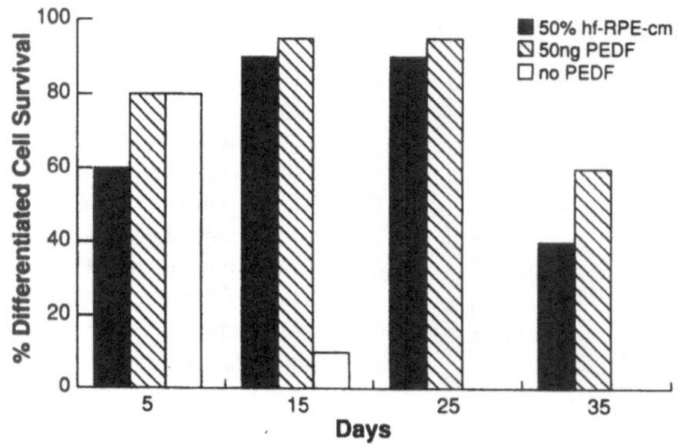

Figure 2. The graph shows the percentage of differentiated Y79 cells surviving with and without addition of PEDF or RPE-CM to attached cultures. The cells were initially treated for 7 days in suspension with either PEDF or hfRPE-CM and then seeded onto poly-D-lysine-coated coverslips. Culture A: 50% hfRPE-CM added every 7 days after attachment. Culture B: Medium supplemented with 50 ng/ml PEDF and added to culture every 7 days. Culture C: Only SF-defined medium added to these cultures 7 days post-attachment.

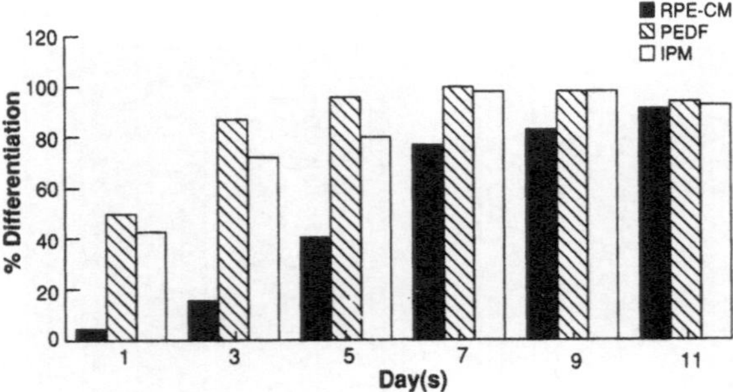

Figure 3. The bar graph represents the percentage of neuronal differentiation in Y79 retinoblastoma cell at days 1, 3, 5, 7, 9 and 11 after treatment for 7 days in suspension with one of the following a) hfRPE-CM, b) HPLC purified PEDF from hfRPE-CM c) human IPM. Under all three conditions greater than 80% neuronal differentiation is seen after 7 days of attachment.

demonstrates its potential for action *in vivo* at critical times of neuronal cell differentiation and, after retinogenesis, in the mature retina where it could promote photoreceptor cell survival.

Because of the difficulties in obtaining and purifying native PEDF from human fetal RPE cultures, we investigated other sources of PEDF production in addition to that found in the interphotoreceptor matrix. Because the vitreous humor is a large extracellular space also found in the eye, we examined it for the presence of PEDF and found, by western blot analysis, substantial amounts of PEDF in the bovine vitreous as compared to the IPM (Fig. 4). Future studies will be necessary to determine both the function(s) and source of PEDF in the vitreous. Thus, both bovine IPM and vitreous are suitable starting materials for the isolation of native PEDF. For this purpose we used the Bio-Rad model 491 prep cell system which fractionates and purifies complex protein samples by continuous-elution electrophoresis using conventional electrophoresis buffers. Approximately 5 μg of PEDF can be purified from the IPM of each eye cup and 150 μg from the vitreous (~ 10 ml for each eye) of each bovine eye (Fig 4). Purified PEDF from both vitreous and IPM is found to be neurotrophic for the Y79 cells. Thus, the bovine vitreous is a useful source of native PEDF for use in both *in vivo* and *in vitro* studies.

Localization of PEDF in Human Fetal RPE cells in vivo by *In Situ* Hybridization

The neurotrophic effects of PEDF on the human Y79 retinoblastoma cell was first demonstrated with conditioned medium obtained from 17 weeks gestation fetal human RPE cells (Tombran-Tink and Johnson, 1989). *In situ* hybridization was therefore performed on specimens of fetal human retina-RPE of similar developmental age. Under conditions of high stringency hybridization, specific binding of a radioactively-labelled antisense probe

M 1 2 3 4 5

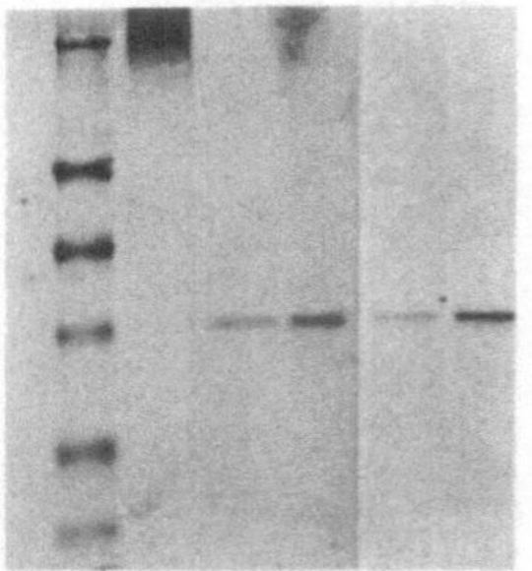

Figure 4. 10 ml of either bovine IPM or vitreous samples were electrophoresed through a cylindrical 10% SDS gel using the Bio-Rad 491 electrophoreses prep cell. Each cylindrical band migrated off the bottom of the gel into an elution collection chamber and fractions containing PEDF were pooled and dialyzed against PBS. 10 μl sample from pooled fractions containing eluted PEDF was analyzed by Coomassie blue and western blot. The PEDF polyclonal antibody detects a 50 kD protein in bovine serum, IPM and vitreous (lanes 1, 2 and 3 respectively). Stronger hybridization signal is seen in the vitreous sample (lane 3). Lanes 4 and 5 represents a Coomassie blue stained 10% SDS gel containing purified PEDF from IPM and vitreous respectively. M: Molecular Weight Standards.

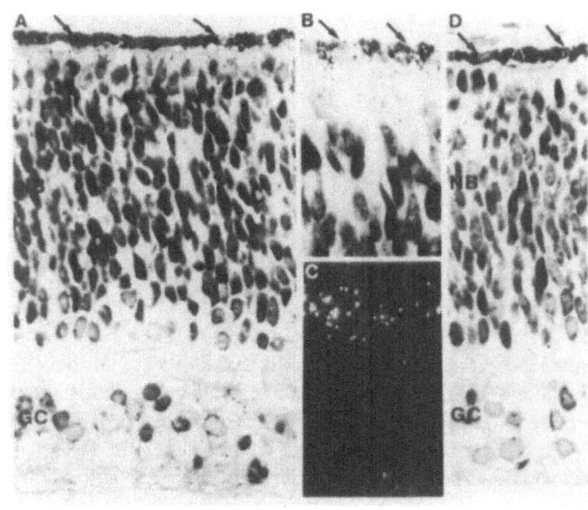

Figure 5. Human fetal eye, at 17 weeks of gestation, fixed in 4% formaldehyde, dehydrated in ethanol and embedded in diethylene glycol distearate. A 264 bp sense and 300 bp antisense probe was generated from the coding region of the PEDF cDNA , ^{35}S-labeled and used for hybridization (Tombran-Tink et al., 1995). Autoradiograms were exposed for 12 days and photographed under brightfield (A, B, and D) or epipolarization illumination (C). A) retinal section incubated with the radioactive antisense probe reveals the concentration of silver grains above background over the retinal pigment epithelium (arrows). The neuroblastic (NB) layer, ganglion cell (GC) layer and other regions of the neurosensory layer were not consistently labeled above background. B) retinal pigment epithelium (arrows) under higher magnification shows more detail, although some silver grains remain obscured by underlying melanin granules. C) Epipolarization microscopy of the same area as that shown in (B) reveals all of the silver grains over the RPE. D) retinal sections incubated with the radioactive sense probe bound only background levels of radioactivity as indicated by the low density of silver grains over the RPE (arrows). A and D: 800x; B and C: 1360x.

was seen only in the RPE cell layer (Fig. 5) (Tombran-Tink et al., 1995). Most of the neurons in the retinal neuroblastic layer, with the exception of the ganglion neurons, are relatively undifferentiated at this developmental stage and were not labeled above background levels as compared to control hybridization using the PEDF sense probe. Thus, in the developing retina as early as 17 weeks of gestation, when most of the retinal neurons are still undifferentiated, PEDF is specifically synthesized and secreted into the adjacent IPM, by the RPE cells, where its inductive potential could be maximized. In the human, RPE cells mature well before the photoreceptors of the neural retina during the fetal period and, as mentioned before, may influence and direct aspects of retinal differentiation either directly by contact or through their secretory products in the IPM (Johnson et al., 1985). These results form an important rationale for our hypothesis that PEDF is a critical neurotrophic factor participating in the process of early retinogenesis.

DOWNREGULATION OF PEDF IN SENESCENT RPE CELL CULTURES

In a recent report, PEDF was shown to be expressed by cultured WI-38 lung fibroblast cells where its expression is linked to the process of senescence (Pignolo et al., 1993). In the human W38 cells, a well defined senescing human fibroblast model, PEDF (called EPC-1) transcripts could only be found in confluent young, quiescent WI-38 cells but not in their non-proliferating senescent counterparts. In a similar vein, the PEDF transcript was undetectable in fully- differentiated Y79 retinoblastoma cells although it is seen in abundance in their rapidly-dividing, morphologically undifferentiated counterparts (Tombran-Tink et al., 1994). It is also of interest that survival of the neuronally- differentiated state is greatly enhanced by the presence of PEDF in the two cell types tested to date which are the Y79 retinoblastoma and cerebellar granule cells. We have also examined the synthesis and

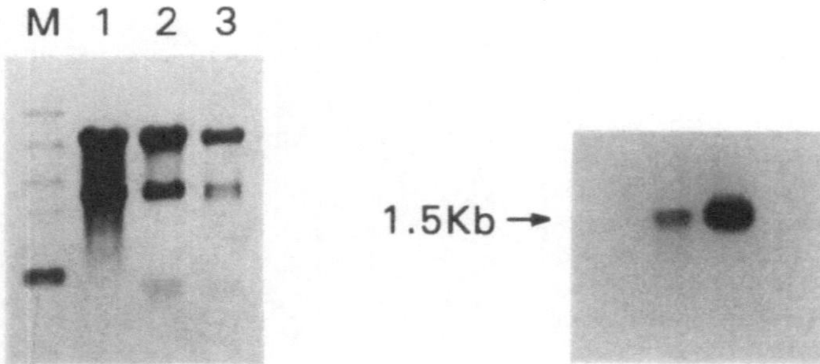

Figure 6. Left panel: Agarose gel containing 5 μ g of total RNA in each lane. Lane 1-3: monkey retina, 1st passage monkey RPE cells and 10th passage monkey RPE cells respectively. The gel is stained with ethidium bromide. Right panel: A northern blot of the gel in panel on the left hybridized with radioactively labelled PEDF cDNA probe. M: molecular weight standards. The weak hybridization seen in the retina RNA may be due to contamination with adhering RPE cells, since, by in situ hybridization the PEDF mRNA is only detected in the RPE cell layer. A strong signal is seen in the first passage RPE cells (lane 2) and no PEDF message detected in tenth passage RPE cells (lane 3)

secretion of PEDF in older RPE cell cultures. Samples of total RNA obtained from monkey retina and from first and 10th passage monkey RPE cells were analyzed by northern blot for the PEDF transcript (Tombran-Tink et al., 1995). Using the PEDF cDNA as a probe, a weak hybridization signal is seen in total retinal tissue while a very strong signal is seen in first-passaged RPE cell cultures. Of importance, is the lack of detectable levels of PEDF message in tenth passage RPE cultures which are non-proliferative and unresponsive to mitogenic agents (Fig. 6). The cultures are operationally defined as "senescent". These findings were supported by data obtained from an examination of the secretion profile of PEDF with sucessive passages of RPE cultures. Using a PEDF polyclonal antibody and western blot analysis, we found that only rapidly proliferating, early passaged RPE cultures (first through fourth passage) secrete PEDF into the surrounding medium. Weak immunoreactivity is seen in slowly-dividing cultures (fifth through eight passage) while no immunoreactivity for the 50 kD PEDF doublet is seen in senescing cultures (between the tenth and fifteenth cell passage) (Fig. 7). Thus, the secretion of PEDF decreases dramatically with sucessive cell passages in an age-related manner and is paralled by the lack of transcription

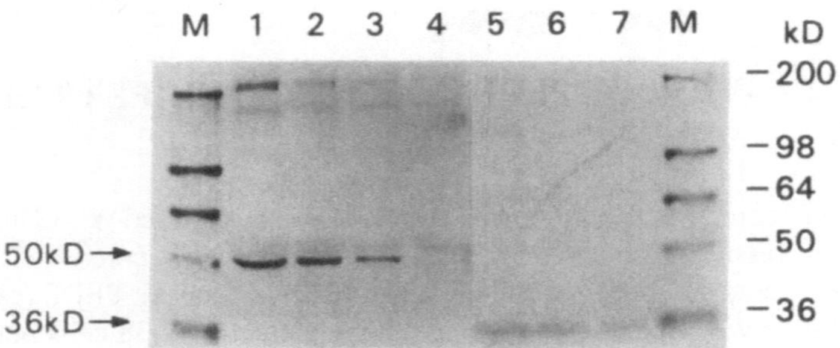

Figure 7. Lanes 1-5 represents a Western blot of samples of RPE cell-conditioned medium (CM) after 1st, 2nd, 5th and 10th passages respectively. The PEDF polyclonal antibody recognizes a prominent 50 kD band in CM from the 1st and 2nd passages, a weak signal in CM of 5th passage cultures and no signal with sucessive cell passages. A 36 kD protein is seen in cell lysates of 2nd passage RPE cell extracts, adult bovine RPE and retinal extracts respectively (lanes 5-7). M: molecular weight standards.

of the PEDF's message in "senescent" RPE cell cultures. In RPE and retina/RPE cell lysates, a 36 kD protein and not the secreted 50 kD species is detected with the PEDF polyclonal antibody. Either a non-glycosylated form of PEDF exists in the cytosol or a small amount of PEDF reenters the cell as the 36 kD species. Alternatively, after uptake the 50 kD species could be cleaved intracellularly resulting in the 36 kD protein that is seen by western blot analysis (Fig. 7). Although further work will have to be done to explain these results, our findings, along with those of Pignolo et al., (1993) raises the interesting possibility that the presence of PEDF could 1) influence a commitment to neuronal differention, 2) promote cell-survival of well-differentiated, mature neurons and 3) the lack thereof could be a general marker for terminal differentiation and/or senescence in such different cell types as RPE cells, retinal neurons and fibroblasts.

MOLECULAR CLONING AND SEQUENCING OF THE PEDF cDNA AND GENOMIC DNA

Tryptic peptides obtained from HPLC-purified PEDF (Tombran-Tink et al., 1992) were microsequenced and analyzed for protein homology/identity with other known proteins. Eight peptides were analyzed and the data indicated that the 50 kD PEDF protein is unique but share sequence homology with members of the serine protease inhibitor gene family known as serpins. From the peptide sequence and codon usage data oligodeoxynucleotides were constructed and used to isolate a full length 1.5 kb, cDNA clone from a human fetal eye Charon BS cDNA library (Steele et al., 1992). Sequence analysis revealed a 1503 base sequence with a long ORF encoding 418 amino acids, a typical ATG start codon and a polyadenylation site. All previously determined peptide sequences align perfectly and confirm the translated product of the clone (Steele et al., 1992). The cDNA sequence shares 27% homology with human α-1-antitrypsin, the prototype of the serpin gene family. Although this percentage is not high, it is well within the range of similarity shared by other serpins.

Serpins are single-chain proteins that range in molecular weight from 45-100 kD, typically containing approximately 400-500, residues and whose structures show a high degree of 3-D architectural similarity. The acronym denotes a superfamily of SERine proteinase INhibitors. Most inhibitory serpins contain a conserved reactive domain or binding loop at the carboxy terminus and which can adopt a variety of conformations. They are variously glycosylated and while the majority have been isolated from plasma, unglycosylated forms are shown to exit in cytosolic fractions while others have been identified in the extracellular matrix (Remold O'Donnel et al., 1992). Serpins play critical regulatory roles in a wide range of vital physiological processes and while for most, the primary function is regulation of proteolytic events associated with a myriad of biochemical pathways, others have diverged from the classical protease inhibitory function and have evolved other roles. These include hormone transporters, zymogen activators, cytokine response modifiers, neurotrophic agents, complement activators, blood coagulators and modifiers of processes involving phagocytosis, and tumor invasion and metastasis (Shapira and Patston, 1991; Pickup et al., 1988). Clinical studies reveal a large number of genetic deficiencies associated with mutations of serpins principally with lung and liver diseases and blood coagulation defects (Huber and Carrel 1989; Engh et al., 1993; Crystal 1991). PEDF lacks homology in the carboxy terminus reactive domain thought to be necessary for protease inhibition and therefore, must have diverged in function with evolution as many other non-inhibitory serpins have done.

Using the PEDF cDNA as a probe, several genomic clones have been isolated and sequenced to reveal a 16 kb PEDF gene whose translated sequence is distributed among 8 exons interrupted by 7 introns (Tombran-Tink et al., 1994). This genomic organization is consistent with a number of serine protease inhibitors and substantiates evidence that PEDF is a member of the serpin gene family. To date, however, no inhibitory activity can be associated with this protein.

CHROMOSOMAL LOCALIZATION OF PEDF

Two well-characterized human x mouse somatic cell hybrid mapping panels were analyzed by southern blot and PCR, for the localization of the human PEDF gene (Tombran-Tink et al., 1994). Data obtained from both the multichromosomal and monochromosomal mapping panels assigned PEDF to human chromosome 17. DNAs from a deletion mapping panel of eight somatic cell hybrids containing specific fragments of human chromosome 17 (Guzzetta et al., 1992) were analyzed by PCR to further sublocalize the PEDF gene. Data obtained from this study confirmed the initial assignment of PEDF and sublocalized it to the short arm of chromosome 17 at 17p13.1-pter. Fluorescent in situ hybridization, using a fragment of the PEDF gene, further confirms the map position to the telomeric region of the short arm of the chromosome (Tombran-Tink et al.,1994).

This chromosomal region is associated with a number of cancer-related loci including, the Li-Fraumeni syndrome, the p53 tumor suppressor gene, the medulloblastoma gene, breast and colorectal tumors. At least one other serpin, the human plasmin α_2-plasmin inhibitor gene (PLI) has been mapped to this region allowing for the interesting possibility of a “gene cluster” of serpins in this area of the genome. Whether PEDF is involved in any tumor-related events is yet to be examined. Most importantly, Greenberg et al., (1994) has provided evidence indicating this region as the locus for one autosomal dominant RP (ADRP) in a large South African family, making PEDF a candidate gene for mutational analysis for the pathogenesis of ADRP in this family. Our mapping also is of interest since it places PEDF close to the telomere of chromosome 17 where it could be involved in cellular aging. Eukaryotic chromosomes end with highly conserved, simple tandem hexamer repeats and, as somatic cells replicate, telomeric sequences are lost. It is thought that telomere shortening activates a complex cascade of molecular events leading to cellular senescence (Allsopp et al., 1992; Harley et al., 1990; Hastie et al., 1990, Greider 1990). The positioning of the PEDF gene in close proximity to the chromosome telomere allows it to be in a location where it might be directly affected by telomere shortening. This phenomenon may influence PEDF’s expression in senescing human fibroblasts and RPE cells.

SUMMARY

RPE cells secrete a unique 50 kD protein, PEDF, *in vitro* and *in vivo*. The protein contains trophic activity both in the induction of neuronal cell differentiation as well as survival of the mature neuron. Because of its presence in fetal and adult IPM, it may participate in early retinogenesis and the survival of mature retinal neurons after development. It is tempting to speculate that its accessibility to nearby photoreceptor neurons may allow it to participate in mechanisms involved in the rescuing of apotosing or injured photoreceptor neurons during degenerative processes caused by exogeneous and endogeneous insults. Transplantation and transgenic studies will further elucidate the function of PEDF in the retina.

REFERENCES

Bogenman E, Lochrie MA, Simon MI, (1988) Cone cell-specific genes expressed in retinoblastoma. Science 240:76-78

Bok D (1985) Retinal photoreceptor-pigment epithelium interactions. Invest. Ophthalmol. Vis. Sci. 26:1659-1694

Campochairo PA, Sugg R, Grotendorst G, Hjelmeland LM (1989) Retinal pigment epithelial cells produce PDGF- like proteins and secrete them into their medium. Exp Eye Res 49:217-227

Connor T, Roberts A, Sporn M, Davis J, Glaser B (1988) RPE cells synthesize and release transforming growth factor- β , a modulator of endothelial cell growth and wound healing. Invest. Ophthalmol Vis. Sci. 29:307- 313

Crystal RG (1991) α -1-antitrypsin, emphysema and liver disease. J. Clinical Invest. 85:1343-1352

Cunha-Vaz JG (1979) The blood-ocular barriers. Surv Opthalmol 23:279-296

Elner UM, Schaffner T, Taylor, Glagou KS (1981) Immunophagocytic properties of retinal pigment epithelial cells. Science 211:74-76

Engh RA, Schulze AJ, Huber R, Bode W (1993) Serpin Structures. Behring Inst. Mitt. 93:41-62

Fassio JB, Jumblatt MM, Barr CC, Georghegan TE, Eiferman RA, Schultz GS (1988) Invest. Ophtalmol. Vis. Sci. 29:242-250

Feeney-Burns L (1988) The pigments of the retinal pigment epithelium. Curr Top Eye Res 2:120-178

Greenberg J, Goliath R, Beighton P, Ramesar R. (1994) A new locus for autosomal dominant retinitis pigmentosa on the short arm of chromosome 17. Hum Mol Genet 3(6):915-918

Greider CW (1990) Telomeres, telomerase and senescence. BioEssays 12:363:369

Guzzetta V, Franco B, Trask BJ, Zhang H, Saucedo- Cardena O, de Oca-Luna RM, Greenberg F, Chinault AC, Lupski JR, and Patel PI (1992) Somatic cell hybrids, sequence-tagged sites, simple repeat polymorphisms and yeast artificial chromosomes for physical and genetic mapping of proximal 17p. Genomics 13:551-559

Harley CB, Futcher AB, Greider CW (1990) Telomeres shorten during ageing of human fibroblast. Nature 345:458-460

Hastie ND, Dempster M, Dunlop MG, Thompson AM, Green DK, Allshire RC (1990). Telomere reduction in human colorectal carcinoma and with ageing. Nature 346:866:868

Hageman GS, Johnson LV (1991) Structure, composition and function of the retinal interphotoreceptor matrix. In Progress in Retinal Research. N. Osborne and G.J. Chader, editors. Pergamon Press, Oxford, England. 207-250

Hewitt AT, Lindsey JD, Carbott D, Adler R (1990) Photoreceptor survival-promoting activity in interphotoreceptor matrix preparations: characterization and partial purification. Exp Eye Res 50:79-88

Huber R, Carrell RW (1989) Implications of the three-dimensional structure of α -1-antitrypsin for structure and function of serpins. Biochemistry 28:8951- 8966

Immel J, Steinberg RH (1986) Spatial buffering of K^+ by the retinal pigment epithelium. J. Neurosci. 6:3197-3204

Kyritsis AP, Tsokos M, Triche TJ, Chader GJ (1984) Nature (London) 307:471-473

Miller S, Steinberg RH (1982) Potassium transport across the frog retinal pigment epithelium. J Membr Biol 67:199-209

Misfeldt D, Hamamoto ST, Pitelka DT (1976) Transepithelial transport in cell culture. Proc Natl Acad Sci (USA) 76:1212-1216.

Pfeffer B (1990) Improved methodology for cell culture of human and monkey retinal pigment epithelium. In progress in Retinal Research. N. Osborne and G.J. Chader editors. Pergamon Press, Oxford UK. 251-291

Pickup DJ, Ink BS, Hu W, Ray CA, Joklik WK (1988) Hemorrhage in lesions caused by cowpox virus is induced by a viral protein that is related to plasma protein inhibitors of serine proteinase. Proc. Natl. Am. Sci. USA 83:7698-7702

Pignolo RJ, Cristofalo VJ, Rotenberg MO (1993) Senescent WI-38 cells fail to express EPC-1, a gene induced in young cells upon entry into the G_0 state. J. Biol. Chem. 268:8949-8957

Plouet J (1988) Implication of the binding of acidic and basic fibroblast growth factors (FGF) to photolysed rhodopsin in visual transduction.Invest Ophthalmol. Vis. Sci. 293:106-114

Remold O'Donnel E, Chin J, Alberts E (1992) Sequence and molecular characterization of human monocyte/neutrophil elastase inhibitor. Proc. Natl. Acad. Sci. USA 89:5635-5639

Shapira M, Patston PA (1991) Serine protease inhibitors (Serpins) TCM 1, 146-151

Steele FR, Chader GJ, Johnson LV, Tombran-Tink J (1993) Pigment epithelium derived factor (PEDF): neurotrophic activity and identification as a unique member of the serine protease inhibitor (SERPIN) gene family. Proc Natl Acad Sci (USA) 90:1526-1530

Steinberg RH, Wood I (1974) Pigment epithelial cell ensheathment of cone outer segments in the retina of the domestic cat. Proc Royal Soc Lond B 187:461-478

Taniwaki T, Becerra SP, Chader GJ, Schwartz JP (1994) Neurotrophic effects of pigment epithelium-derived factor on cerebellar granule cells in culture. J Neurochem suppl) 25:50

Tombran-Tink J, Chader GJ, Johnson LV (1991) PEDF: a pigment epithelium-derived factor with potent neuronal differentiative activity. Exp Eye Res 53:411-414

Tombran-Tink J, Johnson LV (1989) Neuronal differentiation of retinoblastoma cells induced by medium conditioned by human RPE cells. Invest Ophthalmol Vis Sci. 30:1700-1707

Tombran-Tink J, Li A, Johnson MA, Johnson LV, Chader GJ (1992) Neurotrophic activity of interphotoreceptor matrix on human Y79 retinoblastoma cells. J Comp Neurol 317:175-186

Tombran-Tink J, Pawar H, Swaroop R, Rodriguez I, Chader GJ (1994) Localization of the gene for pigment epithelium-derived factor to chromosome 17p13.1 and expression in cultured human retinoblastoma cells. Genomics 19:266-272

Tombran-Tink J, Rodriguez IR, Mazuruk K, Shivaram S, Li A, Chader G (1994) Structural Analysis of the gene for pigment epithelium differentiation (PEDF) factor. Invest. Ophthal. Vis. Sci. 35:1312

Tombran-Tink J, Shivaram S, Chader G, Johnson LV, Bok D (1995) Expression, secretion and age-related downregulation of pigment epithelium-derived factor, a serpin with neurotrophic activity. J. Neurosci. (in press)

Young RW, Bok D (1969) Participation of the retinal pigment epithelium in the rod outer segment renewal process. J Cell Biol 42:392-403

NITRIC OXIDE IN THE RETINA

Potential Involvement in Retinal Degeneration and Its Control by Growth Factors and Cytokines

O. Goureau, F. Becquet, and Y. Courtois

Unité de Recherches Gérontologiques
Institut National de la Santé et de la Recherche Médicale
Association Claude Bernard
29 rue Wilhem, 75016 Paris, France

GENERAL INTRODUCTION

Nitric oxide (NO), an atmospheric gas, is now known to be enzymatically synthesized in a tightly regulated manner by a number of tissues and cell types. Over the past 5 years, significant progress has been made elucidating the mechanism of NO synthesis and the functions of NO in different biological systems. NO is produced by cells, and serves a wide variety of functions in different cells, ranging from vascular endothelia, immune cells, neurons and glia, hepatocytes and smooth muscle cells (reviewed in: 1-3). The functions of NO appear very diverse, having actions on vascular tone, neurotransmission (2), immune cytotoxicity (3,4), and many others. Three isoforms of NO synthase (NOS) have been identified as being responsible for this synthesis in the presence of oxygen, NADPH and flavins, and represent three distinct gene products (1). Two of the enzyme types are continuously present and, thus are termed constitutive NOS. The first, termed NOS-I is found in the cytosol of central and peripheral neurons (2), and the second (NOS-III) was originally expressed by the vascular endothelium. Small amounts of NO are generated by these two iso-enzymes when they are activated by the calcium/calmodulin complex. In contrast, NOS-II, or inducible NOS, is expressed in many cell types after challenge by immunological or inflammatory stimuli (3). This isoform, the activity of which is independent of calcium and calmodulin, generates large amounts of NO over longer periods which are dependent on the presence of the stimuli.

Over the past 2 years, the NO pathway has also been studied in the retina and both the inducible and the constitutive isoforms have been identified. Different techniques were used to describe them. Historically, the first one is the NADPH-diaphorase which is almost but not always associated with NOS activity (1,2). The different cofactors requirements help to discriminate biochemically between the type I, II and III. With the characterization of their sequences, specific polyclonal antibodies have been prepared which are used in the most recent studies. Molecular biology techniques also allow the quantification and localization

Degenerative Diseases of the Retina, Edited by Robert E. Anderson et al.
Plenum Press, New York, 1995

of mRNA specific to each isoforms. However, despite the rapidly evolving situation, there are still very little important molecular and protein data on NOS in the retina or in other parts of the eye.

EXPRESSION OF DIFFERENT NOS ISOFORMS IN THE RETINA.

Constitutive NOS (NOS-I, NOS-III) in the Retina

NOS enzymatic activity was demonstrated in crude extracts of the retina and in isolated rod outer segments (ROS) (5,6). Enzymatic properties indicate a relationship between the retinal enzyme and the brain constitutive isoform (type I). NADPH-diaphorase studies also indicate the presence of NOS in the retina (7,8). NOS-I isoform was clearly identified in some amacrine cells, in the inner nuclear layer and in photoreceptors of the retina from different species by immunohistochemistry (7,9). Recently, NOS-I was identified in the retina of lower vertebrates (10) and cloned from the human retina (11). The presence of NOS-III was not directly demonstrated, however, the large NADPH-diaphorase staining observed in the vascular endothelium of choroid and retina (7, 8) could correspond to this isoform.

NO present in this tissue could serve as a neuromodulator of synaptic transmission in the retina, via a modification of electrical coupling in horizontal cells (12), activation of cGMP-gated conductance in bipolar cells (13) and modulation of a cGMP-gated conductance in ganglion cells (14). Furthermore, NO has been shown to modulate the photoresponse, via a direct modulation of the dark voltage and the light response in frog ROS (15) and by affecting different currents in tiger salamander ROS (16).

Inducible NOS (NOS-II) in the Retina

In the retina, we have demonstrated that retinal Müller glial (RMG) cells can express the NOS-II isoform after endotoxin and cytokines stimulation (17). Bovine (18), human (19) and murine (20,21) retinal pigmented epithelial (RPE) cells , also contain NOS-II. In rat and bovine RPE cells, NOS-II mRNA and enzyme activity are induced by synergistic co-operation between interferon γ (IFNγ) and lipopolysaccharide (LPS) and can be potentiated by tumor necrosis factor α (TNFα), while co-stimulation with IFNγ and interleukin-1β is essential for NO production in human RPE cells (Table 1).

Cytokines and growth factors are key elements in the regulation of NOS-II induction (3). In order to understand their potential either as inducers or as inhibitors of NOS, we have investigated the role of several growth factors in the production of NO in RPE or RMG cells *in vitro* (20,22). We have demonstrated, as summarized in Table 1, that fibroblast growth

Table 1. Comparative role of growth factors and cytokines on nitrite production in RPE and RMG cells from different species

Cell Type:	RMG cells	RPE cells	
Species:	Rat	Bovine	Human
Inducers:	LPS/IFNγ/TNFα	LPS/IFNγ	IL1β/IFNγ
FGF-1 / FGF-2	Potentiation	Inhibition	Potentiation
TGFβ	Inhibition	Potentiation	Inhibition
EGF	N.D	None	None

N.D = non determined.
None = no effect.

Table 2. Correlation between NO production and decrease of bovine RPE cell phagocytosis after stimulation by LPS/IFNγ

Addition (48 hrs)	Nitrite (μM)	Phagocytosis (% control)
Control	1.46 ± 0.7	100
LPS + IFNγ	17.4 ± 0.9	76.8 ± 5.1*
LPS + IFNγ + L-NMMA	2.70 ± 1.5	95.2 ± 4.4

Confluent RPE cells were incubated for 48 hours in medium alone or with LPS (1μg/ml) and IFNγ (100U/ml) with or without L-NMMA (0.5mM). Nitrite release was determined in culture supernatant by Griess reaction (18). RPE cell phagocytosis was estimated by RIA technique (24) after incubation of RPE cells with ROS membranes in fresh medium for 3 hours. Values are means ± SEM for three different experiments, each done in triplicate. * $p<0.001$, significantly different from control.

factors (FGFs) and transforming growth factor beta (TGFβ) have opposing actions on the regulation of the production of NO by these cells.

In bovine RPE cells, FGF-1 and FGF-2 inhibit the induction of NOS at the transcriptional level (23), while it is not an inhibitor of NOS-II induction in rat or in human RPE and rat RMG cells despite the fact that these cells have FGF receptors. This is consistent with an efficient control of the expression of NOS at different levels, since cells must control tightly the formation of high amounts of NO. The opposite data was observed with TGFβ which frequently acts as an immunosupressor signal, suggesting that the regulation of NOS activity in each cell depends on the status of the transduction signal activated by the cytokine network. Since several stimuli act simultaneously or sequentially upon cells, it is not surprising that cell and specie specific signals can be obtained.

The role of high levels of NO produced by NOS-II in the retina is poorly understood. A beneficial antimicrobial, antitumoral and antiviral effect of NO could be considered. However, we could speculate that it would perturbate the neurotransmission processes involving NO described above (12-14), as well as phototransduction (15-16). Recently, we reported that NO could affect RPE cell phagocytic activity *in vitro*, as evaluated by the rod outer segments (ROS) ingestion in a radio-immunoassay (24). Results in Table 2 demonstrate that RPE cells producing large amounts of nitrite (stable end-product of NO) after LPS/IFN γ treatment showed a significant decrease in phagocytic activity (20% less than in control).

Furthermore, addition of the potent inhibitor of NO synthase, L-NMMA, which inhibited nitrite release by LPS/IFN γ-stimulated RPE cells, restored normal phagocytosis of ROS membranes, demonstrating that endogenous NO is the compound involved in LPS/IFN γ-decreased RPE phagocytic activity. This inhibitory effect has also been observed with addition of exogenous NO using a NO donor (24). We suggest that the production of large amounts of NO in the retina, in addition to its beneficial effect (host defense), could also be deleterious by perturbing ROS membrane phagocytosis by RPE cells. This inhibition could lead to the accumulation of ROS debris between photoreceptors and retinal pigmented epithelium, and ultimately result in photoreceptor degeneration.

EXPRESSION OF NOS-II IN DIFFERENT RETINAL PATHOLOGIES

During the last two years, we have investigated the possibility that NO could be one of the mediators involved in the pathogenesis of retinal pathologies or degeneration.

Recently, we demonstrated that NOS-II is expressed *in vivo* in human retina as a result of viral infection (25). Indeed, NOS-II has been detected in cytomegalovirus (CMV)-infected retina from AIDS patients by immunohistochemistry and was localized to CMV-infected glial cells (astrocytes and RMG cells). The role of NO in viral infections of the retina could be beneficial via its antimicrobial and antiviral effects, but also detrimental through its potential to damage tissue.

We recently demonstrated (26) that intraperitoneal injections of NOS inhibitor, N^G-nitro-L-arginine methyl ester (L-NAME), inhibits clinical inflammation in the anterior and in the posterior part of the eye induced by foot pad injection of LPS. This result demonstrating that NO is involved in endotoxin-induced uveitis in Lewis rats was recently confirmed by other studies (27).

NO and Light-Induced Retinal Degeneration

It has been proposed that light induced oxygen free radicals may be mediators of retinal photic injury (28). Furthermore, several reports have demonstrated the protective effect of free radical scavengers in the light damage model (28,29). However, the putative action of NO has never been investigated in retinal degenerative diseases. We have investigated the possible involvement of NO in light-induced photoreceptor degeneration by intraperitoneal injection of L-NAME, an inhibitor of NOS, into albino Fisher rats maintained in constant light for 7 days. By measuring the photoreceptor nuclear layer thickness, we found that L-NAME partially protects against the degeneration of photoreceptors and acts to maintain their morphological organization, with a more pronounced effect observed in the superior hemisphere (Figure 1). The protective effect of L-NAME was dose dependent and maximal protection (up to 35%) was obtained at 100mg/kg of L-NAME administered every other day (30).

We also demonstrated using RT-PCR that constant illumination induced NOS-II in the retina *in vivo* and that this induction correlated with photoreceptor degeneration. A representative RT-PCR experiment, shown in Figure 2, reveals the presence of NOS-II mRNA, 3 days after illumination which increases after 5 and 7 days. Non reverse transcribed

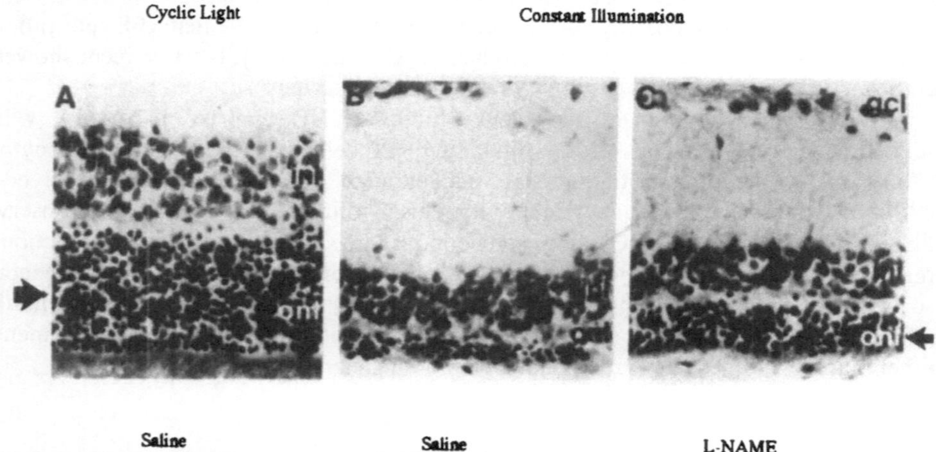

Figure 1. Effect of L-NAME treatment on photoreceptor nuclear layer from rats exposed to constant light. Photographs of the superior part of the retinas from rats exposed to cyclic light (A) or to constant light for 7 days (B,C), and either treated with saline solution (A,B) or with 100mg/kg of L-NAME (C). Adapted from (30).

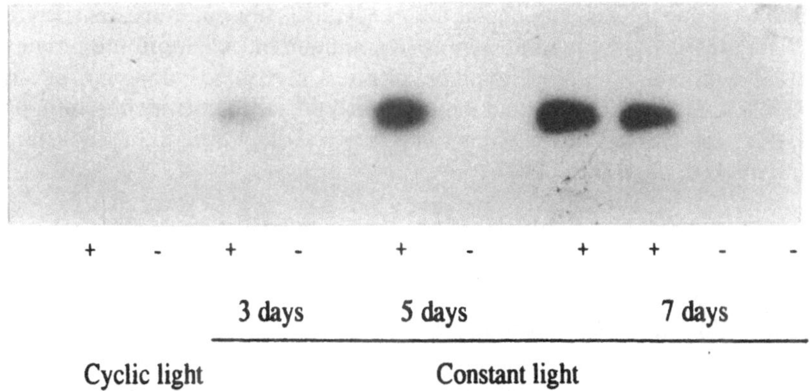

Figure 2. Induction of iNOS mRNA in the retina during constant illumination. One microgramm of total RNA extracted from the retina at the time indicated after constant illumination, was used for each RT-PCR, and iNOS PCR product was identified using specific hybridization probe.

RNA (-) submitted to the PCR amplification steps, did not express NOS-II mRNA, excluding a contamination by genomic DNA. No significant difference in the expression of GAPDH PCR product, used as an internal control, between different RNA extracts was observed, confirming that similar amounts of RNA were transcribed in each sample (data not shown).

The cellular type(s) which is (are) the primary source of NO in the retina, namely resident cell types (RPE or RMG cells) or infiltrating cells (macrophages), are currently under investigation by attempting to localize NOS-II mRNA and protein respectively by *in situ* hybridization and by immunohistochemistry. The involvement of NOS-I and NOS-III has not been investigated, but their role should not be excluded.

A number of different pathways could be involved in the NO-induced light degenerative process : NO might act by eliciting a cGMP increase in the photoreceptor cells leading to excess calcium influx (15,16). NO could ADP-ribosylate certain photoreceptor proteins and modulate their activity (31, 32). NO could also act as a free radical capable of combining with oxygen derivatives resulting in the production of the peroxynitrite anion, $ONOO^-$, which rapidly releases the hydroxyl radical (33, 34). Alternatively, it is possible than an excess of NO might impair the capacity of RPE cells to ingest photoreceptor outer segments, as previously reported and lead to photoreceptor destruction. It will be very interesting to determine which of these molecular mechanisms and cellular pathways are involved in the rescue of photoreceptor degeneration by L-NAME.

Potential Role of NO in Other Retinal Diseases

In our previous study (24), we demonstrated also that, in addition to its effect on phagocytosis, NO was able to inhibit the proliferation of these cells in vitro. This was obtained either with the addition of a NO donor or with endogenous NO produced upon stimulation with LPS and cytokines. It is thus possible that retinal pathologies where RPE cells proliferate as in proliferative vitreoretinal diseases (35), the production of NO may participate in the control of these cells. It will be of great interest to determine the nitrate/nitrite content within the vitreous of these patients as a function of the evolution of their pathology.

This will fit with the dual effect of NO which has been described in many other systems such as in brain ischemia (36). An important feature is to distinguish between the respective role of the NOS isoforms and the possibility to regulate specifically their expression. In addition, NO is only one of the free radicals present in tissues specifically in the retina. The presence of superoxide anion is an important feature in the retina, since it could combine with NO to give rise to peroxynitrite, a very toxic compound, as suggested above (34). The effect of NO is a function of the redox environment where it is produced (35). Thus, it is probable that the combination between these different free radicals induced by inflammation is toxic to the cells.

Growth Factors Rescue and NO

Several growth factors have the property, if injected in the eye, to protect the retina against degeneration in different pathologies. For instance, FGF-1 is efficient in partially maintaining the integrity of the retina in RCS rat (37) and in light-induced photoreceptor degeneration (38). Other growth factors or neurotrophic factors also have a protective action in the latter model (39) and even in ischemic-reperfusion experiments (40). As mentionned above the treatment with radical scavengers has a similar potential (28,29). These data seem to indicate that common neurotoxic mediators may be involved in all these pathologies and that NO may be one of them. One can hypothesize that FGF by its capacity to down-regulate NOS-II expression and thus NO release, is partially able to protect the retina. In other words, by using L-NAME to inhibit the production of NO, we overcome the effect of FGF. However this hypothesis does not take into account that various growth and neurotrophic factors or cytokines do not use the same receptors but form a network which share several pathways in the transduction machinery. Thus, different responses to cytokines, growth factors and neurotrophins can be expected in the treatment of retinal degeneration in various species. A better understanding of the signalisation machinery will be the key to future treatment of the retinal pathologies.

ACKNOWLEDGMENTS

We thank J.C. Jeanny, Y. de Kozak, and J.M. Phillipe for helpful suggestions, L. Jonet and E. Jacquemin for technical assistance and H. Coèt and A. Haslé for photographic work. This work was supported by the Institut National de la Santé et de la Recherche Médicale (INSERM) and by contributions from the "Fondation pour la Recherche Médicale" and from the "Asssociation Française de la Retinite Pigmentaire".

REFERENCES

1. Nathan, C.F., and Xie, Q. 1994. Nitric oxide synthases: Roles, tolls, and controls. *Cell* 78:915-918.
2. Bredt, D.S., and Snyder, S.H. 1994. Nitric oxide: a physiologic messenger molecule. *Ann. Rev. Biochem.* 63:175-195.
3. Nussler, A.K., and Billiar, T.R. 1993. Inflammation, immunoregulation, and inducible nitric oxide synthase. *J. Leukoc. Biol.* 54:171-178.
4. Nathan, C.F., and Hibbs, J.B. Jr 1991. Role of nitric oxide synthesis in macrophage antimicrobial activity. *Curr. Opinion Immunol.* 3: 65-70.
5. Venturini, C.M., Knowles, R.G., Palmer, R.M.J., and Moncada, S. 1991. Synthesis of nitric oxide in the bovine retina. *Biochem. Biophys. Res. Comm.* 180:920-925.
6. Goureau, O., Lepoivre, M., Mascarelli, F., and Courtois, Y. 1992. Nitric oxide synthase activity in bovine retina, In: *Structures and functions of retinal proteins* (Rigaud, J.L., and INSERM, eds) vol. 221, pp395-398. J. Libbey Eurotext Ltd, London.
7. Yamamoto, R., Bredt, D.S., Snyder, S.H., and Stone, R.A. 1993. The localization of nitric oxide synthase in the eye and related cranial ganglia. *Neuroscience* 54:189-200.

8. Osborne, N.N., Barnett, N.L., and Herrera, A.J. 1993. NADPH diaphorase localization and nitric oxide synthetase activity in the retina and anterior uvea of the rabbit eye. *Brain Res.* 610:194-198.
9. Koch, K., Lambrecht, H., Haberecht, M., Redburn, D., and Schimdt, H.H.H.W. 1994. Functional coupling of a calcium/calmodulin-dependent nitric oxide synthase and a soluble guanyl cyclase in vertebrate photoreceptor cells. *EMBO J.* 13:3312-3320.
10. Lieppe, B.A., Stone, C., Koistinaho, J., and Copenhagen, D.R. 1994. Nitric oxide synthase in Müller cells and neurons of salamander and fish retina. *J. Neurosci.* 14:7641-7654.
11. Park, C., Pardhasaradhi, K., Gianotti, C., Villegas, E., and Krishna, G. 1994. Human retina expresses both constitutive and inducible isoforms of nitric oxide synthase mRNA. *Biochem. Biophys. Res. Comm.* 205:85-91.
12. Miyachi, E.I., Murakami, M., and Nakaki, T. 1990. Arginine blocks gap junctions between retinal horizontal cells. *Neuroreport* 1:107-110.
13. Shiells, R., and Falk, G. 1992. Retinal on-bipolar cells contain a nitric oxide -sensitive guanylate cyclase. *Neuroreport* 3:845-848.
14. Ahmad, I., Leinders-Zufall, T., Kocsis, J.D., Shepherd, G.M., Zufall, F., and Barnstable, C.J. 1994. Retinal ganglion cells express a cGMP-gated cation conductance activatable by nitric oxide donors. *Neuron* 12:155-165.
15. Nöll, G.N., Billek, M., Pietruck, C., and Schmidt, K.F. 1994. Inhibition of nitric oxide synthase alters light responses and dark voltage of amphibian photoreceptors. *Neuropharmacology* 33:1407-1412.
16. Kurenny, D.E., Moroz, L.L., Turner, R.W., Sharkey, K.A., and Barnes, S. 1994. Modulation of ion channels in rod photoreceptors by nitric oxide. *Neuron* 13:315-324.
17. Goureau O., Hicks, D., Courtois, Y., and de Kozak, Y. 1994. Induction and regulation of nitric oxide synthase in retinal Müller glial cells *J. Neurochem.* 63:310-317.
18. Goureau, O., Lepoivre, M., and Courtois, Y. 1992. Lipopolysaccharide and cytokines induce a macrophage-type of nitric oxide synthase in bovine retinal pigmented epithelial cells. *Biochem. Biophys. Res. Comm.* 186: 854-859.
19. Goureau, O., Hicks, D., and Courtois, Y. 1994 Human retinal pigmented epithelial cells produce nitric oxide in response to cytokines *Biochem. Biophys. Res. Commun.* 198:120-126.
20. Sparrow, J.R., Nathan, C.F., and Vodovotz, Y. 1994. Cytokine regulation of nitric oxide oxide synthase in mouse retinal pigment epithelial cells in culture. *Exp. Eye Res.* 59:129-139.
21. Liversidge, J., Grabowski, P., Ralston, S., Benjamin, N., and Forrester, J.V. 1994. rat retinal pigment epithelial cells express an inducible form of nitric oxide synthase and produce NO in response to inflammatory cytokines and activated T cells. *Immunolgy* 83:04-409.
22. Goureau, O., Lepoivre, M., Becquet, F., and Courtois, Y. 1993. Differential regulation of inducible nitric oxide synthase by basic fibroblast growth factors and transforming growth factor β in bovine retinal pigmented epithelial cells : Inverse correlation with cellular proliferation *Proc Natl. Acad. Sci. USA* 90:4276-4280.
23. Goureau, O., and Courtois, Y. 1994. Fibroblast growth factors inhibit inducible nitric oxide synthase expression in bovine retinal pigmented epithelial cells . In : *The Biology of Nitric Oxide*, (Moncada, S., Feelisch, M., Busse, R., and Higgs, E.A., eds) Vol.4, pp105-108, Portland Press Ltd, London.
24. Becquet, F., Courtois, Y., and Goureau, O. 1994. Nitric oxide decreases in vitro phagocytosis of photoreceptor outer segments by bovine retinal pigmented epithelial cells. *J. Cell. Physiol.* 159:256-262.
25. Digheiro, P., Reux, I., Hauw, JJ., Fillet, AM., Courtois, Y. and Goureau, O. 1994. Expression of inducible nitric oxide synthase in cytomegalovirus-infected glial cells of retinas from AIDS patients. *Neurosci. Letters*, 166:31-34.
26. Bellot, J., Goureau, O., Thillaye, B., Chatenoud, L., and de Kozak, Y. 1994. Inhibition of endotoxin-induced uveitis by nitric oxide synthase inhibitor; effect on intraocular TNF and NO synthesis".In : *Advances in Ocular Immunology*, (Nussenblatt, R.B., Whitcup, S.M., Caspi, R., and Gery, I., eds) pp225-228, Elsevier Science B.V., New York.
27. Mandai M., Yoshimura, N., Yoshida, M., Iwaki, M. and Honda Y. 1994. The role of nitric oxide synthase in endotoxin-induced uveitis : effect of N^G-nitro L-arginine. *Invest. Ophthalmol. Vis. Sci.* 35:3673-3681.
28. Organisciak, D.T., Darrow, R.M., Bicknell, I.R., Jiang, Y.L., Pickford, M., and Blanks, J. 1991. Protection against retinal light damage by natural and synthetic antioxidants. In: *Retinal Degenerations*. (Anderson, R.E., Hollyfield, J.G., and La Vail, M.M., eds) pp189-201, Boca Raton, CRC Press.
29. Davidson, P.C., and Sterneberg, P. Jr. 1993. Potential retinal phototoxicity. *Am. J. Ophthalmol.* 116:497-501.
30. Goureau, O., Jeanny, JC., Becquet, F., Harthmann, MP., and Courtois, Y. 1994. Protection against light-induced retinal degeneration by an inhibitor of NO synthase inhibitor. *Neuroreport* 5:233-236.

31. Ehret-Hilberer, S., Nullans, G., Aunis, D., and Virmaux, N. 1992. Mono ADP-ribosylation of transducin catalyzed by rod outer segment extract. *FEBS Lett.* 309:394-398.
32. Zoche, M., and Koch, K.W. 1995. Purified retinal nitric oxide synthase enhances ADP-ribosylation of rod outer segments proteins. *FEBS Lett.* 357:178-182.
33. Beckman, J.S., and Crow, J.P. 1993. Pathological implications of nitric oxide, superoxide and peroxynitrite formation. *Biochem. Soc. Trans.* 21:330-334.
34. Stamler, J.S. 1994 Redox Signaling : Nitrosylation and related target interactions of Nitric Oxide. *Cell* 78:931-936
35. Malecaze, F., Mathis, A., Arne, J.L., Raulais, D., Courtois, Y., and Hicks D. 1991. Localization of acidic fibroblast growth factor in proliferative vitreoretinopathy membranes. *Current Eye Res.* 10, 719-729.
36. Dawson D.A. 1994 Nitric Oxide and focal ischemia: multiplicity of actions and diverse outcome *Cerebrovasc. Brain Metab.Rev.* 6:299-324
37. Faktorovitch, E.G., Steinberg, R.G., Yasumura, D., and La Vail, M. 1990. Photoreceptor degeneration in inherited retinal distrophy delayed by basic fibroblast growth factor. *Nature* 347:83-86
38. Faktorovitch, E.G., Steinberg, R.G., Yasumura, D., Matthes, M.T., and LaVail, M. 1992. Basic fibroblast growth factor and local injury protect photoreceptors from light damage in the rat. *J. Neurosc.* 12,3554-3567
39. La Vail, M., Unoki, K., Yasumara, D., Matthes, M.T., Yancopoulos G.D., and Steinberg R.G. 1992. Multiple growth factors, cytokines, and neurotrophins rescue photoreceptors from the damaging effects of constant light. *Proc. Natl. Acad. Sci. USA* 89, 11249-11253
40. Unoki K., LaVail M. 1994. Protection of the rat retina from ischemic injury by brain-derived neurotrophic factor, ciliary neurotrophic factor and basic fibroblast growth factor. *Invest. Ophtalmol. Vis. Sci.* 35, 907-915.

8

IN VITRO EXPRESSION OF EPIDERMAL GROWTH FACTOR RECEPTOR BY HUMAN RETINAL PIGMENT EPITHELIAL CELLS

Yusuf K. Durlu and Makoto Tamai

Department of Ophthalmology,
Tohoku University, School of Medicine
1-1 Seiryo-machi, Aoba-ku
Sendai, Miyagi 980, Japan

INTRODUCTION

Growth factors regulate cell proliferation and differentiation (1). Epidermal growth factor (EGF) is a polypeptide mitogen that stimulates division of various cell types *in vitro* and, in particular, epithelial cells *in vivo* (1,2). It has been previously reported that the addition of EGF to culture medium increases DNA synthesis and number of cultured human retinal pigment epithelial (RPE) cells after 48 hr (3).

The EGF receptor (EGFR) is an oncogene protein that has intrinsic tyrosine kinase activity and mediates its action by EGF and EGF-like molecules (transforming growth factor-alpha [TGF-alpha], pox virus growth factors, and amphiregulin) (2). In several malignant diseases, EGFR appears to be overexpressed, as noted by its increased autophosphorylation activity (4,5).

When cultured, RPE cells lose their highly specialized and differentiated nature, for example, the failure to express cellular retinaldehyde-binding protein and the loss of melanin granules (6). The activity of retinyl ester synthetase in RPE cells that converts all-*trans* retinol to retinyl esters of fatty acids is rapidly lost in conventional culture media, and insulin or insulin-like growth factor type 1 is required to maintain its activity (7). In this study, we looked for the expression of EGFR in cultured human RPE cells to evaluate the transformative and undifferentiated characteristics of this neuroectodermal cell type *in vitro* .

MATERIALS AND METHODS

Human RPE Cell Culture

Human RPE cell cultures were established using 36-, 38- and 65-year-old donor eyes within 12 hours after death. In brief, after removing anterior segment and vitreous, the eyecups

Degenerative Diseases of the Retina, Edited by Robert E. Anderson et al.
Plenum Press, New York, 1995

were incubated with trypsin (0.05%)/ethylenediaminetetraaminicacid (EDTA) (0.53 mM) (Gibco) solution in Hank's balanced salt solution (HBSS) (Bio-Whittaker), without calcium and magnesium (trypsin/EDTA solution), for 15 min at 37°C in 5% CO_2. Following incubation, trypsin/EDTA solution was aspirated, and the neural retina was separated from the RPE layer with use of a spatula under visualization with a dissecting microscope. The eyecups were washed twice with Dulbecco's phosphate-buffered saline, without calcium and magnesium ($DPBS^-$). Further incubation of the eyecups was made for 10 min in trypsin/EDTA solution at 37°C in 5% CO_2. At the end of the incubation, trypsin/EDTA solution was gently aspirated from the eyecups. The eyecups were filled with 10% fetal bovine serum in minimum essential medium (10% FBS/MEM, from Gibco). By performing pipetting under the dissecting microscope, we could collect RPE in 10% FBS/MEM. Following centrifugation at 1000 rpm for 5 min at 4°C, the supernatant was discarded, and the RPE pellet was resuspended and incubated with trypsin/EDTA solution for 5 min at 37°C. Thereafter, an aliqout of 10 μl was taken, and the total number of RPE cells was counted with use of a Burke-Turk hemocytometer. Incubation was stopped by adding 5 times the volume of ice-cold 10% FBS/MEM. Centrifugation was performed at 1000 rpm for 5 min at 4°C. The pellet was resuspended and washed twice with 10% FBS/MEM. Human RPE cells were seeded at 37°C in 5% CO_2 in modified polysterene dishes (Becton-Dickinson) at a density of $1x10^4$ cells/cm^2 in 10% FBS/MEM containing 10 ng/ml of basic-fibroblastic growth factor (b-FGF) mutein CS23 (which is a generous gift of Takeda Chemical Industries, Ltd., Japan) and heparin (17 U/ml, Lederle). It has been previously reported that the serine was substituted for the cysteine residue at 70 and 88 of b-FGF mutein CS23 to increase the stability of this protein (8). The medium was changed every 3 to 4 days. Passages between numbers 5 and 10 were used for immunocytochemistry studies and immunoblotting technique.

Immunocytochemistry

At the indicated passage, the medium was aspirated, and the plates were washed twice with $DPBS^-$. Following incubation with trypsin/EDTA solution at 37°C for 10 min, the RPE cells were recovered from the plate by pipetting. Five times the volume of ice-cold 10% FBS/MEM was added, and the solution was centrifuged at 1000 rpm for 5 min at 4°C. After washing the pellet twice with MEM, RPE cells were seeded to glass chamber slides (Nunc) in MEM at a density of $3x10^4$/cm^2 at 37°C in 5% CO_2. Following incubation of the chamber slides at 37°C for 3-4 hr, the medium was aspirated. These glass slides and cryosections of the human retina prepared 7-9 μ thick using a 48-year-old donor eye were air-dried, and fixed with cold(-20°C) acetone for 10 min at room temperature (RT). The preincubation was made with 0.3% hydrogen peroxide including 0.1% sodium azide in phosphate-buffered saline (PBS) for 20 min at RT to remove endogeneous peroxidase activity (9). The slides were washed twice with Tween-80 (0.05%) containing PBS (Tw-PBS). Incubations were performed at RT for 2 hr with the following antibodies at a dilution factor 1/50, in Tw-PBS containing 2% bovine serum albumin (BSA): anti-EGFR (activated and unactivated [monoclonal, clone:Z025, Zymed] and activated [monoclonal, clone:Z026, Zymed]) and anti-EGFR (polyclonal rabbit IgG raised against human recombinant EGFR extracellular domain, Austral Biologicals Inc.). Incubation with anti-cytokeratin pan (a monoclonal antibody mixture of the following clones: C-11, PCK-26, CY-90, Ks1A3, M20, A53-B/A2, supplied as prediluted from Sigma) was performed at RT for 2 hr at a dilution factor 1/10, in Tw-PBS-2% BSA. At the end of the incubation, the glass slides were washed 5 times with Tw-PBS. Then the slides were incubated with the peroxidase-labeled second antibody (goat anti-mouse or goat anti-rabbit, depending on the first antibody, Dako), 1/100 dilution, in Tw-PBS-2% BSA for 30 min at RT. Following incubation with the second antibody, the slides were washed 5 times with Tw-PBS then with PBS twice. The color development was achieved by 3-amino-9-ethylcarbazole (Dako) for 15 min at RT. Control slides were made by omitting the first antibody in the incubation mixture containing Tw-PBS-2% BSA. The reaction was stopped by transferring the slides to distilled water. Thereafter, they were counterstained with Mayer's hematoxyline and mounted with glycerol.

Immunoblotting

The immunoblotting was performed against EGFR and cytokeratin antigens. In brief, after reaching the confluent state of the RPE cultures, the medium was aspirated and the plates were washed twice with ice-cold DPBS$^-$ containing protease inhibitors (2.5 mM EDTA [Dojindo Laboratories], 1 mM phenylmethylsulfonylfluoride [Sigma], 2 μg/ml leupeptin [Boehringer-Mannheim], and 4 μg/ml pepstatin A [Sigma]). The RPE cells were recovered by using a scraper, then they were centrifuged at 1000 rpm for 5 min at 4°C and washed twice with the same buffer. The final pellet was resuspended (0.3-0.4 mg/ml protein concentration) and sonicated (10 strokes, 1-sec interval) on ice in the same buffer and divided into several portions. The protein assay was made (10), and the sample was lyophilized and kept frozen at -80°C until its use. The lyophilized sample described was vortexed in sodium dodecyl sulfate (SDS) sample buffer, sonicated (3 strokes, 1-sec interval) on ice and treated for 1 hr at 37°C. Next, 14 μg of this sample and 5 μl of prestained protein standard (Bio-Rad) was subjected to SDS/10% polyacrylamide gel electrophoresis, according to the method of Laemmli (11). The electrophoretic transfer of the proteins to the nitrocellulose membrane was carried out with a buffer consisting of 0.025 M Tris, pH 8.3/0.192 M glycine/20% (vol/vol) methanol, 0.25% SDS by a Bio-Rad Trans-Blot apparatus overnight at RT (12). The nitrocellulose membrane was incubated with 3% gelatin in Tw-PBS for 1 hr at 37°C to block nonspecific binding followed by incubation overnight at 4°C with the first antibody (anti-EGFR [polyclonal], 1/1000 dilution, and prediluted anti-cytokeratin pan [monoclonal], 1/100 dilution), in 2% BSA/Tw-PBS. After washing with Tw-PBS five times, incubation for 1 hr was made with the second antibody (alkaline phosphatase-conjugated goat anti-rabbit IgG or goat anti-mouse IgG, depending on the first antibody, Cappel, 1/5000 dilution) in 2% BSA/Tw-PBS at 37°C. The membrane was washed with Tw-PBS five times and then with Tris-buffered saline (pH 9.5) twice. The color was developed with a solution containing 10 ml of 100 mM NaCl, 5 mM $MgCl_2$, 100 mM Tris/HCl, pH 9.5, 66 μl of NBT solution (50 mg/ml nitroblue tetrazolium in 70% dimethyl formamide), and 33 μl of BCIP solution (50 mg/ml 5-bromo-4-chloro-3-indolyl-phosphate, p-toluidine salt in dimethyl formamide) for 10 min at RT. The reaction was stopped by washing the membrane with distilled water.

RESULTS

The cultured human RPE cells were achieved from donor eyes, using the appropriate isolation technique described, and were free of other retinal or choroidal cells (Fig. 1 A and B). The proliferation of cultured human RPE cells was observed within 2-3 weeks from the initial seeding using 10% FBS containing 10 ng/ml b-FGF with heparin. At passage 5 and 9, most of the proliferated human RPE cells had lost their pigmentation (Fig. 1 C and D). We found the addition of b-FGF to be essential, as the human RPE cells became more epitheloid shaped in b-FGF and heparin-containing medium (Fig. 1 C and D).

The cultured human RPE cells isolated from three donors exhibited EGFR immunoreactivity by using a monoclonal antibody against the synthetic 30 residue region of EGFR (Fig. 2 A) and also a polyclonal antibody against human recombinant EGFR (data not shown). However, we could find no immunoreactivity to the activated EGFR by a monoclonal antibody against tyrosine-phosphorylated human EGFR (Fig. 2 B). The sample

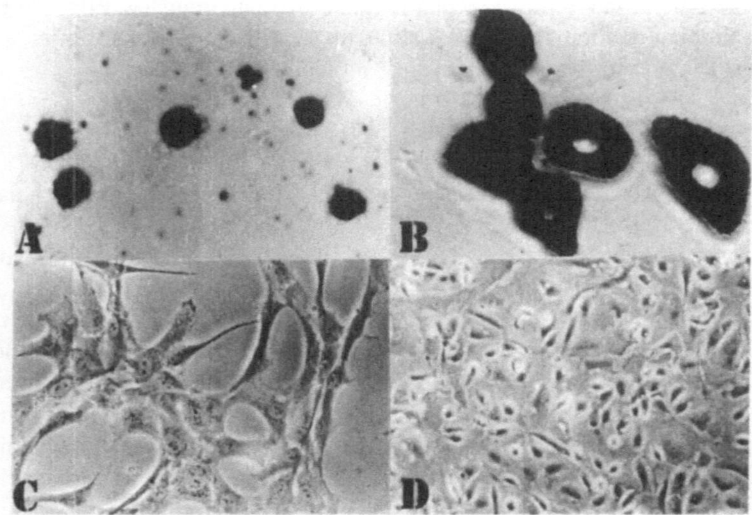

Figure 1. Cultured human RPE cells isolated from the 38-year-old donor eye. (A) primary culture, X240 (B) passage 1, X240 (C) passage 5, X240 (D) passage 9, X120.

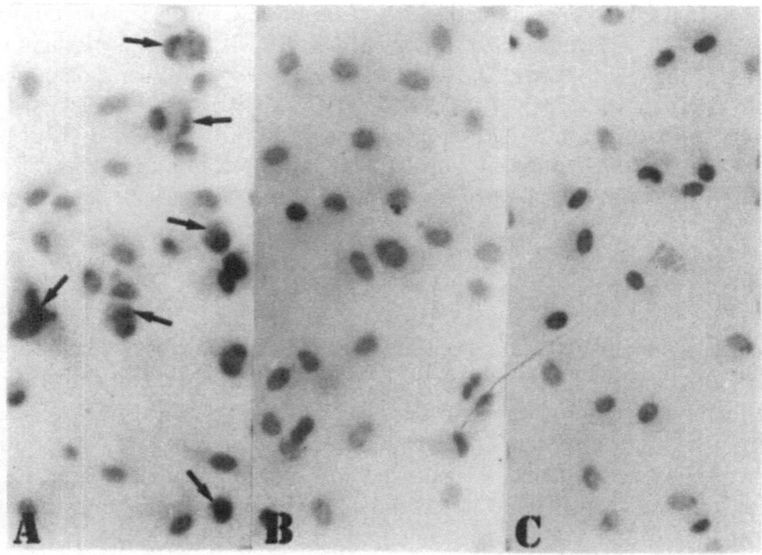

Figure 2. (A) The immunocytochemistry of the epidermal growth factor receptor (EGFR), (B) activated EGFR of the cultured RPE isolated from the 36-year-old donor eye at passage 8, X240. (C) Control where first antibody omitted, X240. Counterstained with Mayer's hematoxyline. Arrows indicate EGFR immunoreactive cells at A.

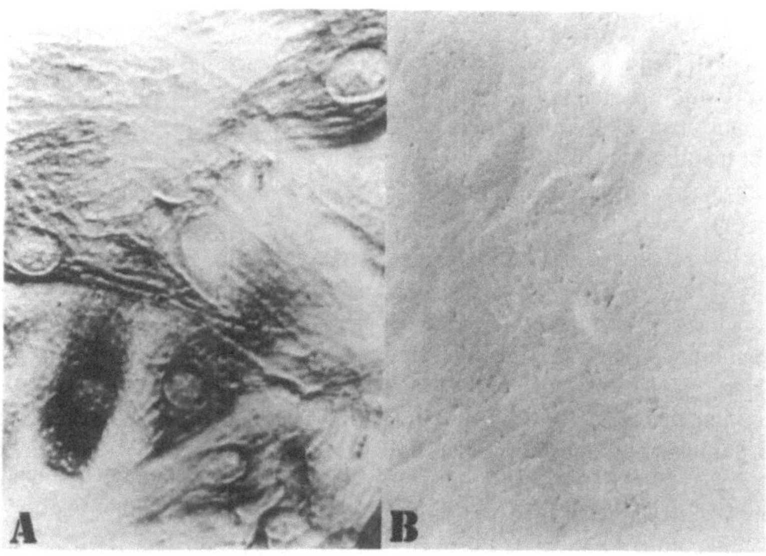

Figure 3. (A) Cytokeratin immunocytochemistry of the cultured human RPE cells from the 65-year-old donor eye at passage 7, X480. (B) Control where first antibody omitted, X480. (Nomarski optics).

without first antibody showed no staining (Fig. 2 C). The cryosections of the retina from the 48-year-old donor eye failed to show EGFR immunoreactivity at RPE (data not shown) using the immunohistochemical technique described.

The cytokeratin immunoreactivity of the cultured human RPE cells isolated from three donors also was found positive (Fig. 3 A) where the control lacked reaction product (Fig. 3 B).

The immunoblotting of the cultured human RPE cell extract against EGFR showed one band at 170 kDa (Fig. 4). The immunoblotting using the same sample against cytokeratin stained two bands with molecular weights 45 kDa and 40 kDa (Fig. 4).

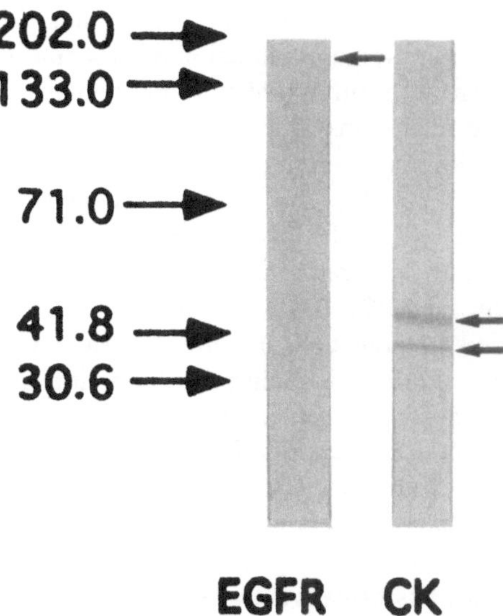

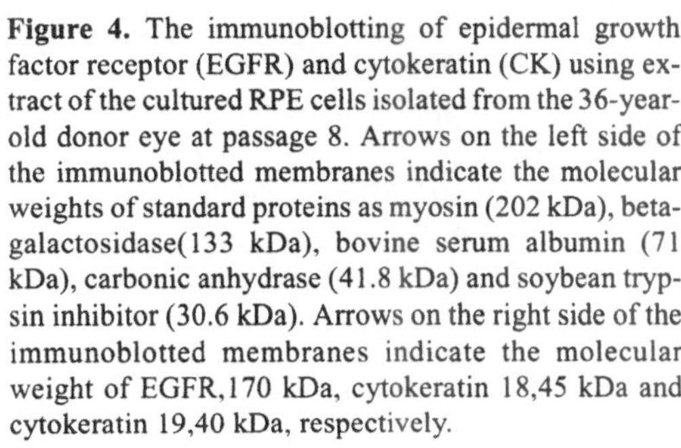

Figure 4. The immunoblotting of epidermal growth factor receptor (EGFR) and cytokeratin (CK) using extract of the cultured RPE cells isolated from the 36-year-old donor eye at passage 8. Arrows on the left side of the immunoblotted membranes indicate the molecular weights of standard proteins as myosin (202 kDa), beta-galactosidase(133 kDa), bovine serum albumin (71 kDa), carbonic anhydrase (41.8 kDa) and soybean trypsin inhibitor (30.6 kDa). Arrows on the right side of the immunoblotted membranes indicate the molecular weight of EGFR,170 kDa, cytokeratin 18,45 kDa and cytokeratin 19,40 kDa, respectively.

DISCUSSION

EGF acts through a cell surface receptor, EGFR, which has three functional domains (2). The EGF-binding domain of EGFR is located on the external cell surface, which might be secreted or truncated from the transmembrane domain of the receptor, although the physiological nature of this process is not well-known (13,14). The cytoplasmic domain of EGFR has intrinsic tyrosine kinase activity, which has a central role in the regulation of cell proliferation (1).

It has been reported that EGF is a potent mitogen of cultured human RPE cells (3). The synthesis of EGF and TGF-alpha by neural retinal cells may indicate that these secreted growth factors may be used by the secreted cells themselves (autocrine effect) and/or by the RPE cells (paracrine effect) (15). High EGFR activity was found in Muller cells of the retina when they were cultured (16). The presence of high affinity TGF-alpha/EGF-receptors in membrane homogenates of the bovine neural retina has been reported, although isolated bovine RPE homogenates showed no detectable receptor activity (15). We also could find no immunoreactivity of EGFR at the human RPE using cryosections of donor eye retina. However, our results showed the expression of EGFR by cultured human RPE cells (Figs. 2 A and 4). Why cultured human RPE cells failed to express activated (tyrosine-phosphorylated) EGFR remains unknown (Fig. 2 B). The immunoblotting of EGFR using the human RPE cell extract stained one band at 170 kDa (Fig. 4), which is in agreement with the molecular weight of EGFR reported in other tissues and cell lines (2).

In our study, cytokeratin was used to determine whether the cultured cells that expressed EGFR were epithelial in origin. We have shown that the cells used in this study also expressed cytokeratin, as detected by immunocytochemistry (Fig. 3 A) and immunoblotting (Fig. 4). Our immunoblotting results using cytokeratin antibody demonstrated that the stained bands were cytokeratin 18 and 19. Cytokeratin 18 has a molecular weight of 45 kDa, which represents the cytoskeleton of simple epithelial cells (17,18). Cytokeratin 19, which has a molecular weight of 40 kDa, is reportedly observed primarily in proliferating human RPE cells (17,18). We also found cytokeratin to be a useful marker of RPE. As the human RPE cells lost some of their differentiating characteristics, such as pigmentation and phenotype after subsequent passages (Figs. 1 C and D, 2, and 3), it might have been difficult without cytokeratin to identify the origin of the cells in culture.

In vivo, RPE cells lack mitosis. However, RPE cells have been implicated in the proliferation and induction of new fibrovascular membrane formations in certain diseases, such as proliferative vitreoretinopathy (19), proliferative diabetic retinopathy (20), and age-related macular degeneration (21). In addition, RPE cells participate in ocular wound healing, as seen in hyperpigmented retinal scars after cryotherapy (22) and laser photocoagulation (23). Through their oncogene receptor EGFR, EGF and TGF-alpha may play a part in activating the proliferation of RPE cells in pathological conditions. In the Royal College of Surgeons rat, retinal neovascularization and vascular transformations are stimulated by diseased RPE (24). Angiogenesis is promoted by EGF and TGF-alpha by their binding to several target cells *in vitro* (25). However, it has been shown that TGF-alpha is more potent than EGF in inducing the formation of new blood vessels *in vivo* (25). In proliferative diseases of the retina, EGF and/or TGF-alpha may be secreted by RPE cells for their autocrine effect to express EGFR; otherwise, the supply of these growth factors from adjacent neural retinal cells to RPE cells may contribute in a paracrine fashion to the induction of this oncogene protein by RPE cells.

In conclusion, we think that the expression of EGFR may be related to the mitotic and transformative changes of human RPE *in vitro*, as shown in our study, and in certain ocular diseases, which needs to be proved. Additional studies are necessary to understand the mechanisms involved in the expression of EGFR by cultured human RPE cells.

ACKNOWLEDGMENTS

This study was supported in part by the Ministry of Education, Science and Culture of Japan and Mishima Sai-ichi fund.

REFERENCES

1. Carpenter G. and Cohen S.; Epidermal growth factor. J Biol Chem 265(14): 7709-7712, 1990.
2. Carpenter G.; Receptors for epidermal growth factor and other polypeptide mitogens. Ann Rev Biochem 56: 881-914, 1987.
3. Leschey K.H., Hackett S.F., Singer J.H. and Campochiaro P.A.; Growth factor responsiveness of human retinal pigment epithelial cells. Invest Ophthalmol Vis Sci 31: 839-846, 1990.
4. Libermann T.A., Nusbaum H., Razon N., Kris R., Lax I., Soreq H., Whittle N., Waterfield M.D., Ullrich A. and Schlessinger J.; Amplification, enhanced expression, and possible rearrangement of the EGF-receptor gene in primary human brain tumors of glial origin. Nature (London) 313:144-147, 1985.
5. Ekstrand A.J., James C.D., Cavenee V.K., Seliger B., Petterson R.F. and Collins V.P.; Genes for epidermal growth factor receptor, transforming growth factor alpha and their expression in human gliomas in vivo. Cancer Res 51(8):2164-2172, 1991.
6. Campochiaro P.A. and Hackett S.F.; Corneal endothelial cell matrix promotes expression of differentiated features of retinal pigment epithelial cells: Implication of laminin and basic fibroblast growth factor as active components. Exp Eye Res 57: 539-547, 1993.
7. Edwards R.B., Adler A.J. and Claycomb R.C.; Requirement of insulin or IGF-1 for the maintenance of retinyl ester synthetase activity by cultured retinal pigment epithelial cells. Exp Eye Res 52:51-57, 1991.
8. Seno M., Sasada R., Iwane M., Sudo K., Kurokawa T., Ito K. and Igarashi K.; Stabilizing basic fibroblast growth factor using protein engineering. Biochem Biophys Res Comm 151(2):701-708, 1988.
9. Hoppenreijs V.P.T., Pels E., Vrensen G.F.J.M., Felten P.C. and Treffers W.F.; Platelet-derived growth factor: Receptor expression in corneas and effects on corneal cells. Invest Ophthalmol Vis Sci 34(3):637-649, 1993.
10. Lowry O.H., Rosenbrough N.J., Farr A.L. and Randall, R.J.; Protein measurement with the folin phenol reagent. J Biol Chem 193:265-275, 1953.
11. Laemmli U.K.; Cleavage of structural proteins during the assembly of the heads of bacteriophage T4. Nature (London) 227:680-685, 1970.
12. Towbin H., Staehelin T. and Gordon J.; Electrophoretic transfer of proteins from polyacrylamide gels onto nitrocellulose sheets: procedure and some applications. Proc Natl Acad Sci U.S.A.81:4683-4687, 1979.
13. Weber W. and Gill G.N.; Production of an epidermal growth factor receptor-related protein. Science 224: 294-297, 1984.
14. Soderquist A.M., Stoscheck C and Carpenter G.; Similarities in glycosylation and transport between the secreted and plasma membrane forms of the epidermal growth factor receptor in A-431 cells. J Cell Physiol 136:447-454, 1989.
15. Fassio J.B, Brockman E.B., Jumblatt M., Greaton C., Henry J.L., Geoghegan T.E., Barr C. and Schultz G.S.; Transforming growth factor alpha and its receptor in neural retina. Invest Ophthalmol Vis Sci 30(9): 1916-1922, 1989.
16. Roque R.S., Caldwell R.B. and Behzadian M.A.; Cultured Muller cells have high levels of epidermal growth factor receptors. Invest Ophthalmol Vis Sci 33(9): 2587-2595, 1992.
17. Hunt R.C. and Davis A.A.; Altered expression of keratin and vimentin in human retinal pigment epithelial cells in vivo and in vitro. J Cell Physiol 145:187-199, 1990.
18. Fuchs U., Kivela T and Tarkkanen A.; Cytoskeleton in normal and reactive human retinal pigment epithelial cells. Invest Ophthalmol Vis Sci 32(13):3178-3186, 1991.
19. Baudouin C., Fredj-Reygrobellet, Brignole F., Negre F., Lapalus P., Gastaud P.; Growth factors in vitreous and subretinal fluid cells from patients with proliferative vitreoretinopathy. Ophthalmic Res 25: 52-59, 1993.

20. Hiscott P., Gray R., Grierson I. and Gregor Z.; Cytokeratin-containing cells in proliferative diabetic retinopathy membranes. Br J Ophthalmol 78(3):219-222, 1994.
21. Das A., Puklin J.E., Frank R.N. and Zhang N.L; Ultrastructural immunocytochemistry of subretinal neovascular membranes in age-related macular degeneration. Ophthalmology 99(9):1368-1376, 1992.
22. Lincoff H., Kreissig I., Jakobiec F. and Iwamoto T.; Remodeling of the cryosurgical adhesion. Arch Ophthalmol 99:1845, 1981.
23. L'Esperance F.A.: The ocular histopathologic effect of krypton and argon laser radiation. Am J Ophthalmol 68:263, 1969.
24. Roque R.S. and Caldwell R.B.; Pigment epithelial cell changes precede vascular transformations in the dystrophic rat retina. Exp Eye Res 53:787-798, 1991.
25. Schreiber A.B., Winkler M.E. and Derynck R.; Transforming growth factor-alpha: A more potent angiogenic mediator than epidermal growth factor. Science 232:1250-1253, 1986.

9

NEW RETINAL DEGENERATIONS IN THE MOUSE

Thomas H. Roderick,[1] Bo Chang,[1] Norman L. Hawes,[1] and John R. Heckenlively[2]

[1] The Jackson Laboratory
Bar Harbor, Maine
[2] The Jules Stein Eye Institute
Los Angeles California

For many years we followed with great interest the comparative advances of genetics in the mouse[1] and in human.[2] Relative frequencies of dominant and recessive traits were similar, mapping of genes proceeded at similar rates, and comparative mapping produced surprises through the discovery of the large number and size of mouse and human homologous chromosomal segments retained since the separation of the species 65 million years ago. What was different between mouse and human was the relative frequency of genetic eye disorders, that is, they were relatively frequent in human, and rare in mouse. Recognizing this was because of bias in ascertainment, because mice do not refer themselves for visual diagnosis and treatment, we began a systematic program to find and characterize mouse eye disorders. The Jackson Laboratory, having the largest collection of mouse mutant stocks and genetically diverse inbred strains was an ideal place to look for genetically determined eye variations and disorders. We have not been disappointed. Through ophthalmoscopy, electroretinography and histology, we have discovered disorders affecting all aspects of the eye including the lid, cornea, iris, lens, and retina, resulting in cornea disorders, cataracts, retinal degenerations and glaucoma. Additional studies have shown predisposition to certain eye problems in aged mice of specific stocks or strains.

Precise mapping of these mutations along with the extensive knowledge of the homologous regions of mouse and human chromosomes, make it possible to predict mapping locations of human homologs. Some of these are good models and actual genetic homologs of human eye disorders. Having homologs of human eye disorders in a genetically well characterized experimental mammal such as the mouse, can lead to understanding of primary actions of genes, developmental timing and anatomical or physiological etiologies, pre-clinical development, as well as providing a tool for experimenting with diet or other environmental conditions in hopes of ameliorating disease. In this report we review salient findings on the known retinal degenerations, *pcd*, *nr*, *rd1* and *Rd2*, and summarize information on newly discovered retinal degenerations *mnd*, *rd3*, *Rd4*, and *rd5*.

Degenerative Diseases of the Retina, Edited by Robert E. Anderson et al.
Plenum Press, New York, 1995

FOUR PREVIOUSLY KNOWN RETINAL DEGENERATIONS IN THE MOUSE

Purkinje Cell Degeneration (*pcd*)

Two neurologic disorders with slow progressive photoreceptor degeneration have been reported in the mouse. The first, Purkinje cell degeneration (*pcd*), is an autosomal recessive mutation that arose on the C57BR/cdJ strain. Homozygotes show a moderate ataxia beginning at 3 to 4 weeks, and then there is rapid degeneration of nearly all cerebellar Purkinje cells beginning at 15 to 18 days. This is accompanied by a slower degeneration of photoreceptor cells of the retina and mitral cells of the olfactory bulb.[3,4] Complete degeneration of photoreceptor cells proceeds slowly over a year.[5,6]

Nervous (*nr*)

The second, nervous (*nr*), arose on the BALB/cGr strain. At 2 or 3 weeks homozygotes are obviously smaller with hyperactive ataxic behavior. By 23 to 50 days, there is a 90% loss of Purkinje cells, subsequent to which most Purkinje cells degenerate

Degeneration of the photoreceptors in the retina is observable at 13 days, and large whorls of outer segment membranes are present at 3 weeks. Outer segments eventually completely disappear.[3,7]

Retinal Degeneration -1 (*rd1*)

A third retinal degeneration and one of major importance, is *rd1* (formerly *rd*, identical with rodless retina, *r* .[8,9] This mutation has been found among several commonly used laboratory inbred strains, as well as in inbred strain MOLD/Rk, a strain derived recently from feral mice of *Mus musculus molossinus* in Japan.[10] The *rd1* gene on mouse chromosome 5 codes for the ß-subunit of cyclic GMP-phosphodiesterase, and has been mapped to human chromosome 4p16.[11,12] Through linkage analysis using recombinant inbred strains, a close genetic association of the *rd* gene was found with an endogenous xenotropic murine leukemia virus, *Xmv-28*.[13] In *rd1* homozygous mice there is no known defect other than that of the retina. The eyes appear normal up to about 10 days when the outer segments and the rod cells degenerate rapidly; and by day 35 of age, they have disappeared. Figure 1 shows the typical retinal outer segment loss seen in *rd1* homozygotes.

Retinal Degeneration -2 (*Rd2*)

A fourth retinal degeneration is retinal degeneration-slow, formerly *rds*. Later it was found to be partially dominant and so renamed *Rds*, and it has since been renamed *Rd2* in the series because its slow progression is no longer unique. This mutant gene was first found in strain O20/A and mapped near H2 on chromosome 17.[14] This gene in homozygous condition causes onset retinal degeneration, at 7 days but the progression is relatively slow. The outer nuclear layer degenerates slowly and is gone by nine months in the peripheral retina and 12 months in the center.[15,16] In heterozygotes, the outer segments are abnormal, with a very slow rate of photoreceptor cell loss. An mRNA encoded by Rd2 is specific to retinal photoreceptors[17], and the normal product of the gene is peripherin.[18]

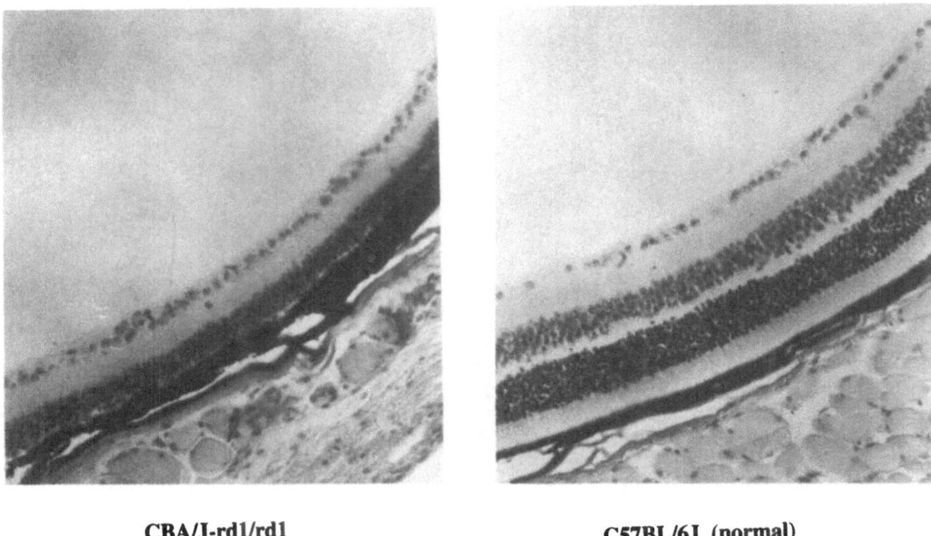

Figure 1. Retinal degeneration at 1 month of age for *rd1* homozygotes (left) and normal mice.

A longer description of these mutations and their effects with further citations is found in the Mouse Genome Database[1]

FOUR NEW RETINAL DEGENERATION MUTATIONS

Motor Neuron Degeneration (*mnd*)

Motor neuron degeneration was originally discovered in strain B6Kb2/Rn, a congenic strain for markers on chromosome 17. Like *pcd* and *nr*, it is a neurologic disorder characterized by gradual hind limb paresis beginning at 5 months and progressing to paralysis by 14 months.[19,20,21] It is a good model for human Batten disease and its expression is clearly recessive.[22] Using indirect ophthalmoscopy, we found early arteriolar attenuation, venous dilation, and a granular appearance to the retinal pigment epithelium by 6 weeks.[23] By 7 months, indirect ophthalmoscopy showed tiny scattered drusen through the retinal pigment epithelium. Histology showed photoreceptor cells reduced in the peripheral retina by 2 months and absence in the entire retina by 6 months. The ERG was undetectable with single flash methods at 6 months. Ultrastructural studies showed in all layers of retina distinct cytoplasmic lysosome-like inclusions characteristic of curvilinear profiles observed in lysosomes of neurons from humans with ceroid lipofuscinosis. The disease process in homozygous mnd mice is similar to Spielmeyer-Vogt disease in children. In both mouse and human, the retinal degeneration occurs before neuromuscular dysfunction. The course of the retinal disease and neurologic defect in mice is different. The retinal degeneration was nearly complete by 6 months, but the motor neuron abnormalities began around 6 months. Mapping of the human disorder is tentatively to chromosome 16. The *mnd* location on the proximal end of mouse chromosome 8 has nearby homologous segments for human chromosomes 8, 13, 19, and 21.[24] At this point, we can only say *mnd* is a good model of Batten disease, but may not be a genetic homolog.

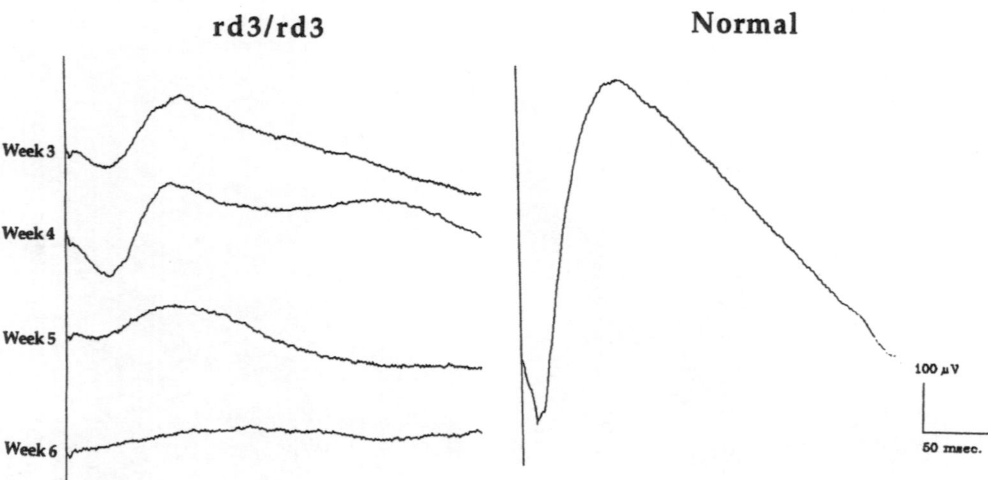

Figure 2. The electroretinogram to the left is that of RBF/DnJ homozygous for *rd3*, compared with the normal on the right.

Retinal Degeneration -3 (*rd3*)

We discovered retinal degeneration -3 (*rd3*) in strain RBF/DnJ and related strains, all carrying Robertsonian chromosomes with ancestry in a stock of mice derived from the Valle de Poschiavo in Switzerland.[25] We had originally imported the stock to The Jackson Laboratory for chromosomal studies. In homozygotes, photoreceptor degeneration starts by 3 weeks. The rod photoreceptor loss is complete by 5 weeks, although remnant cone cells are seen through 7 weeks. Figure 2 shows the electro retinograms with time course for the homozygous rd3 compared with normal. Ultrastructural analysis indicates that photoreceptor death occurs by apoptosis.[26 This is the only mouse homozygous retinal degeneration so far known in which photoreceptor cells are initially normal. We mapped the locus to position 87 on mouse chromosome 1 about 10cM distal to *Akp1*. In that location it is one of the most distal loci known on chromosome 1. Homology mapping suggests the homologous human locus should be on chromosome 1q.

Fig. 2 shows the electroretinographs of affected and normal mice; This ERG is the maximal response from a 10 µ second single-flash stimulation (intensity 73 ft -Lamberts -s) The wave form of *rd3* homozygotes was 25% normal and largest at four weeks but undetectable by 7 weeks.

Retinal Degeneration -4 (*Rd4*)

We discovered retinal degeneration-4 (*Rd4*) in a stock of mice with a chromosomal inversion, In(4)56Rk, induced with 1000r cesium.[27,28] In this dominantly expressed condition, retinal plexiform layers begin to reduce at 10 days of age and show a total loss of photoreceptor cells at 6 weeks. Recordable electroretinograms (ERG) are poor at 3 to 6 weeks and barely detectable after 6 weeks of age. Retinal vessel attenuation, pigment spots and optic atrophy appeared in the fundus at 4 weeks of age. Inversion In(4)56 Rk, which spans most of mouse chromosome 4, is homozygous lethal. In all crosses mice carrying the inversion always exhibit this retinal degeneration.

Although females and males were examined to characterize the retinal degeneration, only the males were followed for linkage analysis In all crosses typing for the inversion was

Table 1. Mouse strains with different retinal degenerations

	Phenotype evaluation			Chromosome	
	Retinal vessels attenuate and disappear by (months)	Photoreceptor cells degenerate and disappear by (months)	b-wave of ERG decreases and is flat by (months)	Mouse	Human
C3H/HeJ rd1/rd1	2	1	1	Chr.5	4p16
C3H-Rd2/Rd2	14	12	9	Chr. 17	6p
C57BL/6J-pcd/pcd	14	13	10	Chr. 13	5q?
C3H-nr/nr	11	10	8	Chr. 7	8p?
C57BL/6J-mnd/mnd	8	6	5	Chr. 8	16?
RBF/DnJ rd3/rd3	3	3	2	Chr. 1	1q32
In-56 Rd4/+	2	2	1	Chr. 4	8q or 1p
C57BL/6J-rd5/rd5	8	8	5	Chr. 7	11p15

by examining histological sections of testes for a high frequency of first meiotic anaphase bridges.[29] In numerous crosses *Rd4* always was associated with the inversion. If *Rd4* were in the central 4/5 region of the inversion, it should easily recombine with it. The simplest explanation, postulating no more than the two breakpoint mutations, is that *Rd4* is located at one of the breakpoints of the inversion, perhaps part of a large deletion which makes the inversion homozygous lethal. Comparative mapping of the breakpoint regions of the inversion would place a homolog of *Rd4* on human chromosomes 1p or 8q.

Retinal Degeneration -5 (*rd5*)

Retinal degeneration -5, *rd5*, was found in our systematic examination of mutant mice from the Mouse Mutant Resource of The Jackson Laboratory.[30] It was found in C57BL/6-*tub*, now a subline of C57BL/6J wherein a mutation to tubby (*tub*) occurred in 1977.. Indirect opthalmoscopy revealed arteriolar attenuation, venous dilation and a granular appearance in the retinal pigment epithelium by six weeks of age. By five months there was retinal vessel attenuation and loss of pigment epithelium. At this time patches of pigment deposits are clear. From 3 to 7 months, biomicrosopy showed fine drusen through the retinal pigment epithelium. Electroretinography revealed deterioration parallel with the direct observation. Hearing tests showed that *rd5* homozygotes were virtually deaf at 90 to 100 dB by 5 to 6 months of age, whereas heterozygotes were normal. Linkage studies place *rd5* close to *tub* on chromosome 7. but we cannot yet rule out the possibility that *rd5/rd5* is a manifestation of *tub/tub*. Comparative mapping would place a homolog on human llp15, the region where one study has located a human Usher Syndrome Type I.[31]

Table 1 summarizes the age of onset and progression of retinal degenerations for all eight mutations. Figures 3 and 4 show schematic representations of changes in visual cell structure of the eight mutations from their onset to complete degeneration. Although the end point in retinal degeneration is very similar in appearance among the mutations, there is a richness in variation of onset and progression, offering opportunities for study and experiments to ameliorate the conditions. Table 1 further provides potential locations of human homologs derived from comparative mapping of large conserved chromosomal segments in mouse and human.

In surveying the literature on mouse and human retinal genetic disorders, it is surprising that no X-linked forms have yet been described in the mouse, and yet so many are known for the human X chromosome[1,2]

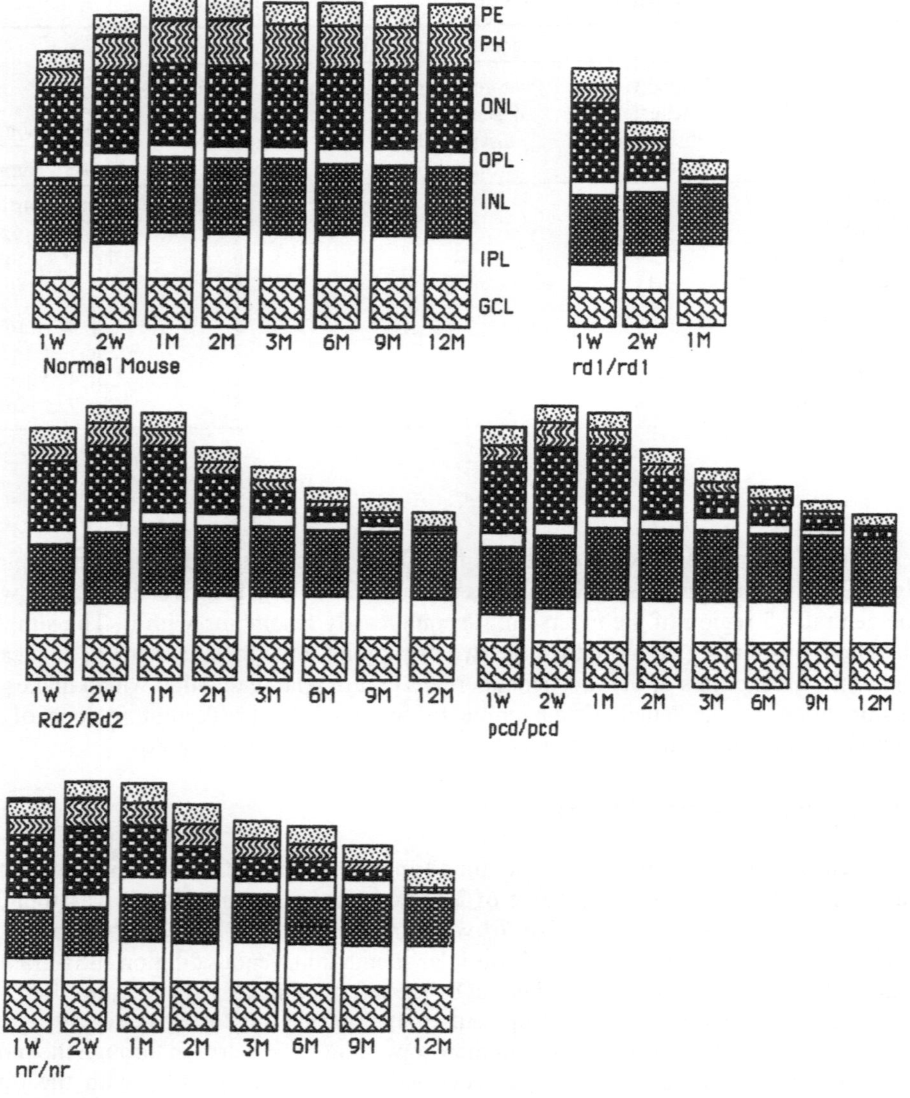

Figure 3. Schematic representation showing changes in visual cell structure of the retina of mutants *rd1*, *Rd2*, *pcd* and *nr*, as well as normal mice from 1 week until the age of 12 months. GCL, ganglion cell layer; IPL, inner plexiform layer; INL, inner nuclear layer; OPL, outer plexiform layer; ONL, outer nuclear layer; PH, photoreceptors; PE, pigment epithelium.

The variety of backgrounds on which retinal mutations have been found is important to note. Some are associated with other naturally occurring mutations (*nr*, *pcd*, *mnd*, and perhaps *tub*) and are no doubt in each a part of a syndrome of effects; some were found as mutations in widely different inbred strains including some recently derived from feral populations (*rd1*, *rd3*), one from a commonly used inbred strain (*Rd2*), and one in the breakpoints of an inversion induced by irradiation (*Rd4*). There are differences in expression, i.e. dominance and recessiveness, although as for retinal degeneration each

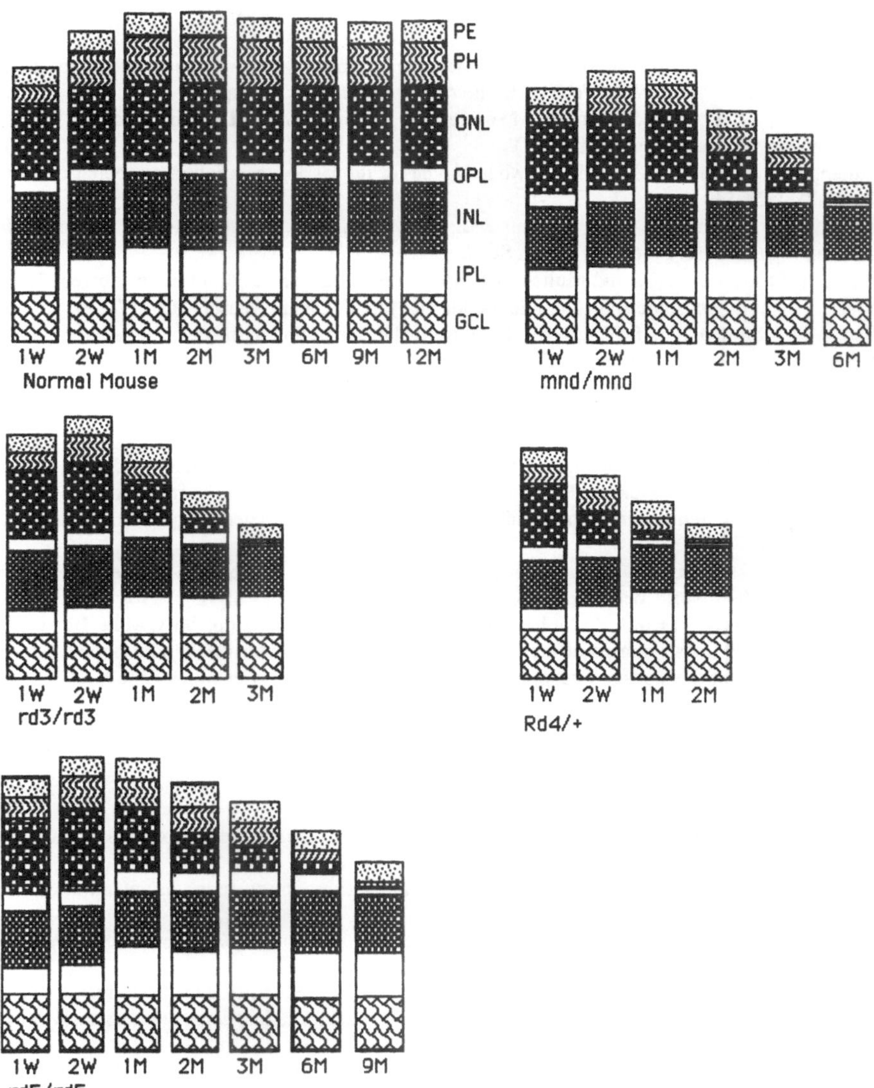

Figure 4. Schematic representation showing changes in visual cell structure of the retina of mutants *mnd*, *rd3*, *Rd4*, and *rd5*, as well as normal mice from 1 week until the age of 12 months. GCL, ganglion cell layer; IPL, inner plexiform layer; INL, inner nuclear layer; OPL, outer plexiform layer; ONL, outer nuclear layer; PH, photoreceptors; PE, pigment epithelium.

is fully penetrant and within mutants the expression is very consistent. Clearly wherever there is genetic variation, there is an opportunity to find genetic effects on the retina.

ACKNOWLEDGMENTS

This work was supported by National Eye Institute Grant R01 EY07758. The distribution of mice to other investigators was supported by a grant from The Foundation Fighting Blindness.

REFERENCES

1. Mouse Locus Catalog, MLC of MGD, Mouse Genome Informatics Project, The Jackson Laboratory, Bar Harbor, ME, 1995 World Wide Web (URL:http://www.informatics.jax.org/)
2. McKusick, V.A., 1995. Online Mendelian Inheritance in Man. World Wide Web (http://gdbwww.gdb.org/omim/omimq/1)
3. Mullen, R.J., and LaVail, M.M.,1974. Two new types of retinal degeneration in cerebellar mutant mice. Nature 258:528-530
4. Mullen, R.J., Eicher, E.M., and Sidman, R.L.,1976. Purkinje cell degeneration, a new neurological mutation in the mouse. Proc. Natl. Acad. Sci. USA .73:208-212.
5. LaVail, M.M., Blanks, J.C., and Mullen, R.J., 1982. Retinal degeneration in the *pcd* cerebellar mutant mouse I. Light microscopic and autoradiographic analysis. J. Comp. Neurol. 212:217-230.
6. Blanks, J.C., Mullen, R.J., LaVail,M.M., 1982. Retinal degeneration in the *pcd* cerebellar mutant mouse II. Electron microscopic analysis. J. Comp. Neurol. 212:231-246;
7. Sidman, R.L., and Green, M.C., 1970. "Nervous," a new mutant mouse with cerebellar disease. In: M Saboudry, ed. Les Mutants Pathologiques Chez l'Animal. Paris: Centre National de la Recherche Scientifique, pp.69-79.
8. Keeler, C, 1966. Retinal degeneration in the mouse is rodless retina. J. Hered. 57:47-50;
9. Pitler, S.J., Keeler, C.E.,Sidman, R.L., and Baehr, W., 1993. PCR analysis of DNA from 70-year old sections of rodless retina demonstrates identity with the mouse rd defect. Proc. Natl. Acad. Sci. USA 90:9616-9619.
10. MATRIX, Mouse Genome Database, Mouse Genome Informatics Project, The Jackson Laboratory, Bar Harbor, ME., 1995 World Wide Web (URL:http://www.informatics.jax.org/).
11. Bateman, B., Klisak, I., Kojis, T., Mohandas, T., Sparkes, R.S., Li, T., Applebury, M.L., Bowes, C., and Farber, D.B., 1992. Assignment of the ß-subunit of rod photoreceptor cGMP phosphodiesterase gene PDEB (homolog of the mouse *rd* gene) to human chromosome 4p16. Genomics 12: 601-603
12. Danciger, M., Bowes, C., Kozak, C.A., Lavail, M., and Farber, D.B., 1990. Fine mapping of a putative *rd* cDNA and its cosegregation with *rd* expression. Invest. Ophthalmol. &Vis. Sci. (Suppl.) 31:1427-1432
13. Bowes, C., Li, T., Frankel, W.N., Danciger, M., Coffin,J.M., Applebury, M.L., and Farber, D.B., 1993. Localization of a retroviral element within the *rd* gene coding for the ß subunit of cGMP phosphodiesterase. Proc. Natl. Acad. Sci. USA 90:2955-2959).
14. Van Nie, R., Ivanyi, D., and Demant, P., 1978. A new H-2-linked mutation, *rds*, causing retinal degeneration in the mouse. Tissue Antigens 12:106-108.
15. Jansen, H.G., and Sanyal, S., 1984. Development and degeneration of retina in mutant mice: electron microscopy. J. Comp. Neurol. 224:71-84.
16. Sanyal, S., DeReuter, A., and Hawkins, R.K., 1980. Development and degeneration of retina in *rds* mutant mice: light microscopy. J. Comp. Neurol. 194:193-207.
17. Travis, G.H., Brennan, M.B., Danileson, P.E., Kozak, C.A., and Sutcliffe, J.G. 1989. Identification of a photoreceptor-specific mRNA encoded by the gene responsible for retinal degeneration slow (*rds*). Nature 338:70-73.
18. Connell, G., Bascom, R., Molday, L., Reid,D., McInnes, R.R., and Molday, R.S., 1991. Photoreceptor peripherin is the normal product of the gene responsible for retinal degeneration in the rds mouse. Proc. Natl. Acad. Sci. USA 88:723-726.
19. Messer, A, and Flaherty L., 1986. Autosomal dominance in a late onset motor neuron disease in the mouse. J. Neurogenet. 3:345-355;
20. Messer, A., Stominger, N.L., and Mazurkiewicz, J.E., 1987. Histopathology of the late-onset motor neuron degeneration (*Mnd*) mutant in the mouse. J. Neurogenet. 4:201-213.
21. Messer, A., Plummer, J., Maskin, P., Coffin, J.M., and Frankel, W.N., 1992. Mapping of the motor neuron degeneration (*Mnd*) gene, a mouse model of amyotrophic lateral sclerosis (ALS). Genomics 13:797-802.
22. Bronson, R.T., Lake, B.D., Cook, S., Taylor, S., and Davisson, M.T., 1993. Motor neuron degeneration (*mnd*) of mice is a model of neuronal ceroid lipofuscinosis (Batten's disease) Ann. Neurol. 33:381-385.
23. Chang, B., Bronson, R.T , Hawes, N.L., Roderick, T.H., Peng, C., Hageman, G.S., and Heckenlively, J.R., 1994. A retinal degeneration in motor neuron degeneration: A mouse model of ceroid lipofuscinosis. Investigative Ophthal. & Vis. Sci. 35:1071-1076.
24. Hillyard, A.L., Davisson, M.T., Doolittle, D.P., Guidi, J.N., Maltais, L.J., and Roderick, T.H., 1993. Locus map of mouse (Mus musculus/domesticus). In: Genetic Maps, Nonhuman Vertebrates, Book 4, pp. 4.110-4.142. 6th Edit., Ed. S.J. O'Brien. Cold Spring Harbor Laboratory Press, Cold Spring Harbor, NY.
25. Chang, B., Heckenlively, J.R., Hawes, N.L., and Roderick, T.H., 1993. New mouse primary retinal degeneration (*rd-3*). Genomics 16:45-49.

26. Linberg, K.A ., Fariss, R.N., Heckenlively, J.R , Peng, C., Bowes, C., Farber, D.B., and Fisher, S.K., 1994. Structural changes in the developing retina of the *rd-3* mouse. Investigative Ophthalmology & Vis.Sci., 35(4):1610.
27. Heckenlively, J.R., Chang, B., Hawes, N.L., and Roderick, T.H., 1992. Two new mouse retinal primary degenerations. Investigative Ophthalmol & Vis. Sci. 33(4):1063.
28. Roderick, T.H., Chang,B., Hawes, N.L., and Heckenlively,J.R., 1996. A new dominant retinal degeneration (*Rd4*) associated with a chromosomal inversion in the mouse (in preparation)
29. Roderick, T.H., 1983. Using inversions to detect and study recessive lethals and detrimentals in mice, pp.135-167. In: F.J deSerres, and W. Sheridan (eds.), Utilization of Mammalian Specific Locus Studies in Hazard Evaluation and Estimation of Genetic Risk. Plenum Publ. Corp.
30. Heckenlively, J.R., Chang, B. , Peng, C., Erway, L.C., Hawes, N.L., Hageman, G.S., and Roderick, T.H., 1995. A mouse model (*rd5*) for Usher Syndrome; linkage mapping suggests homology to Usher Type I reported at human chromosome 11p15. Proc. Natl. Acad. Sci USA, (in press)
31. Smith, R.J.H., Lee, E.C., Kimberling, W.J., Daiger, S.P., Pelias, M.Z., Keats, B.J.B., Jay, M., Bird, A., Reardon, W., Guest, M., Ayyagari, R. and Hejtmancik, J.F., 1992. Localization of two genes for Usher Type I to chromosome 11. Genomics 14:995-1002.

CHORIORETINAL INTERFACE

Age-Related Changes in Rodent Retina

P. M. Leuenberger,* Y. Gambazzi, and E. Rungger-Brändle

Electron Microscopy
University Eye Clinic
Geneva, Switzerland

SUMMARY

Ultrastructural features related to ageing that specifically concern the basal aspect of the retinal pigment epithelium (RPE) and underlying Bruch's membrane (BM), were studied in rat and rabbit. Disintegration of the regular arrangement of RPE basal infoldings, a massive increase in basal lamina-like deposits, and the presence of unusual collagen polymers are common features in both rodent species. In the rat, minor strain-specific variations exist as to the pattern and severity of particular alterations. A morphometric evaluation revealed that, in the Fischer rat, RPE hemidesmosomes are markedly decreased. Basal infoldings tend to shed which may contribute to the formation of vesicular deposits in senescent BM. Moreover, RPE cells show an increase in coated pits, lysosomes, and intermediate filaments.

INTRODUCTION

Senile macular degeneration (SMD) is the second cause of visual impairment in the elderly. The fact that only very limited therapeutic possibilities are presently at hand, resides in the problematic assessment of the pathophysiological etiology of the various clinical forms (e.g. 2, 11) and, moreover, in the scarcity of adequate animal models. It is, however, generally accepted that changes at the retino-choroidal interface are prerequisites for the formation of SMD (3). Other factors such as exogenous and endogenous toxic effects and a genetic predisposition may also contribute to its development.

Given the functional importance of the retino-choroidal interface in transport mechanisms and retinal adhesion and integrity, we investigated ultrastructural features

* Correspondence: Dr. P.M. Leuenberger, Clinique d'Ophtalmologie, Hôpital cantonal universitaire, 22, rue Alcide-Jentzer, CH-1211 Genève 14. Fax: ++41 22 38 28 382.

Degenerative Diseases of the Retina, Edited by Robert E. Anderson et al.
Plenum Press, New York, 1995

of both the basal aspect of retinal pigment epithelium (RPE) and Bruch's membrane (BM) in ageing specimens of rat and rabbit. Both species display common morphological characteristics, although species differences exist. By comparing two rat strains (Fischer and Wistar), we also found strain specific senile patterns. An ubiquitous, ultrastructurally well defined feature associated with the senescent RPE is disintegration of the regularly arranged basal infoldings attached to the basal lamina of BM, accompanied by a decrease in the ultrastructurally well defined attachment sites, the hemidesmosomes. Furthermore, we describe cell-associated changes such as an increase in coated pits, lysosomes, and intermediate filaments as well as extracellular matrix-associated changes observed as a massive accumulation of basal lamina-like material and unusual collagen polymers intercalated between RPE and BM.

MATERIALS AND METHODS

Animals

Albino rats of both the Fischer and Wistar strain were aged to 28-31 months. The animals were kept in a 12 hrs light - dark cycle at 110 and 20 Lux. For overall morphology, two month-old Wistar rats served as control. Morphometric measurements were done on two groups of 12 animals each of the Fischer strain that had been aged to 16 months (control) and 28 months respectively. Albino rabbits of the New Zealand and Alaska strain were between 3 and 4 years old.

Electron Microscopy

Rats were sacrificed 3-4 hrs after onset of light, after premedication with pentobarbital and according to the laws on animal experimentation of the cantons of Geneva and Basel. Eyes were fixed by immersion in 2.5% glutaraldehyde, 0.1M cacodylate buffer (pH 7.2), supplemented with 0.2% tannic acid. After rinsing in 0.1M cacodylate buffer (pH 7.4), the tissue pieces were postfixed in 1% osmium tetroxide in 50mM cacodylate buffer (1hr, 4°C) stained "en bloc" with 2% uranyl acetate in 50mM maleate buffer (1hr, 4°C), dehydrated through a series of increasing concentrations of ethanol, and embedded in Spurr resin.

Thin sections were conventionally stained with uranyl acetate and lead citrate and observed in a Philips EM 300 electron microscope.

Morphometric Procedures

Seven eyes of both senile and control Fischer rats were chosen arbitrarily and 3 tissue blocks per eye were selected randomly. Thin sections were rapidly screened and the first grid square displaying the entire height of the RPE with adjacent Bruch's membrane was taken as start point for series of up to 15 micrographs. Overlaps were avoided. Pictures were taken at a magnification of 6000x. After conversion to a positive film, morphometric analysis was done on a light screen at a final magnification of 17000x. The morphometric method used was essentially as described (14) using a transparent test grid placed over the projected micrograph. The number of coated pits in the plasmalemma of the basal infoldings and the number of hemidesmosomes were counted as a function of Bruch's membrane length (μm). Maximal penetration of the basal infoldings was expressed as a function of the mean height of RPE cells. For statistical analysis, the Student's t-test was used.

RESULTS

1. Morphological Characteristics

Prominent age-related alterations in both rat and rabbit involve both cytoplasm and underlying extracellular matrix (ECM) of RPE cells and BM. As exemplified for the *Fischer strain*, the cytoplasm of young RPE cells (Fig. 1a) is characterized by the presence of numerous lipid droplets and smooth endoplasmic reticulum with clustered free ribosomes. Regularly oriented basal infoldings abut onto the basal lamina of Bruch's membrane, the electron dense layer of which measures about 40nm in thickness. In the aged RPE (Fig. 1b), ribosomes are associated with the endoplasmic reticulum. Lipofuscin granules and primary lysosomes (Fig. 5a) are frequent and, in some of the cells, intermediate filaments are hypertrophied (not shown). The basal infoldings are heavily disorganized and many of them do not end on the original basal lamina. Hemidesmosomes are enlarged and electron dense (Figs. 1b and 2b). By contrast, electron density of the lamina densa belonging to the basal lamina is strongly decreased. Instead, such material now forms extracellular invaginations extending deeply into the epithelium (Figs. 2b and 3). The presence of associated hemidesmosomes indicates that the plasma membrane lining these extracellular pouches has maintained its basal characteristics.

In the senile Fischer rat, the basal lamina-like material contains electron dense fibrillar structures (Figs. 2-4) of distinct periodicity probably representing unusual collagen polymers. One such type are electron dense fibrils displaying 130nm periodicity (fibrous long-spacing collagen, FLSC). They form higher order linear or stellar aggregates (Figs. 2a and 3). Another type of polymers are microfibrils of 10-15nm diameter (Fig. 2b) that, in longitudinal and cross-sections, display staining irregularities suggesting a helicoidal nature. Microfibrillar collagen and FLSC may be physically contiguous as are FLSC and collagen fibers of normal (66nm) periodicity. Fig. 4 shows an extended fiber sheet, where FLSC and normal collagen are associated both laterally and longitudinally. Such mixed polymers are found intercalated between entirely normal collagen fibers of BM and pure FLSC close to the basal plasmalemma of RPE.

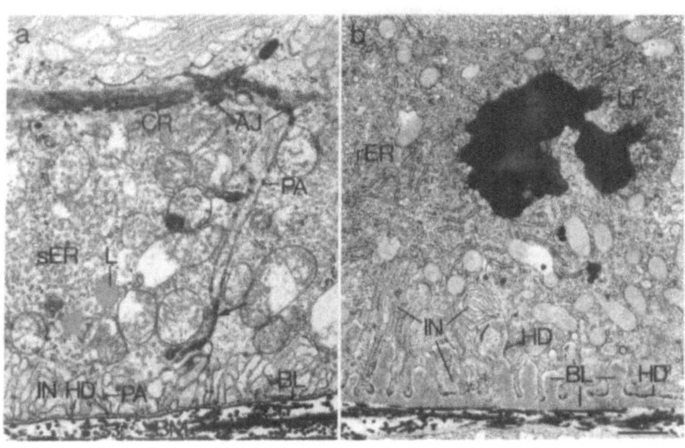

Figure 1. Interface of RPE-Bruch's membrane (BM) in young (a) and aged (b) albino rats. a, Two month-old specimen of the Wistar strain showing regular arrangement of basal infoldings (IN) with numerous hemidesmosomes (HD) and puncta adherentia (PA). The electron-dense portion of the basal lamina (BL) is well delimited to about 40nm thickness. Note the presence of smooth endoplasmic reticulum (sER) and lipid droplets (L). Arrow points to lamellar material in the intercellular space. b, Thirty-one month-old specimen of the Fischer strain. Basal infoldings are irregular. The lamina densa of BL is thickened and of decreased electron density. Hemidesmosomes are enlarged. The cytoplasm is filled with rough endoplasmic reticulum (rER) and lipofuscin (LF). CR, Contractile ring; AJ, adherens junction. Bar, 1μm.

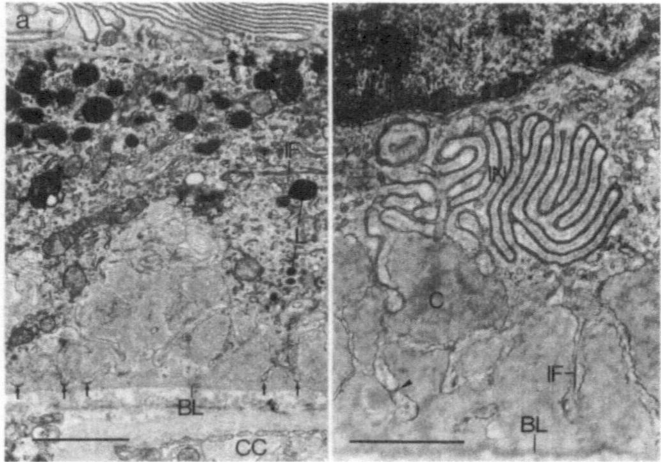

Figure 2. Unusual collagen polymers in the RPE-Bruch's membrane interface of aged Fischer rats. a, Fibrous long spacing collagen (fC) displaying 130 nm periodicity (small arrows). b, Continuity (arrowheads) between microfibrils (mC) and fibrous-long-spacing collagen fibers. BL, basal lamina-like material; HD hemidesmosomes; IN, infoldings; N, nucleus. Bar, 1μm.

Using the same fixation procedure, senile rats of the *Wistar strain* do not display FLSC aggregates as observed in the Fischer strain. However, large deposits of basal lamina-like material and abnormal collagen polymers other than FLSC between RPE and BM are abundant (Fig. 5). Hemidesmosomes at basal infoldings in contact with the former basal lamina are scarce (a). Such cytoplasmic basal extensions tend to vesiculate (b).

Similar to what has been observed in the rat, senile alterations at the RPE-BM interface of the *rabbit* are also accompanied by an increase in basal lamina-like material and abnormal collagen polymers (not shown). In addition and unlike in rats, RPE cells were observed to locally protrude into BM (Fig. 6) in areas where the structure of the basal lamina appeared to be interrupted, though intercellular junctional complexes were well preserved in these regions.

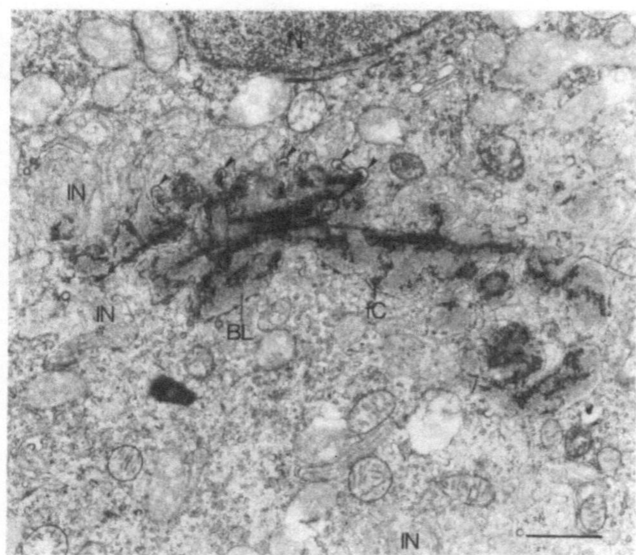

Figure 3. Oblique section through the basal aspect of RPE in an aged Fischer rat showing an extracellular invagination filled with basal lamina-like deposit (BL) and fibrous long-spacing collagen (fC). Note numerous coated pits (arrowheads) associated with collagen fibers. IN, infoldings; N, nucleus. Bar, 1μm.

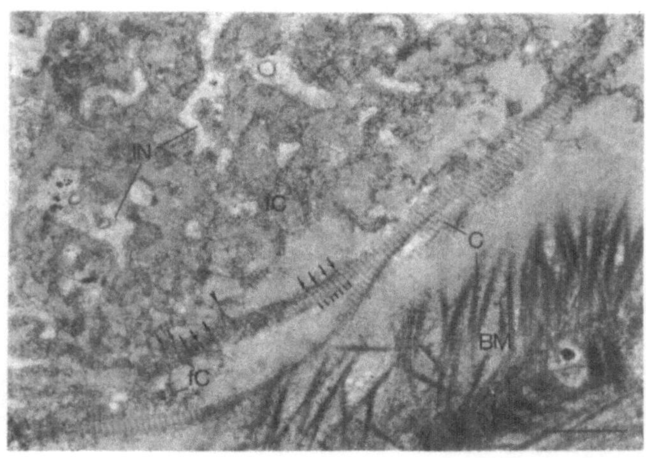

Figure 4. Oblique section through Bruch's membrane (BM) and adjacent basal infoldings (IN) of an aged Fischer rat. A large fiber sheet shows continuity between collagen of normal periodicity (C, 66 nm) and fibrous long-spacing collagen (fC, small arrows). A lateral array in the 130 nm collagen fiber sheet is typical for a transition zone (arrowhead) between the two polymers. Note that collagen fibers close to the cell surface are of abnormal, those within Bruch's membrane of normal periodicity. Bar, 1μm.

2. Morphometric Evaluation of Senile Alterations in the Fischer Rat

Table 1 summarizes some of the parameters considered in the stereological study on age-related changes linked to the basal aspect of RPE. The epithelial cell volume remains constant with increasing age but the volume of basal lamina-like material intercalated between BM and basal RPE plasmalemma increases more than threefold. The presence of deep extracellular invaginations (cf. Fig. 3) into the RPE is reflected in an increase of the relative height of the cell compartment occupied by basal infoldings. Similar observations have been reported previously (5). The number of hemidesmosomes per unit length of BM is drastically decreased, while the number of coated pits within the basal plasmalemma is

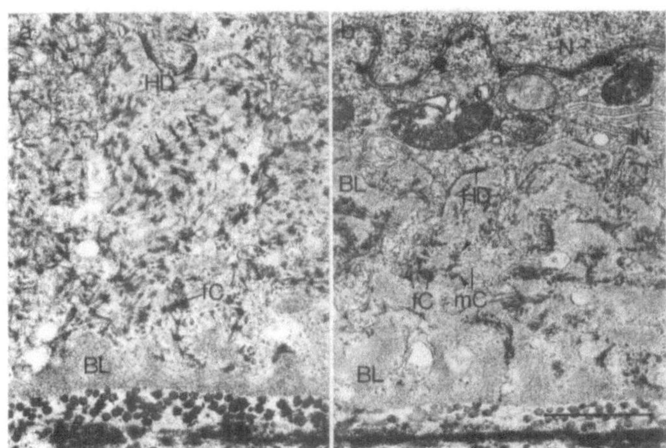

Figure 5. Typical senile alterations at RPE-Bruch's membrane interface in the Wistar rat. a, Decreased epithelial thickness and presence of large amounts of extracellular deposits in between the infoldings. Arrows point to few hemidesmosomes still in contact with the original basal lamina (BL). The cytoplasm is filled with lysosomes (L). b, High magnification of infoldings some of which are in contact with the former basal lamina but tend to vesiculate (arrowhead). The bulk of infoldings (IN) have no contact with extracellular matrix. IF, intermediate filaments. N, nucleus; C, aberrant polymers of collagen; CC, choriocapillary. Bars, 1μm.

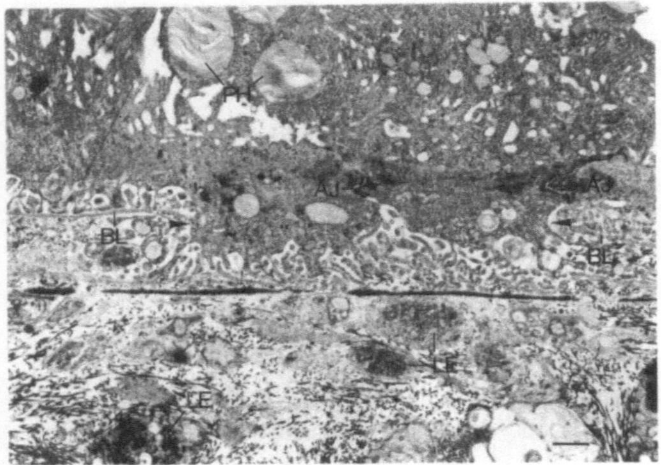

Figure 6. View of basal aspect of RPE and Bruch's membrane in a 4 year-old rabbit. Mount of 2 consecutive electron micrographs. Basal lamina (BL) continuity is interrupted and 2 adjacent epithelial cells are displaced basally (arrowheads). Note the presence of leucocytes (LE) in the choroidea. PH, photoreceptor outer segments. AJ, adhaerens junction. Bar, 2.5μm.

increased. Interestingly, coated pits in the senile rat are frequently associated with FLSC (cf. Fig. 3), a situation that is virtually never observed in the young rat.

Taken together, these evaluations show that a disproportional increase in extracellular basal lamina-like material, intercalated between RPE and BM, is correlated with a general decrease in hemidesmosomal attachment sites. Moreover, endocytic or secretory activity of the RPE is disturbed as judged by the number of coated pits often associated with collagen polymers within the basal plasmalemma.

DISCUSSION

In this study, we describe age-related alterations at the chorio-retinal interface in rat and rabbit that specifically concern both basal infoldings of RPE and adjacent basal lamina of BM. These alterations appear to be ubiquitous and resemble those described earlier in human (3), rat (4, 5) and mouse (9). However, pronounced species- and strain-specific differences exist as to the severity of particular changes. The massive increase in volume and thickness of extracellular basal lamina-like material in between RPE and BM appears to be correlated with a decrease in both the hemidesmosomes and basal infoldings in contact with the original basal lamina leading to a reduction of physical contact between RPE and

Table 1. Stereologic comparison of parameters linked to RPE - Bruch's membrane

	Young rats (n = 7)	Old rats (n = 7)	p
V_C (μm^3)	80.89 ± 1.26	80.63 ± 1.76	$p > 0.9$
V_{BL}/V_C (%)	3.25 ± 0.33	11.45 ± 1.42	$p < 0.01$
H_{IN}/H_C (%)	20.05 ± 1.13	38.35 ± 1.74	$p < 0.01$
HD/L_{BM} (n/μm)	2.78 ± 0.1	1.66 ± 0.15	$p < 0.01$
CP/L_{BM} (n/μm)	0.33 ± 0.08	0.56 ± 0.01	$p < 0.01$

V_C, cell volume; V_{BL}, volume of basal lamina-like material; H_{IN}, height of cytoplasmic area occuped by basal infoldings; H_C, RPE cell height; HD, hemidesmosomes; L_{BM}, length of Bruch's membrane; CP, coated pits.

BM. It might be speculated that reduction of specialized attachment sites ultimately lead to RPE detachment, a general feature observed in human SMD. Nevertheless, intercellular cohesion between the epithelial cells remains preserved and it is conceivable that reattachment of RPE, regularly observed in human, is based on the preservation of intercellular junctions and apical-basal cell polarity.

Given the continuous interactions between epithelial cells and underlying extracellular matrix, we cannot decide by ultrastructural criteria, which of the observed senile alterations, whether associated with RPE or extracellular matrix, represents the primary cause triggering the pathological cascade. Clearly, senescent RPE cells display distinct morphological features indicating alterations in the synthetic and catabolic machinery, such as the presence of rough endoplasmic reticulum and a high number of primary lysosomes and coated pits, the latter being often associated with fibrous components that may represent aberrant collagen polymers. Moreover, an increased intermediate filament content may be indicative of a general pathological state of RPE cells reminiscent of that of gliotic cells.

We believe that the senile modifications in the chorio-retinal interface described above decrease sensibly the metabolically active exchange surface between the choriocapillaris and outer retina, a situation which, ultimately, may contribute to impaired photoreceptor function. Both, alterations at the chorio-retinal interface and reduced retinal sensitivity have also been observed in the mouse homozygous for the mutant *pearl* (*pe/pe*, 15). Moreover, the transport of metabolites and other essential molecules to the retina is likely to be impaired by the massive deposits of basal lamina-like material intercalated between RPE and BM. The altered electron density of this material, as compared to normal BL of younger animals, is indicative of altered composition and it is therefore likely that the molecular sieve capacities of the basal lamina (7, 10) are modified.

Within the basal lamina-like deposits, we have observed fibrous components with periodicity similar to that of collagen. Similar polymers have also been described in senescent human (3, 6) and rat (4) retina. We ignore whether several or only one type of collagen are contained in the polymers described here. If only one type of collagen were involved, it ought to have the capacity to polymerize into lateral and linear aggregates of different periodicity. Indeed, we have observed physical continuity not only between microfibrils and FLSC but also between FLSC and normal fibers. The peculiar local distribution of FLSC and microfibrils close to the RPE plasma membrane on the one hand and collagen polymers of normal periodicity close to and within BM on the other, with mixed aggregates in between, suggests that external factors determine polymerization. Non-enzymatic cross-linking by glycation (8), proteoglycans (12), pH, and oxygen pressure may modulate collagen polymerization (for discussion, see 13). In vitro observations of aberrant collagen polymerization at acid pH (1) argue in favor of an acidic extracellular microenvironment close to the RPE plasma membrane in the ageing retina. Acidity itself could be a consequence of hypoxia close to the RPE or of altered molecular filter capacity of the massive extracellular matrix deposits.

During senescence, reduction of physical contact at the chorio-retinal interface may progressively continue. We have observed that basal infoldings in contact with the former basal lamina, tend to shed and vesiculate. This may finally lead to the formation of vesicular deposits within BM that are widely observed in several mammalian species.

ACKNOWLEDGMENTS

This study was supported by the Swiss National Science Foundation, grants 32-9489.88 and 31-32047.91.

REFERENCES

1. Doyle, B.B., Hukins, D.W.L., Hulmes, D.J.S., Miller, A., and Woodhead-Gallway, J. (1975) Collagen polymorphism; its origins in the amino acid sequence. J. Mol. Biol. **91**, 79-99.
2. Ferris, F.L. (1983) Senile macular degeneration: review of epidemiologic features. Am J. Epidemiol. **118**, 132-151.
3. Hogan, M.J. (1972) Role of the retinal pigment epithelium in macular disease. Trans. Am. Acad. Ophthal. Otolaryng. **76**, 64-80.
4. Katz, M.L. and Robison, W.G. (1984) Age-related changes in the retinal pigment epithelium of pigmented rats. Exp. Eye Res. **38**, 137-151.
5. Katz, M.L. and Robison, W.G. (1985) Senescence and the retinal pigment epithelium: Alterations in basal plasma membrane morphology. Mech. Ageing Dev. **30**, 99-105.
6. Killingsworth, M.C. (1987) Age-related components of Bruch's membrane in the human eye. Graefes Arch. Clin. Exp. Ophthalmol. **225**, 406-412.
7. Lyda, W., Eriksen, N., and Krisna, N. (1957) Studies of Bruch's membrane. Flow and permeability studies in a Bruch's membrane-choroid preparation. Am. J. Ophthalmol. **44**, 362-370.
8. Malik, N.S., Moss, S.J., Ahmed, N., Furth, A.J., Wall, R.S., and Meek, K.M. (1992) Ageing of the human corneal stroma; structural and biochemical changes. Biochim. Biophys. Acta **1138**, 222-228.
9. Mishima, H. and Kondo, K. (1981) Ultrastructure of age changes in the basal infoldings of aged mouse retinal pigment epithelium. Exp. Eye Res. **33**, 75-84.
10. Pino, R.M., Essner, E., and Pino, L.C. (1982) Location and chemical composition of anionic sites in Bruch's membrane of the rat. J. Histochem. Cytochem. **30**, 245-252.
11. Sarks, S.H. (1976) Ageing and degeneration in the macular region: a clinico-pathological study. Brit. J. Ophthal. **60**, 324-341.
12. Scott, J.E. (1990) Proteoglycan:collagen interactions and subfibrillar structure in collagen fibrils. Implications in the development and ageing of connective tissues. J. Anat. **169**, 23-35.
13. van der Schaft, T.L., de Brujin, W.C., Mooy, C.M., Ketelaars, D.A.M., and deJong, P.T.V.M. (1991) Is basal laminar deposit unique for age-related macular degeneration? Arch. Ophthalmol. **109**, 420-425.
14. Weibel, E.R. (1979) Stereological methods. Volume 1: Practical methods for biological morphometry. Academic Press: London.
15. Williams, M.A., Pinto, L.H., and Gherson, J. (1985) The retinal pigment epithelium of wild type (C57BL/6J +/+) and pearl mutant (C57BL/6J *pe/pe*) mice. Invest. Ophthalmol. Vis. Sci. **26, 657-669.**

11

FRACTIONATION OF INTERPHOTORECEPTOR MATRIX METALLOPROTEINASES

James J. Plantner, Timothy A. Quinn and George J. Dwyer

Department. of Ophthalmology
Case Western Reserve University
10900 Euclid Avenue, Cleveland, Ohio 44106-5068

ABSTRACT

Metalloproteinases (MPs) present in the interphotoreceptor matrix (IPM) may play a role in both the maintenance of normal retinal homeostasis and the progression of various degenerative retinal diseases. We wished to extend our previous studies of these enzymes and begin their characterization by first obtaining them in purified form.

IPM was obtained from fresh bovine eyes. The material was fractionated by a combination of gel filtration, DEAE ion exchange and gelatin- affinity chromatographies. Fractions were monitored by SDS-PAGE carried out on both plain gels and those preloaded with gelatin or casein (zymograms) and by quantitative analysis for enzyme activity Combination of the procedures described above provided a partial fractionation of the MPs as detected by zymography. A previously unrecognized 52 kDa caseinase was also revealed, possibly identical with MMP-3 (stromelysin).

INTRODUCTION

The interphotoreceptor matrix (IPM) is the extracellular material which fills the subretinal space. It is bounded by the retinal pigment epithelium (RPE), the photoreceptor cells and the Müller glial cells. The IPM is involved in adhesion between the retina and RPE, transport of metabolites, and phagocytosis of shed rod outer segment tips [1]. Our laboratory originally showed the presence of metalloproteinases (MPs) and their inhibitors in bovine IPM [2,3], and we have recently shown that three of the matrix metalloproteinases, MMP-2, 3 and 9, are present in bovine and human IPM and vitreous [4]. Jones, et al [5] have also shown the presence of MMP-2 and one member of the tissue inhibitor of metalloproteinases family (TIMP-1) in bovine IPM.

The MPs are a large family of endoproteinases found in bacteria, plants and animals [6-8]. The matrix metalloproteinases (MMPs), found in animals, have been especially well

Degenerative Diseases of the Retina, Edited by Robert E. Anderson et al.
Plenum Press, New York, 1995

studied. All members of this class are found extracellularly and share the need for a divalent cation, usually zinc, at the active site. Most cells which elaborate an extracellular matrix will secrete these enzymes; although not all cells produce all members of the group [9]. Because of their location and substrate specificity, the MPs have been shown to take part in a number of important processes involving remodeling and turnover of extracellular matrices [6-8]. Besides these beneficial functions, however, they have also been implicated in a number of pathological states, including: arthritis, osteoporosis, periodontitis, metastasis and tumor invasion [6-8, 10]. All of these processes require degradation of extracellular matrix components. In the pathological conditions, it is likely that an imbalance in the production of active enzyme vs. inhibitor occurs [7,8]. Proper control of the production and activity of all members of this family of enzymes and inhibitors is thus essential for maintaining an extracellular matrix.

A number of ocular tissues have been shown to secrete MMPs in culture, including cornea [11], trabecular meshwork [12] and retinal pigment epithelium (RPE) [13, 14]; and it has been suggested that an imbalance of MMPs and their inhibitors may be involved in glaucoma [12], keratoconus and corneal ulceration [11]. Studies have demonstrated the efficacy of a synthetic MMP inhibitor to impede the progression of corneal ulcers secondary to alkali burns [15].

A number of retinal diseases involve neovascularization and/or abnormal cellular proliferation at one time or another during their course, including: retinopathy of prematurity [16], diabetic retinopathy[16], proliferative vitreoretinopathy [17], and age-related macular degeneration[18]. During the course of each disease, abnormal, invasive tissue becomes associated with the retina or RPE. For cellular migration or neovascularization to occur, cells must move through, or proliferate into, an extracellular matrix. To gain this mobility, the matrix must be changed, or "remodeled", requiring the action of hydrolytic enzymes [10]. Indirect evidence of MMP involvement in the pathogenesis of these retinal diseases is present in the literature [19, 20] and TIMP-3 has been shown to be involved in simplex retinitis pigmentosa [21] and in the autosomal dominant macular degeneration of Sorsby's retinal dystrophy [22]. The present studies describe our initial attempts to separate and purify the MPs and their inhibitors present in bovine IPM.

METHODS

Preparation of Interphotoreceptor Matrix

IPM was prepared by the method of Adler and Severin [23], as described previously [24]. Briefly, after removal of the cornea, lens, aqueous and vitreous humor from fresh bovine eyes, the vitreal surface of the attached retina was washed with cold 0.05M HEPES, pH 7.3, containing: 3 mM PMSF, 5 μgm/ml benzamidine, 3 μgm/ml pepstatin A and 2 μgm/ml leupeptin (Buffer). The wash was discarded. All remaining steps were performed at 4°. The retina was removed, placed into Buffer (2 ml/retina) and gently agitated for about 30 min. After agitation, the retinas were filtered to obtain the crude retina-IPM in the filtrate. Additional cold Buffer was added to the eyecup and gently pipetted against the apical surface of the RPE a number of times to produce the RPE-IPM. The crude IPMs were then centrifuged at 48000xg for 30 min to remove remaining debris (mostly rod outer segments). The supernatant fluids were concentrated to the equivalent of about 5 eyes/ml by ultrafiltration using Amicon (Beverly, MA) YM-10 membranes. All material was stored at -20°C.

Fractionation

Gel Filtration. IPM was fractionated using a 1.6 cm x 95 cm column of Sephacryl S-500 (Pharmacia, Uppsala), superfine, eluted with 0.05M Na phosphate, pH 7.5, 0.1M NaCl. Elution was followed by measuring absorbance at 280nm and proteolytic activity (see below). Fractions were pooled as shown in Fig. 1A.

Ion Exchange Chromatography. Pool B obtained from gel filtration of retinal-IPM (see Fig. 1A) was dialysed against 10mM Tris-acetate, pH 7.5, applied to a 2.5 cm x 60 cm column of DE52 (Whatman, Hillsboro, OR) and washed with the same buffer. Elution then proceeded using a linear gradient composed of 1L each of the Tris-acetate buffer and 0.5M NaCl in this buffer, followed by a wash of this buffer containing 1.0M NaCl. Fractions were monitored for conductivity, UV absorbance and proteolytic activity (see below) and pooled as shown in Fig. 2A.

Gelatin affinity chromatography Pools were applied to a 10 ml column of gelatin-agarose and washed with 50 mM Tris-HCl, pH 7.5; 2mM CaCl2; 0.02% Brij-35, containing 0.5M NaCl until no more 280nm absorbing material was eluted. At that point elution was continued with this buffer containing 1.0M NaCl and 10% DMSO. Individual fractions were monitored for UV absorbance and gelatinase and caseinase activity on zymograms (see below).

Gel Electrophoretic Techniques

Polyacrylamide Gel Electrophoresis. Samples were denatured with (final concentrations): SDS, 2.5%, with or without ß-mercaptoethanol, 0.36M and EDTA, 0.5mM, by incubating overnight at 4°. Electrophoresis was performed as described by Laemmeli. Gels were stained with Coomassie blue.

Visualization of Forms by Zymography. The presence of multiple forms of the metalloproteinases was demonstrated by zymogram analysis [25]. Samples were subjected to electrophoresis on substrate-loaded gels containing added gelatin or casein (1.0 mg/ml) as described above. Samples were denatured without ß-mercaptoethanol or EDTA. Following electrophoresis, the gels were washed with 2.5% Triton X-100 for one hour prior to incubation in a solution containing 50mM Tris-HCl, pH 7.2; 5mM CaCl2; 1μM ZnCl2; and 0.02% Na azide at 37°C. The gels were then stained with Coomassie blue. Clear areas indicated proteolytic activity. EDTA (10 or 50mM) or 1,10-phenanthroline (10mM) were occasionally included in both the Triton X-100 wash and in the incubation buffer to ensure that the activity observed in the zymograms was due only to metalloproteinases (data not shown).

Measurement of Enzymatic Activity

Azoalbumin. Incubations contained azoalbumin (0.4 mg); 0.1M Tris-HCl, pH 7.2; toluene, 0.01 ml; along with samples and any additives in a final volume of 0.27 ml [26]. Following incubation at 37° C (usually for about 3 days), TCA was added to 6.5%. The precipitate which formed during subsequent incubation at 4° C was removed by centrifugation. Aliquots of the supernatant fluid were transferred to multiwell plates, made alkaline by the addition of NaOH, and A450nm measured using an ELISA reader. Incubations containing thermolysin were also prepared to measure the presence of endogenous inhibitors.

^{3}H-Gelatin. Incubations contained ^{3}H-labeled, heat-denatured, type I collagen (DuPont NEN, Wilmington, DE), 0.04 mg/ml; 0.05M Tris-HCl, pH 7.2; 5mM CaCl2; 0.1M NaCl and 0.02% Na Azide; along with samples and any additives in a final volume of 0.2 ml. After incubation overnight at room temperature, 10µl of 50 mg/ml milk powder in 0.5M EDTA was added, followed by 40µl of a solution of 50% TCA in 2% tannic acid. After mixing, standing on ice to allow precipitation and centrifugation, an aliquot of the supernatant solution was counted. Incubations containing partially purified IPM as an enzyme source were also prepared to measure the presence of endogenous inhibitors.

Ac-Peptide. Incubations contained 0.1µM substrate (Ac-pro-leu-gly-thioester-leu-gly-OEt) [27]; 0.12mM DTNB; 12mM CaCl2; 60mM HEPES, pH 7.0; along with samples and any additives in a final volume of 0.2 ml in a multiwell plate. Samples were incubated at room temperature, and A410nm was periodically measured. Incubations containing partially purified IPM as an enzyme source were also prepared to measure the presence of endogenous inhibitors. NOTE: This material is not a substrate for common proteinases (e.g., trypsin, papain) or bacterial collagenase [27].

All other biological materials were obtained from Sigma (St. Louis, MO).

RESULTS AND DISCUSSION

Gel Filtration of Retinal-IPM

When retinal-IPM was fractionated by gel filtration on Sephacryl S-500, two peaks of 280 nm-absorbing material were seen (Fig. 1A) as previously described [2, 3, 24]. A small peak at the void volume contains the mucin-type glycoproteins [24] along with some low molecular weight material evidently trapped by the mucins (Fig. 1B, lane C). The major peak contains the interphotoreceptor retinoid binding protein (IRBP) along with other, lower molecular weight materials (Fig. 1B, lane D). Unfractionated retinal-IPM exhibits no gelatinase (Fig. 1C, lane A) or caseinase (Fig. 1D, lane A) activity nor activity toward azoalbumin or Ac-peptide (data not shown). However, after gel filtration, numerous forms are revealed (Fig. 1C, lanes C-E), especially in pool B (lane D), the most prominent being at about 65 kDa, 80-90 kDa and 130 kDa. A high molecular weight caseinase activity is also revealed in pool B (Fig. 1D. lane D). For comparison, RPE-IPM, whether unfractionated or subjected to gel filtration exhibits only two bands of gelatinase activity, at about 65 kDa and 90 kDa (Fig. 1C and D, lane B) [4]. Metalloproteinase and MP inhibitor activity as measured by azoalbumin are also evident in many fractions (Fig. 1A). Thus the rather innocuous procedure of gel filtration is sufficient to separate the MPs of retinal-IPM from at least some of the inhibitory restraint present in the unfractionated state.

DEAE Cellulose Ion Exchange Chromatography of Pool B

Fractionation on DEAE cellulose provided a significant separation of the components of pool B (Fig. 2 A and B). Fractions were pooled on the basis of MP activity or inhibition of MP activity as assayed by the various techniques. For reference, IRBP is present in pool

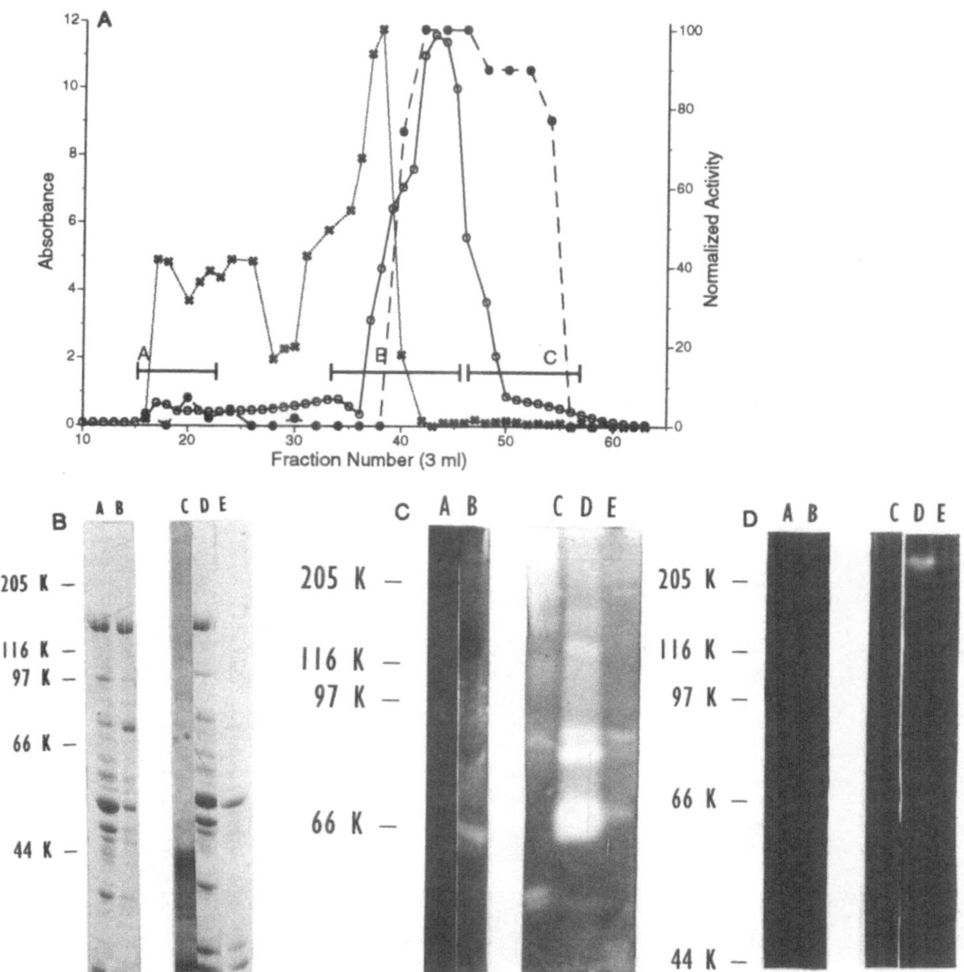

Figure 1. Gel filtration of retinal IPM. Retina-IPM (130 mg protein) was applied to a column of Sephacryl S-500, as described in Methods. (A) UV absorbance (O) was measured on the fractions (≈ 3 ml), and aliquots (0.1 ml) of individual fractions were dialysed vs. water prior to analysis for proteinase activity (x) and proteinase inhibitor activity (●) as described in Methods using azoalbumin as the substrate and thermolysin as the enzyme to be inhibited. Remaining material was then pooled as indicated by the bars. (B) SDS-PAGE — Both unfractionated retinal-IPM (5 µl) and RPE-IPM (5 µl) (lanes A and B) and pools A, B and C (45 µl each) (lanes C-E) were subjected to SDS-PAGE under reducing conditions as described in Methods. Zymography — Both retinal-IPM (60 µl) and RPE-IPM (40 µl) (lanes A and B) and pools A, B and C (60 µl each) (lanes C-E) were subjected to SDS-PAGE under nonreducing conditions in gels impregnated with either (C) gelatin or (D) casein as described in Methods. Following incubation for 2.5 days, they were stained with coomassie blue.

D (Fig. 2 B, lane D). The large 260 nm-absorbing peak centered near fraction 210 (Fig 2A) contained no MP activity. A single peak of Ac-peptide-cleaving activity (data not shown) coincides with the major peak of gelatin-cleaving activity (data not shown) in pool B, and may coincide with the 60-65 kDa band seen in the gelatin zymogram (Fig. 2C, lane B). The other higher molecular weight gelatinases present in pool B from gel filtration (Fig. 1 C,

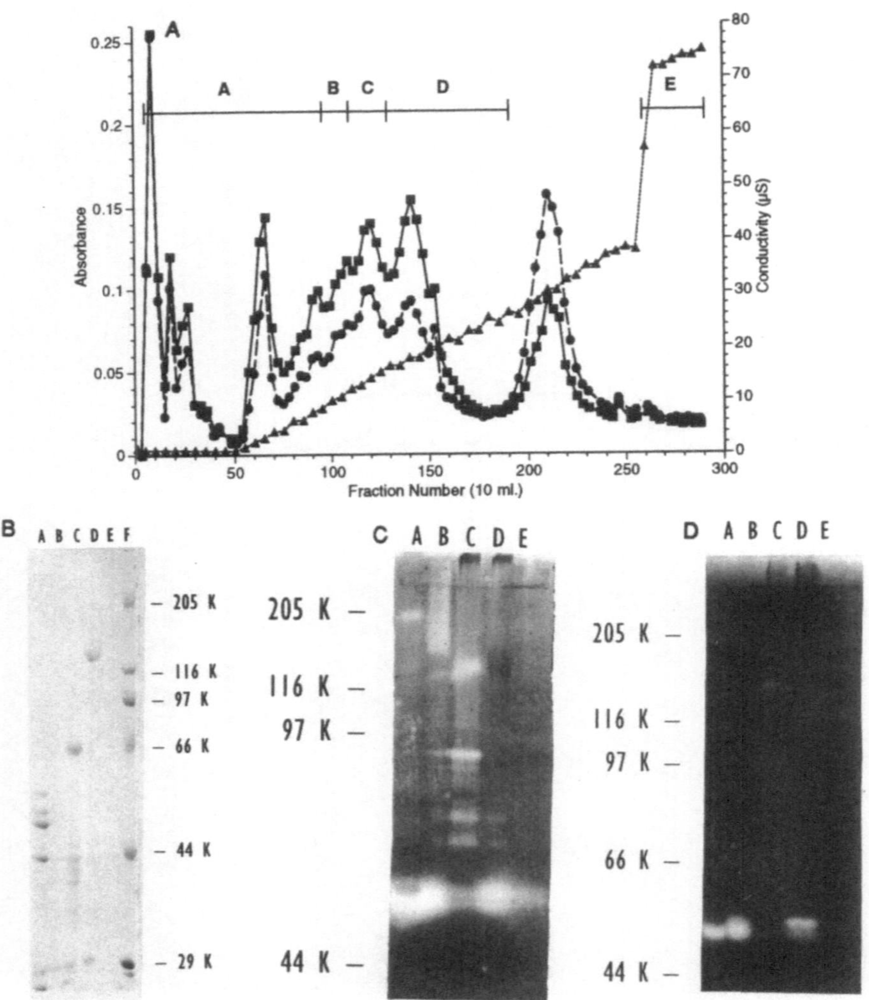

Figure 2. DEAE Cellulose ion exchange chromatography. Pool B, obtained from gel filtration of retina-IPM (Fig. 1), was subjected to chromatography on DE52 as described in Methods. Fractions were monitored for the following: (A) Absorbance at 280 nm (■) and 260 nm (●) and conductivity (▲), (B) SDS-PAGE — Pools A to E (15 μl each) (lanes A-E) were subjected to SDS-PAGE under reducing conditions as described in Methods. (Lane F contained standards.) Zymography—Pools A to E (30 μl each) (lanes A-E) were subjected to SDS-PAGE under nonreducing conditions in gels impregnated with either (C) gelatin or (D) casein as described in Methods. Following incubation for 2.5 days, they were stained with coomassie blue.

lane D) are present in pool C (Fig. 2C, lane C). A lower molecular weight caseinase is now evident in pools A, B and D (Fig. 2D) migrating at about 50 and 52 kDa.

An inhibitor of the Ac-peptide-cleaving activity was found in pool E (data not shown), and two inhibitors of azoalbumin-cleaving activity are present in pools B and C (data not shown). These two inhibitor assays measure two different types of MP. Ac-peptide uses partially purified IPM as an enzyme source, while the azoalbumin assay uses the bacterial MP, thermolysin. There appears to be no cross-reactivity between the two types of inhibitors.

Gelatin-agarose affinity chromatography of pools A, B and C from DEAE cellulose

Affinity chromatography with gelatin as the ligand can be used to separate gelatinases from other MPs as well as other proteins. This procedure has thus far been applied to the material in DEAE cellulose pools A, B and C (Fig 3). In each case, the caseinase activity did not adhere to the column (Fig. 3 C, lanes A, B and D). This was also true for the major 60 kDa gelatinase activity (Fig. 3 B, lanes A, B and D), although this activity was somewhat retarded and eluted slightly behind the bulk of the nonadherent protein (data not shown). Pool B2 contained no measurable MP activity. It should be noted that pool C now contains 60 kDa gelatinase activity, possibly reflecting the removal of an inhibitor, or auto-activation upon standing. The 70 and 90 kDa gelatinases of pool C did adhere to the column and were eluted with the DMSO in pool C2 (Fig 3B, lane C). This pool contains no caseinase activity and little extraneous protein.

CONCLUDING REMARKS

These studies have extended our characterization of the MPs present in bovine IPM. The combination of procedures used here has enabled us to obtain two of the gelatinases (70 kDa and 90 kDa) free from most other extraneous protein (pool C2) and other MP activity. These two forms correspond to gelatinases A and B (MMP-2 and 9), respectively, which we have shown to be present in retinal-IPM [4]. The major 60-65 kDa gelatinase is probably the activated form of gelatinase A [6-8]. The presence of mostly the activated forms of the enzymes would be expected in fresh IPM as opposed to stimulated secretions of cultured cells which would be expected to contain mostly the unactivated, precursor forms. Why this form does not adhere to the gelatin affinity matrix is not clear at this time.

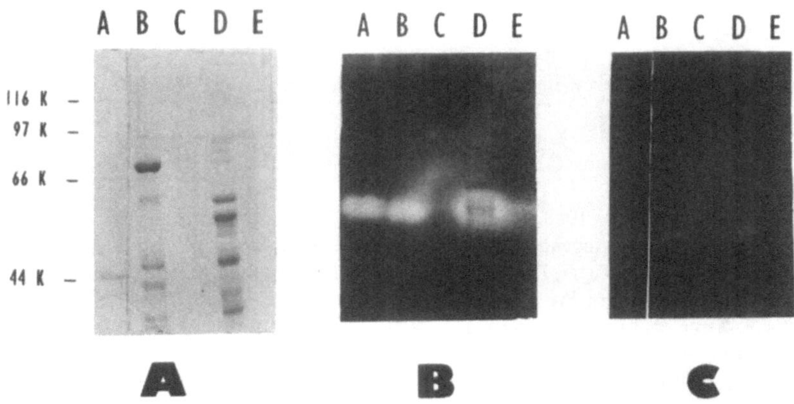

Figure 3. Gelatin-agarose affinity chromatography. Pools A, B and C, obtained from DE52 chromatography (Fig. 2), were subjected to affinity chromatography on gelatin-agarose as described in Methods. Individual fractions were pooled based on whether they were retained by the column and eluted by DMSO (A2, C2) or not retained and obtained in the initial wash (A1, B1, C1). (A) SDS-PAGE — Pools B1 (8 μl), C1 (8 μl), C2 (11 μl), A1 (8 μl) and A2 (11 μl) (lanes A-E, respectively) were subjected to SDS-PAGE under reducing conditions as described in Methods. Zymography — Pools B1 to A2 (19 μl each) (lanes A-E as in part A) were subjected to SDS-PAGE under nonreducing conditions in gels impregnated with either (B) gelatin or (C) casein as described in Methods. Following incubation for 2.5 days, they were stained with coomassie blue.

As the retinal-IPM has been further fractionated, additional forms of MP have been revealed. Starting with unfractionated retinal-IPM, which exhibits no activity using any of the assay systems described here, gel filtration revealed many forms of gelatinase, some of which are probably fragments or aggregates of others, along with a very minor, high molecular weight caseinase. Ion exchange chromatography has revealed a major, 50 kDa caseinase which corresponds to Stromelysin 1 (MMP-3), which we have identified as being present in retinal-IPM as well [4]. We are currently attempting to separate the caseinase from the activated gelatinase A.. Pool B1 (Fig. 3, lane A) contains both enzymes, but little extraneous protein. These additional forms have been revealed through a combination of removal of inhibitors and activation of latent, zymogen forms. It is possible that other forms may be found as we fractionate this material and RPE-IPM further.

ACKNOWLEDGMENTS

This work was supported in part by NIH EY10184, Eppley Foundation for Research, Ohio Lions Eye Research Foundation and The Retinal Institute of Mt. Sinai Medical Center.

REFERENCES

1. Hageman, G. and Johnson, L., 1991, Structure, composition and function of the retinal interphotoreceptor matrix, in Prog. Ret. Res., (N. Osborne and G.J. Chader, eds.), 10:207-69
2. Plantner, J.J., 1992, The presence of neutral metalloproteolytic activity and metalloproteinase inhibitors in the interphotoreceptor matrix, *Curr. Eye Res.* 11:91-101.
3. Plantner, J.J. and Drew, T.A., 1994, Polarized distribution of metalloproteinases in the bovine interphotoreceptor matrix, *Exp. Eye Res.* 59:577-585.
4. Plantner, J.J. and Quinn, T.A., Matrix Metalloproteinases in the Interphotoreceptor Matrix and Vitreous, submitted to Invest. Ophthalmol. Vis. Sci. for publication.
5. Jones, R.E., Moshyedi, P., Gallo, S., Tombran-Tink, J., Arand, G., Reid, D.A., Thompson, E.W., Chader, G.J. and Waldbillig, R.J., 1994, Characterization and novel activation of 72-kDa metalloproteinase in retinal interphotoreceptor matrix and Y-79 cell culture medium, *Exp. Eye Res.* 59:257-269.
6. Birkedal-Hansen, H., 1988, From tadpole collagenase to a family of matrix metalloproteinases, *J. Oral Pathol.* 17:445-451.
7. Matrisian, L.M., 1992, The matrix-degrading Metalloproteinases. *BioEssays*, 14:455-463.
8. Woessner, Jr., J.F., 1991, Matrix metalloproteinases and their inhibitors in connective tissue remodeling, *FASEB J.* 5:2145-2154.
9. Kolkenbrock, H., Orgel, D., Hecker-Kia, A., Noack, W. and Ulbrich, N., 1991, The complex between a tissue inhibitor of metalloproteinases (TIMP-2) and 72-kDa progelatinase is a metalloproteinase inhibitor, *Eur. J. Biochem.* 198:775-781.
10. Moscatelli, D. and Rifkin, D.B., 1988, Membrane and matrix localization of proteinases: a common theme in tumor cell invasion and angiogenesis, *Biochim. Biophys. Acta* 948:67-85.
11. Fini, M.E., Yue, B.Y.J.T. and Sugar, J., 1992, Collagenolytic/Gelatinolytic Metalloproteinases in Normal and Keratoconus Corneas, *Curr. Eye Res.* 11:849-862.
12. Alexander, J.P., Samples, J.R., Van Buskirk, E.M. and Acott, T.S., 1991, Expression of Matrix Metalloproteinases and Inhibitor by Human Trabecular Meshwork, *Invest Ophthalmol Vis Sci.* 32:172-180.
13. Alexander, J.P., Bradley, J.M., Gabourel, J.D. and Acott, T.S., 1990, Expression of Matrix Metalloproteinases and Inhibitor by Human Retinal Pigment Epithelium, *Invest Ophthalmol. Vis. Sci.* 31:2520-2528.
14. Hunt, R.C., Fox, A., Al Pakalnis, V., Sigel, M.M., Kosnosky, W., Choudhury, P. and Black, E.P., 1993, Cytokines Cause Cultured Retinal Pigment Epithelial Cells to Secrete Metalloproteinases and to Contract Collagen Gels, *Invest. Ophthalmol. Vis. Sci.* 34:3179-3186.
15. Wentworth, J.S., Paterson, C.A. and Gray, R.D., 1992, Effect of a Metalloproteinase Inhibitor on Established Corneal Ulcers After an Alkali Burn, *Invest. Ophthalmol. Vis. Sci.* 33:2174-2179.
16. Patz, A., 1980, Studies on retinal neovascularization, *Invest. Ophthalmol. Vis. Sci.* 19:1133-8.
17. Machemer, R., 1988, Proliferative vitreoretinopathy (PVR): A personal account of its pathogenesis and treatment, *Invest. Ophthalmol. Vis. Sci.* 29:1771-1783.

18. Green, W.R., McDonnell, P.J. and Yeo, J.H., 1985, Pathologic features of senile macular degeneration, *Ophthalmol.* 92:615-627.
19. Taylor, C.M., Weiss, J.B., Kissun, R.D., and Garner, A., 1986, Effect of oxygen tension on the quantities of procollagenase-activating angiogenesis factor in the vitreous humour of oxygen treated kittens, *Brit. J. Ophthalmol.* 70:162-168.
20. Caldwell, R.B., Slapnick, S.M., and Roque, R.S., 1989, RPE-associated extracellular matrix changes accompany retinal vascular proliferation and retino-vitreal membranes in a new model for proliferative retinopathy: the dystrophic rat, In "Inherited and Environmentally Induced Retinal Degenerations," Alan R. Liss, Pp. 393-407.
21. Jones, S.E., Jomary, C. and Neal, M.J., 1994, Expression of TIMP3 mRNA is elevated in Retinas Affected by Simplex Retinitis Pigmentosa, *FEBS Lett.* 352:171-174.
22. Weber, B.H.F., Vogt, G. Pruett, R.C., Stohr, H. and Felbor, U., 1994, Mutations in the Tissue Inhibitor of Metalloproteinases-3 (TIMP3) in Patients with Sorsby's Fundus Dystrophy, *Nature Genetics* 8:352-356.
23. Adler, A.J. and Severin, K.M., 1981, Proteins of the Bovine Interphotoreceptor Matrix: Tissues of Origin, *Exp. Eye Res.* 32:755-769.
24. Plantner, J.J., 1992, High Molecular Weight Mucin-like Glycoproteins of the Bovine Interphotoreceptor Matrix, *Exp. Eye Res.* 54:113-125.
25. Heussen, C. and Dowdle, E.B., 1980, Electrophoretic Analysis of Plasminogen Activators in Polyacrylamide Gels Containing Sodium Dodecyl Sulfate and Copolymerized Substrates, *Anal. Biochem.* 102:196-202.
26. Plantner, J.J., 1991, A microassay for proteolytic activity, *Anal. Biochem.* 195:129-131.
27. Weingarten, H., Martin, R. and Feder, J., 1985, Synthetic Substrates of Vertebrate Collagenase, *Biochem,* 24:6730-6734.

PEPTIDES FROM RHODOPSIN INDUCE EXPERIMENTAL AUTOIMMUNE UVEORETINITIS IN LEWIS RATS

Grazyna Adamus,[1] Henry Ortega,[1] Lundy Campbell,[1] Anatol Arendt,[2] and Paul A. Hargrave[2]

[1] R. S. Dow Neurological Sciences Institute
1120 NW 20th Avenue, Portland, Oregon 97209

[2] Department of Ophthalmology
University of Florida
Gainesville, Florida 32610

Experimental autoimmune uveoretinitis (EAU) is an organ-specific, T cell mediated autoimmune disease, which is characterized by destruction of the photoreceptor cells of the retina. EAU serves as a model for human intraocular inflammatory diseases, like uveitis, which are major causes of visual impairment. The animal disease can be induced by immunization with certain retinal proteins. The antigens used in most studies for EAU induction have been S-antigen (arrestin) and interphotoreceptor retinoid binding protein (IRBP)(Gery, Michizuki and Nussenblatt, 1986). We, and other investigators, have reported that rhodopsin is similarly uveitogenic in rats, guinea pigs, and in monkeys (Adamus, et al., 1992; Marak, et al., 1980; Meyers-Elliott and Sumner, 1982; Moticka and Adamus, 1991; Schalken, et al., 1988b; Schalken, et al., 1989; Wong, et. al., 1977). Rhodopsin is the major protein of photoreceptor cells and has a molecular weight of 40,000 daltons. It is the photoreceptor protein that initiates the visual transduction process (Hargrave and McDowell, 1993).

In the 1970's, it was first reported that immunization with a crude homogenate of unsolubilized rod cell outer segments produced pathological damage to the retina. Subsequently, animals injected with purified rhodopsin/opsin have been shown to develop a primarily posterior, nongranulomatous choroiditis (Marak, et al., 1980). Studies in the 1980's have suggested that rhodopsin is more immunogenic and pathogenic than opsin (Schalken, et al., 1988a). However, our experiments have not confirmed this report (Moticka and Adamus, 1991). The development of both cellular and antibody responses against rhodopsin/opsin have been measured. The formation of B and T cell responses in Lewis rats during the disease induction shows a unique pattern of specificities. The animals produced anti-rhodopsin antibodies mostly specific for the IV-V loop region (174-203) and T immunocompetent cells that recognized primarily the C-terminal region (324-348). The immunopathology

Degenerative Diseases of the Retina, Edited by Robert E. Anderson et al.
Plenum Press, New York, 1995

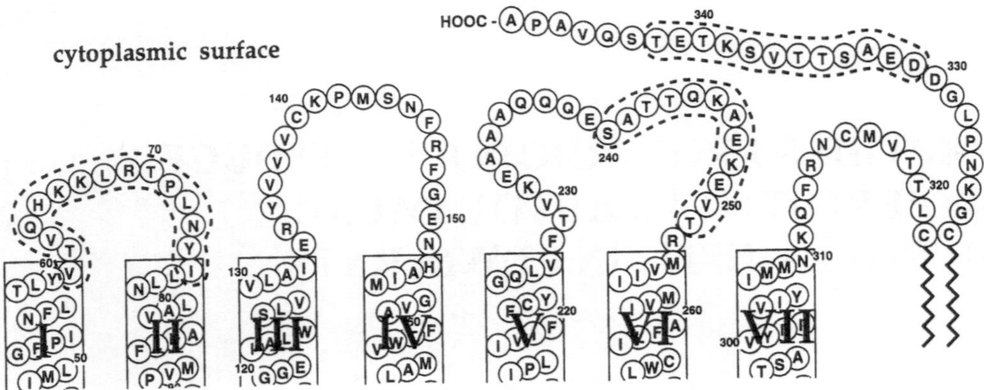

Figure 1. A topographic model of rhodopsin's cytoplasmic surface in the disc membrane of retinal rod cells, showing uveitogenic regions.

seen in rhodopsin-induced uveitis is T cell mediated, based on our observations that the disease can be transferred by $CD4^+$ T cells (Moticka and Adamus, 1991).

We have used synthetic peptides to define pathogenic determinants for rhodopsin (Adamus, et al., 1994; Adamus, et al., 1992) and this is the topic of our current report. Immunization of animals with rhodopsin synthetic peptides, that cover hydrophilic, exposed regions of the protein, resulted in the display of some different T cell specificities than immunization with whole antigen. Furthermore, immunization with peptides induces specificities that are not present after rhodopsin immunization e.g., T cells specific against residues 230-252 and 2-32. Three distinct uveitogenic epitopes were identified on rhodopsin's conserved, cytoplasmic surface within the carboxyl terminus (324-348), loop I-II (61-75), and loop V-VI (230-252) and are shown in Figure 1.

Female Lewis rats (Harlan-Sprague-Dawley Laboratories) in groups of 3 animals (150-180 g) were immunized with rhodopsin synthetic peptides (100 mg per animal). Antigens were emulsified with an equal volume of complete Freund's adjuvant (CFA), containing Mycobacterium tuberculosis H37Ra (Sigma). At the same time rats received 1 μg pertussis toxin intraperitoneally. Control animals received saline instead of peptide. Peptides were synthesized by solid phase techniques using Fmoc derivatives of amino acids with an automatic synthesizer (Applied Biosystems Model 431A, Foster City, CA). Seventeen days after immunization, the animals were killed and spleens were removed aseptically. The eyes were removed and fixed in 10% buffered neutral formaldehyde solution and submitted to routine histological examination. The severity of EAU in each eye was assessed using blind-coded samples and graded from 0 (no EAU) to 5+ (most severe) as described elsewhere (Adamus, et al., 1992). All procedures adhered to the ARVO Resolution on the Use of Animals in Research.

Proliferative responses of lymphocytes were determined by measuring incorporation of [^{3}H]-thymidine as previously described (Adamus, et al., 1992). Purified T cells from spleens of immunized animals were cultured with or without stimulant (1 μg of concanavalin A, 5 μg, 10 μg of synthetic peptide). Results are presented as stimulation indices (SI) calculated as mean cpm of cultures with stimulant divided by mean cpm of the unstimulated control culture. Stimulation was considered positive if the SI of injected rats was equal to or greater than twice the background (SI=2).

The nine synthetic peptides derived from rhodopsin's hydrophilic, exposed sequences vary in their immunological activities. The most immunopathogenic peptide was 230-251 (V-VI loop) which caused severe inflammation of the retina. Histologically confirmed uveitis

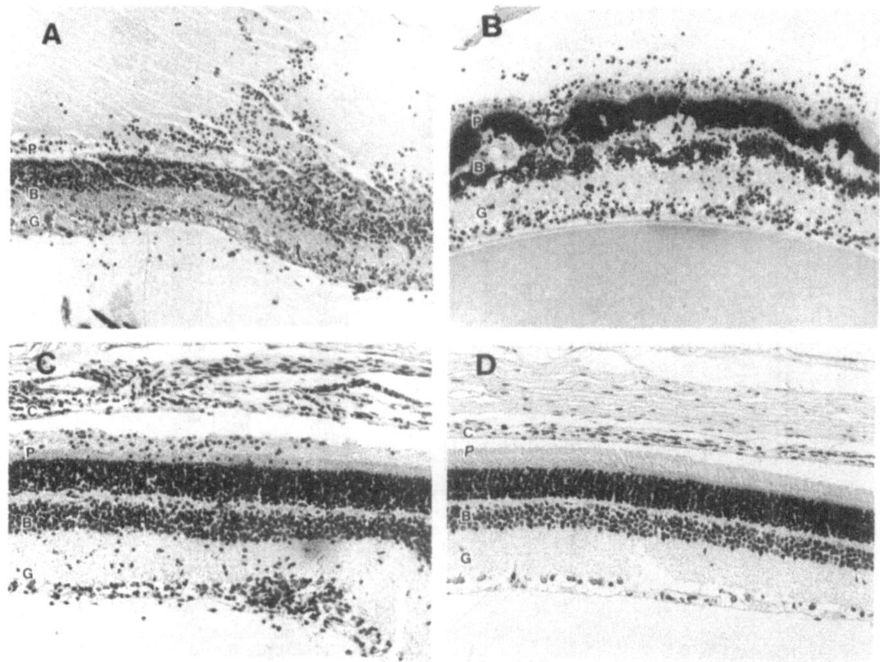

Figure 2. Representative micrographs of the retinas of Lewis rats immunized with peptides derived from loop V-VI sequence. Destruction of photoreceptor cell layer following immunization with (A) peptide 230-252; (B) peptide 240-251; (C) peptide 240-251 (Ala 250); (D) saline instead of peptide. The eyes were collected 18 days postimmunization. The inflammation was similar to that found in animals immunized with intact bovine rhodopsin. Normal retina consists of P - photoreceptor cell layer including outer and nuclear cell layers, B - bipolar cell layer, G - ganglion cell layer, and C - choroid.

occurred 14 days after injection, and active inflammation was present for at least the next 10 days. At first, mild mononuclear cell infiltration with foci of inflammatory cells in the retina was observed. Later, infiltration of the photoreceptor cell layer, outer nuclear layer, and inner layer of the retina was characteristic. Control animals injected with saline in complete Freund's adjuvant did not develop any histological changes (Figure 2). The cellular responses determined by *in vitro* lymphocyte proliferation assay indicated strong proliferative activities against immunizing peptide with a peak of proliferative activity on day 15-16 after immunization. Lymphocytes sensitized against this peptide did not recognize the intact rhodopsin molecule in culture but sensitized lymphocytes could transfer the disease.

By application of a series of 12-amino acid long overlapping peptides that covered rhodopsin sequence 230-255, only a single peptide - the one containing sequence SATTQKAEKEVT (240-251) - has been found to contain a potent uveitogenic site. The elimination of serine from its amino-terminus suppressed the pathogenic features suggesting importance of this residue for pathogenic T cell activation. Proliferative activities against these overlapping peptides are shown in Figure 3. Peptide 240-251 stimulates vigorous proliferative response and is the only peptide that contains the uveitogenic site. Injection of Lewis rats with this peptide induced infiltration of the photoreceptor cell layer, and inner layer of the retina (Figure 2C), which finally led to complete destruction of the photoreceptor cell layer. A few other peptides exhibited proliferative activities without the disease induction. Peptide 243-254 stimulated in vitro the most vigorous proliferative response but did

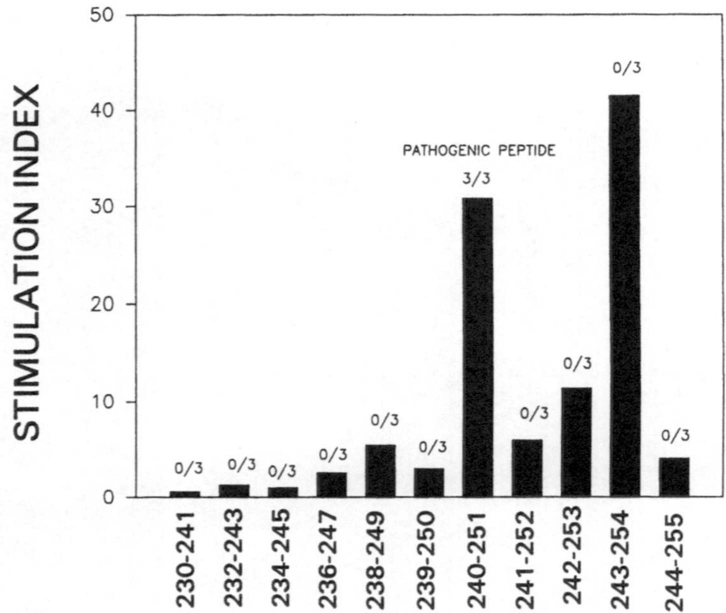

Figure 3. Proliferative responses of splenic T cells to V-VI loop overlapping peptides from animals injected with 100 μg of the appropriate peptide. Data are presented as stimulation indices. Unstimulated cultures of cells from those animals incorporated 714-2162 cpm. Peptides were tested *in vitro* at 2.5 μM. Numbers over the bars (0/3, 3/3) are the number of animals showing uveitogenic response out of a total of 3 tested.

not induce EAU. The proliferation and pathogenicity data together demonstrate the presence of pathogenic and proliferative sites within the sequence of loop V-VI. A proliferative site is distinct from, but overlapping, the uveitogenic site. These indicate that rhodopsin, in a manner similar to that of other retinal autoantigens, contains separate and distinct T cell epitopes that are responsible for the active induction of EAU and lymphocyte proliferation (Gregerson, et al., 1989; Kotake, et al., 1991).

The role of each of the amino acids contained in the uveitogenic site (240-251) was further examined by testing the activity of peptide analogs in which individual residues were substituted by alanine. Certain residues were found to be important for the site's activities. Similarly to studies performed with other uveitogenic peptides, the effects of substitution with alanine differ according to the chemical properties of the amino acids being substituted. Three residues were found essential for both the uveitogenicity and immunogenicity: lysine 245, glutamic acid 247, and lysine 248. Peptides with alanine substitutions at these positions did not elicit autoimmune uveoretinitis. Threonines at positions 242 and 243 are apparently important in the binding site. Some analogs of peptide 240-251 were found to be immunogenic without being immunopathogenic. Threonine at position 251 is important for uveitogenicity. The alanine substitution in this position failed to induce EAU. Figure 4 presents important residues involved in pathogenicity of loop V-VI of rhodopsin.

Figure 4. Scheme representing uveitogenic site, peptide 240-252. Amino acids involved in pathogenicity of loop V-VI are boxed.

The proliferation and pathogenicity data together demonstrate the presence of pathogenic and proliferative sites within the sequence 240-251. A proliferative site is distinct from, but overlapping, the uveitogenic site. The identification of regions of the molecule which are responsible for the pathogenicity of the potential autoantigen is important in understanding the immune mechanisms by which ocular inflammation occurs. Loop V-VI is a good example of a cryptic determinant. Sensitized lymphocytes were found to be capable of inducing the disease in naive animals, yet failed to respond to rhodopsin in culture. Similarly to other determinants studied that induced experimental EAU (Gregerson, et al., 1990; Lipham, et al., 1991) this previously cryptic region is uveitogenic when injected as a peptide. The pivotal event that initiates autoimmune disease may also involve induction of a strong T cell response against some cryptic determinants. Once initiated, the activated T cells can cause tissue destruction.

ACKNOWLEDGMENTS

This work was supported in part by NIH grants EY06225 and EY06226, and the Oregon Lions Sight and Hearing Foundation.

REFERENCES

Adamus, G., Dugger, D., Arendt, A., and Hargrave, P.A. (1994) Characterization of an immunopathogenic epitope in rhodopsin that induces experimental autoimmune uveitis. *Reg. Immunol.* 6:44-46.

Adamus, G., Schmied, J.L., Hargrave, P.A., Arendt, A., and Moticka, E.J. (1992) Induction of experimental autoimmune uveitis with rhodopsin synthetic peptides in Lewis rats. *Curr. Eye Res.* 11:657-667.

Gery, I., Michizuki, M., and Nussenblatt, R.B. Retinal specific antigens and immunopathogenic process they provoke. Vol. 5 of Progress in Retinal Research. Oxford: Pergamon Press, 1986. 75-109.

Gregerson, D.S., Fling, S.P., Obritsch, W.F., Merryman, C.F., and Donoso, L.A. (1989) Identification of T cell recognition sites in S-antigen: Dissociation of proliferative and pathogenic sites. *Cell. Immunol.* 123:427-440.

Gregerson, D.S., Fling, S.P., Obritsch, W.F., Merryman, C.F., and Donoso, L.A. (1990) A new perspective of S-antigen from immunochemical analysis. *Curr. Eye Res.* 9 (Suppl.):145-153.

Hargrave, P.A., and McDowell, J.H. (1993) Rhodopsin and Phototransduction. *International Rev. Cyt.* 137B:49-97.

Kotake, S., Redmond, T.M., Wiggert, B., Vistica, B., Sanui, H., Chader, G., and Gery, I. (1991) Unusual immunologic properties of the uveitogenic interphotoreceptor retinoid-binding protein-derived R23. *Invest. Ophthalmol. Vis. Sci.* 32:2058-2064.

Lipham, W.J., Redmond, T.M., Takahashi, H., Berzofsky, J.A., Wiggert, B., Chader, G.J., and Gery, I. (1991) Recognition of peptides that are immunopathogenic but cryptic. Mechanisms that allow lymphocytes sensitized against cryptic peptides to initiate pathogenic autoimmune processes. *J. Immunol.* 146:3757-3762.

Marak, G.E., Shichi, H., Rao, N.A., and Wacker, W.B. (1980) Patterns of experimental allergic uveitis induced by rhodopsin and retinal rod outer segments. *Ophthalmic Res.* 12:165-176.

Meyers-Elliott, R.H., and Sumner, H.L. (1982) Experimental uveitis induced by products of activated lymphocytes: intraocular effects of rhodopsin-induced lymphokines. *Cell. Immunol.* 66:240-253.

Moticka, E.J., and Adamus, G. (1991) Specificity of T and B cell responses to bovine rhodopsin in Lewis rats. *Cell. Immunol.* 138:175-184.

Schalken, J.J., VanVugh, A.H.M., Winkens, H.J., Bovee-Geurts, P.H.M., DeGrip, W.J., and Broekhuyse, R.M. (1988a) Experimental autoimmune uveoretinitis in rats induced by rod visual pigment: Rhodopsin is more pathogenic than opsin. *Graefe's Arch. Clin. Exp. Ophthalmol.* 226:255-261.

Schalken, J.J., Winkens, H.J., VanVugt, A.H.M., Bovee-Geurst, P.H.M., DeGrip, W.J., and Broekhuyse, R.M. (1988b) Rhodopsin-induced experimental autoimmune uveoretinitis: Dose dependent clinicopathological features. *Exp. Eye Res.* 47:135-145.

Schalken, J.J., Winkens, H.J., Vugt, A.H.M.V., Grip, W.J.D., and Broekhuyse, R.M. (1989) Rhodopsin-induced experimental autoimmune uveoretinitis in monkeys. *Br. J. Ophthalmol.* 73:168-172.

Wong, V.G., Green, W.R., and McMaster, P.R.B. (1977) Rhodopsin and blindness. *Tr. Am. Ophthal. Soc.* 75:272-284.

13

RETINAL DEGENERATION IN RATS INDUCED BY VITAMIN E DEFICIENCY

E. El-Hifnawi, H. -J. Hettlich, and C. Falk

Department of Ophthalmology
University of Lübeck
Ratzeburger Allee 160, D-23538 Lübeck, Germany

INTRODUCTION

The biological function of vitamin E as a lipid antioxidant has been a topic of investigations for over fifty years. Since the discovery of antioxidant activity, several functions of vitamin E have been elucidated and its role in various vertebrate organs has been examined and discussed. Numerous studies have demonstrated that α-tocopherol (vitamin E) is related to reproduction and that its deficiency results in sterility (White et al. 1978). Furthermore, the role of vitamin E in oxygen metabolism and membrane function as a protector against free radical oxidation has been documented in several studies (Lucy 1978; McCay et al. 1978).

It is well known that the vertebrate retina is particularly vulnerable to damage by activated oxygen radicals due to its high concentration of polyunsaturated fatty acids (Anderson and Andrews 1982; Wiegand et al. 1986). In bovine rod outer segments, vitamin E normally occurs in relatively large amounts (Dilley and McConnell 1970), therefore, it is possible that α-tocopherol may protect the retina from degeneration. Previous evidence suggests that lipid peroxidation plays a major role in the retinal degeneration following e.g. constant illumination (Wiegand et al. 1983) or vitamin E deficiency (Hayes 1974; Robison et al. 1979; Riis et al. 1981). In rabbit eyes, cataract formation has also been reported as a result of nutritional vitamin E deficiency (Devi et al. 1965).

Some symptoms of inflammation are due to increased fragility of the lysosomal membrane and the release of their lytic enzymes, leading to cellular and extracellular injuries (Hayasaka 1974). Amemiya (1981) observed that rats deficient in vitamin E developed a disruption of the RPE lysosomal membrane resulting in infiltration of lytic enzymes into the subretinal space. These alterations became more prominent with increasing duration of vitamin E deficiency. From these findings, the author assumed that disc degeneration of the photoreceptor outer segments may be due to enhanced activity of the lysosomal enzymes in the RPE and because of the effects these released enzymes have on the photoreceptor outer segments. Experiments on RCS rats, an animal model for hereditary retinal dystrophy, demonstrated a defect in the RPE lysosomal membrane of this strain of rat (El-Hifnawi and Kühnel 1987a, b;

Degenerative Diseases of the Retina, Edited by Robert E. Anderson et al.
Plenum Press, New York, 1995

El-Hifnawi et al. 1994). Extracellular release of lytic enzymes, e.g. acid phosphatase and cathepsin D, occurred at a very early stage of the disease and increased as the animals grew older. Our recent findings on RCS rats have shown that a daily administration of vitamin E caused the stabilization of the RPE lysosomal membrane. Furthermore, an increase in the survival rate of the photoreceptor cells and pigment epithelium was observed (El-Hifnawi et al. 1993). The experiments undertaken in this study were to examine the retinas of rats following deprivation of vitamin E over various periods of time. Particular emphasis was placed on comparing the results of nutritional factors with those related to inherited factors.

MATERIAL AND METHODS

Three groups of 9 month old non-dystrophic animals were obtained from the RCS-rdy$^+$ congeneic strain and fed a vitamin E deficient diet ad libitum for various periods of time (35, 75 and 95 days). Control animals were obtained from the same strain and fed a normal diet. All animal groups were maintained in a 12 hr light - 12 hr dark environment at a room illumination of approximately 12-18 foot-candles from overhead fluorescent lamps. The illumination levels within the cages varied considerably, depending upon the position of the cage. However, it never exceeded 15 foot-candles. The temperature of the room was controlled at 21 ± 1°C. The treatment of the animals in this study was in conformance with the ARVO resolution on animal use in research. The rats were sacrificed by injecting an i.p. overdose of phenobarbital (Nembutal® 0.1 ml/100 g body weight). It is well established that disc shedding and the phagocytic activity of the retinal pigment epithelium, is greatest during the first two hours of light exposure (LaVail 1976). For this reason, and because of the increase in lytic enzyme activity during this time, the animals were sacrificed within this period and the eyes enucleated immediately upon death. In order to facilitate identification of the individual eye quadrants, they were marked with a surgical marking pen. The eyes were enucleated and prepared for electron microscopic and immunohistochemical examination.

Fixation for electron microscopy was done using 4% glutaraldehyde in 0.1 M cacodylate buffer at a pH of 7.4. The cornea was slit horizontally, and fixation continued for 30 min. at room temperature. Tissue samples were obtained in such a way that each section included a full length of retina from the optic nerve head to the ora serrata and into some portion of the cornea. Samples were rinsed several times in the buffer solution mentioned above with the addition of 8.5% sucrose for a period of 1 hr. The specimens were post-fixed for 2 hrs. in a Dalton solution (1955), containing 1% osmium tetroxide. Samples were dehydrated through a graded series of alcohols and then flat embedded in Durcupan ACM^R (FLUKA). Oriented semi-thin sections were cut at 1-1.5 μm with glass knives on a Reichert OmU2a ultramicrotome, mounted on glass slides and stained following the same procedure described by Richardson et al. (1960). Ultrathin sections from selected retinal regions were cut with a diamond knife and double stained with uranyl acetate and lead citrate (Reynolds 1963). Electron microscopic examination of the ultrathin sections was carried out on a TEM Zeiss 9S.

For immunohistochemical examination, the enucleated eyes were fixed immediately and left overnight in 4% buffered neutral formalin at 4°C. Following standard histological procedures, the specimens were embedded in paraffin wax. Oriented serial sections (4-5 μm) were cut, so as to include the whole length of the retina passing through the optic nerve head and extending forward to the ora serrata. The sections were mounted on glass slides which had been additionally cleaned with Extran® (MERCK) and coated with 2% 3- Aminopropyltriethoxysilane (SIGMA) for better adhesion. Following deparaffinization, the presence of cathepsin D was demonstrated immunohistochemically by the alkaline phosphatase-antialkaline phosphatase (APAAP) technique. Nonspecific protein binding was eliminated by a 30 min. incubation at room temperature in 20% normal swine serum in TBS buffer pH 7.6,

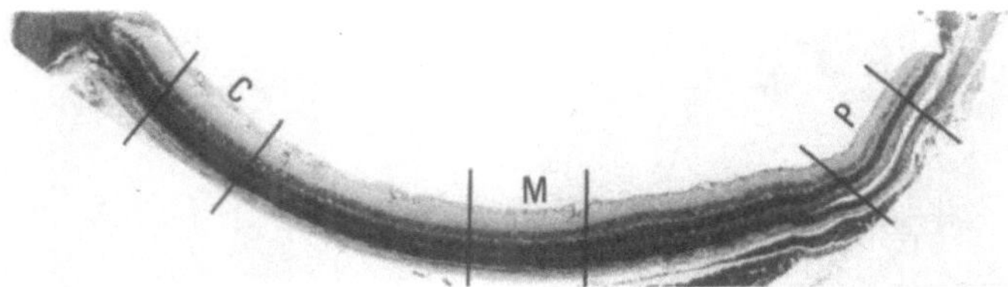

Figure 1. Low power light micrograph showing the subdivision of each eye quadrant prior to semi-thin and ultrathin sectioning. Periphery (P); Midperiphery (M); Center (C) X25.

obtained from DAKO. The polyclonal antibody (anti-cathepsin D Code No. A561), obtained from DAKO, was diluted to 1:100 and incubated for 30 min. at room temperature. After each step, sections were washed three times in TBS buffer. All incubations were carried out in humidified chambers in order to prevent evaporation. Immunohistochemical labelling was then performed using the APAAP technique. The visualization of the enzyme was carried out with the new fuchsin stain. The nuclei were counterstained with Mayer's hemalum and mounted in glycergel. Negative control staining was performed by omitting either the primary polyclonal antibody or subsequent stages of the labelling procedure. No positive staining was seen in the negative control preparations.

RESULTS

In rats fed a vitamin E deficient diet for 35 days, beginning 9 months after birth, the photoreceptor outer segments were disorganized and an abnormal localization of cathepsin D was detected on the distal regions (Fig. 2). The extracellular distribution of cathepsin D in the subretinal space is due to instability of the lysosomal membranes in the RPE cells. Following a prolonged duration on the deficient diet, the extracellular enzyme activity became more pronounced. In control animals, cathepsin D was restricted to the retinal pigment epithelium (RPE) and no reaction product was found extracellularly. Compared to normal rats (Fig. 3), the outer nuclear layer (ONL) showed signs of degeneration in the form of pyknotic nuclei and a reduction in thickness. These alterations were restricted to the far periphery of the retina. The rest of the neural retina - bipolar, ganglion cells - and RPE appeared to be morphologically unaffected.

In rats fed a vitamin E deficient diet for 75 days, the alterations became more prominent. The retina was reduced in thickness as a result of photoreceptor cell degeneration. The outer nuclear layer contained numerous pyknotic cell nuclei corresponding to the degree of outer segment degeneration. In the midperiphery and center of the retina, the outer segments of the photoreceptor cells were extremely thin, unusually long and disorganized. When viewed at light microscopic level, it was difficult to distinguish morphological damage in the RPE and inner segments. However, when examined ultrastructurally, disruption of the RPE cell membranes and loss of their cell organelles could be seen. In these areas, the RPE cell layer had lost its original thickness and become atrophic. Non-phagocytized discarded outer segment material was found in close proximity to the RPE cells (Figs. 4a, 5). Vacuolation of the outer segment discs and inner segments were apparent (Fig. 5). However, the periphery was more affected than the other retinal regions, e.g. midperiphery and center (Fig. 4).

In rats fed a vitamin E deficient diet for 95 days, the degenerative changes became more prominent than in 75 days (Fig. 4). All the retinal layers were affected, in some areas the photoreceptor cells had almost disappeared, consisting of only 1-3 nuclear rows (Figs. 4c, d, 6). In regions in which the photoreceptor cell layer had been greatly reduced in

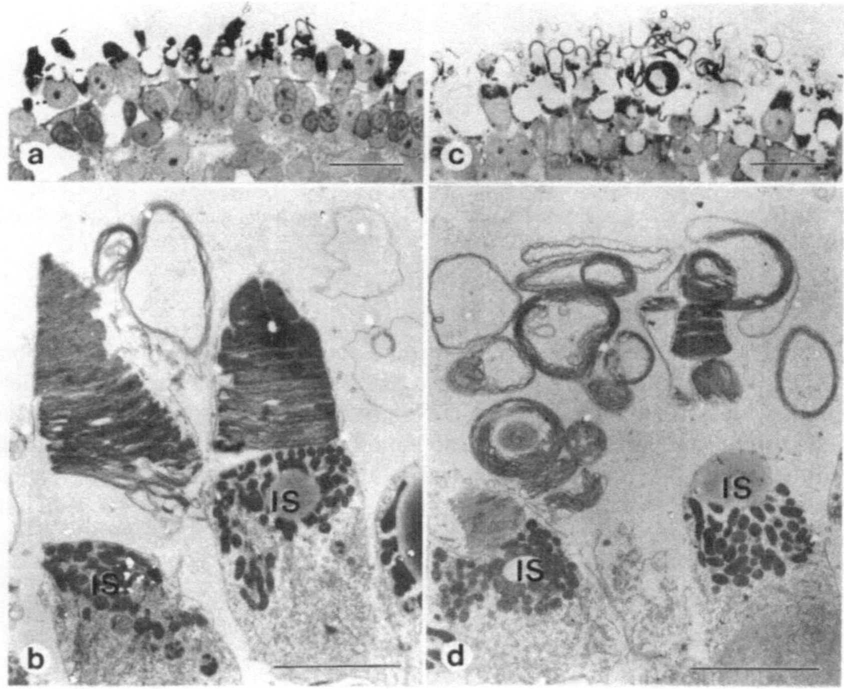

Figure 2. Paraffin section of 9 month old normal rat retina fed a vitamin E deficient diet for 35 days. Abnormal distribution of cathepsin D activity on the distal portion of photoreceptor outer segments is visualized as red reaction product (arrowheads). X950.

thickness (Fig. 7), the RPE cells showed dramatic changes in the form of extreme thinning, loss of cell organelles and partial disruption of the cell membranes. Numerous lipid-laden lysosomes were seen in the cytoplasm of the RPE cells (Fig. 8). In the far periphery of the retina, a focal loss of the inner and outer segments was present and their remnants were located in the subretinal space, next to the pigment epithelium (Fig. 7). In other areas,

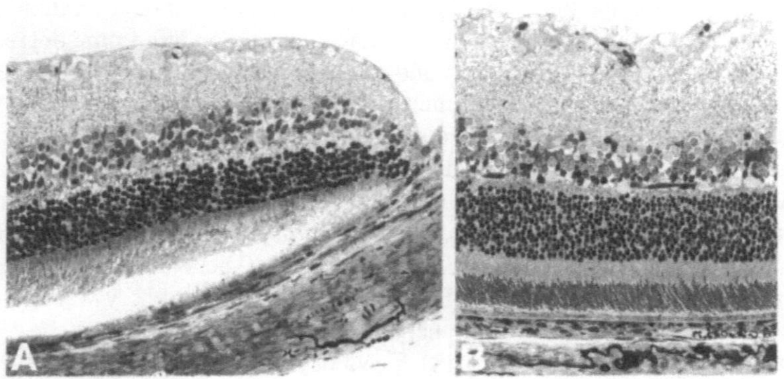

Figure 3. Light micrographs illustrating the periphery (A) and midperiphery (B) of the retina of a 9 month old rat fed a normal diet. A-B, X190.

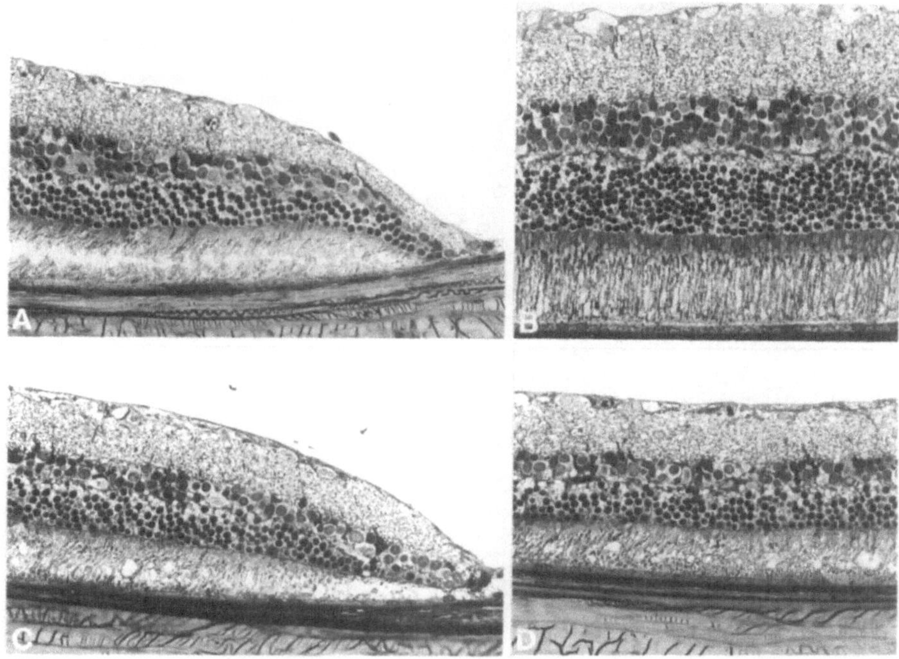

Figure 4. Micrographs showing the retina of rats fed a vitamin E deficient diet for 75 days (A, B) and 95 days (C, D) at the same magnification. With prolonged deficiency, a reduction in retinal thickness is evident due to loss of photoreceptor cells. (A, C periphery; B, D midperiphery) A-D, X320.

vacuolation and loss of cytoplasmic ground substance were evident in the perikarya and inner segments of the photoreceptor cells (Fig. 6). The outer segments were extremely short, their discs disorganized and vacuolated. Due to the loss of photoreceptor cell nuclei, the outer limiting membrane was interrupted (Fig. 6). Occasionally, free cells, which could be identified as resembling photoreceptor cells, were located next to the apical portion of RPE cells (Fig. 8). Signs of morphological alterations were found in the choriocapillaris and Bruch's membrane in the form of focal narrowing and condensation of osmiophilic material, as can be seen in Fig. 8. Degenerative changes were also seen in the innermost retinal layers.

DISCUSSION

Experimental vitamin E deficiency has elicited a multitude of pathologic entities in a variety of tissues. In ocular pathology, including human and animals, cataract and retinal degeneration has been well proven (Devi et al. 1965; Riis et al. 1981; Katz et al. 1982). The present study demonstrates that vitamin E deficiency induces degenerative changes in the retina and retinal pigment epithelium. These alterations become increasingly prominent with prolonged duration on the diet. Our findings are in accordance with those of several laboratories, in that experimentally induced vitamin E deficiency leads to the degeneration of photoreceptor outer segments and retinal pigment epithelium (Hayes 1974; Robison et al. 1979). It is worth noting that retina and rod outer segments have a greater percentage of their fatty acids as docosahexaenoic acid than most other tissues (Wiegand et al. 1986). The

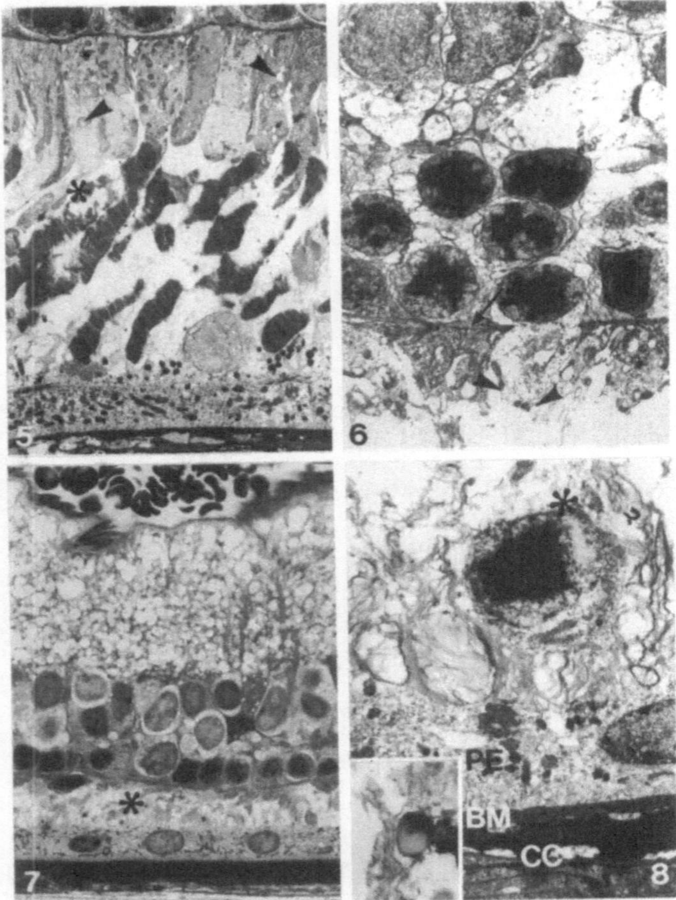

Figure 5. Electron micrograph illustrating the midperipheral retina of a rat fed a deficient diet for 75 days starting 9 months after birth. Note the disorganization and vacuolation of the photoreceptor outer segment discs (*) and the loss of cell organelles and distended mitochondria in the inner segments (arrowheads). X3,250.

Figure 6. Electron micrograph of the photoreceptor cells from a rat fed a vitamin E deficient diet for 95 days. The thickness of the outer nuclear layer has decreased to 2-3 nuclei. A loss of cell organelles and disruption of the inner segment cell membrane are clearly apparent (arrowheads). Arrows indicate the outer limiting membrane. X3,400.

Figure 7. Light micrograph from the periphery of the retina showing changes in photoreceptor cells and pigment epithelium as a result of vitamin E deficiency for 95 days. The photoreceptor cell layer has almost disappeared and consits only of 1 row. Structures reminiscent of inner and outer segments (*). X1,000.

Figure 8. Electron micrograph showing the pigment epithelium and choriocapillaris of a rat fed a vitamin E deficient diet for 95 days. Asterisk indicates a cell nucleus probably originating from a photoreceptor cell lying close to the pigment epithelium (PE). Signs of alterations are evident in the Bruch's membrane (BM) and choriocapillaris (CC). Inset: higher magnification of lipid-laden lysosomes in the apical portion of the pigment epithelium. X7,800., Inset: X17,000.

evidence to date, suggests that vitamin E, as an effective radical scavenger normally found in the retina (Dilley and McConnell 1970), plays a major role in protecting the highly unsaturated lipids of retinal membranes from free radical induced peroxidation (Robison et al. 1979, Wiegand et al. 1986). Katz et al. (1982) observed that albino rats deficient in selenium and vitamin E, showed a loss of photoreceptor cells, marked disruption of outer segment membranes and an increase in lipofuscin granules in the pigment epithelium. Lipofuscin accumulation in the rat RPE cells, as a result of vitamin E deficiency, has been well documented (Robison et al. 1979, 1980; Bieri et al. 1980). This accumulation is not confined to the RPE cells, but is also demonstrated in melanocytes and fibroblasts of the choroidal stroma in rats fed a vitamin E deficient diet (Herrmann et al. 1984). However, dietary vitamin A levels influence the extent of lipofuscin accumulation in the retinal pigment epithelium following vitamin E deficiency (Robison et al. 1982).

Instability of the lysosomal membrane has been demonstrated in different tissues under various pathological conditions, e.g. high concentration of retinol (Dewar et al. 1975), light exposure (Weissmann and Dingle 1961), vitamin E deficiency (Amemiya 1981). In the present study, however, the release of cathepsin D from the RPE lysosomes into the subretinal space occurs and increases with prolonged duration of the deficiency. The findings presented in this paper are in accordance with those of Amemiya (1981) in that lytic enzymes were found on the disc membrane of photoreceptor outer segments in vitamin E deficient rats. He concluded that the disc degeneration of photoreceptor outer segments might be due to the enhanced activity of lysosomal enzymes in the retinal pigment epithelium and the effect their release has on the photoreceptor outer segments. Our recent immunohistochemical studies have clearly shown that cathepsin D in normal human and rat retinas is restricted only to the RPE cells, no extracellular reaction product was found (El-Hifnawi 1995; El-Hifnawi et al. in press). In hereditary retinal dystrophy, cathepsin D reaction product was found extracellularly in the subretinal space and on the photoreceptor outer segments (El-Hifnawi et al. 1994). From these findings, we concluded that photoreceptor cell degeneration may be due to the release of lytic enzymes from the retinal pigment epithelium. In previous studies, we were able to demonstrate that the oral administration of acetyl salicylic acid had a positive effect on the survival rate of photoreceptor cells (El-Hifnawi and Kühnel 1989; El-Hifnawi et al. 1992). The survival of photoreceptors corresponds with the RPE, which was in markedly better condition in comparison to untreated animals. Similar results were obtained by administrating alpha-tocopherol (El-Hifnawi et al. 1993). Based on our earlier and present findings, it is plausible to conclude that lipid peroxidation may play a role in retinal blindness caused by nutritional and inherited factors. Further investigations are in progress to substantiate this assumption.

ACKNOWLEDGMENTS

This work was supported by the Deutsche Forschungsgemeinschaft (DFG) under grant Hi 434/3-3

REFERENCES

Amemiya, T., 1981, Photoreceptor outer segment and retinal pigment epithelium in vitamin E deficient rats. An electron microscopic histochemical study, *Graefes. Arch. Clin. Exp. Ophthalmol.* 216:103-109.

Anderson, R.E. and Andrews, L.D., 1982, Biochemistry of retinal photoreceptor membranes in vertebrates and invertebrates. In: Visual Cells in Evolution, J.A. Westfall, editor. Raven Press, New York 1-22.

Bieri, J.G., Tolliver, T.J., Robison, W.G., Jr. and Kuwabara, T., 1980, Lipofuscin in vitamin E deficiency and the possible role of retinol, *Lipids* 15:10-13.

Dalton, A.J., 1955, A chrome-osmium fixative for electron microscopy, *Anat. Rec.* 121:281.

Devi, A., Raina, P.L. and Singh, A., 1965, Abnormal protein and nucleic acid metabolism as a cause of cataract formation induced by nutritional deficiency in rabbits, *Br. J. Ophthalmol.* 49:271-275.

Dewar, A.J., Barron, G. and Reading, H.W., 1975, The effect of retinol and acetylsalicylic acid on the release of lysosomal enzymes from rat retina in vitro, *Exp.Eye Res.* 20:63-72.

Dilley, R.A. and McConnell, D.G., 1970, Alpha-tocopherol in the retinal outer segment of bovine eyes, *J. Membr. Biol.* 2:317-323.

El-Hifnawi, E. and Kühnel, W., 1987a, The role of lysosomes in hereditary retinal dystrophy in RCS rats. In: Advances in the Biosciences vol 62. Research in Retinitis pigmentosa. E. Zrenner, H. Krastel, and H.-H. Goebel, editors. Pergamon Journals Ltd. Great Britain 381-395.

El-Hifnawi, E. and Kühnel, W., 1987b, Lysosomal instability in the retinal pigment epithelium of RCS rats, *Invest. Ophthalmol. Vis. Sci.* 28:346 Abstract.

El-Hifnawi, E. and Kühnel, W., 1989, The effect of anti-inflammatory drug administration on the course of retinal dystrophy in RCS rats, *Prog. Clin. Biol. Res.* 314:343-355.

El-Hifnawi, E., Kühnel, W., Orün, C., Haug, H. and Laqua, H., 1992, Die Wirkung von Cyclooxygenasehemmern auf den Verlauf der hereditären Netzhautdystrophie bei RCS-Ratten, *Ann. Anat.* 174:251-258.

El-Hifnawi, E., Evers-Püschel, C. and El-Hifnawi, A., 1993, Effect of vitamin E on the survival of photoreceptor cells in RCS rats, *Invest. Ophthalmol. Vis. Sci.* 34:741 Abstract.

El-Hifnawi, E., Kühnel, W., El-Hifnawi, A. and Laqua, H., 1994, Localization of lysosomal enzymes in the retina and retinal pigment epithelium of RCS rats, *Ann. Anat.* 176:505-513.

El-Hifnawi, E., 1995, Localization of cathepsin D in rat ocular tissues. An immunohistochemical study, *Ann. Anat.* 177:11-17.

El-Hifnawi, E., BenEzra, D., Reichenbach, A. and Hettlich, H.-J., Distribution of cathepsin D in human ocular tissue: An immunohistochemical study, *Ann. Anat.* In press.

Hayasaka, S., 1974, Distribution of lysosomal enzymes in the bovine eye, *Jpn. J. Ophthalmol.* 18:233-239.

Hayes, K.C., 1974, Retinal degeneration in monkeys induced by deficiencies of vitamin E or A, *Invest. Ophthalmol.* 13:499-510.

Herrmann, R.K., Robison, W.G., Jr. and Bieri, J.G., 1984, Deficiencies of vitamins E and A in the rat: lipofuscin accumulation in the choroid, *Invest. Ophthalmol. Vis. Sci.* 25:429-433.

Katz, M.L., Parker, K.R., Handelman, J., Bramel, T.L. and Dratz, E.A., 1982, Effects of antioxidant nutrient deficiency on the retina and retinal pigment epithelium of albino rats: a light and electron microscopic study, *Exp. Eye Res.* 34:339-369.

LaVail, M.M., 1976, Rod outer segment disc shedding in relation to cyclic lighting, *Exp.Eye Res.* 23:277-280.

Lucy, J.A., 1978, Structural interactions between vitamin E and polyunsaturated phospholipids. In: Tocopherol, Oxygen and Biomembranes, C. de Duve and O. Hayaishi, editors. Elsevier/North-Holland Biomedical Press, Amsterdam 109-120.

McCay, P.B., Fong, K., Lai, E.K. and King, M.M., 1978, Possible role of vitamin E as a free radical scavenger and singlet oxygen quencher in biological systems which initiate radical-mediated reactions. In: Tocopherol, Oxygen and Biomembranes, C. de Duve and O. Hayaishi, editors. Elsevier/North-Holland Biomedical Press, Amsterdam 41-57.

Reynolds, E.S., 1963, The use of lead citrate at high pH as an electron-opaque stain in electron microscopy, *J. Cell Biol.* 17:208-212.

Richardson, K.C., Jarett, L. and Finke, E.H., 1960, Embedding in epoxy resins for ultrathin sectioning in electron microscopy, *Stain Technol.* 35:313-323.

Riis, R.C., Sheffy, B.E., Loew, E., Kern, T.J. and Smith, J.S., 1981, Vitamin E deficiency retinopathy in dogs, *Am. J. Vet. Res.* 42:74-86.

Robison, W.G., Jr., Kuwabara, T. and Bieri, J.G., 1979, Vitamin E deficiency and the retina: photoreceptor and pigment epithelial changes, *Invest. Ophthalmol. Vis. Sci.* 18:683-690.

Robison, W.G., Jr., Kuwabara, T. and Bieri, J.G., 1980, Deficiencies of vitamins E and A in the rat. Retinal damage and lipofuscin accumulation, *Invest. Ophthalmol. Vis. Sci.* 19:1030-1037.

Robison, W.G., Jr., Kuwabara, T. and Bieri, J.G., 1982, The roles of vitamin E and unsaturated fatty acids in the visual process, *Retina* 2:263-281.

Weissmann, G. and Dingle, J., 1961, Release of lysosomal protease by ultraviolet irradiation and inhibition by hydrocortisone, *Exp.Cell Res.* 25:207-210.

White, A., Handler, P., Smith, E.L., Hill, R.L. and Lehman, I.R., 1978, *Principles of Biochemistry*, 6th Ed., 1371.

Wiegand, R.D., Giusto, N.M., Rapp, L.M. and Anderson, R.E., 1983, Evidence for rod outer segment lipid peroxidation following constant illumination of the rat retina, *Invest. Ophthalmol. Vis. Sci.* 24:1433-1435.

Wiegand, R.D., Joel, C.D., Rapp, L.M., Nielsen, J.C., Maude, M.B. and Anderson, R.E., 1986, Polyunsaturated fatty acids and vitamin E in rat rod outer segments during light damage, *Invest. Ophthalmol. Vis. Sci.* 27:727-733.

14

RETINAL PIGMENT EPITHELIAL CELLS CULTURED FROM RCS RATS EXPRESS AN INCREASED MEMBRANE CONDUCTANCE FOR CALCIUM COMPARED TO NORMAL RATS

O. Strauß[1] and M. Wienrich[2]

[1] Institute for Clincal Physiology
University Clinic Benjamin-Franklin
Freie Universität Berlin
Hindenburgdamm 30, 12200 Berlin, Germany

[2] CNS Pharmacology
Boehringer Ingelheim,
Binger Straße, 55216 Ingelheim am Rhein, Germany

INTRODUCTION

The Royal College of Surgeon (RCS) rat is an animal model for Retinitis pigmentosa (1) with a still unknown genetical defect. In order to find a candidate gene which also serves as candidate for human retinitis pigmentosa, the functional defect in the retina of the RCS rat has been intensively studied. It has been shown, that the defect is located in the retinal pigment epithelium (RPE). Using cultured RPE cells or *in situ* preparations of the posterior eye (2-7), it has been demonstrated that the RPE of RCS rats is unable to ingest shed outer segments of the photoreceptors. The RPE is able to bind the outer segments of photoreceptors to the apical membrane but fails to incorporate them (2). Several lines of evidence show that in the RPE of RCS rats the second messenger system is disturbed which regulates the phagocytosis of photoreceptor outer membranes (5-8). Normal phagocytosis of photoreceptor outer membranes is regulated by calcium and the phosphoinositol second messenger system (9-11). Light stimulates the onset of phagocytosis of photoreceptor outer membranes which is associated with an increase of the phosphoinositol lipid concentration in the cytosol of the RPE (10). An increase of the cytosolic free calcium, which leads to activation of protein kinase C, represents the "off" signal for phagocytosis (9). Both mechanisms contribute to the fine regulation of phagocytosis of photoreceptor outer membranes in the RPE. The unspecific phagocytosis of for example latex microspheres is not regulated by this second messenger system (12). It has been reported for RCS rats that the second messenger system

Degenerative Diseases of the Retina, Edited by Robert E. Anderson et al.
Plenum Press, New York, 1995

of the RPE cells is disturbed. An abnormal phosphoinositol generation (8) and a reduced cAMP production has been reported in RCS RPE cells (6). As consequence to these findings the RPE in RCS rats expresses an altered growth factor responsiveness (7, 13) and protein phosphorylation (5). In addition, the regulation of photoreceptor membrane phagocytosis may also be misregulated.

Many functions of the RPE are related to the ion conductance of the RPE cell membrane. Potassium and chloride conductances of the RPE membrane are involved in epithelial transport and maintaining the ion homeostasis in the subretinal space (14). Using the patch-clamp technique (15) on cultured or freshly isolated RPE cells a variety of ion conductances of the cell membrane have been described. Inwardly and outwardly rectifying potassium channels are dominating the membrane conductances in all investigated species (16-21). In addition, volume-activated, calcium-dependent or cAMP-activated chloride channels were found in RPE cells (22-24). Voltage-dependent calcium conductances were observed in cultured and freshly isolated RPE cells from the rat (25, 26). Human and monkey RPE cells were also able to express voltage-dependent sodium channels (27, 28).

The aim of this study was to compare characteristics of the RPE cell membrane conductances from normal non-dystrophic and RCS rats. Cultured RPE cells have been shown to be a useful model to study electrical properties of the RPE (19-21, 23, 26). In addition, cultured RPE cells of the RCS rat display their functional defect also in cell culture (2-7). Therefore, we compared the ion conductance of the membrane of cultured RPE cells from non-dystrophic and RCS rats using the whole-cell configuration of the patch-clamp technique (15). We found an increased membrane conductance for calcium in RPE cells from RCS rats compared with non-dystrophic rats (29).

METHODS

RPE cells of both RCS and non-dystrophic rats were cultured using the method of Edwards (30). In short, eyes were incubated in Puck's saline F without Mg^{2+} and Ca^{2+} containing 0.1% trypsin for 30 minutes at 37°. After the anterior parts of the eye including the vitreous and the retina were removed, sheets of the RPE were gently brushed off using a fine pair of forceps. The sheets of the RPE were collected in Ham's F10 culture medium containing 20% FBS, 100µg/ml kanamycin and 50µg/ml gentamycin. After the cells had been dissolved into a cell suspension by gentle pipetting, the cell suspension was plated out into petri-dishes equipped with glass cover-slips and maintained at 37° C and 5% CO in air. After 24 hours the cells had settled down and started to spread out resulting in confluent monolayers of pigmented epitheloid-shaped cells after 9-16 days in culture.

Figure 1. Whole-cell currents in cultured cells of the rat retinal pigment epithelium. A. Pattern of electrical stimulation: the cells were electrically stimulated via the patch pipette using a series of voltage-steps positive and negative from the holding potential at 1s intervals. From the holding potential, which was adjusted to -45mV, nine voltage-steps of 50 msec duration and 10 mV increasing amplitude were performed to depolarize the cell. This was followed by nine voltage-steps of 50 msec duration and 10 mV increasing amplitude to hyperpolarize the cell. B. Top: Whole-cell currents in a cultured RPE cell of non-dystrophic rats: the electrical stimulation of a control cell by the stimulation protocol described above, led to voltage-dependent outward and inward currents (currents without capacitance compensation). Bottom: Current/voltage plot of the steady-state currents shown above. The steady-state current amplitudes were plotted against their step-potentials of the stimulation protocol. C. Top: Whole-cell currents in a cultured RPE cell from RCS rats: the electrical stimulation led voltage-dependent outward and inward currents (currents without capacitance compensation). Bottom: Current/voltage plot of the currents shown above. The steady-state currents were plotted against the step potentials of the stimulation protocol.

For patch-clamp recordings the cover-slips with cultured RPE were placed into a perfusion chamber mounted onto the stage of an inverted microscope. Membrane currents were measured using the conventional whole-cell configuration of the patch-clamp technique (15). During the whole-cell configuration the cells were superfused using a saline containing the salts of the Ham's F10 culture medium (mM): 0.3 $CaCl_2$, 3 KCl, 0.6 $MgCl_2$, 130 NaCl, 14 $NaHCO_3$, 1 Na_2HPO_4, 33 HEPES, 5.5 glucose; adjusted to pH = 7.2 with Tris.

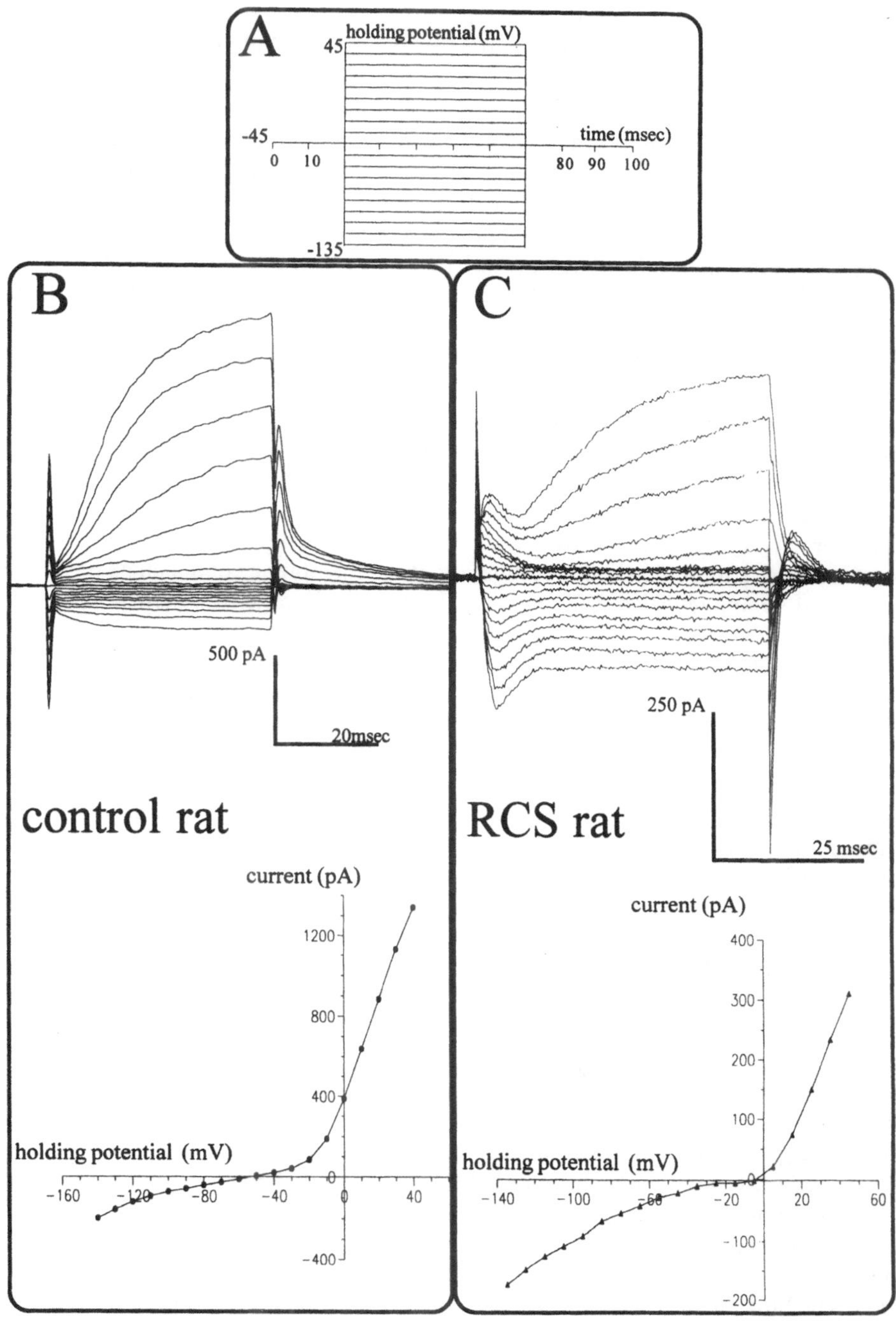

The patch-pipette contained a solution which mimicks the intracellular ion composition (mM): 0.5 $CaCl_2$, 5.5 EGTA/KOH, 100 KCl, 2 $MgSO_4$, 10 NaCl, 10 HEPES; adjusted to pH = 7.2 with Tris.

RESULTS

In order to compare an overview of voltage-dependent membrane currents present in cells from non-dystrophic and RCS rats, both types of cells were electrically stimulated using a series of voltage-steps positive and negative from the holding potential of -45 mV (fig. 1A). From the holding potential, nine voltage-steps of 50 msec duration and 10 mV increasing amplitude were performed to depolarize the cell. This was followed by nine voltage-steps of 50 msec duration and 10 mV increasing amplitude to hyperpolarize the cell. At a first glance, the membrane currents induced by this stimulation were not different in RCS and non-dystrophic rats (fig. 1B and 1C). Depolarization of cells from both RCS and normal rats led to voltage-dependent outward currents. These currents activated in a time-dependent manner and reached maximal current amplitudes after 50msec. Hyperpolarization led in 30% of the cells cultured from both RCS and non-dystrophic rats to an inward current which activated in a nearly time-independent manner. The inward and outward currents were reversibly blockable by blockers of voltage-dependent potassium channels (21). In addition, removing intra- and extracellular potassium eliminated the voltage-dependent currents in RPE cells from RCS and non-dystrophic rats (21). Thus, the membrane conductance in RPE cells from RCS and non-dystrophic rats seemed to be dominated by potassium conductances. An outwardly rectifying conductance displaying the characteristics of the delayed rectifier potassium channel and a second conductance expressing the characteristics of the inward rectifier potassium channel.

However, a more detailed investigation of the membrane conductance of RPE cells from both rat strains revealed differences in the electrical properties. With - 38.9 ± 9 mV (mean ± SD, n = 31) the cells of the RCS rats showed a significant ($p < 0.05$) more positive resting potential (figure 1 B, C) than cells from non-dystrophic rats (- 47 ± 14 mV; mean ± SD; n = 32; ref. 29). In addition, the activation threshold of the delayed rectifier (figure 1 B, C) of - 7.3 ± 2 mV (mean ± SEM; n = 13) was shifted in RPE cells from RCS rats to more positive values compared with cells from non-dystrophic rats (- 29 ± 2 mV, mean ± SEM; n = 11, ref. 29). This indicates the presence of an additional conductance in RPE cells from RCS rats. This additional conductance must be a conductance for an ion with a Nernst potential more positive than that for potassium and shifted the activation threshold of the delayed rectifier currents to a more positive value.

Therefore, we investigated more closely the outwardly directed delayed rectifier currents. For this purpose we used another stimulation protocol which depolarized the cell from a holding potential of -45mV with nine voltage-steps of 50 msec duration and 10 mV increasing amplitude (fig. 2a). In cells from both rat strains the currents of the delayed rectifier were reversibly blocked by the potassium channel blocker barium(fig. 2b). In contrast to cells from non-dystrophic rats, we observed in RPE cells from RCS rats in the presence of 10mM barium not only a reduction of the outward current but an additional inward current (fig. 2c). Thus, the additional membrane conductance in RPE cells from RCS rats led to large inward currents in the presence of barium. Since it is known that barium passes through L-type calcium channels better than calcium itself, we tested the hypothesis that the additional inwardly directed membrane conductance was a calcium conductance. The additional currents in the presence of barium in cells from RCS rats were reversibly blocked by the calcium channel blockers verapamil (30μM, ref. 29) and diltiazem (30μM, ref. 29). The blocker for L-type calcium channels nifedipine (1μM) reversibly reduced the inward current in the presence of barium to 12 ± 2 % of control (mean ± SEM, n = 3); recovery to 61 ± 6% of control (mean ± SEM; n = 3). Thus, the additional membrane conductance in RPE cells from RCS rats showed characteristics of L-type calcium channels.

DISCUSSION

In this study we described functional differences between retinal pigment epithelial cells from RCS and non-dystrophic rats (29).

The membrane conductance of RPE cells is dominated by potassium conductances. This has been shown in various species using cultured or freshly isolated cells and by investigating the membrane conductance using the patch-clamp technique or in micropuncture studies (16-21). In patch-clamp studies two main potassium conductances have been reported: one outwardly directed conductance with characteristics of the delayed rectifier and a second inwardly directed conductance with the characteristics of the inward rectifier potassium channel (17-21). In our hands, cultured RPE cells from the RCS rat and from non-dystrophic rats displayed the same two potassium conductances. Thus, the membrane conductance of RPE cells from RCS rats seemed not to be different compared with cells from non-dystrophic rats or with cells from other species including man (16-21).

However, a closer investigation of the outwardly rectifying potassium channel revealed differences between the RPE cells from RCS and non-dystrophic rats. In the presence of barium an additional inward current was observed in RPE cells from RCS rats. This inward current was blocked by the Ca^{2+}-channel blockers verapamil, diltiazem (29) and nifedipine which indicates this additional inward current to be a barium current through L-type calcium channels. Thus, the RPE cells from RCS rats express an increased membrane conductance for calcium compared to normal rats. In recent studies it has been shown that freshly isolated (24) or cultured RPE cells from non-dystrophic rats also express L-type calcium channels (25, 26). However, these studies were performed under conditions where potassium currents were totally blocked. In our investigation we could detect barium currents through L-type channels in the presence of potassium currents in cells from RCS rats, whereas these barium currents were never observed under these conditions in RPE cells from non-dystrophic rats. In cultured cells from non-dystrophic rats the calcium current amplitudes were too small to be observed in the presence of potassium currents under control conditions. In addition, the increased membrane conductance for calcium in cells from RCS rats led to differences in the properties of the resting cell. RPE cells from RCS rats showed a significant more positive resting potential and the activation threshold of the outwardly rectifying potassium conductance was shifted to more positive values.

The increased calcium conductance in RPE cells from RCS rats possibly has two consequences for the interaction of photoreceptors and the RPE. First, the role of potassium conductances in the RPE may be affected by the increased calcium conductance. One role of potassium conductances in the RPE is to determine the resting potential of the cell (15). It is known from studies using the electroretinogram that the onset or offset of light leads to hyper- or depolarizations of the RPE (31). This is normally compensated for by the potassium conductance of the RPE cell membrane. As we could show the activation threshold of the outwardly rectifying potassium conductance was shifted to more positive values in the RPE of RCS rats. In non-dystrophic rats, the RPE depolarized by 10-15 mV in response to the removal of light from the eye (31) which in turn activates the outward rectifier potassium channels for compensation. In the RPE of RCS rats the outward rectifier cannot by activated by removal of light from the eye because the activation threshold is shifted to a more positive value. In addition, the depolarization by 10-15 mV would lead to an influx of calcium into the cell through L-type calcium channels in cells from RCS rats which would not occur in

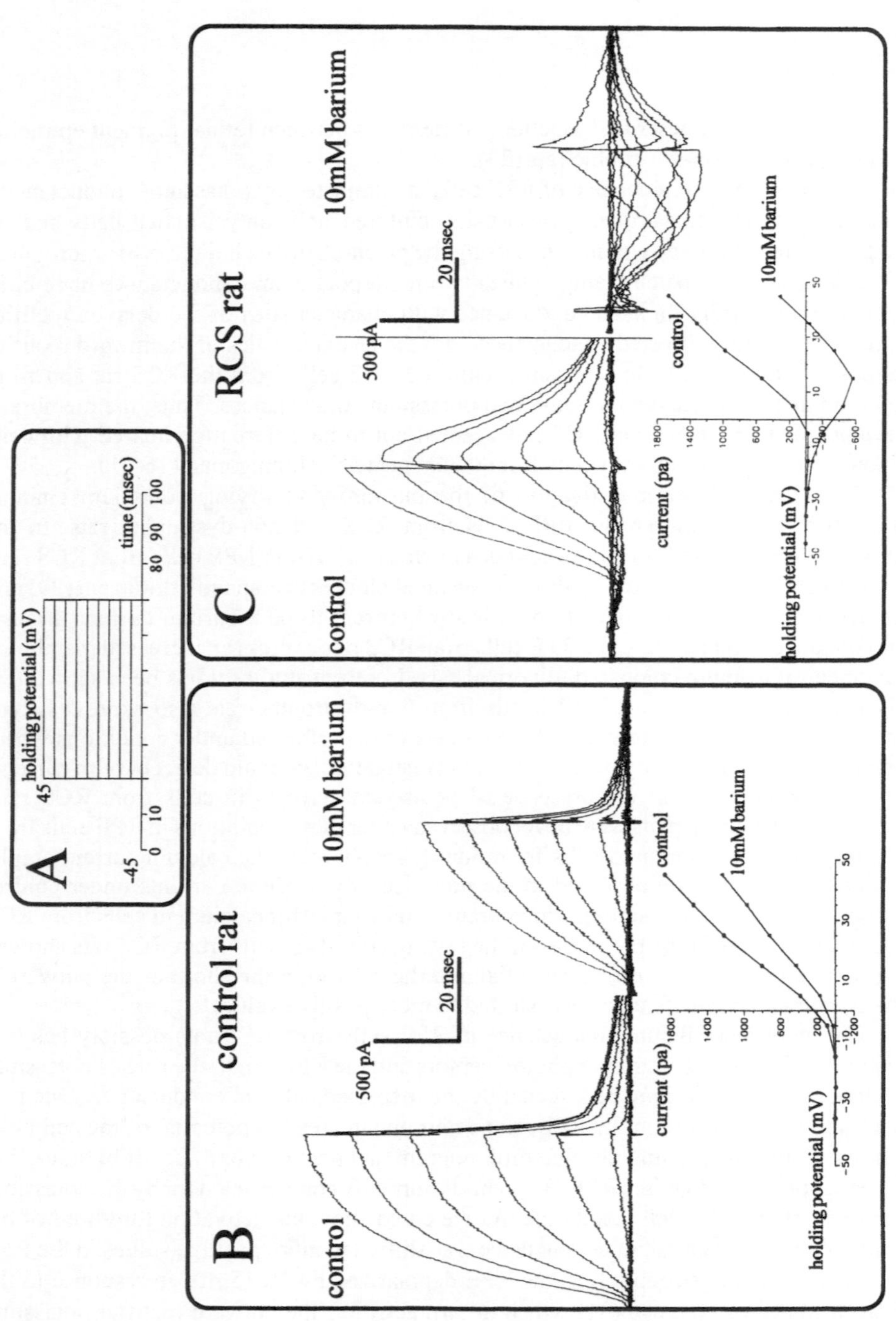
A
holding potential (mV)
time (msec)
B
control rat
control
10mM barium
500 pA
20 msec
current (pa)
holding potential (mV)
C
RCS rat
control
10mM barium
500 pA
20 msec
current (pa)
holding potential (mV)

Figure 2. Effect of barium on the voltage-dependent outward currents in RPE cells from RCS and non-dystrophic rats. A. Pattern of electrical stimulation to induce voltage-dependent outward currents: From a holding potential of -45 mV the cells were depolarized by nine voltage-steps of 50 msec duration and 10mV increasing amplitude at 1s intervalls. B. Effect of barium on the voltage-dependent outward currents in RPE cells from non-dystrophic rats. Top: Application of 10 mM barium in the bath solution led to a reduction (right panel) of the control outward currents (left panel) induced by the electrical stimulation. Bottom: Current/voltage plot summarizing the experiment shown above. The maximal current amplitudes were plotted against their step-potentials of the stimulation protocol. Barium led to a decrease (triangles) of the control currents (circles). C. Effect of barium on the voltage-dependent outward currents in RPE cells from RCS rats. Top: Application of 10 mM barium to the bath solution (right panel) reduced the voltage-dependent outward currents induced by the stimulation protocol. In contrast to cells from non-dystrophic rats, an additional voltage-dependent inward current was observed in the presence of barium in RPE cells from RCS rats. Bottom: Current/voltage plot summarizing the experiment shown above. The maximal current amplitudes were plotted against their step-potentials of the stimulation protocol. In the presence of barium (triangles) not only the current amplitudes of the outward currents (circles) were reduced, but an additional inward current was observed in the presence of barium in RPE cells from RCS rats.

RPE cells from non-dystrophic rats. Thus, light causes different responses in the RPE cells from RCS rats compared to non-dystrophic rats which might affect their functional integrity. Second, the increased calcium conductances in RPE cells from RCS rats may change the second messenger system which regulates the phagocytosis in these cells. Using cultured RPE cells it has been shown that the RPE in RCS rats is unable to phagocytose shed photoreceptor discs (2-7). These RPE cells are able to bind the disc membranes but are unable to ingest them (2). An increase of the cytosolic free calcium and activation of protein kinase C represent the off signal for phagocytosis in RPE cells (9). Here we report that the RPE cells from RCS rats express an increased membrane conductance for calcium which might lead to an influx of calcium into the cell when the RPE cells are depolarized with every extinction of light. Thus, every extinction of light could generate an off signal for phagocytosis in the RPE of RCS rats. In addition, the regulation of calcium stimulating agonists in the RPE of RCS rats could be altered. The phagocytosis by the RPE is regulated by the calcium/inositol-phosphate second messenger system (9, 10). Several studies suggest that there is a disturbed second messenger system in the RPE of RCS rats, which may lead to a disturbed regulation of phagocytosis (6, 7, 8, 13). The increased calcium conductance in RCS RPE cells could result in a changed increase of the cytosolic free calcium in response to agonists that activate the calcium second messenger system. Thus, regulation of phagocytosis by calcium stimulating agonists in the RPE of RCS rats could be altered.

In summary, we could demonstrate that cultured cells from RCS rats express an increased membrane conductance for calcium compared to cells from non-dystrophic rats. This observation can be related to the functional defect of the RPE in RCS rats which leads to the retinal degeneration in this animal model. This observation may help to find a candidate gene to elucidate the genetical defect in RCS rats which may also serve as candidate gene for human Retinitis pigmentosa.

ACKNOWLEDGMENTS

This work was supported by the DFG grant Wi/992 and the German Retinitis pigmentosa Society (DRPV). The authors thank Prof. M. Wiederholt, PD Dr. H. Kettenmann and Dr. T. Weiser for helpful discussions. The expert technical assistance of H. Morhardt and E. Fietkau is greatfully acknowledged.

REFERENCES

1. Merin, S. and Auerbach, E., 1976, Retinitis pigmentosa, *Surv. Ophthalmol* 20: 303-346.
2. Chaitin, M.H. and Hall, M.O., 1983, Defective ingestion of rod outer segments by cultured dystrophic rat pigment epithelial cells, *Invest.Ophthalmol.Vis.Sci.* 24:812-820.
3. Gregory, C.Y. and Hall, M.O., 1992, The phagocytosis of ROS by RPE cells is inhibited by an antiserum to RPE plasma membranes, *Exp.Eye Res.* 54: 843-851.
4. Hall, M.O. and Abrams, T.A., 1991, RPE cells from normal rats do not secrete a factor which enhances the phagocytosis of ROS by dystrophic rat RPE cells, *Exp.Eye Res.* 52: 461-464.
5. Heth, C.A. and Schmidt, S.Y, 1992, Protein phosphorylation in retinal pigment epithelium of Long-Evans and Royal College of Surgeon rats, *Invest.Opthalmol.Vis.Sci.* 33: 2839-2847.
6. Heth, C.A. and Marescalchi, P.A., 1992, Generation of inositol triphosphate during rod outer segment phagocytosis is abnormal in cultured RCS retinal pigment epithelium, *Invest.Ophtalmol.Vis.Sci.* 33: 1203.
7. McLaren, M.J., Holderby, M., Brown, M.E. and Inana, G., 1992, Kinetics of ROS binding and ingestion by cultured RCS rat RPE cells: modulation by conditioned media and bFGF, *Invest.Ophthalmol.Vis.Sci.* 33: 1207.
8. Gregory, C.Y., Abrams, T.A. and Hall, M.O., 1992, cAMP production via the adenyl-cyclase pathway is reduced in RCS rat RPE, *Invest.Ophthalmol.Vis.Sci.* 33: 3121-3124.

9. Hall, M.O., Abrams, T.A. and Mittag, T.W., 1991, ROS ingestion by RPE cells is turned off by increased protein kinase C activity and increased calcium, *Exp.Eye Res.* 52: 591-598.
10. Rodriguez de Turco, E.B., Gordon, W.C. and Bazan, N.G., 1992, Light stimulates in vivo inositol lipid turnover in frog retinal pigment epithelial cells at the onset of shedding and phagocytosis of photoreceptor membranes. *Exp.Eye Res.* 55: 719-725.
11. Salceda, R., 1992, Zymosan induced $^{45}Ca^{2+}$ uptake by retinal pigment epithelial cells. *Curr.Eye Res.* 11: 195-201.
12. Heth, C.A. and Marescalchi, P.A., 1994, Inositol triphosphate generation in cultured rat retinal pigment epithelium, *Invest.Ophthalmol.Vis.Sci.* 35: 409-416.
13. Leschey, K.H., Hackett, S.F., Singer, J.H. and Campochiaro, P.A., 1990, Growth factor responsiveness of human retinal pigment epithelial cells, *Invest.Ophthalmol.Vis.Sci.* 31: 839-846.
14. Steinberg, R.H., 1985, Interactions between the retinal pigment epithelium and the neural retina. *Doc.Opthalmol 60*: 327-346.
15. Hamill, O.P., Marty, A., Neher, E., Sakmann, B. and Sigworth, F.J., 1981, Improved patch-clamp techniques for high-resolution current recording from cells and cell-free patches, *Pflüger's Arch.* 391: 85-100.
16. Hughes, B.A. and Steinberg, R.H., 1990, Voltage-dependent currents in isolated cells from the frog retinal pigment epithelium, *J.Physiol. (Lond)* 428: 273-297.
17. Wen, R., Lui, G.M. and Steinberg, R.H., 1993, Whole-cell currents in fresh and cultured cells of the human and monkey retinal pigment epithelium, *J.Physiol. (Lond)* 465: 121-147.
18. Fox, J.A. and Steinberg, R.H., 1992, Voltage-dependent currents in isolated cells of the turtle retinal pigment epithelium, *Pflüger's Arch.* 402: 451-460.
19. Tao, Q., Rafuse, P.E. and Kelly, M.E.M., 1994, Potassium currents in cultured rabbit retinal pigment epithelial cells, *J.Membrane Biol.* 141: 123-138.
20. Strauss, O., Richard, G. and Wienrich, M., 1993, Voltage-dependent potassium currents in cultured human retinal pigment epithelial cells. *Biochem.Biophys.Res.Commun.* 191: 775-781.
21. Strauss, O., Weiser, T. and Wienrich, M., 1994, Potassium currents in cultured cells of the rat retinal pigment epithelium, *Comp.Biochem.Physiol. 109A*: 975-983.
22. Botchkin, L.M. and Matthews, G., 1993, Chloride current activated by swelling in retinal pigment epithelium cells, *Am.J.Physiol.* 265: C1037-C1045.
23. Ueda, Y. and Steinberg, R.H., 1994, Chloride currents in freshly isolated rat retinal pigment epithelial cells, Exp.Eye Res. 58: 331-342.
24. Hughes, B.A. and Segawa, Y., 1993, c-AMP activated chloride currents in amphibian retinal pigment epithelial cells, *J.Physiol. (Lond)* 466: 749-766.
25. Ueda, Y. and Steinberg, R.H., 1992, Voltage-operated calcium channels in fresh and cultured rat retinal pigment epithelial cells, *Invest.Ophthalmol.Vis.Sci.* 34: 3408-3418.
26. Strauss, O. and Wienrich, M., 1994, Ca^{2+}-conductances in cultured rat retinal pigment epithelial cells, *J.Cell.Physiol.* 160: 89-96.
27. Botchkin, L.M. and Matthews, G., 1994, Voltage-dependent sodium channels develop in rat retinal pigment epithelium in culture, Pro.Natl.Acad.Sci.USA 91: 4564-4568.
28. Wen, R., Lui, G.M. and Steinberg, R.H., 1994, Expression of a tetrodotoxin-sensitive Na^{+}-current in cultured human retinal pigment epithelial cells, *J.Physiol. (Lond)* 476: 187-196.
29. Strauss, O. and Wienrich, M., 1993, Cultured retinal pigment epithelial cells from RCS rats express an increased calcium conductance compared with cells from non-dystrophic rats, *Pflüger's Arch.* 425: 68-76.
30. Edwards, R.B., 1977, Culture of rat retinal pigment epithelium, *In Vitro* 13: 301-304.
31. Griff, E.R., 1991, Potassium-evoked responses from the retinal pigment epithelium of the toad Bufo marinus, *Exp.Eye Res.* 53: 219-228.

15

EFFECT OF SUGARS ON PHOTORECEPTOR OUTER SEGMENT ASSEMBLY

Monica M. Stiemke* and Joe G. Hollyfield

Cullen Eye Institute
Baylor College of Medicine
One Baylor Plaza
Houston, Texas 77030

INTRODUCTION

During embryonic development, the rudimentary layers that will become the pigment epithelium (PE) and the neural retina are brought into close proximity upon the collapse of the optic vesicles. At this developmental stage, the retina is morphologically undifferentiated and photoreceptor outer segments have not yet begun to form. The close apposition of the PE and photoreceptors prior to the time when outer segments first appear raises the possibility that the PE could be a source of signals that induce or regulate photoreceptor development and outer segment elaboration. Also, the observation that outer segment development is limited in the absence of the PE in most species suggests that interactions between these two cell types may be of fundamental importance for the structural and functional maturation of photoreceptors [1, 2, 3, 5, 13, 16, 19, 20, 26].

The effects of the PE on photoreceptor survival has been known for several years. Retinal detachment studies indicate that the cone and rod outer segments degenerate very quickly after retinal detachment and the degree of recovery of cell morphology and function are negatively correlated with the duration of the detachment [4, 7, 11, 12, 17]. In the Royal College of Surgeons (RCS) rat, it has been shown that a defect in the PE results in photoreceptor degeneration unless growth factors [8] or subretinal fluid [23] are injected into the subretinal space or a PE transplant is performed [18]. In all cases, the rescue extends beyond the limits of the transplant suggesting that a diffusible factor from the PE is responsible for the rescue. While a great deal of data indicate that the PE or "factors" supplied by the PE are of paramount importance to the development and/or survival of photoreceptors, the nature of these interactions is poorly understood.

Recent studies [21, 22] utilizing *Xenopus laevis* retinal rudiments have demonstrated that intact undifferentiated retinas when placed in culture in the presence of an apposing PE

*Present address for Monica M. Stiemke is The University of Texas Health Science Center at Houston, Department of Ophthalmology, Houston, Texas 77030.

Degenerative Diseases of the Retina, Edited by Robert E. Anderson et al.
Plenum Press, New York, 1995

are able to elaborate organized outer segment membranes forming stacked saccules and the amount of opsin is equivalent to that present at approximately stage 41 of in vivo development after 3 days of culture. In the absence of a closely apposed PE, extensive amounts of disorganized outer segment membrane are synthesized as whorl-like structures, forming a mat-like array at the outer retinal surface. The amount of opsin produced is approximately one-half of that present when the retinas are cultured in the presence of the PE. This model of retinal development allows for exploration of the interactions between the PE and the photoreceptor cells that are possibly involved in maturation and differentiation of the photoreceptor cells, as well as outer segment membrane organization. This system can be exploited to determine what factors are able to alter the organization of the membranes and the synthesis of the proteins therein.

HYPOTHESIS

What is lacking from the cultures maintained in the absence of the PE that does not allow the proper organization of the photoreceptor outer segment membranes? The hypothesis tested in this study is that a specific carbohydrate-lectin interaction is necessary for the correct assembly of outer segment membranes. In the absence of the PE, the proper carbohydrate is missing because it was synthesized by the PE or the sugar is not able to exert its organizational effect because of dilution into the culture medium. Addition of the proper saccharide to the culture media may allow proper formation of the outer segment membranes. To test this hypothesis, we added a series of exogenous carbohydrates to the culture medium and assayed their effect on the organization of nascent outer segment membranes in retinas that had the PE removed prior to in vitro development. Retinas were removed from stage 33/34 embryos, when the first outer segments can be detected on occasional photoreceptors [22]. Therefore, the vast majority of the outer segments are formed in the absence of the PE, yet in the presence of an exogneous carbohydrate. In the presence of some sugars, outer segment membranes were able to form stacked flattened discs that were aligned with individual inner segments. These are termed permissive sugars. Those that had no effect on the formation of the outer segment structure are termed non-permissive. Sugars were tested at concentrations from 50 mM to 5 μM. Control conditions included retinas that matured with and without an adherent PE in the absence of any added carbohydrate.

Light microscopy was used to monitor the assembly of outer segment discs of control retinas, as well as retinas allowed to develop with exogenous carbohydrate. Tissues were fixed and processed following standard procedures [21]. Criteria used to score the morphologic appearance of the outer segment membranes were: organization of outer segment membranes (flattened, stacked, membranous leaflets which could be localized to individual photoreceptor cell bodies) and consistency of appearance between retinal rudiments incubated in any particular sugar.

OBSERVATIONS

Nascent photoreceptor outer segments produced in the presence of a closely apposed PE elaborated highly organized membranes composed of stacked saccules sclerad to individual photoreceptor inner segments (Figures 1a and 1b). A previous study has shown that most of the membranes immunolabel with an anti-opsin antibody that is specific for the principal rod outer segment in *Xenopus laevis* [22]. Under control conditions in the absence of exogenous sugars and the PE, outer segment membranes are produced, however they are

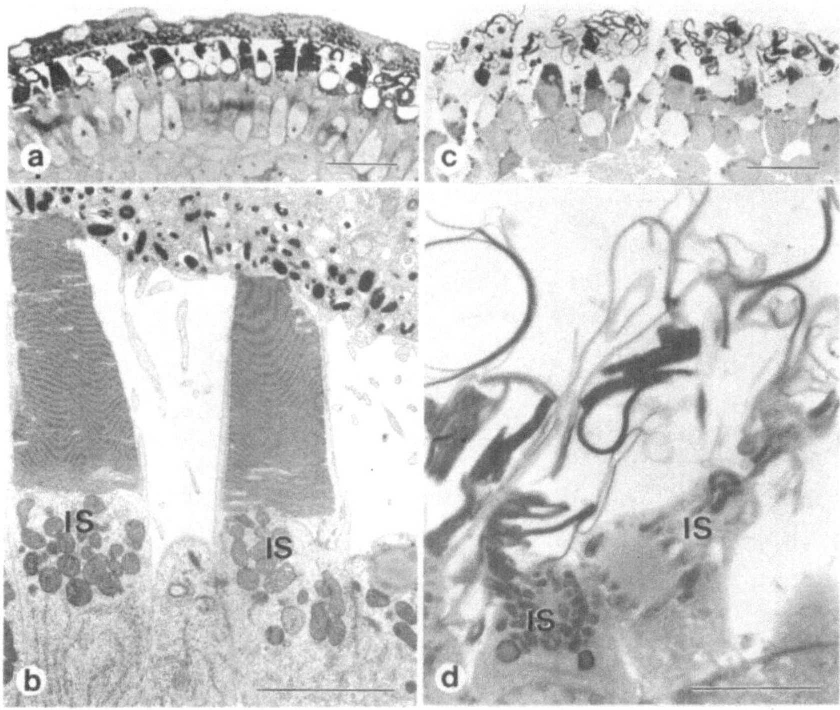

Figure 1. Light and electron micrographs of retinal rudiments removed from stage 33/34 *Xenopus* embryos and placed into culture for 3 days under cyclic lighting conditions. a) and b) are control retinas cultured in the presence of the PE. Note the highly organized outer segment membranes composed of flattened stacked saccules localized to individual inner segments. c) and d) are control retinas cultured in the absence of the PE. Note the highly disorganized outer segments that appear to be membranous whorls seemingly not localized to any particular inner segment. IS = inner segment. Scale bars = 1 μm in a and c and 0.5 μm in b and d. (From Stiemke *et al.*, *Dev. Brain Res.* 80:285-289.)

disorganized and are present as an expansive membranous mat that is neither disc-like nor associated with any particular photoreceptor cell(s) (Figures 1c and 1d).

In the absence of the PE, N-acetyllactosamine, N-acetylgalactosamine, lactose, galactose, 3′-N-acetylneuramin-lactose, 6′-N-acetylneuramin-lactose and lactulose when added to the culture media had a significant effect on both rod and cone outer segment organization compared to control retinas maintained without these sugars (Table 1). Outer segments developing in the presence of each of these sugars are composed of arrays of highly ordered, flattened membrane saccules that are localized proximal to individual photoreceptor inner segments (Figure 2a and 2b). Most outer segments of rods are elongated and cylindrically shaped. Cone outer segments are tapered and shorter than their rod counterparts (Figure 2a). The appearance of both rod and cone outer segments are very similar to those seen in Figures 1a and 1b in which the retinas were cultured with an apposing PE. With increasing dilution, the organizational effect of the permissive sugars appeared to be reduced, with the exception of galactose in which retinas had more organized outer segments in lower sugar concentrations (Table 1).

Melibiose, glucose, fucose, N-acetylglucosamine, methyl mannopyranoside, mannose, mannan, maltose and chitobiose were non-permissive sugars. At all concentrations

Table 1. Outer segment organization in the presence of exogenous sugars

Sugar	Concentration (mM)	Improvement in outer segment organization relative to control w/o PE
N-acetyllactosamine	5	+ + +
	0.5	+ + +
	0.05	+
	0.005	0
N-acetylgalactosamine	5	+ +
	0.5	+ + +
	0.05	+ +
	0.005	0
Lactose	50	+ +
	20	+ +
	5	+ + +
	0.5	+ +
	0.05	+ +
	0.005	0
Galactose	50	–
	20	–
	5	+ +
	0.5	+ +
	0.05	+ +
	0.005	+ +
3′-N-acetylneuramin-lactose	5	NA
	0.5	+ +
	0.05	+ +
	0.005	0
6′-N-acetylneuramin-lactose	5	NA
	0.5	+ +
	0.05	+ +
	0.005	0
Lactulose	5	+ +
	0.5	+ +
	0.05	+ +
	0.005	+
Melibiose	5	–
	0.5	+
	0.05	+
	0.005	+
Glucose	5	+
	0.5	+
	0.05	+
Fucose	50	0
	20	0
	5	+
N-acetylglucosamine	50	0
	20	0
	5	0
	0.5	+
	0.05	0

Table 1. (*Continued*)

Sugar	Concentration (mM)	Improvement in outer segment organization relative to control w/o PE
Methyl mannopyranoside	50	0
	20	0
	5	0
Mannose	50	–
	20	0
	5	0
Mannan	50	0
	20	0
	5	0
Maltose	5	0
	0.5	0
	0.05	0
Chitobiose	50	–

Grading Scale (Improvement in outer segment organization relative to control retinas allowed to differentiate without a PE)

+ + + best organization of outer segment membrane discs; very similar to retinas allowed to differentiate in the presence of the PE (See Fig. 2a and 2b)

+ + very good organization; much more disc stacking than control retinas without a PE, yet somewhat less than controls with a PE

+ slightly more organization of outer segment discs than control retinas maintained without a PE; some evidence of disc stacking with substantial amounts of whorled membrane

0 same level of organization as control retinas maintained without a PE; whorls of outer segment disc membranes with little evidence of organized stacking (See Fig. 2c and 2d)

– less organization than control retinas maintained without a PE

NA data not available.

From Stiemke *et al.*, *Dev. Brain Res.* 80:285-289.

tested, these non-permissive sugars had little to no effect on rod outer segment organization (Table 1). Illustrated in Figures 2c and 2d are retinas from cultured with 0.5 mM maltose and 5 mM methyl mannopyranoside, respectively. The outer segments have been elaborated as a membranous mat-like array, virtually identical to those which develop in the absence of any added sugar (compare to Figures 1c and 1d).

POSSIBLE MECHANISMS

It is unlikely that this improvement in outer segment disc assembly in the embryonic retina is due to an osmotic effect since lactose at concentrations as low as 0.05 mM was able to sustain this effect, while other sugars when provided at the same or higher concentrations failed to produce a similar effect. Lactose (4-O-β-D-galactopyranosyl-β-D-glucose) is a disaccharide containing galactose and glucose. Since glucose did not provide a significant permissive organizational effect at any concentration and galactose supported outer segment

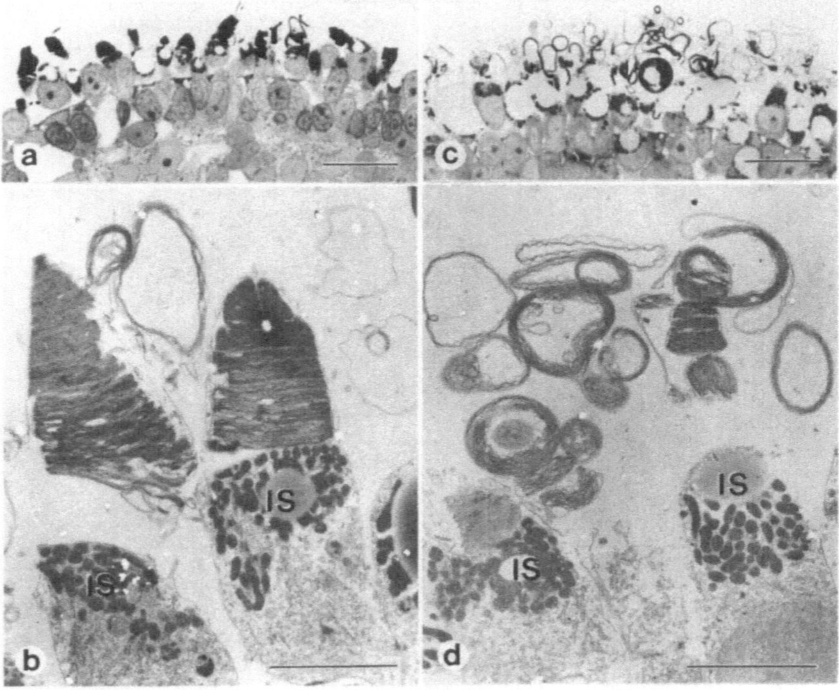

Figure 2. Light and electron micrographs of retinal rudiments from stage 33/34 *Xenopus* embryos that had been maintained in vitro for 3 days under cyclic lighting conditions: a) retina cultured with 0.5 mM N-acetyllactosamine b) retina cultured with 5 mM lactose c) retina cultured with 0.5 mM exogenous maltose, and d) retina cultured with 5 mM methyl mannopyranoside. Retinas allowed to develop in the presence of N-acetyllactosamine and lactose (a and b, respectively) elaborate outer segment membranes that are composed of highly ordered, flattened membrane saccules that were localized to particular photoreceptor cell bodies. While those cultured in maltose and methyl mannopyranoside (c and d, respectively) produced membranes that are disorganized and are similar in appearance to control retinas that differentiated without a PE. IS = inner segment. Scale bars = 1 μm in a and c and 0.5 μm in b and d.(From Stiemke *et al.*, *Dev. Brain Res.* 80:285-289.)

organization at even the lowest concentration tested, it is unlikely that the individual monosaccharides which comprise lactose function by serving as an energy source.

Data from several labs, in addition to our own, have indicated that carbohydrates are necessary for outer segment formation and organization. The precise mechanism by which the carbohydrates are functioning is not known, but it could be through direct interactions of sugars and lectins on outer segment membranes or by interactions with the extracellular matrix surrounding the membranes. Previous studies have shown that tunicamycin, an antibiotic that prevents the formation of N-linked oligosaccharides during protein glycosylation via the lipid intermediate pathway, significantly alters membrane morphogenesis in adult *Xenopus* retinas. Rather than a compact array of stacked outer segment discs being formed, the newly synthesized membranes form tubulo-vesicular profiles that appear to contain opsin [6, 9, 25]. These data suggest that the lack of sugar moieties on glycoproteins within the retina may be responsible for the disorganization of outer segment membrane saccules. If post-translational trimming of oligosaccharides is inhibited with castanospermine, nascent disc morphology is identical to control conditions, suggesting that post-translational removal of oligosaccharides is not essential for normal disc morphogenesis [10].

Data from this laboratory indicates that specific mono- and disaccharides when added to the culture media in which retinas without an adherent PE are maintained are able to permit organization of outer segment membranous discs into a compact and organized array.

The close apposition of stacked membrane leaflets in an organized outer segment suggests that some mechanism for the stabilization of large expanses of sheet-like membrane domains exists because organized membranes with folds at the disc periphery are at a much higher energy state than disorganized vesiculated membranes [24]. Lactose, galactose and substituted forms of these sugars in some way were able to permit assembly of membranes into this high energy state. These observations suggest that an endogenous β-galactoside specific lectin or other lectin-like molecule is present in the developing Xenopus retina and that an endogenous carbohydrate ligand may be necessary for the initial stages of normal photoreceptor outer segment formation.

While there is precedence for the existence of carbohydrate binding sites in the retina and IPM of the *Xenopus*, the mechanism by which exogenous permissive sugars are acting in the culture system described here is unknown. Several possibilities may be operative:

1. the sugars may form bridges between lectins localized to the developing outer segment membranes, thus permitting apposing membranes to come into proper register and organize.

   While it is intellectually appealing to suggest this first hypothesis, there is no established or known mechanism by which this could occur, since only mono-valent saccharides were used.
2. the sugars may expose an essential reactive site by binding lectins that were inhibiting cross-linking of apposing membranes by interacting with the reactive site.

   Rather than bridging lectins on apposing membranes, the exogenous sugars are more likely to inhibit binding or cross-linking. This could be possible by the sugars binding to a lectin and exposing a reactive site that had previously been masked, thus exposing essential reactive sites that are necessary for proper membrane organization.
3. the sugars may bind to an outer segment membrane lectin(s) and induce a conformational change of the protein(s) that may be required for proper stability of the newly formed membranes via interactions with the outer segment cytoskeleton.

   The exogenously added sugar may bind to a membrane bound β-galactoside-specific receptor and induce a conformational change within the receptor that is transmitted to the cytoskeletal architecture or some other structural components in the outer segment. In the event that the binding site of the receptor is vacant, normal assembly of the outer segment into compact, flattened, stacked discs is prevented because the proper conformation of structural proteins has not been achieved.
4. the sugars may organize the developing interphotoreceptor matrix (IPM), allowing this extracellular matrix to act as a mechanically restrictive microenvironment into which outer segment membranes are elaborated, thus indirectly supporting normal outer segment organization.

   The IPM may provide a structural template into which outer segment membranes are elaborated and within that template the proper physical constraints are provided such that outer segments form membranous saccules of the proper size and structure. Previous studies have shown that the PE is required for the assembly and organization of the IPM, yet it is not required for the synthesis of the normal amount of wheat germ agglutinin (WGA) binding proteins in the developing

Xenopus IPM [14]. In the absence of the PE, however, only limited amounts of WGA binding sites are present in the immediate vicinity of outer segments during their initial formation [15]. Perhaps, the addition of exogenous sugars allows the IPM components that are synthesized to become restricted to the photoreceptor outer segments which in turn provide structural support for the nascent photoreceptor outer segment membrane organization.

Although the mechanism through which the exogenously added permissive sugars are acting to support formation of organized, stacked, flattened membranous saccules is unknown at this point in time, it is clear that lactose, galactose and substituted forms of these sugars are very effective in supporting the normal organization of nascent membranes produced by *Xenopus* photoreceptors in the absence of the PE. We have proposed several mechanisms that may be acting in our culture system, each of which will be tested in future studies.

ACKNOWLEDGEMENTS

This work was supported in part by NEI Grants EY06341, EY02363, The Retina Research Foundation, The Retinitis Pigmentosa Foundation and The Knights Templar Eye Foundation.

REFERENCES

1. Adler R (1986) Developmental predetermination of the structural and molecular polarization of photoreceptor cells. Dev. Biol., 117:520-527.
2. Adler R (1987) Nature and nurture in the differentiation of retinal photoreceptors and neurons. Cell Differ., 20:183-188.
3. Adler R, Politi L (1989) Expression of a "survival crisis" by normal and rd/rd mouse photoreceptor cells in vitro. In: LaVail MM, Anderson RE, Hollyfield JG (eds) Inherited and environmentally induced retinal degenerations, Alan R. Liss, Inc., New York, pp 169-181.
4. Anderson DH, Guérin CJ, Erickson PA, Stern WH, Fisher SK (1986) Morphological recovery in the reattached retina. Invest. Ophthalmol. Vis. Sci., 27:168-183.
5. Caffé AR, Visser H, Jansen HG, Sanyal S (1989) Histopathic differentiation of neonatal mouse retina in organ culture. Curr. Eye Res., 8:1083-1092.
6. Defoe DM, Besharse JC, Fliesler SJ (1986) Tunicamycin-induced dysgenesis of retinal rod outer segment membranes. I. Quantitative freeze-fracture analysis. Invest. Ophthalmol. Vis. Sci., 27:1595-1601.
7. Erickson PA, Fisher SK, Anderson DH, Stern WH, Borgula GA (1983) Retinal detachment in the cat:the outer nuclear and outer plexiform layers. Invest. Ophthalmol. Vis. Sci., 24:927-942.
8. Faktorovich EG, Steinberg RH, Yasumura D, Matthes MT, LaVail MM (1990) Photoreceptor degeneration in inherited retinal dystrophy delayed by basic fibroblast growth factor. Nature, 347:83-86.
9. Fliesler SJ, Rayborn ME, Hollyfield JG (1985) Membrane morphogenesis in retinal rod outer segments: inhibition by tunicamycin. J. Cell Biol., 100:574-587.
10. Fliesler SJ, Rayborn ME, Hollyfield JG (1986) Inhibition of oligosaccharide processing and membrane morphogenesis in retinal rod photoreceptor cells. Proc. Natl. Acad. Sci., 83:6435-6439.
11. Guérin CJ, Anderson DH, Fariss RN, Fisher SK (1989) Retinal reattachment of the primate macula. Invest. Ophthalmol. Vis. Sci., 30:1708-1725.
12. Guérin CJ, Lewis GP, Fisher SK, Anderson DH (1993) Recovery of photoreceptor outer segment length and analysis of membrane assembly rates in regenerating primate photoreceptor outer segments. Invest. Ophthalmol. Vis. Sci., 34:175-183.
13. Lahav M (1987) In vitro model of retinal photoreceptor differentiation. Trans. Am. Ophthalmol. Soc., 85:600-638.
14. Lahiri D, Hollyfield JG (1992) Development of WGA-binding domains in the IPM of Xenopus laevis embryos. Invest. Ophthalmol. Vis. Sci. (Suppl.), 33:815.

15. Lahiri D, Rayborn ME, Hollyfield JG (1991) Development of the IPM in Xenopus laevis embryos. Invest. Ophthalmol. Vis. Sci. (Suppl.), 32:1217.
16. LaVail MM, Hild W (1971) Histotypic organization of the rat retina in vitro. Z. Zellforsch, 114:557-579.
17. Lewis GP, Erickson PA, Anderson DH, Fisher SK (1991) Opsin distribution and protein incorporation in photoreceptors after experimental retinal detachment. Exp. Eye Res., 53:629-640.
18. Li L, Turner JE (1991) Optimal conditions for long-term photoreceptor cell rescue in RCS rats: The necessity for healthy RPE transplants. Exp. Eye Res., 52:669-679.
19. Politi LE, Lahav M, Adler R (1988) Development of neonatal mouse retinal neurons and photoreceptors in low density cell culture. Invest. Ophthalmol. Vis. Sci., 29:534-543.
20. Roof D, Hayes A, Adamian M, Marescalchi P, Heth C (1991) Photoreceptor development in rat neural retina-retinal pigment epithelium co-cultures. Invest. Ophthalmol. Vis. Sci., 32:1148.
21. Stiemke MM, Hollyfield JG (1994) Outer segment disc membrane assembly in the absence of the pigment epithelium: The effect of exogenous sugars. Dev. Brain Res., 80:285-289.
22. Stiemke MM, Landers RA, Al-Ubaidi MR, Hollyfield JG (1994) Photoreceptor outer segment development in *Xenopus laevis*: Influence of the pigment epithelium. Dev. Biol., 162:169-180.
23. Takeda Y, Yamaguchi K, Yamada K, Matthes MT (1993) Photoreceptor rescue of Royal College of Surgeons (RCS) rat retina by human subretinal fluid. Invest. Ophthalmol. Vis. Sci., 34:1097.
24. Travis GH, Sutcliffe JG, Bok D (1991) The *retinal degeneration slow (rds)* gene product is a photoreceptor disc membrane-associated glycoprotein. Neuron, 6:61-70.
25. Ulshafer RJ, Allen CB, Fliesler SJ (1986) Tunicamycin-induced dysgenesis of retinal rod outer segment membranes. II. A scanning electron microscopy study. Invest. Ophthalmol. Vis. Sci., 27:1587-1594.
26. Watanabe T, Raff MC (1990) Rod photoreceptor development in vitro: intrinsic properties of proliferating neuroepithelial cells change as development proceeds in the rat retina. Neuron, 2:461-467.

16

REGULATORY INFLUENCES ON THE GLYCOSYLATION OF RHODOPSIN BY HUMAN AND BOVINE RETINAS

Edward L. Kean, Jermin Ju, and Naiqian Niu

Department of Ophthalmology
Case Western Reserve University
Cleveland, Ohio 44106

ABSTRACT

The retina has assembled oligosaccharide chains on rhodopsin that are unlike those found in other asparagine-linked, complex class of glycoproteins in terms of their abridged size and limited number of sugars. We have examined the kinetic properties of two glycosyltransferases of the retina whose action would be required in order to synthesize oligosaccharide chains of rhodopsin which would be similar in structure to those found in other molecules of this type. We have examined the kinetics of N-acetylglucosaminyltransferase II (GlcNAc-transferase II), required for branching, and galactosyltransferase, required for extending the oligosaccharide chain. Golgi-enriched preparations from human retina, bovine retina and rat liver were used as the enzyme sources and rhodopsin, opsin, and the oligosaccharide isolated from rhodopsin used as acceptors. From an evaluation of the V_{max}/K_m ratios it was observed that bovine and human retinas have very limited abilities compared to the rat liver to carry out the transfer of GlcNAc and galactose to these acceptors. In keeping with the presence of a relatively high concentration of galactosylated isomers found in human rhodopsin, human retinas had up to 13-fold greater activity of galactosyltransferase than did bovine retina. It was also apparent that the glycosyltransferases were not appreciably influenced by the polypeptide portion of the molecule. A further aspect of metabolic regulation was revealed with the observation that galactosylation of rhodopsin blocked the addition of GlcNAc and thus prevented branching of the oligosaccharide chain. It is suggested that these properties contribute to the assembly of the abridged structures that have been observed in the oligosaccharides of rhodopsins of all species thus far examined. The identification of the substrates and products of the reactions was carried out by chromatographic means.

Degenerative Diseases of the Retina, Edited by Robert E. Anderson et al.
Plenum Press, New York, 1995

INTRODUCTION

The major oligosaccharide chains of bovine (1, 2) , frog (3) and human rhodopsins (4), while identical in structure, are unusual when compared to other asparagine-linked glycoproteins in terms of their abridged size and lack of branching. This situation must be reflected in the activities and properties of the glycosyltransferases of the retina that are concerned with their biosynthesis. A key enzymatic step required for branching to occur would be the action of GlcNAc-transferase II catalyzing the transfer of a residue of N-acetylglucosamine to the terminal unsubstituted mannose residue of the major rhodopsin oligosaccharide (5, 6). This could lead to the assembly of multiantennary oligosaccharides, structures that are not found on rhodopsin. A key step for chain extension would be via the action of galactosyltransferase catalyzing the addition of galactose to the GlcNAc residue attached to the α 1,3-linked mannose. These reactions are depicted in Figure 1.

Galactosylation of rhodopsin has been observed in frog (3) and human rhodopsin (4), but whether such a modification is present on bovine rhodopsin is in dispute (1, 2, 7). Galactosylation could lead to the formation of sialic acid-containing oligosaccharides. That the enzymatic machinery is present in the retina for such a process was shown by Kean (8, 9) who demonstrated the activation of sialic acid in retina nuclei, and by O'Brien and Muellenberg (10) who showed the presence of retinal glycoprotein sialyltransferases. Sialic acid-containing isomers have been detected only in trace amounts in human rhodopsin (4), have not been seen in bovine rhodopsin (1,2), but have been observed in frog rhodopsin (3).

In the present studies, the kinetic properties of the glycosyltransferases were examined using Golgi-enriched preparations from human retina, bovine retina and rat liver using as acceptors rhodopsin, opsin, and the oligosaccharide isolated from rhodopsin. The V_{max}/K_m ratios obtained from these studies provide an index for evaluating the relative efficiencies of the retinal enzymes using rat liver as a reference tissue. Other influences on glycosylation were examined such as that of the protein portion of the molecule, and the influence of galactosylation on the GlcNAc-transferase reaction. An evaluation of the ability of the retina to carry out these important glycosyltransferase reactions may aid in understanding the

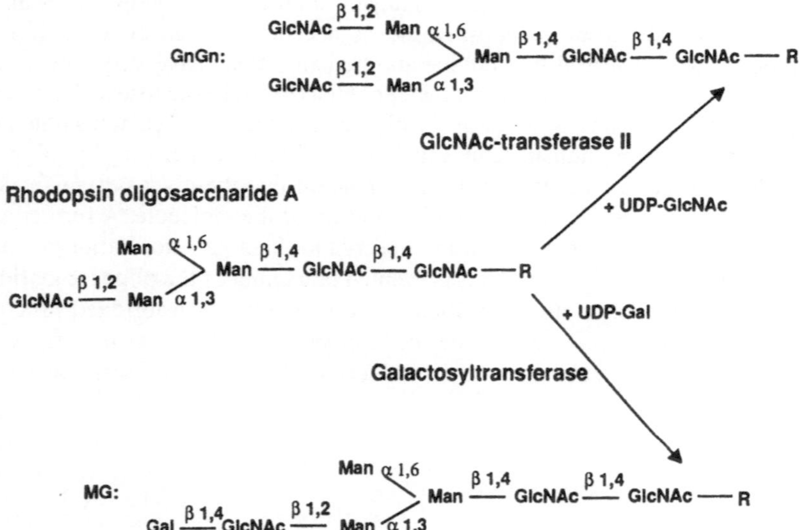

Figure 1. Galactosyltransferase and GlcNAc-transferase II action on the major oligosaccharide of rhodopsin.

metabolic environment of the cell that results in the formation of the abridged oligosaccharides observed in rhodopsins of all species thus far examined.

METHODS

Enzymes

Golgi-enriched fractions from bovine and human retinas were prepared as described previously (11).

Incubation Conditions for GlcNAc-Transferase Activity

Bovine Retina Enzyme. Golgi-enriched preparations from bovine retinas, prepared as described previously (11, 12), were incubated for 60 min at 37°C in the presence of 0.1 M Mes buffer, pH 6.5, 0.016% TX-100, 10 mM ATP, 22 mM $MnCl_2$, 2.5 mM UDP[^{3}H]GlcNAc (2.7 x 10^7 dpm/μmol) and rhodopsin, opsin or oligosaccharide as acceptors in a total volume of 0.04 ml. When using the oligosaccharide as acceptor, the detergent was omitted from the incubations.

Human Retina Enzyme. The same conditions were used as with bovine retina, except for the following: 90 min incubations; 7.5 mM $MnCl_2$; 1.0 mM ATP.

Rat Liver Golgi Preparation. Golgi-enriched fractions were prepared from rat liver (male, Sprague-Dawley) as described by Morre', Cheetham and Nyquist (13). Incubations were carried out for 10 min at 37°C with enzyme, buffer, metal ion, detergent, 10 mM ATP, 4.0 mM UDP[^{3}H]GlcNAc (1.6 x 10^7 dpm/μmol) and acceptors in a total vol of 0.04 ml.

Assay for Enzymatic Activity

High voltage paper electrophoresis in borate buffer (pH 9.4) was carried out as described previously (11, 12).

Oligosaccharide Production and Product Identification

Oligosaccharides were prepared enzymatically from bovine rhodopsin using peptide-N-glycosidase-F as described previously (14, 15). The oligosaccharides used as acceptors, and the oligosaccharide products of the reactions were analyzed by high pH anion exchange chromatography (HPAEC) using pulsed amperometric detection (PAD) (Dionex Corp.) and scintillation spectrometry of the [^{3}H]-labeled oligosaccharides, as described previously (11, 12, 15, 16). The column was calibrated with standard oligosaccharides.

Incubation Conditions for Galactosyltransferase Activity

Human Retina Enzyme. Incubations were carried out as described previously (11, 16). In short, incubations were performed at 37°C for 60 min in the presence of 60 mM MOPSO (3-[N-Morpholino]-2-hydroxypropanesulfonic acid), pH 7.3, 2 mM ATP, 12 mM $MnCl_2$, 0.1% Emulphogene, 0.24 mM UDP-[^{3}H]Gal (sp. act., 1.0 - 1.6 x 10^8 dpm/μmol), rhodopsin, opsin, or oligosaccharide, and enzyme (22 μg protein) in a total volume of 0.05 ml. When using the oligosaccharide as acceptor, ATP was omitted.

Bovine Retina Enzyme. Incubations were performed for 60 min at 37° C under conditions similar to those described previously (16) in the presence of 60 mM MOPSO, pH 6.3, 3 mM ATP, 9 mM $MnCl_2$, 0.1% Emulphogene, 0.1 mM UDP[^{3}H]-Gal (sp. act; 1.62 - 2.3 x 10 ^{8} dpm/µmol), enzyme (25 µg protein when using rhodopsin or opsin as acceptors, and 55 µg protein when using oligosaccharide as acceptor). Rhodopsin was present over a range from 5.8 µM - 34.8 µM; opsin, from 5.8 µM - 58 µM; the oligosaccharide, from 22.1 µM - 133 µM. The total volume of the incubation mixtures was 0.05 ml.

Rat Liver Golgi. The incubations were performed for 10 min at 37°C in a manner similar to that described previously (11, 16) in the presence of 40 mM cacodylate buffer, pH 6.6, 18 mM $MnCl_2$, 6 mM ATP, 0.016% TX-100, 0.2 mM UDP-[^{3}H]Gal (sp. act., 3.2 - 7.8 x 10^7 dpm/µmol), rhodopsin, opsin, or oligosaccharides, and enzyme (12 µg protein) in a total volume of 0.05 ml.

RESULTS

Kinetic Properties of GlcNAc-Transferases

Shown in Table I are the results of analyses of the kinetics of the transfer of GlcNAc to the various acceptors as catalyzed by the preparations from the human and bovine retinas and from rat liver

The relative efficiencies of the reactions are provided by the V_{max}/K_m ratios. From such an analysis, using rhodopsin, opsin, the oligosaccharide as acceptors, the bovine retina

Table 1. Kinetic properties of GlcNAc-transferases of human and bovine retinas and rat liver

Tissue/substrate	Apparent Km (µM)	Apparent V_{max} (pmol $mg^{-1}min^{-1}$)	V_{max}/K_m
Bovine Retina			
Rhodopsin	23.7 ± 11	32.2 ± 7.9	1.49 ± 0.36
Opsin	18.5 ± 7.9	23.9 ± 0.5	1.41 ± 0.58
Oligosaccharide	116 ± 2.1	155 ± 13	1.34 ± 0.13
Rat liver			
Rhodopsin	49.7 ± 10	6,890 ± 112	142 ± 26
Opsin	12.7 ± 5.1	4,010 ± 525	334 ± 91
Oligosaccharide	150 ± 83	27,000 ± 10,900	190 ± 33
Human Retina			
Rhodopsin	—	—	0.329 ± 0.006*
Opsin	—	—	0.431 ± 0.023*
Oligosaccharide	—	—	0.357 ± 0.151*

Two separate kinetic studies were carried out for each determination in which the effect on the rate of the reaction of 5 to 6 variations in acceptor concentration, in duplicate, were performed. Golgi-enriched fractions of each of the tissues were used. The apparent kinetic constants were calculated by computerized analysis of Lineweaver-Burk plots of the data presented as the average and range.

*Saturation with acceptor could not be achieved. The value of V_{max}/K_m was obtained from the slope of the plot of V(pmol/mg/min) versus S (µM), assuming that [S] is much less than K_m.

Table 2. Kinetic properties of the galactosyltransferases of human and bovine retinas and rat liver

Tissue/acceptor	Apparent K_m (μM)	Apparent V_{max} ($mg^{-1}min^{-1}$)	V_{max}/K_m
Human retina			
Rhodopsin	20.7 ± 6.0	42.2 ± 18	1.95 ± 0.31
Opsin	10.9 ± 1.7	74.0 ± 22	6.65 ± 0.95
Oligosaccharide	544 ± 35	186 ± 9.0	0.335 ± 0.005
Rat liver			
Rhodopsin	32.5 ± 0.50	1,180 ± 145	36.2 ± 3.9
Opsin	20.5 ± 5.6	2,840 ± 138	148 ± 34
Oligosaccharide	2,320 ± 270	12,400 ± 1,240	5.37 ± 0.09
Bovine retina			
Rhodopsin	26.0 ± 6.1	5.68 ± 1.6	0.216 ± 0.011
Opsin	37.9 ± 1.1	18.9 ± 2.1	0.497 ± 0.041
Oligosaccharide	161 ± 28	18.2 ± 0.15	0.117 ± 0.022

The kinetic data were collected and presented in the same manner as described in Table I.

preparations were shown to be 100 to 200 fold less active than rat liver, and human retina 400 to 800 fold less active. The major factor influencing these evaluations was the relatively low apparent V_{max} values of reactions with the retina enzymes (12).

Kinetic Properties of the Galactosyltransferases

In Table II is a summary of kinetic data of the galactosyltransferase reactions by preparations of human and bovine retina and from rat liver (16).

From an analysis of the V_{max}/K_m ratios, as with the GlcNAc-transferase described above, the galactosyltransferases of human retina using rhodopsin, opsin or the oligosaccharide isolated from rhodopsin as acceptors were about 20-fold less active than those of rat liver Golgi. Bovine retina was even less active compared to rat liver using these acceptors. Thus, not only was low galactosyltransferase activity displayed by the retina, consistent with previous studies (11), but a difference in this response was seen between bovine and human retinas. The human retina exhibited from 3 fold to 13-fold greater galactosyl transferase activities toward rhodopsin, opsin, and the oligosaccharide, than the bovine retina enzyme.

Lack of Influence of the Polypeptide Matrix on the Glycosyltransferases

The possibility was examined that the low GlcNAc-transferase and galactosyltransferase activities of the retina reflected a directing influence of the protein matrix of the acceptor molecules on glycosylation as has been demonstrated previously(17, 18). Such an effect was not demonstrated in the present studies. As seen in Tables I and II, the V_{max}/K_m ratios of the GlcNAc-transferases and galactosyltransferases using the isolated oligosaccharide as acceptor were similar to or lower than those obtained with intact rhodopsin or opsin. Thus, removal of the protein from the visual pigment did not result in an enhanced catalytic efficiency of the retina or liver enzymes in glycosylating the oligosaccharide, i.e., the limited ability of the holo-visual pigment to act as an acceptor of galactose was not due to a major directing influence of the peptide backbone.

The conformation of the molecule did appear to influence glycosylation since opsin was a better acceptor of galactose than rhodopsin using bovine and human retina as well as rat liver. This effect was not so evident concerning the activity of GlcNAc-transferase.

Relationship between Galactosyltransferase and GlcNAc-Transferase Activities

Among several control mechanisms that have been described for glycoprotein biosynthesis, Bendiak and Schachter (6) demonstrated a relationship between galactosylation and GlcNAc addition whereby the addition of a residue of galactose inhibits the activity of GlcNAc-transferase II. This mechanism was also detected concerning the glycosylation of rhodopsin (12). As seen in Table III, galactosylated rhodopsin, either generated in situ, or added exogenously, extensively blocked the subsequent transfer of GlcNAc to rhodopsin.

Product Identification

The major oligosaccharide cleaved from rhodopsin/opsin (oligosaccharide A) served as an acceptor in GlcNAc-transferase and galactosyltransferase experiments as described above. The products of these reactions were separated by HPAEC-PAD. Shown in Fig. 2 is the pattern obtained after transfer of galactose to the oligosaccharide catalyzed by the Golgi-enriched fraction from human retina. A single peak of labeled material was seen having the elution properties expected for that of oligosaccharide MG (Fig. 2B). The products fomed after incubations carried out in the presence of UDP[^{3}H]GlcNAc was identified as GnGn (see Figure 1 for structure) (data not shown). Similar results were

Table 3. Table III. Inhibition of GlcNAc-transferase by galactosylation of rhodopsin: Relative GlcNAc-transferase activity

	Enzyme source					
	Bovine retina		Rat liver		Partially purified GlcNAc-transferase II	
Acceptor	% of control	% inhib.	% of control	% inhib.	% of control	% inhib.
Rhodopsin (control)	100	—	100*	—	100	—
Galactosylated rhodopsin	32	68	3.1*	96.9*	5.1	94.9
Oligosaccharide (control)	—	—	—	—	100	—
Galactosylated oligosaccharide	—	—	—	—	0.2	99.8

Rhodopsin, or the oligosaccharide isolated from rhodopsin, was preincubated for 60 min at 37°C with a milk galactosyltransferase system (Ju and Kean (11)) containing non-radioactive UDP-galactose (2.6 mM), thus forming galactosylated rhodopsin *in situ*. Controls did not contain UDP-galactose. In a second stage incubation, either a Golgi-enriched fraction from retina or a partially purified GlcNAc-transferase II from rat liver was added as indicated, plus [^{3}H] UDP-GlcNAc (2-3 mM, 0.6 - 1.5 x 10^7 dpm/μmol, and the incubation continued for an additional 60 min. *In addition to generating galactosylated product *in situ*, galactosylated rhodopsin synthesized enzymatically (11) was incubated in the presence of the GlcNAc-transferase system and the rat liver Golgi preparation, as above. The incorporation of GlcNAc into the acceptor was measured by borate electrophoresis as described in Methods

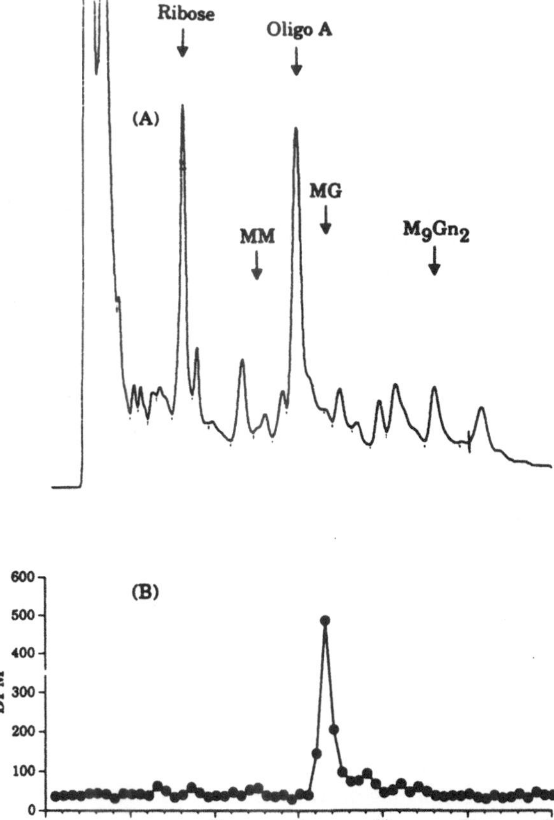

Figure 2. Oligosaccharide analysis by HPAEC-PAD of the galactosylated products formed after incubating with a Golgi-enriched fraction from human retina. The oligosaccharide recovered after digesting rhodopsin with PNGase-F was incubated with the retina enzyme,UDP[^{3}H]galactose and the products recovered as described in Methods and Materials and analyzed by HPAEC-PAD. (A) HPAEC-PAD elution pattern from the column which contained ribose (0.5 nmol) and $Man_9GlcNAc_2$ (0.21 nmol) as internal standards and calibrated also with MM and MG. (B) radioactivity of the fractions from the PAD cell. The abbreviations refer to the following compounds:

M_9Gn_2:
Man $\overset{\alpha2}{—}$ Man ↘α6
Man $\overset{\alpha2}{—}$ Man ↗α3 Man ↘α6
Man $\overset{\alpha2}{—}$ Man $\overset{\alpha2}{—}$ Man ↗α3 Man $\overset{\beta4}{—}$ $(GlcNAc)_2$

Rhodopsin oligosaccharide (A):
Man ↘α6
GlcNAc $\overset{\beta2}{—}$ Man ↗α3 Man $\overset{\beta4}{—}$ $(GlcNAc)_2$

MM:
Man ↘α6
Man ↗α3 Man $\overset{\beta4}{—}$ $(GlcNAc)_2$

MG:
Man ↘α6
Gal $\overset{\beta4}{—}$ GlcNAc $\overset{\beta2}{—}$ Man ↗α3 Man $\overset{\beta4}{—}$ $(GlcNAc)_2$

obtained with the other enzyme preparations and acceptors used in these studies (data not shown).

DISCUSSION

The findings of mutations in the rhodopsin gene, some at the glycosylation sites (21), in patients with retinitis pigmentosa have highlighted the importance of understanding the chemistry and biochemistry of the glycosylation of rhodopsin. We have examined two key enzymatic reactions that could regulate branching and extension of the oligosaccharide chains of this molecule. When compared to rat liver the kinetic studies revealed the very low efficiency of GlcNAc-transferase II of the retina, a key reaction required for oligosaccharide branching. Likewise, the capacity for chain extension by the addition of galactose to the major oligosaccharide was also shown to be greatly limited in the retina. Previously (11), low galactosyltransferase activity of bovine retina was shown also using other acceptors such as ovalbumin and GlcNAc, where less than 0.7% activity as compared to rat liver was detected. An additional aspect of metabolic regulation of the glycosylation of rhodopsin was revealed with the observation that galactosylated rhodopsin is a potent inhibitor of GlcNAc-transferase II, thus blocking branching. The transient galactosylation of rhodopsin described by Smith, et al (19) might exert such an effect. An influence of the peptide backbone on glycosylation was not indicated by these studies since use of the oligosaccharide as acceptor did not bring about enhanced Vmax/Km ratios. An effect due to the conformation of the visual pigment, however, was revealed in the galactosyltransferase reactions with greater activity obtained with opsin as compared to rhodopsin. This was not oserved with GlcNAc-transferase. These relationships may contribute to the virtual inability of human and bovine retina to assemble branched, sialylated oligosaccharides on rhodopsin (1, 2, 4).

How closely these studies showing properties detected in Golgi-enriched fractions of whole retina reflect processes in the photoreceptor cell is not clear. Unlike bovine rhodopsin in which little if any galactosylated species are present, about 18% of the oligosaccharide chains of human rhodopsin are galactosylated (4). The correspondence of this relationship with the much greater efficiency of the human enzyme compared to bovine retina suggests that the observations made with Golgi-enriched preparations from the whole retina might also reflect the enzymatic environment of the photoreceptor cell. However, other considerations in addition to the kinetics of the reactions must play important roles in regulating the metabolism of the photoreceptor cell such that the unique structure of the rhodopsin oligosaccharide is assembled. This clearly must be the case since the photoreceptor cell synthesizes other glycoproteins, such as the interphotoreceptor binding protein (IRBP), whose structure (20) indicates the participation of galactosyltransferase and GlcNAc-transferase II. Differences in compartmentalization of the water soluble IRBP molecule and the hydrophobic, membrane-bound visual pigment could influence the availability and accessibility of the various substrates and cofactors. These relationships in addition to the relatively limited activity of galactosyltransferase and GlcNAc-transferase II toward rhodopsin/opsin shown by the present studies, could contribute to the differences in oligosaccharide structures assembled on the glycoproteins of the photoreceptor cell.

ACKNOWLEDGMENTS

This work was supported in part by NIH Grant EY00393, and the Ohio Lions Eye Research Foundation.

REFERENCES

1. Liang, C.-J., Yamashita, K., Muellenberg, C.G., Shichi, H., and Kobata, A. (1979) Structure of the carbohydrate moieties of bovine rhodopsin, *J. Biol. Chem.* 254:6414-18.
2. Fukuda, M.N., Papermaster, D.S., and Hargrave, P.A. (1979). Rhodopsin Carbohydrate. Structure of small oligosaccharides attached at two sites near the NH2 Terminus, *J. Biol. Chem.* 254:8201-7.
3. Duffin, K.L., Lange, G.W., Welply, J.K., Florman, R., O'Brien, P.J., Dell, A., Reason, A.J., Morris, H.R. and Fliesler, S.J. (1993) Identification and oligosaccharide structure analysis of rhodopsin glycoforms containing galactose and sialic acid, *Glycobiol.* 3:365-380
4. Fujita, S., Endo, T., Ju, J., Kean, E.L. and Kobata, A. (1994) Structural studies of the N-linked sugar chains of human rhodopsin, *Glycobiol.* 4:633-640.
5. Schachter, H., Narasimhan, S., Gleeson, P., and Vella, G. (1983) Enzymatic control of oligosaccharide branching during synthesis of membrane glycoproteins, *GANN Monograph on Cancer Research* 29:177-195
6. Bendiak, B., and Schachter, H. (1987) (b) Control of Glycoprotein Synthesis, Kinetic mechanism, substrate specificity, and inhibition characteristics of UDP-N-acetylglucosamine:α-D-mannoside β1-2 N-acetylglucosaminyltransferase II from rat liver, *J. Biol. Chem.* 262:5784-5790
7. Kean, E.L., Hara, S., Mizoguchi, A., Matsumoto, A. and Kobata, A. (1983) The enzymatic cleavage of rhodopsin by the retinal pigment epithelium. II. The carbohydrate composition of the glycopeptide cleavage product, *Exp. Eye Res.* 36: 817-825.
8. Kean, E.L. (1969) Sialic acid activating enzyme in ocular tissue, *Exp. Eye Res.* 8: 44-54.
9. Kean, E.L. (1970) Nuclear cytidine 5'-monophosphate synthetase, *J. Biol. Chem.* 245:2301-2308.
10. O'Brien, P.J. and Muellenberg, C.G. (1968) Properties of glycosyltransfer enzymes of bovine retina, *Biochim. Biophys. Acta* 167:268-273.
11. Ju, J. and Kean, E.L. (1992) In vitro galactosylation of rhodopsin and opsin: kinetics, properties and characterization, *Exp. Eye Res.* 55:589-6047.
12. Ju, J. and Kean, E.L. (1994) Retinal GlcNAc-transferases and the glycosylation of rhodopsin, *Exp. Eye Res.* 59:565-576.
13. Morre', D.J., Cheetham, R.D., and Nyquist, S.E. (1972). A simplified procedure for isolation of Golgi apparatus from rat liver, *Prep. Biochem.* 2:61-9.
14. Plantner, J.J., Le, M.-L., and Kean, E.L. (1991) Enzymatic deglycosylation of bovine rhodopsin, *Exp. Eye Res.* 53:269-274.
15. Prasad, A.V.K., Plantner, J.J., and Kean, E.L. (1992) Effect of enzymatic deglycosylation on the regenerability of bovine rhodopsin, *Exp. Eye Res.* 54:913-920.
16. Kean, E.L., Ju, J., and Niu, N. (1995) Galactosylation of rhodopsin by the human retina, *Current Eye Res.*, 14:413-419.
17. Shao, M.-C., and Wold, F. (1988) The effect of the protein matrix on glycan processing in glycoproteins, *J. Biol. Chem.* 263:5771-5774
18. Yet, M.-G., Shao, M.-C. and Wold, F. (1988) Effect of the protein matrix on glycan processing in glycoproteins, *FASEB J.* 2:22-31.
19. Smith, S.B., St Jules, R.S., and O'Brien, P.J. (1991) Transient hyperglycosylation of rhodopsin with galactose, *Exp. Eye Res.* 53:525-37.
20. Taniguchi, T., Adler, A.J., Mizuochi, T., Kochibe, N., and Kobata, A. (1986) The structures of the asparagine-linked sugar chains of bovine interphotoreceptor retinol-binding protein, *J. Biol. Chem.* 261:1730-1736
21. Dryja, T.P., Hahn, L.B., Cowley, G.S., McGee, T.L., and Berson, E.L. (1991) Mutation spectrum of the rhodopsin gene among patients with autosomal dominant retinitis pigmentosa, *Proc. Nat'l Acad. Sci. USA*, 88:9370-9374.

17

THE DOLICHOL PATHWAY IN THE RETINAL PIGMENT EPITHELIUM OF THE EMBRYONIC CHICK

Edward L. Kean and Naiqian Niu

Department of Ophthalmology
Case Western Reserve University
Cleveland, Ohio 44106

ABSTRACT

Retinal pigment epithelium (RPE) cells from the embryonic chick, maintained in culture, were incubated in the presence of [2-^{3}H]-mannose and the oligosaccharide-lipids isolated and analyzed. After mild acid hydrolysis and reduction with $NaBH_4$, the oligosaccharide pattern was examined by HPLC. The full array of intermediates of the dolichol pathway was observed with the glucose-containing species, Glc_1, Glc_2 and $Glc_3Man_9GlcNAc_2$-P-P-dolichol, being the major oligosaccharide-lipids detected. Additional evidence supporting the formation of glucose-containing oligosaccharide-lipids by the RPE was obtained when analyzed by high pH anion exchange chromatography, comparing the products synthesized by the RPE with those formed by Madin-Darby canine kidney cells.

INTRODUCTION

In the recent period a great deal of attention has been directed to the importance of glycoconjugates in normal tissue and in disease states. There is, however, little information available concerning glycoprotein biosynthesis by the cells of the retinal pigment epithelium (RPE), the cell layer which engages in many important functions with the neural retina. Research from a variety of laboratories over the past 25 years has demonstrated that the biosynthesis of the core region of the oligosaccharide chains of asparagine-linked glycoproteins proceeds by a series of reactions known as the dolichol pathway (see review, (1)). This process involves the participation of some 16 separate enzymatic reactions that lead to the formation of $Glc_3Man_9GlcNAc_2$-P-P-dolichol, the activated oligosaccharide-lipid which transfers its oligosaccharide to asparagine residues on nascent, developing, glycoprotein chains. We have previously examined the capacity of the retina to carry out the synthesis of the various intermediates of the pathway (2), and studies on control mechanisms of its initial

Degenerative Diseases of the Retina, Edited by Robert E. Anderson et al.
Plenum Press, New York, 1995

reactions (3-6). As an extension of these investigations, the formation of the various oligosaccharide-lipid intermediates of the dolichol pathway by RPE cells from the embryonic chick maintained in cell culture has also been examined.

MATERIALS AND METHODS

Cell Culture; Labeling

RPE cells of the embryonic chick were maintained in cell culture as described previously (7). RPE cells from the embryonic chick were grown in 150 cm^2 plastic Corning culture flasks until confluent. After removal of the media, the monolayer of cells was rinsed with Dulbecco's phosphate buffered saline without metals ($DPBS^-$). A solution of [2-^{3}H]-mannose (12.5 μCi/ml, 28 Ci/mmol) in 10 ml $DPBS^-$ was added to the cells and incubated at 37° C for 1 hr. with gentle shaking, after which the medium was removed and the cells rinsed with 10 ml $DPBS^-$. The cells were scraped from the flask and transferred to a glass centrifuge tube, aided with $DPBS^-$ to a total volume of 2 ml from which the oligosaccharide-lipids were isolated as follows.

Oligosaccharide-Lipid Isolation

The oligosaccharide-lipid fraction was isolated essentially as described by D'Souza et al (8). In short, this involved the following procedures. A solution of 4 ml of chloroform:methanol (C/M) 1:1 v/v) was added to the cell suspension and the mixture vortexed vigorously. After centrifuging at 250 x g for 5 min, the upper phase and the particulate interface which was formed were retained, and the lower phase discarded. The upper phase and interface pool was extracted with 2 ml chloroform after which 4 ml of methanol was added. After centrifugation, the resultant pellet was washed with 50% methanol followed by methanol. The oligosaccharide lipid fraction was obtained by extracting the pellet with a mixture of C/M/water (10:10:3, v/v/v).

Oligosaccharide-Lipid Analysis by HPLC

After evaporating an aliquot of the 10:10:3 extract to dryness, the material was subjected to mild acid hydrolysis (0.2 ml n-propanol, 0.4 ml 0.02 N HCl, 100° C, 20 min), reduction with $NaBH_4$, followed by ion exchange chromatography as described previously (2). The product was dissolved in acetonitrile-water (1:1, v/v) and analyzed by HPLC on a Varian model 5000 liquid chromatograph using a column of Absorbosphere silica (3 μm, 0.46 cm x 15 cm; Alltech, Avondale, PA) as described previously (2). Elution was carried out with a gradient using acetonitrile (solvent A) and 10% acetonitrile in water (solvent B) between a mixture of 63 A and 37 B and a mixture of 42 A and 58B, each containing also 0.05% diaminobutane, for a total elution time of 100 min, at a flow rate of 1 ml/min.

Oligosaccharide-Lipid Analysis by Dionex Chromatography

After mild acid hydrolysis as above, the liberated [^{3}H]-oligosaccharides were also analyzed by high pH anion exchange chromatography (HPAEC) on a CarboPac PA-1 anion exchange column (4 mm x 250 mm) (Dionex Corp., Sunnyvale, CA). After equilibrating the column with a mixture of 99 parts of 0.1M NaOH and 1 part of 0.5 M NaOAc in 0.1 M NaOH, the sample (20 μl in water) was injected. Elution was carried out using a linear gradient from 1 % to 15% of 0.5 M NaOAc in 0.1 M NaOH at a flow rate of 1 ml/min for

40 min. This was followed by a linear gradient from 15% to 20% of 0.5 M NaOAc in 0.1 M NaOH for 10 min. The radioactivity of the fractions was measured by scintillation spectrometry. The analyses were performed in the presence of ribose and the oligosaccharide, $Man_9GlcNAc_2$, as internal standards. In addition, the column was calibrated with the major oligosaccharide isolated from bovine rhodopsin (oligosaccharide A). This material was obtained after incubating rhodopsin with peptide N-glycosidase, as described previously (9). The elution of the standards was followed by pulsed amperometric detection (PAD), as described previously (9). The structures of the various standards are shown in the legends to the figures.

MDCK Cell Incubation

The major oligosaccharide-lipid formed by Madin-Darby canine kidney (MDCK) cells is $Glc_3Man_9GlcNAc_2$-P-P-dolichol (8, 10). MDCK cells were therefore used as a source of this compound to aid in exploring for the production of the glucose-containing oligosaccharide-lipids by the RPE cell. MDCK cells were obtained from the American Type Culture Collection (Rockville, MD) and grown in Dulbecco's Modified Eagles Medium containing high glucose (4,500 mg/l) and L-glutamine (Gibco), 10% fetal bovine serum (Hyclone), and penicillin (100 units/ml) and streptomycin (100 µg/ml). Large-scale growths of the cells were made, and after removing the growth medium and washing the cells with $DPBS^-$, the cells were incubated with [2-^{3}H]-mannose in the same manner as described in Methods for the production of ^{3}H-labeled intermediates of the dolichol pathway by the RPE cells. The MDCK cells were quite stable for several hours in this environment as were the RPE cells. The oligosaccharide-lipid fraction was isolated and the oligosaccharides examined by HPAEC in the same manner as described above.

RESULTS

HPLC analysis of the Oligosaccharides

Shown in Fig. 1 is the elution pattern obtained by HPLC on the Varian liquid chromatograph of the oligosaccharides liberated from the C/M/W 10:10:3 oligosaccharide-lipid fraction which was formed after incubating the RPE cells as described above.

As seen here, many of the oligosaccharide-lipids of the dolichol pathway observed in other systems were detected in the RPE after labeling with [^{3}H]mannose. The predominant types were the glucose-containing oligosaccharide-lipid intermediates. Calibration of the HPLC column was based upon the elution pattern of standard oligosaccharides as described previously (2), and the the elution pattern of the oligosaccharide-lipids synthesized by Thy-1 cells (11). A preliminary account of some of the early observations concerning this effect has been made previously (2).

MDCK Cells

Additional evidence was obtained concerning the formation of the glucose-containing oligosaccharide-lipids by the RPE cells by carrying out the analysis using the Dionex chromatograph. The compound, [^{3}H]$Glc_3Man_9GlcNAc_2$-P-P-dolichol, has been shown to be the major oligosaccharide-lipid formed by MDCK cells under these incubation conditions (8, 10). The products formed by the MDCK cells were therefore used as a reference source.

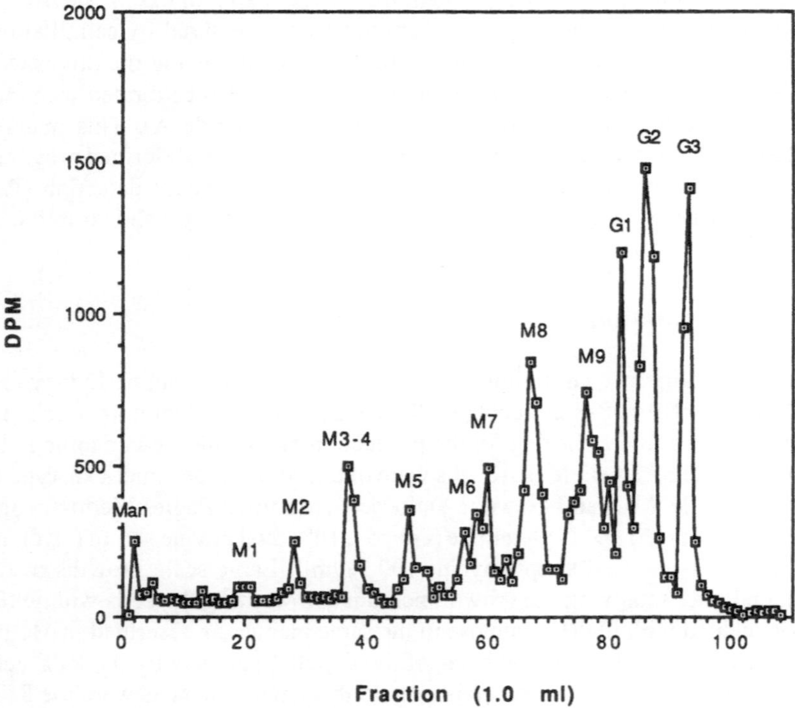

Figure 1. HPLC separation of the oligosaccharides derived from the oligosaccharide-lipids synthesized by the RPE of the embryonic chick. RPE cells were incubated with 2-[^{3}H]mannose and the oligosaccharides were analyzed on the Varian liquid chromatograph as described in Materials and Methods. The radioactivity of 1.0 ml fractions was determined by scintillation spectrometry in the presence of 4 ml of EcoLume (ICN, Costa Mesa, CA) is indicated on the ordinate. The following abbreviations refer to the elution positions of standard oligosaccharides: G3, G2 and G1, $Glc_nMan_9GlcNAc_2$; M1-9, $Man_nGlcNAc_2$. (See legend to Figure 2 for structures.)

The analyses shown in Fig. 2 clearly demonstrate the similarity of the major components from the MDCK cells and the RPE cells, i.e., the glucose-containing oligosaccharide-lipids (8, 9). However, resolution by this technique of the three glucose-containing intermediates from one another was not obtained.

CONCLUDING REMARKS

Similar to most other cell types which have been examined, the present studies demonstrated that the major intermediates of the dolichol pathway which are formed by the RPE of the embryonic chick are the glucose-containing oligosaccharide-lipids. This is in contrast to the non-glucosylated species, $Man_9GlcNAc_2$-P-P-dolichol, which was shown to be accumulated by the bovine retina (2). In addition to the synthesis of the many intermediates of the dolichol pathway demonstrated by the present work, several aspects of metabolic regulation of the dolichol pathway which have been shown by the retina and other tissues (3-6) have also detected in cell-free preparations from the RPE (Kean, unpublished experiments). These observations all indicate that an active capacity for glycoprotein biosynthesis exists in the RPE, an aspect of the metabolism of this tissue which thus far has been only limitedly explored.

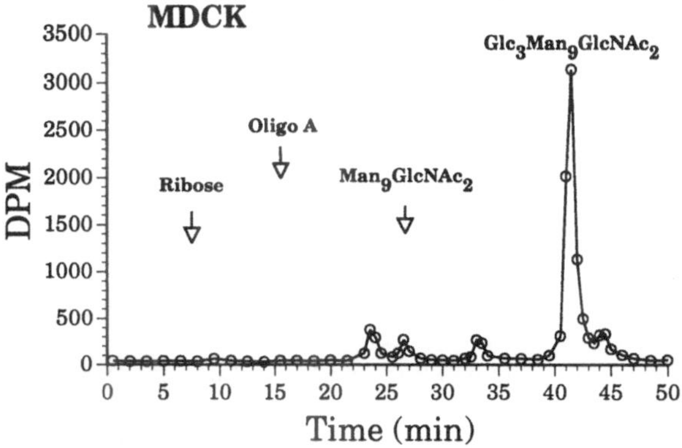

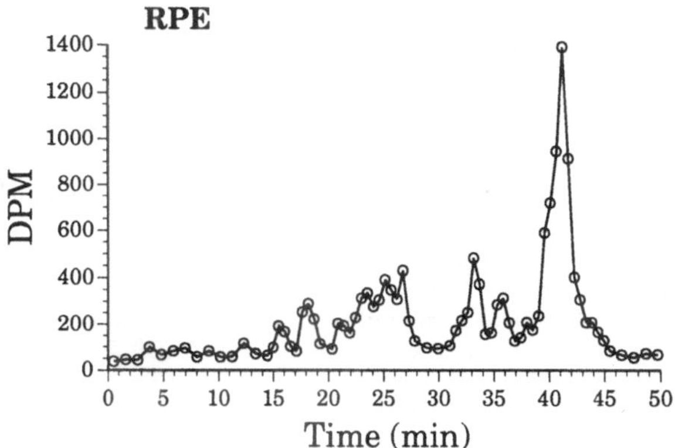

Figure 2. Analysis by HPAEC of the oligosaccharides derived from oligosaccharide-lipids synthesized by embryonic chick RPE cells and MDCK cells. MDCK cells and RPE cells were incubated with 2-[^{3}H] mannose and the oligosaccharides analyzed on a Dionex liquid chromatograph (Sunnyvale, CA), as described in Methods and Materials. The radioactivity of the fractions, 0.5 ml, collected at a flow rate of 1.0 ml/min was determined in the presence of 3 ml of EcoLume. The samples were analyzed in the presence of internal standards, ribose (0.7 nmol) and $Man_9GlcNAc_2$ (0.16 nmol). The retention time of the standards and of oligosaccharide A from bovine rhodopsin was determined by pulsed amperometric detection. The following are the structures of the standard oligosaccharides and of $Glc_3Man_9GlcNAc_2$:

$Man_9GlcNAc_2$:

Man $\xrightarrow{\alpha2}$ Man ↘α6
Man ↘α6
Man $\xrightarrow{\alpha2}$ Man ↗α3
Man ↘ $\xrightarrow{\beta4}$ $(GlcNAc)_2$
Man $\xrightarrow{\alpha2}$ Man $\xrightarrow{\alpha2}$ Man ↗α3

Rhodopsin oligosaccharide (A):

Man ↘α6
Man $\xrightarrow{\beta4}$ $(GlcNAc)_2$
GlcNAc $\xrightarrow{\beta2}$ Man ↗α3

$Glc_3Man_9GlcNAc_2$:

Man $\xrightarrow{\alpha2}$ Man ↘α6
α2—
Man ↘α6
Man Man ↗α3
Man $\xrightarrow{\beta4}$ $(GlcNAc)_2$
Glc $\xrightarrow{\alpha2}$ Glc $\xrightarrow{\alpha3}$ Glc $\xrightarrow{\alpha3}$ Man $\xrightarrow{\alpha2}$ Man $\xrightarrow{\alpha2}$ Man ↗α3

ACKNOWLEDGMENTS

This work was supported in part by NIH Grant EY00393 and the Ohio Lions Eye Research Foundation.

REFERENCES

1. Kornfeld, R., and Kornfeld, S. (1985) Assembly of asparagine-linked oligosaccharides. *Annu. Rev. Biochem.* 54, 631-664.
2. Plantner, J. J., and Kean, E. L. (1988) The dolichol pathway in the retina: Oligosaccharide-lipid biosynthesis. *Exp. Eye Res.* 46, 785-800.
3. Kean, E. L. (1982) Activation by dolichol phosphate-mannose of the biosynthesis of N-acetylglucosaminylpyrophosphoryl polyprenols by the retina. *J. Biol. Chem.* 257: 7952-7954.
4. Kean, E. L. (1985) Stimulation by dolichol phosphate-mannose and phospholipids of the biosynthesis of N-acetylglucosaminyl pyrophosphoryl- dolichol. *J. Biol Chem.* 260, 12561-12571.
5. Kean, E. L. (1987). Metabolic regulation of the initial reactions of the dolichol pathway: activation and deactivation of the biosynthesis of GlcNAc-P-P-dolichol. *Chemica Scripta.* 27, 127-134.
6. Kean, E. L., Rush, J. S., and Waechter, C. J.(1994) Activation of GlcNAc-P-P-dolichol synthesis is stereospecific and requires a saturated α-isoprene unit. *Biochem.* 33, 10508-10512.
7. Kean, E. L., Hara, S., and Lentrichia, B. B. (1981) Binding of neoglycoproteins by the chick pigment epithelial cell in culture. *Vis. Res.* 21, 133-135.
8. D'Souza, C. D., Sharma, C. B., and Elbein, A. D. (1992) Biosynthesis of lipid-linked oligosaccharides. *Anal. Biochem.* 203, 211-217.
9. Prasad, A. V. K., Plantner, J. J., and Kean, E. L. (1992) Effect of enzymatic deglycosylation on the regenerability of bovine rhodopsin. *Exp. Eye Res.* 54, 913-920.
10. Pan, Y. T., and Elbein, A. D. (1990) Control of N-linked oligosaccharide synthesis: cellular levels of dolichol phosphate are not the only regulatory factor. *Biochem.* 29, 8077-8084.
11. Kean, E. L., and DeBrakeleer, D. J.(1986) Stimulation by dolichol-P-mannose of GlcNAc-lipid biosynthesis by membranes from class E Thy-1 negative mutant mouse lymphoma cells which are defective in dolichol-P-mannose biosynthesis. *Arch. Biochem. Biophys.* 250, 146-152.

18

MORPHOLOGICAL AND BIOCHEMICAL STUDIES OF THE RETINAL DEGENERATION IN THE VITILIGO MOUSE

A Model with Perturbed Retinoid Metabolism

Sylvia B. Smith and Barbara N. Wiggert

Department of Cellular Biology and Anatomy and Ophthalmology
Medical College of Georgia
Augusta, Georgia and
Laboratory of Retinal Cell and Molecular Biology
National Eye Institute, National Institutes of Health
Bethesda, Maryland

INTRODUCTION

The vitiligo mouse has been studied since the mid 1980's for the depigmentary condition of its skin and fur. As such, it is a promising model for the human skin disease vitiligo (1). Vitiligo may occur in isolation or in combination with other disorders, including retinal degeneration (Vogt-Koyanagi-Harada syndrome) (2). In 1988, Dr. Richard Sidman and co-workers provided a preliminary report in Mouse News Letter that the vitiligo mouse had a slow progressive retinal degeneration (3). We obtained breeding pairs from Dr. Sidman and set out to characterize the retinal degeneration in this mouse and to determine the etiology of the disease with the hope of eventually designing strategies to treat the disorder. This chapter will review our findings about the morphologic, electrophysiologic and biochemical characteristics of this mutant with particular emphasis on promising results from studies of retinoid metabolism.

MORPHOLOGICAL AND ELECTROPHYSIOLOGICAL CHARACTERIZATION OF PHOTORECEPTOR CELLS AND RPE

The first full-length study of the retinal degeneration in the vitiligo mouse established the rate of photoreceptor cell (PRC) loss and quantified rhodopsin levels between the ages 2 and 52 weeks (4). Rows of PRCs are lost at a rate of about one per month in the posterior retina beginning at 8 weeks postnatally. Outer segments in this region become disrupted rather early in the disease and macrophage-like cells are present in the subretinal space. The neural retina is frequently separated from the RPE. Subsequent studies revealed a marked

Degenerative Diseases of the Retina, Edited by Robert E. Anderson et al.
Plenum Press, New York, 1995

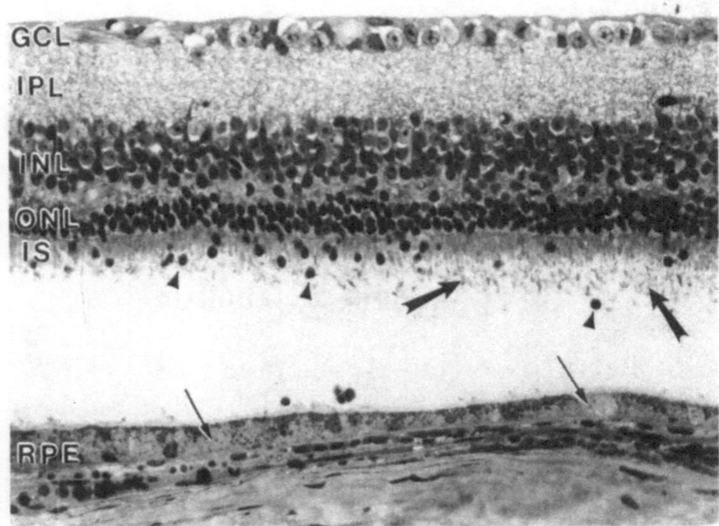

Figure 1. Photomicrograph of posterior region of retina of 8 month vitiligo mouse. Although the ganglion cell layer (GCL), inner plexiform layer (IPL) and the inner nuclear layer (INL) appear normal, the photoreceptor cells in the outer nuclear layer (ONL) are reduced to about 2-3 rows. Arrow heads point to displaced cells, presumably photoreceptor cell nuclei. Inner segments (IS) are present, but the outer segments severely fragmented (indented arrow). The retinal pigment epithelium (RPE) is lightly pigmented (thin arrows point to areas of RPE with minimal pigmentation). Magnification = 630X.

gradient in cellular loss between the posterior and peripheral regions with PRC loss occurring at a much slower rate in the peripheral retina (5). Even as late as 17 months, vitiligo mice have multiple rows of PRCs in the periphery which have well-formed outer segments that interdigitate with the retinal pigment epithelium (RPE). Although there is a definite gradient of degeneration along the posterior-peripheral axis, there appears to be no preferential loss of cells in the superior quadrant versus the inferior quadrant (14 month old animals had 4.03 ± 0.86 rows in superior quadrant, 4.51 ± 1.24 in inferior quadrant).

In addition to changes in PRCs, there are obvious differences in pigmentation of RPE cells between vitiligo and C57BL/6 controls. The pigmentation is uneven; some cells are completely devoid of pigment while adjacent cells are slightly or heavily pigmented (6,7). There appears to be no correlation between RPE pigmentation and PRC loss (4, I. Nir, unpublished observations). Figure 1 is a photomicrograph showing the reduced number of PRCs and sparse pigmentation of RPE in the posterior retina of an 8 month vitiligo mouse.

Electrophysiologic studies confirmed the slow-progressive nature of the degeneration (8). Scotopic ERGs indicated that at 4 weeks the mean of the maximum b-wave amplitude was about 230μV which was similar to controls. With increasing age, all components of the ERG decreased, by 18 weeks the mean V_{max} was 75μV. The gradual decrease in b-wave amplitude correlates highly with the loss of photoreceptor cells and rhodopsin.

FUNCTIONAL STUDIES OF RPE AND PHOTORECEPTOR CELLS

The primary cellular site of the retinal degeneration in the vitiligo mouse is not known, but the two likely cells are the PRC and the RPE. It is well established that these two cells function cooperatively in maintaining retinal homeostasis. With that in mind, we have designed experiments to test specific PRC or RPE functions in this mutant.

Phagocytosis Studies

Phagocytosis of shed outer segment disks by the RPE occurs in the vitiligo mutant (9). There is a measurable peak of phagocytosis that appears during the first two hours of light onset. The number of phagosomes in vitiligo mice is less than controls. Even as early as 4 weeks, the number is about 50% the number in controls. Between 4 and 8 weeks the number of phagosomes is reduced markedly in vitiligo mice to about 16% that of controls; between 8 and 20 weeks the number of phagosomes does not diminish. The observation that phagosomes are present is significant because it distinguishes the vitiligo mouse retinal degeneration from the retinal dystrophy of the RCS rat, the hallmark of which is virtual absence of phagocytosis of shed outer segment discs by RPE (10).

Rod Outer Segment Renewal

The process of outer segment renewal has been analyzed autoradiographically and biochemically in the vitiligo mouse (11). Within 24 hours of intraperitoneal injection of ^{3}H leucine, there is a distinct band of radioactivity present at the junction of the ROS and RIS in mutant mice that is similar to controls. The displacement of the radioactive band progresses normally in the peripheral regions of the vitiligo retina, but not in the posterior retina. Assessment of incorporation of radioactivity into rhodopsin using SDS-PAGE indicates a progressive displacement of radio-labeled rhodopsin through the RER, but not as complete a progression through the outer segments. The disrupted attachment of outer segments to the RPE may be the source of abnormal outer segment renewal in the posterior region.

STUDIES OF THE VISUAL CYCLE AND RETINOID METABOLISM

The results of the phagocytosis/renewal studies suggested that neither PRC nor RPE were functioning entirely normally in the vitiligo mutant. Hence it became necessary to examine a second biological function to further elucidate which cell might be the initial site of malfunction. Both cells are known to function cooperatively in maintaining the visual cycle (shown in Fig. 2). We have designed several experiments to determine if any portion of this cycle is altered in the vitiligo mutant.

Rhodopsin, Opsin and Dark-Rearing

Rhodopsin levels were quantified in vitiligo mice over the course of the degeneration. The level diminishes with time in a manner that correlates with the loss of PRCs (4). The capacity of isolated mutant retinas to synthesize opsin, the apoprotein of rhodopsin was determined using in vitro techniques (12). The isolated retinas are able to synthesize this protein for a surprisingly long period of time. Even as late as one year, mutant retinas synthesized about 20% the amount of opsin as controls. This data suggests that vitiligo PRCs that remain in the periphery have the capacity to synthesize this important protein. In a subsequent study, an attempt was made to interrupt the visual cycle (by eliminating light) in the vitiligo mouse and evaluating the consequences on the rate of PRC degeneration (13). A group of mutant mice were reared in darkness for periods of time up to 28 weeks and eyes were examined histologically. The rate of PRC loss did not differ significantly from the rate

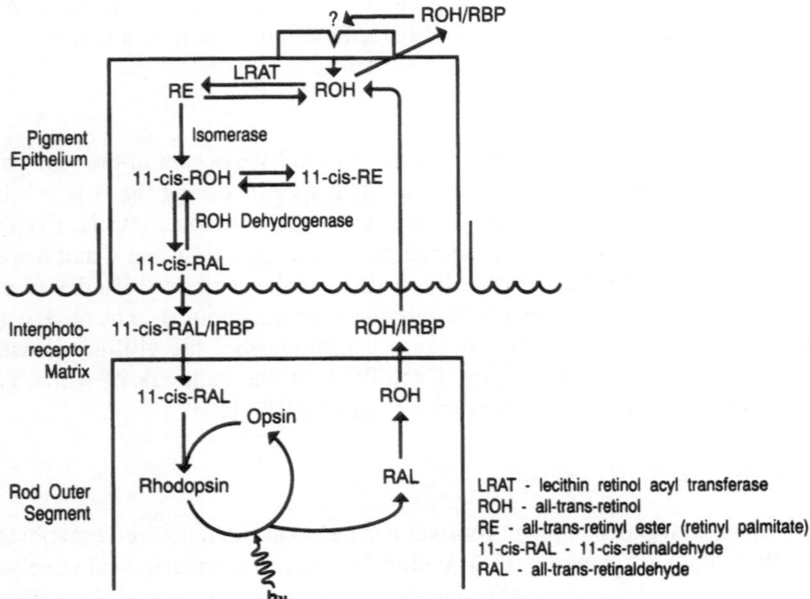

Figure 2. Visual cycle and vitamin A (retinoid) metabolism in the retina. The photoreceptor cell synthesizes opsin which uses 11-cis-retinaldehyde in regeneration of rhodopsin. Following photolysis (hv), there is a cis to trans isomerization and the all-trans-RAL is released. In the photoreceptor cell it is converted to all-trans-retinol and in the presence of IRBP is returned to the RPE. The membranolytic all-trans-retinol is converted in the presence of the enzyme LRAT to the less toxic retinyl ester. In the presence of an isomerase and subsequently a dehydrogenase, the retinyl ester can be converted back to the 11-cis-RAL which, again in company with IRBP, is returned to photoreceptor cells for rhodopsin regeneration. Two forms of vitamin A are elevated in the vitiligo mouse RPE: all-trans-retinol and retinyl palmitate.

determined for vitiligo animals reared in a standard light cycle. Furthermore, retinal detachment from RPE, the displacement of darkly-staining cells into the subretinal space and the influx of macrophage-like cells in the area of the ROS were still present in vitiligo mice reared in darkness. The findings suggest that light deprivation neither retards nor accelerates degeneration of the vitiligo retina.

RETINOID METABOLISM

The studies of rhodopsin and opsin did not reveal a specific cellular defect in the vitiligo mouse, however rhodopsin regeneration is known to require a steady source of vitamin A. The metabolism of vitamin A was examined in whole eyes and then in neural retina separated from the RPE in order to determine if there were any metabolic abnormalities in either of these cells (14).

Levels of Retinoids in Whole Eyes

Retinoid levels in whole eyes were determined by HPLC. Of the various forms of vitamin A shown in Fig. 2, only the level of retinyl palmitate (a retinyl ester) differed between vitiligo and control mice. Figure 3 illustrates the differences. Although levels were similar

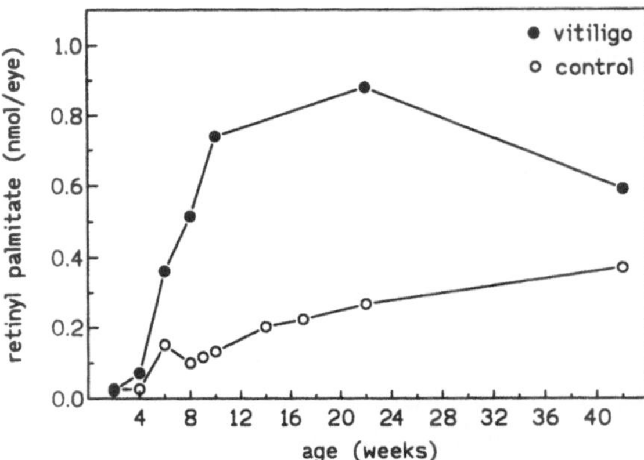

Figure 3. Increase in retinyl palmitate levels in vitiligo mouse eyes. Retinoid levels were determined by HPLC. At each age studied, 12 affected and 12 control animals were used. Four eyes were used per assay and each assay was performed three times. The variation in the data was less than 5%. ANOVA indicated that levels of retinyl palmitate differed significantly between vitiligo and controls mice ($F = 5.57$, $P = 0.0011$). (From Smith et al, 1994, with permission Portland Press, Ltd., London).

at two weeks, by 4 weeks retinyl palmitate levels were greater than controls, they increased sharply at 6 weeks and by 10 weeks levels were more than three-fold greater than controls.

Analysis of Retinoids in RPE and Neural Retina Separately

In the normal eye, retinyl palmitate is present primarily in the RPE. In order to determine whether retinyl palmitate was accumulating in the RPE or in the neural retina in the eyes of affected mice, these tissues were dissected free from each other. The HPLC analysis of retinoids in these two tissues revealed that retinyl palmitate was accumulating in the RPE, rather than the neural retina (Table 1). The levels in affected mice were three fold greater than in controls. Retinyl palmitate levels in the neural retinas did not differ significantly between the two groups. Surprisingly, the analysis of retinoids in the two tissues separately unmasked the elevation of another retinoid, all-trans-retinol (Table 1). In RPE, levels of all-trans-retinol were 4 times greater in the affected animals versus controls. In neural retina, levels of all-trans-retinol were similar between the two groups.

IRBP

The elevation of retinyl palmitate levels in the vitiligo eye precedes the onset of photoreceptor cell loss. Retinyl palmitate is the primary storage form of vitamin A in the retina. Retinyl esterification is considered a detoxifying reaction because it converts retinol, which is known to be membranolytic, into an insoluble relatively non-toxic form (15). One possible explanation for the accumulation of retinyl palmitate is that is a compensatory reaction to an excess amount of all-trans-retinol. All-trans-retinol is a potentially toxic compound. Long-term exposure to membranes could result in a lytic episode that could severely compromise retinal integrity. The vitiligo retina may be shuttling the more toxic compound toward the storage form,

Table 1. Retinyl palmitate and all-trans retinol concentrations in RPE and neural retina of vitiligo and control mice

	Retinyl palmitate (nmol/g)		All-trans-Retinol (nmol/g)	
Tissue	C57BL/6	Vitiligo	C57BL/6	Vitiligo
RPE	0.118 ± 0.007	0.386 ± 0.070*	0.023 ± 0.012	0.091 ± 0.005*
Neural retina	0.014 ± 0.004	0.015 ± 0.003	0.054 ± 0.010	0.048 ± 0.017

retinyl palmitate, to decrease the membranolytic consequences of excessive amounts of the all-trans-retinol. The observation that the all-trans-retinol level was elevated led to experiments to assess levels of the interphotoreceptor retinoid-binding protein, IRBP, which would bind excess retinol in the interphotoreceptor space.

Analysis of IRBP and IRBP mRNA

IRBP levels were determined using the slot-blot analysis technique and data are presented in Fig. 4. Levels of IRBP in control mice ranged between 0.5-1.5μg/mg protein. Levels in affected mice were similar to controls between postnatal day 3 and 4 weeks, but after 4 weeks the IRBP levels increased abruptly and remained elevated through about 12-14 weeks, although the IRBP levels had returned to the control values by 16 weeks. The abrupt elevation of IRBP just after 4 weeks prompted a second set of experiments to examine IRBP levels at each day between 4 and 5 weeks. There was a steady increase in IRBP levels between 4.0 weeks (1.6μg IRBP/mg protein) and the peak at 4.4 weeks when IRBP levels were 4μg/mg protein in vitiligo mice, a value that was 68% greater than controls (1.31μg IRBP/mg protein). Western blotting of IRBP from normal and vitiligo mouse eyes showed no alteration in the size of the protein in the mutant mouse (14).

The marked elevation of IRBP at 4 weeks coincides with the elevation of retinyl palmitate levels in the retina of the affected animal. Interestingly, the IRBP levels remain high for an extended period of time and return to normal levels only by 16 weeks, a time when the number of rows of photoreceptor cell nuclei is reduced by about 50%. This is intriguing because the photoreceptor cell synthesizes IRBP and a decrease in cell number would be expected to result in a decrease in protein synthesis, not the retention of normal levels. One possible explanation for such an elevation could be increased protein synthetic rates. Quantitation of IRBP mRNA using the RT-PCR analysis indicated that the amount of IRBP mRNA in affected animals is not significantly different from that of controls (14). Northern blot analysis also showed that the size and the amount of IRBP mRNA in 8 week old vitiligo mice are identical to that of controls. These data suggest that an increased expression of IRBP mRNA is not the cause of the increased amount of protein. Perhaps, the elevation of IRBP is the result of a profound decrease in turnover of the protein.

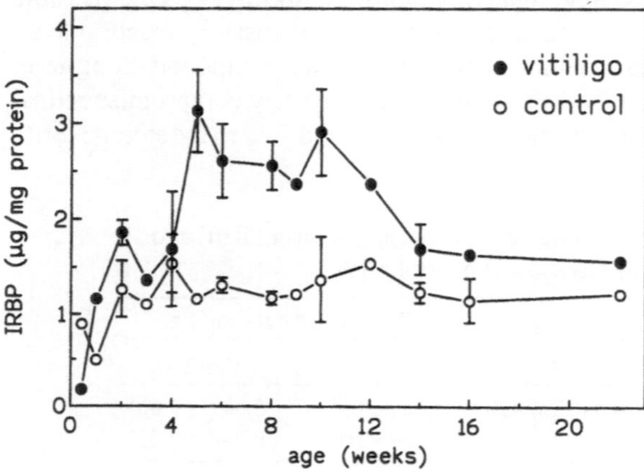

Figure 4. IRBP levels in eyes of vitiligo and control mice. Each point represents the mean ± S.D. of two-four assays. In each assay, two eyes were homogenized in 300μl of Tris-buffered saline, ultracentrifuged at 100,000 g (35000 rev./min at 4°C, 1 hr). The supernatant was analyzed for total protein using slot-blot analysis in which samples were applied in triplicate to Immobilon poly(vinylidene fluoride) membrane. The primary antibody was a 1:150 dilution of goat anti-(bovine IRBP) incubated 4°C overnight, and the secondary antibody was affinity-purified horseradish peroxidase conjugated rabbit anti-goat IgG (H+L) (Kirkegaard-Perry). Purified bovine IRBP was used as the standard at 1-75ng concentration. Quantification was accomplished using an LKB ultrascan XL laser densitometer and 2400 gel scan software package. ANOVA indicated that between 4 and 16 weeks of age, IRBP levels differed significantly from controls ($F = 26.83$, $P = 0.0003$). (From Smith et al, 1994, with permission Portland Press, Ltd., London.)

SYSTEMIC STUDIES

It is well known that free retinol is taken up by intestinal cells, reesterified to a palmitate, incorporated into the chylomicra of the mucosa and transported to the liver through lymph (16). Vitamin A is transported to the RPE via plasma. Given that an alteration in the retinoid metabolism in the RPE of the vitiligo mouse was observed, it was necessary to determine if this represented a systemic effect as well.

Analysis of Retinoids in Liver and Plasma

Using HPLC analysis, retinoid levels were determined in liver and plasma of affected and control animals at 6 weeks of age, a time when there was a significant elevation of retinoid concentration in the eye (14). The mean liver total vitamin A levels in vitiligo mice were approximately 1.7 times greater than and differed significantly from controls. In 6 week mutant mice, the level of vitamin A was 477.43 nmol/g ± 63.5 (n = 6), whereas in controls it was only 279.1 nmol/g ± 104.55 (n = 6). Analysis of esterified vs. unesterified retinoids showed that the elevation was in retinyl palmitate levels. Unlike the RPE, there was no elevation of retinol concentration.

In the vitiligo mouse plasma, the level of retinol averaged 1.04 μmol/L ± 0.10 (n = 4) and in controls, the levels averaged 1.18μmol/L ± 0.17 (n = 4). These values did not differ significantly. Other studies have shown that plasma retinol levels are usually maintained within a normal range despite wide variations in dietary and hepatic vitamin A levels (17), so the lack of alteration in plasma retinol levels in the vitiligo mice was not surprising.

SUMMARY AND FUTURE DIRECTIONS

The vitiligo mouse is an intriguing example of slow-progressive degeneration of the retina. It appears to have an alteration in retinoid metabolism that occurs not only in the RPE, but perhaps in the liver as well. These studies represent the first evidence of biochemical malfunction in this mutant mouse. Studies to determine if there are alterations in the enzymes that control the levels of various forms of Vitamin A must be performed if the precise biochemical lesion is to be determined. The morphologic and biochemical studies described here are timely because information about the genetic defect in this mouse has been reported recently. The vitiligo mutation maps to the *mi* (microphthalmia) allele of mouse chromosome 6 (18, 19) and so it is designated the mi^{vit}/mi^{vit} mouse. The *mi* locus codes for a basic helix-loop-helix DNA transcription factor (20,21). In the case of the vitiligo mouse, the specific mutation is a G to A transition at bp 793 leading to an Asp 222Asn substitution in helix 1 (22). It has not been determined how this single point mutation in a DNA transcription factor results in the slowly-progressing retinal degeneration observed in the vitiligo mutant. Our studies of altered retinoid metabolism and the functions of RPE and PRCs should help in determining how this mutation results in this retinopathy.

ACKNOWLEDGMENTS

This research was supported in part by U.S. Public Health Service Grant EY09682, unrestricted funds from Research to Prevent Blindness awarded to the Department of Ophthalmology, Medical College of Georgia and the Medical College of Georgia Research Institute. The original breeding pairs of vitiligo mice were provided by Dr. R. L. Sidman,

New England Regional Primate Research Center, Southboro, MA (funded by NEI grant EY06859).

REFERENCES

1. Lerner AB, Shiohara T, Boissy RE, Jacabson KA, Lamoreux ML, Moellmann GE. (1986) A mouse model for vitiligo. J. Invest. Dermatol. 87, 299-304.
2 Nussenblatt RB, Palestine AG. (1989) Vogt-Koyanagi-Hirada syndrome. In Retina. SJ Ryan (Ed.) Mosby, St Louis, pp. 723-728.
3. Sidman RL, Neumann P. (1988) Vitiligo: a new retinal degeneration mutation. Mouse News Letter 81, 60.
4. Smith SB. (1992) C57BL/6J-vit/vit mouse model of retinal degeneration: Light microscopic analysis and evaluation of rhodopsin levels. Exp. Eye Res. 55, 903-910.
5. Smith SB. (1995) Evidence of a difference in photoreceptor cell loss in the peripheral versus posterior regions of the vitiligo (C57BL/6J-mi^{vit}/mi^{vit}) mouse retina. Exp. Eye Res. (In press).
6. Boissy RE, Moellmann GE, Lerner AB. (1987) Morphology of melanocytes in hair bulbs and eyes of vitiligo mice. Am. J. Pathol. 127, 380-388.
7. Smirnakis SM, Tang M, Sidman RL. (1991) Abnormalities of pigment epithelium precede photoreceptor cell degeneration in vitiligo mutant mice. Invest. Ophthalmol. Vis. Sci. (Suppl). 31, 298.
8. Smith SB, Hamasaki DI. (1994) Electroretinographic study of the C57BL/6-mi^{vit}/mi^{vit} mouse model of retinal degeneration. Invest. Ophthalmol. Vis Sci. 35, 3119-3123.
9. Smith SB, Cope BK, McCoy JR, McCool DJ, Defoe DM. (1994) Reduction of phagosomes in the vitiligo mouse model of retinal degeneration. Invest. Ophthalmol. Vis. Sci. 35, 3625-3632.
10. Herron WL, Riegel BW, Myers OE, Rubin ML. (1969) Retinal dystrophy in the rat - a pigment epithelial disease. Invest. Ophthalmol. Vis. Sci. 8, 595-604.
11. Smith SB, Defoe DM. (1995) Autoradiographic and biochemical assessment of rod outer segment renewal in the vitiligo (mi^{vit}/mi^{vit}) mouse model of retinal degeneration. Exp. Eye Res. (In press).
12. Smith SB, McCoy JR, Cope BK. (1993) Opsin synthesis in the C57BL/6-mi^{vit}/mi^{vit} mouse model of retinal degeneration. Curr. Eye Res., 12, 711-717.
13. Smith SB, Cope BK, McCoy JR. (1994) Effects of dark-rearing on the retinal degeneration of the C57BL/6-mi^{vit}/mi^{vit} mouse. Exp. Eye Res. 58, 77-84.
14. Smith SB, Duncan T, Kutty G, Kutty RK, Wiggert B. (1994) Increase in retinyl palmitate concentration in eyes and livers and the concentration of interphotoreceptor retinoid-binding protein in eyes of vitiligo mutant mice. Biochem. J. 300, 63-68.
15. Saari JC, Bredberg DL, Farrell DF. (1993) Retinol esterification in bovine retinal pigment epithelium: Reversibility of lecithin:retinol acyltransferase Biochem. J. 291, 697-700.
16. Olson JA. (1969) Metabolism and function of vitamin A. Fed Proc. Fed. Am. Soc. Exp. Biol,.28, 1670-1677.
17. Duncan T, Green JB, Green MH. (1993) Liver vitamin A levels in rats are predicted by a modified isotope dilution technique. J. Nutr. 694-703.
18. Lamoreux ML, Boissy RE, Womack JE, Nordlund JJ. (1992) The vit gene maps to the mi (microphthalmia) locus of the laboratory mouse. J. Hered. 83, 435-439.
19. Tang M, Neumann PE, Kosaras B, Taylor BA, Sidman RL. (1992) Vitiligo maps to mouse chromosome 6 within or close to the mi locus. Mouse Genome, 90, 441-443.
20. Hodgkinson CA, Moore KJ, Nakayama A, Steingrimsson E, Copeland NG, Jenkins NA, Arnheiter H. (1993) Mutations at the mouse microphthalmia locus are associated with defects in a gene encoding a novel basic-helix-loop-helix-zipper protein. Cell, 74, 395-404.
21. Hughes MJ, Lingrel JB, Krakowsky JM, Anderson KP. (1993) A helix-loop-helix transcription factor-like gene is located at the mi locus. J. Biol. Chem. 268-20687-20690.
22. Steingrimsson E, et al (1994) Molecular basis of mouse microphthalmia (mi) mutations helps explain their developmental and phenotypic consequences. Nature Genetics 8:256-263.

19

IMMUNOLOGICAL ASPECTS OF RETINAL TRANSPLANTATION IN RETINAL DEGENERATION RODENTS

Luke Qi Jiang,[1] Duco Hamasaki,[2] Jessica Zuletta,[2] and Marianela Jorquera[1]

[1] The Schepens Eye Research Institute
Department of Ophthalmology, Harvard Medical School
Boston, Massachusetts
[2] Bascom Palmer Eye Institute
University of Miami
Miami, Florida

ABSTRACT

Retinal transplantation holds promise as a treatment for restoring vision in the eye which has been blinded by retinal degenerative disease. We have demonstrated that allogeneic retinal grafts implanted into intraocular spaces of normal eyes are immunologically privileged and this privilege is associated with an active downregulation of systemic immunity. Since the normal eye maintains an immune suppressive intraocular microenvironment, it is important to determine 1) whether the degenerative eye has a similar immunological microenvironment and 2) what impact donor-specific alloimmunity has on the function of retinal allografts. Two sets of experiments were carried out to answer these questions. First, we assayed the systemic immunity in C3H/Hen retinal degenerative mice which received an allogeneic neural retinal graft in either the anterior chamber or subretinal space. We found that these spaces are immunologically privileged sites for allogeneic retinal grafts and this privilege is accompanied by induction of suppression of donor-specific delayed hypersensitivity. Moreover, intraocular retinal allografts in C3H/Hen mice enjoyed a prolonged survival in these immunologically privileged sites. Our results from rd mice suggest that photoreceptor cells may not play an important role in maintaining an immune suppressive intraocular microenvironment. Second, we implanted normal retinal pigment epithelium (RPE) allografts into the subretinal space of retinal degenerative Royal College of Surgeons (RCS) rats and corneal electroretinographic (ERG)s were periodically recorded. The results indicate that the normal RPE allografts can rescue ERG function in the RCS recipients; however, after donor-specific immunization, the rescued ERG function was abolished. The property of immune privilege is maintained in the retinal degenerative eyes of RCS rats and rd mice. However, the emergence of systemic alloimmunity can overcome the immune privilege leading to an immunologically mediated impairment of retinal graft

Degenerative Diseases of the Retina, Edited by Robert E. Anderson et al.
Plenum Press, New York, 1995

function. Thus, systemic immunity directed at donor-specific antigens is a major obstacle for functional retinal transplants.

INTRODUCTION

A common consequence of degenerative retinal disease is blindness caused by incurable loss of retinal cells and destruction of the retina. Progress in studies of retinal transplantation over the past several years has provided hope for using grafted retinal cells to replace those which have been lost as a means of restoring vision to blind eyes. Scientists have attempted to implant either neural retinal or RPE cells into the eyes of different rodent models. The results are encouraging. Both neural retinal and RPE grafts have been shown to survive and even function for a period of time [1-7]. The relatively better survival for the intraocular retinal grafts is due to a hospitable transplantation site in the eye. Today, we know that intraocular spaces of the normal eye are immunologically privileged sites for retinal allografts and this privilege can be attributed to several factors: 1) an active down regulation of systemic immunity, 2) a constitutive immunosuppressive microenvironment, and 3) the blood/eye barrier. Thus, immune privilege is a beneficial factor for the survival of retinal grafts. However, this privilege for retinal allografts is neither permanent nor absolute. We have previously demonstrated that both neural retinal and RPE allografts implanted into a normal (i.e., non-diseased) eye eventually deteriorated, and this deterioration was accompanied by the emergence of systemic alloimmunity [8]. Because therapeutic retinal transplantation eventually will be performed in the diseased eye in which retinal cells have been destroyed by degeneration or other types of damage, it is important to determine the immunological features of the diseased eye.

In this study, retinal transplantation was performed using two different retinal degeneration rodent models, RCS rats and retinal degenerative (rd) mice. Although the primary defect in these models is considered to be mediated by different cell types [9], the pathological consequences in their eyes appear similar. Both the RCS rats and rd mice progressively lose their photoreceptor cells after birth. In addition, it has been reported that abnormal retinal vascularization accompanies deterioration of their retinas [10]. The pathological changes in the eyes of these animals may have an impact on their immunological features. We examined the fate of the retinal grafts implanted into the eyes of these retinal degenerative rodents using either histologic or ERG examinations. An immunological assay for delayed hypersensitivity (DH) was used to measure systemic immunity. In order to observe the impact of systemic immunity on graft survival, we also determined the function of retinal grafts after the host received donor-specific immunization by measuring the alteration of the rescued corneal ERG.

MATERIAL AND METHODS

Animals

Donor retinal tissue for the mouse experiments was obtained from newborn BALB/c mice (aged <24 hours). Adult male C3H/Hen mice (aged 7 to 12 weeks) served as allogeneic recipients. All experimental mice were obtained from our breeding colonies at the University of Miami School of Medicine, Miami, FL, The Schepens Eye Research Institute, Boston, MA, and from the Charles River Laboratories, Wilmington, MA. Mice were maintained in a common room of the vivarium where an overhead fluorescent light provided 12 hour cycles of light and dark. Inoculations, clinical examinations, and enucleations of grafted eyes were performed under anesthesia induced by intramuscular injections of ketamine (Ketalar, Parke Davis) 0.075 mg/g body weight, and xylazine (Rompun, Haver-Lockhart), 0.006 mg/g body weight.

All experimental RCS rats were obtained from the breeding colony at the University of Miami School of Medicine, Miami, FL. The original breeding pairs of RCS rats were supplied by the NIH Genetic Resources Services. The recipient RCS rats ($rdy^{-}p^{-}$) are a pink-eyed, albino inbred strain with retinal dystrophy. Their MHC Class I (*RT1A*) haplotype is u and Class II region remains unknown [11]. For transplantation, host RCS rats aged 20 days were used.

Donor RPE was isolated from (RCSxLong Evans(LE))F1 rats aged about 20 days. Three female LE rats were obtained from Charles River Laboratories (Wilmington, MA) and the male RCS rats from our breeding colony. The LE rats were chosen as the donor source so that the donor/host combination in our experiments would be compatible with previous morphological studies [1-2]. Because the LE rats used in this study are an outbred strain, offspring from the same litter were used as the source for donor RPE tissue.

The experimental rats were maintained in clear plastic cages and fed Lab Chow (Purina, St. Louis, MO). The illumination in the animal room ranged from 4-8 ft/cd and the light was kept on a 12:12 light:dark cycle. Inoculation, clinical examination and enucleation of the eyes were performed under anesthesia induced by intramuscular ketamine (Ketalar, Parke Davis, Shawnee, KS) 0.075 mg/g body weight, and xylazine (Rompun, Haver-Lockhart, Morris Plains, NJ), 0.006 mg/g body weight. For ERG recording, the animals were anesthetized with ketamine-xylazine-urethane (15 mg, 15 mg, and 600 mg/kg body weight, respectively). All experimental procedures conformed to the ARVO Resolution on the Use of Animals in Research.

Preparation of Retinal Tissue

Donor RPE tissue was prepared using a modification of the technique of Chang et al. [12]. Briefly, donor (RCSxLE)F1 rats were decapitated, their eyes immediately enucleated and placed in ice-cold Dulbecco's modified Eagle's medium (DMEM) with 2 mM L-glutamine. The intact eyeballs were incubated for 45 minutes at 37°C in a solution of 2% Dispase (Grade II, Boehringer Mannheim, Indianapolis, IN) in DMEM. After incubation, the eyes were rinsed twice in DMEM and placed in a Petri dish containing fresh DMEM. Under a dissecting microscope, the eyeballs were cut open along the edge of the cornea using microsurgical scissors. After the lens and iris were removed, the RPE and its attached neural retina were gently separated from the eye cup and transferred to another Petri dish containing fresh DMEM. After two changes to fresh medium and incubation for 15 minutes at room temperature, the entire RPE sheet was easily separated from the neural retina. Separated RPE sheets were transferred to a Petri dish containing cold calcium- and magnesium-free Hank's Balanced Salt Solution (CMF-HBSS) and the dish placed on ice. Each RPE sheet was cut in half and one half implanted into the left eye of recipient rats.

For preparation of donor neural retinal tissue, BALB/c newborn mice (<24 hours) were decapitated. Their eyes were immediately enucleated and placed in ice-cold CMF-HBSS. Each eye was cut open along the edge of the cornea using microsurgical scissors. After the lens and iris were removed, the neural retina was gently separated from the eye cup using fine forceps. The retinas were transferred to a new Petri dish containing cold CMF-HBSS. After rinsing twice with fresh medium, the entire neural retina sheet was cut in half and each half used for transplantation.

Implantation of Retinal Grafts into the Anterior Chamber, Subretinal Space or Subconjunctival Space

Recipient mice received general anesthesia. For implantation of retinal tissue into the anterior chamber or subretinal space, the eyelids were kept open and the eyeball held steady with forceps. A 0.3 mm penetrating wound for the anterior chamber was made at the cornea of the eye using a microsurgical knife with a 15° angle (Edward Weck and Company, Inc., Research Triangle Park, NC). For retinal implantations into the subretinal space, a penetrating wound was made through the posterior portion of the wall of the eye while the wound was in the fornix portion of the conjunctiva for implantation into the subconjunctival space. The retinal tissue was then drawn into a glass needle made from a glass bore of a 10 μl micropipetter (diameter of 200 μm). The retinal tissue, along with about 1 μl of HBSS, was slowly injected via the wound into the anterior chamber, subretinal space or subconjunctival space of the eye.

Clinical Examination

Clinical examinations were made to determine changes in the graft size and general appearance of the retinal grafts, including translucency and vascularization. In addition, the host eyes were examined for evidence of intraocular inflammation. Under anesthesia the anterior segment of the eye was examined on designated days after transplantation with a dissecting microscope. To examine the posterior segments of the eye, the pupils were dilated with 0.5% mydriacyl (Alcon, Humacao, RP) and 2.5% phenylephrine hydrochloride, and the fundus was visualized through a contact lens. The eyes bearing retinal grafts were photographed using a dissecting microscope equipped with a 35 mm camera.

Histological Examination

Based on identification of the graft during clinical examination, sections were cut through the graft. Representative sections were selected at three levels and examined. For H&E staining the eyes were fixed with 10% buffered formalin, embedded with paraffin and cut 5 μm thick.

Photographs

Photographs of the fundus were taken with a 35-mm camera system attached to an Olympus dissection microscope (using Kodak Plus-X pan 125 Black-and-White print film). Micrographs of sections stained with H and E technique were taken with a 35-mm camera system attached to an Olympus research microscope (using Kodak Plus-X pan 125 Black-and-White print film).

Delayed Hypersensitivity Response to Retinal Grafts

DH was measured based on ear swelling, as previously described [13]. Briefly, half of the neural retina of newborn BALB/c donor mice was implanted into the anterior chamber of recipient mice. To set positive controls, mice received similar retinal grafts into the subconjunctival space. Normal C3H/Hen mice without retinal grafts served as negative controls. On day 12 after transplantation, ear pinnae of the mouse were challenged and measured for DH responses. A Mitutoyo engineer's micrometer was used to measure the thickness of both ears immediately before challenge. For challenge, 10^6

(BALB/cxC3H/Hen)F1 spleen cells (irradiated) were suspended in 10 μl of HBSS and injected into the dermis of the left ear pinnae. The right ear served as an untreated control. The difference in measured ear thickness after 24 hours was used as a measure of DH intensity. Results were expressed as specific ear swelling = (24 hour measurement - 0 hour measurement) experimental ear - (24 hour measurement - 0 hour measurement) negative control ear x 10^{-3} mm. A two-tailed Student's *t* test was performed on the data presented and significance was assumed to exist if $p<0.05$.

Adoptive Transfer of Capacity to Suppress Alloantigen-Specific DH

Spleens from C3H/Hen mice bearing BALB/c retinal grafts in the AC were collected aseptically. Single cell suspensions were prepared by pressing whole spleens through stainless steel screens (60-mesh). Spleen cells were then washed and resuspended in HBSS. Each naive C3H/Hen recipient received 5×10^7 spleen cells in 100 μl of HBSS via the tail vein. For positive control, a similar number of naive C3H/Hen spleen cells were infused intravenously into naive C3H/Hen recipients. Negative controls received no infusion. Within two hours, all experimental mice received implants in subconjunctival space of neural retina grafts from immature BALB/c mice (aged 8-14 days). Twelve days later, DH reactivity, as described above, was assayed by ear challenge.

ERG Recording

Under anesthesia, ERGs were picked up by a wick Ag:AgCl electrode placed on the cornea. The reference electrode was a needle placed subcutaneously on the head and the animal was grounded using another needle electrode placed subcutaneously in the neck region. The pupils were dilated with 0.2% tropicamide, and the cornea was anesthetized with 0.5% topical proparacaine HCl.

The ERGs were amplified and displayed on an oscilloscope with the half-amplitude bandwidth set at 0.1-10 KHz. The amplified signals were sent to a Texas Instruments 960A minicomputer (Houston, TX) to average 16 responses with an analysis time of 10 msec. The data were permanently stored on digital tapes and the amplitude of the average ERGs measured from a printout of the digital values.

The light for the stimulus was obtained from a 150 watt quartz-halogen lamp bulb. The filament of the bulb was focused by a lens onto a 3-mm diameter fiberoptic bundle. The fiberoptic bundle was brought into a Faraday cage and positioned with the tip 2-3 mm from the cornea of the eye.

The luminance of the unattenuated stimulus was 5.431 log cd/m^2 (Salford Electrical Instruments, Ilford, England) and neutral-density (ND) filters (Eastman Kodak, Rochester, NY) were used to reduce the stimulus intensity. A Uniblitz shutter (Eastman Kodak. Rochester, NY), controlled by a Grass S4 stimulator (Quincy, MA), was used to deliver flashes of 250-msec duration at 5 sec intervals.

The anesthetized animal was strapped to a platform with its head resting in a U-shaped holder. The upper and lower lids were retracted with masking tape to hold open and proptose the eye. After the electrodes were in place and the fiberoptic bundle adjusted to provide uniform illumination of the eye, the animal was dark adapted for 30 min. The ERG recordings were started with a 8.0-log unit ND filter. In the preliminary experiments, this was found to be lower than the stimulus intensity necessary to elicit a small b-wave response from congenic rdy^+p^- RCS rats. In addition, prolonging the dark-adaptation to 1 hr did not change the threshold or amplitude of the ERGs of normal rats. The stimulus intensity was increased in 0.5-log unit steps and responses were recorded from 8.0 to 3.5 log units (ten intensities). A recovery time was provided between each increase in stimulus intensity; 30 sec between filters 8.0 to 7.5 and increasing in 30-sec steps with each increase in intensity. The recovery time between the 4.0 and 3.5 filters was 4.5 min.

A two-tailed student's *t* test was performed on the ERG amplitude data presented and significance was assumed to exist if $p < 0.05$.

RESULTS

Two sets of experiments using rd mice and RCS rats were included in this study.

Retinal Implantation in the Eye of rd Mice

Clinical Findings. Neural retinas obtained from newborn BALB/c mice (age <24 hr) were implanted into the anterior chamber (n=5) or subretinal space (n=5) of the eye of adult C3H/Hen mice. The experiment was repeated using the same number of animals. Periodically after implantation, both the host eye and graft were clinically examined. Under a dissecting microscope, the graft implanted in the anterior chamber appeared as a white translucent mass with well defined edges (Fig. 1). There was no neovascularization on the surface of the retina nor was there evidence of inflammation in the anterior chamber of the eye. For 12 days, the graft showed no significant change in size. An examination of the fundus revealed that the graft implanted into the subretinal space had a similar appearance as the graft in the anterior chamber (Fig. 2). The vessels of the retina were superior to the graft and no neovascularization was detected in the eye. Overall, as demonstrated by clinical examination, BALB/c newborn neural retinal grafts implanted into both the anterior chamber and subretinal space survived well. The appearance of the graft was very comparable to that previously found in normal mice [13].

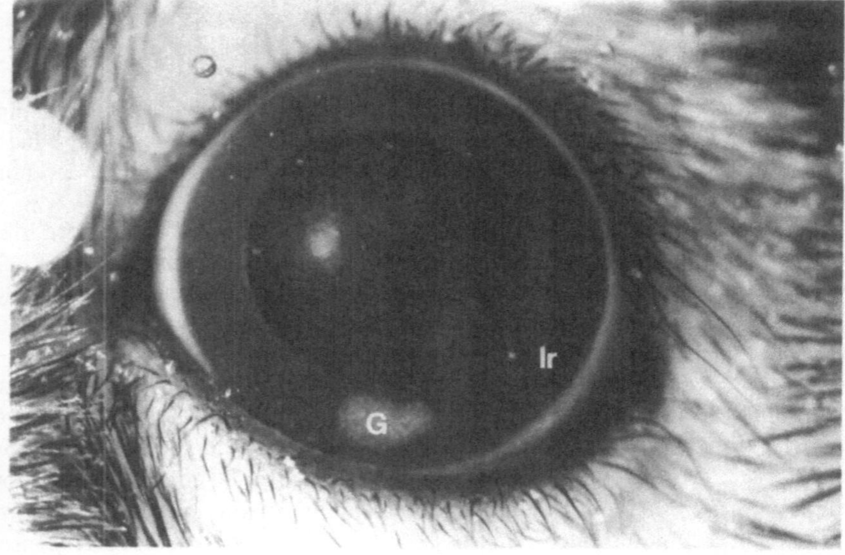

Figure 1. Clinical appearance of BALB/c newborn neural retinal graft implanted into the anterior chamber of C3H/Hen mouse 12 days after transplantation. The graft (G) has a white translucent appearance and rests on the host iris (Ir).

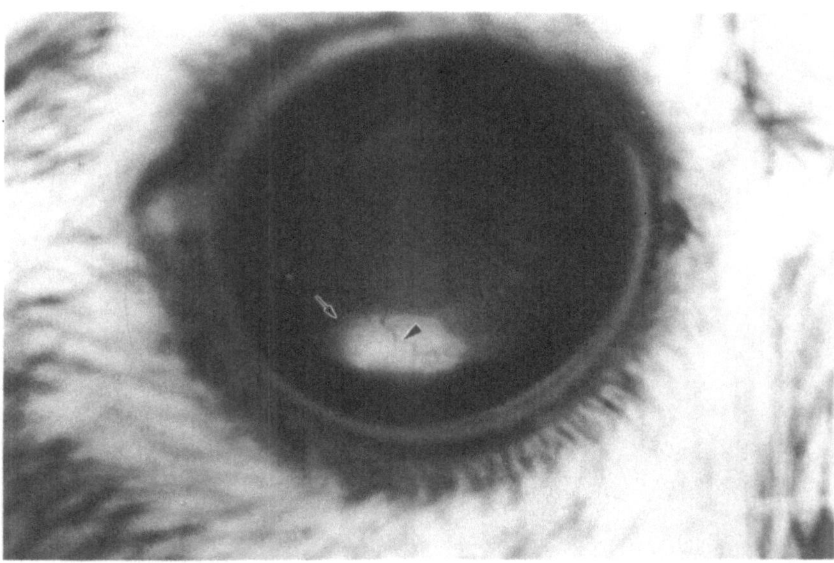

Figure 2. The appearance of the fundus of a C3H/Hen mouse which received BALB/c newborn neural retinal graft into the subretinal space of the eye 12 days prior. The retinal vessels (arrowhead) are superior to the white mass of the retinal graft (arrow).

Histological Findings. The eyes of C3H/Hen mice (n=10) which received newborn BALB/c neural retinal grafts 12 days previously were enucleated and histologically examined. Examination of H&E stained sections (Fig. 3) indicate that the graft consists of differentiated and well-organized cell layers. The photoreceptor cells formed numerous rosettes which were surrounded by large cells that resembled those cells found in the inner nuclear layer of the retina. There were no inflammatory cells infiltrating the graft, anterior chamber or host iris. An examination of the subretinal grafts revealed that the graft contained multiple well-organized cell layers; however, the number of rosettes formed by the photoreceptor cells was significantly reduced in these animals (Fig. 4). There was no evidence of inflammatory cell infiltration in the retinal graft or in the space and tissue surrounding it. The histological appearance of the subretinal graft in rd mice is different from the appearance of the graft in either the anterior chamber of rd mice or subretinal space of normal mice [13].

Immunological Examination. In our previous studies on normal mice, we have demonstrated that allogeneic retinal grafts implanted into the eye can induce suppression of donor-specific DH in their recipients [13]. In order to determine whether this also occurs in the degenerative mouse, we implanted newborn BALB/c neural retinas into either the anterior chamber or subretinal space of adult C3H/Hen mice. For positive controls, similar grafts were implanted into the subconjunctival space of the host eye, while negative control mice did not receive a graft (n=5/group). Twelve days later, all mice received ear challenge with irradiated spleen cells of (BALB/cxC3H)F_1 mice, and 24 and 48 hrs later ear swelling was measured to determine DH. As shown in Figure 5, grafts implanted into the subconjunctival space (positive control) induced a significant DH in their recipients. In contrast, no significant DH was induced by grafts implanted into either the anterior chamber or subretinal space. These results suggest that allografts

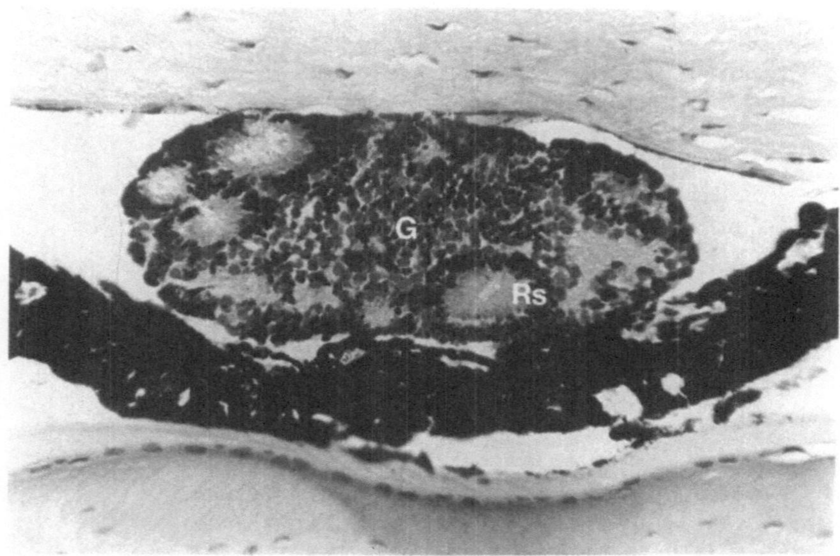

Figure 3. Histological appearance of newborn BALB/c neural retinal graft implanted into the anterior chamber of the eye of a C3H/Hen mouse 12 days previously. The graft (G) contained different cells layers and photoreceptor cells formed rosettes (Rs). There was no evidence of inflammatory cell infiltration.

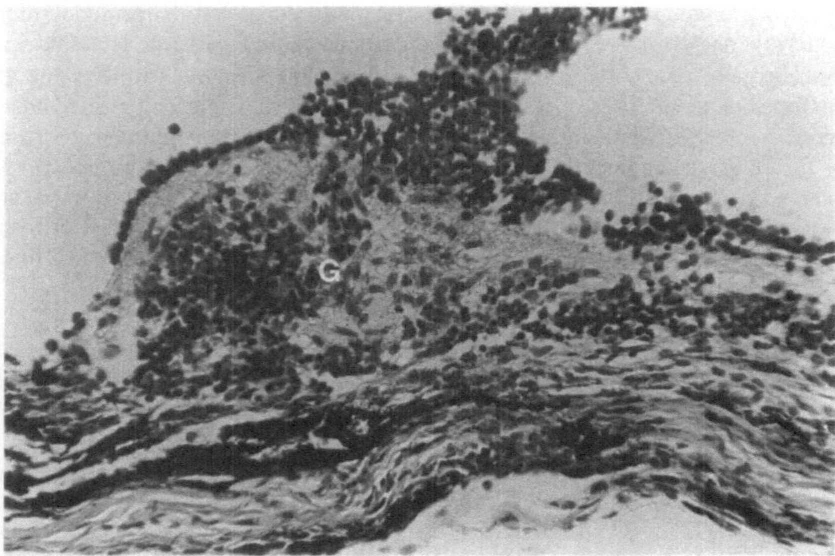

Figure 4. Histological appearance of newborn BALB/c neural retinal graft implanted into the subretinal space of the eye of a C3H/Hen mouse 12 days previously. Although the graft (G) contained organized cell layers, the rosettes (formed by photoreceptor cells) could not be clearly detected. There is no evidence of inflammatory cell infiltration.

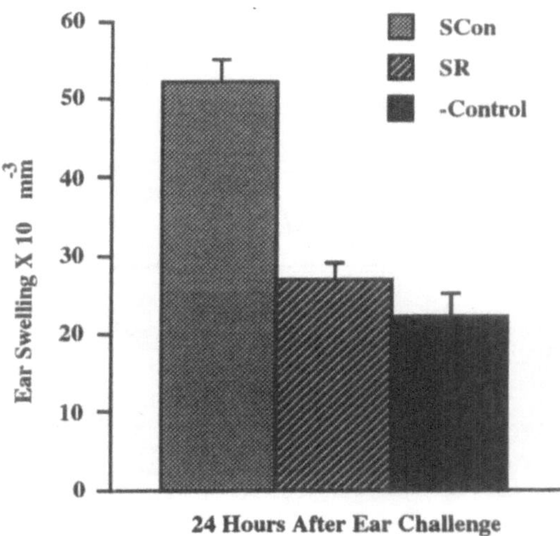

Figure 5. DH response in C3H/Hen mice receiving BALB/c retinal allografts. Newborn neural retinal allografts of BALB/c mice were implanted into the anterior chamber or subretinal space (test groups) 12 days prior. Similar grafts were implanted into the subconjunctival space (positive control). Negative control mice received no graft. Twenty-four hours after ear challenge with irradiated spleen cells of (BALB/cxC3H)F_1 mice, ear swelling was measured. Ear swelling in the positive control group was significantly higher than in either of the test groups or the negative control group ($p < 0.05$).

implanted into the retinal degenerative eye are capable of inducing suppression of donor-specific DH.

We next used an adoptive transfer assay to confirm the generation of donor-specific suppressor T cells in recipients of anterior chamber and subretinal grafts. As Figure 6 indicates, naive C3H/Hen mice which received splenic cells from C3H/Hen mice bearing either anterior chamber or subretinal grafts, mounted insignificant DH after implantation with subconjunctival grafts. Therefore, when retinal allografts were implanted into the eye of retinal degenerative mice, they retained the capacity to induce the generation of donor-specific suppressor T cells in their recipients.

Retinal Implantation in the RCS Model. We have shown in our previous studies that normal RPE from (RCSxLE)F_1 rats implanted into the subretinal space of RCS rats can rescue corneal ERG function [6]. The criteria for determining rescued function was: 1) the grafted eye displayed a significantly larger photoreceptor potential (negative wave) ampli-

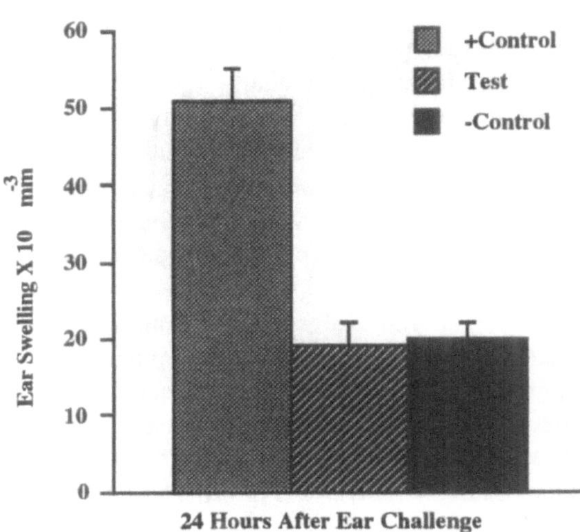

Figure 6. Adoptive transfer of suppression of DH induced by newborn BALB/c retinal allografts in C3H/Hen recipient mice. For test groups, naive mice received spleen cells (50×10^6) intravenously from mice bearing either anterior chamber or subretinal space neural retinal allografts. One hour later, recipients received BALB/c retinal allografts into the subconjunctival space. Twelve days after implantation their ears were challenged with (BALB/c x C3H/Hen)F1 spleen cells and ear swelling responses measured 24 hours later. Negative controls as described in Figure 3. Responses of positive control group were significantly greater than those of either test group or negative control groups ($p < 0.05$).

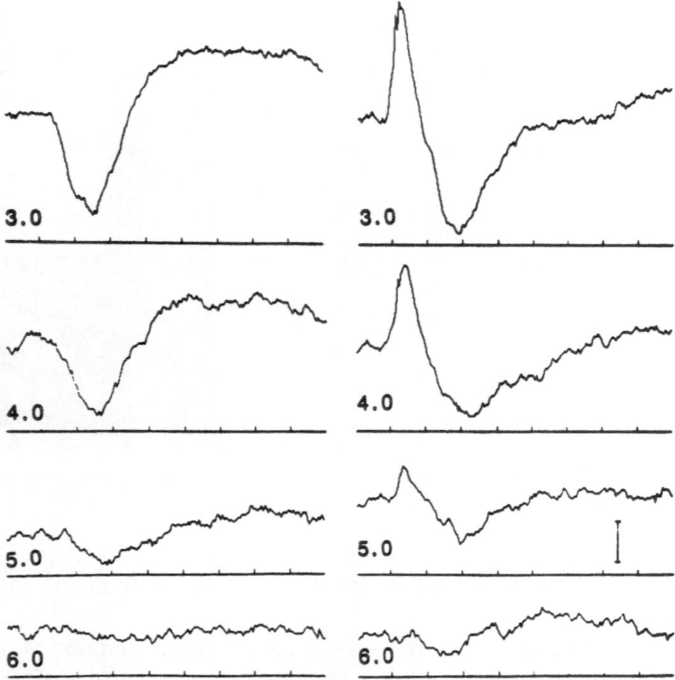

Figure 7. Corneal ERGs recorded from the right eye (left column) and left eye (right column) of a 60 day old RCS rat which received RPE allografts into the subretinal space of the left eye at 20 days of age. The number to the left of each ERG represents the value of the ND filter used to reduce stimulus intensity. The full intensity stimulus was 5.431 log cd/m^2. Calibration: 20 μV; 200 msec.

tude compared to the non-grafted eye, or 2) the existence of a b-wave in the grafted eye, but not in the non-grafted eye. This rescued function was sustained for up to 16-17 weeks at which time the experiment ended [6]. However, the amplitude of the rescued ERG declined during this period, indicating that deterioration and impairment of function of RPE graft had gradually taken place.

In our previous retinal transplantation studies in mice, we found that allogeneic retinal grafts were eventually rejected by conventional alloimmunity [8]. In order to determine the impact of alloimmunity induced by (RCSxLE)F1 grafts on rescued ERGs, we randomly selected 14 recipients which had a functional rescue of ERGs as described above (Fig.7). Six recipients were challenged with an injection of (RCSxLE)F1 spleen cells into the subcutaneous space while the remaining 8 recipients received a subcutaneous injection of HBSS and served as controls. The corneal ERG in both eyes were recorded periodically for the next 3 to 8 weeks. Three weeks after donor-specific immunization, 5 of 6 animals had no significant difference in the amplitude of negative wave between the left and right eye (Fig. 8). In contrast, during a similar time period, all non-challenged animals retained significant rescued ERG. These rescued ERGs were followed for 16-17 weeks after HBSS injection. Although the ERG responses showed a decline in both eyes, the amplitude of the negative wave in the grafted (left) eye remained significantly greater than that recorded in the untreated (right) eye (Fig. 9). These results suggest that a donor-specific alloimmunity can abolish the rescued ERG and, may indeed have a negative impact on the rescue of corneal ERG function in RCS rats.

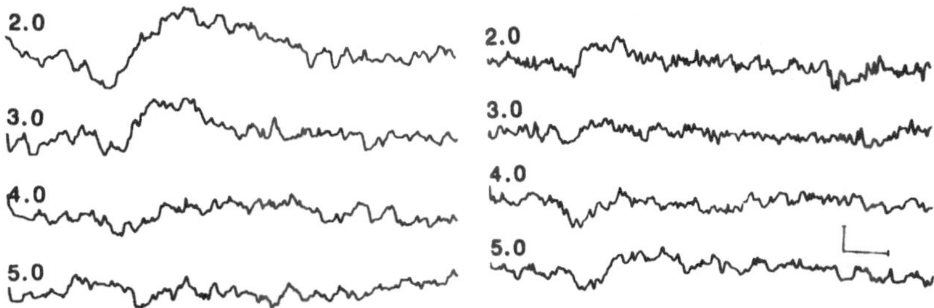

Figure 8. Corneal ERGs recorded from the right eye (left column) and left eye (right column) of the same RCS rat, as shown in Figure 7, 21 days after challenge with donor-specific splenic cells. The number to the left of each ERG represents the value of the ND filter used to reduce stimulus intensity. The full intensity stimulus was 5.431 log cd/m^2. Calibration 12.5 μV; 200 msec.

DISCUSSION

The eye, an extension of the brain, is an immunologically privileged site in which a grafted foreign tissue enjoys prolonged survival. We have previously demonstrated that retinal allografts implanted into the intraocular spaces of the eye, including the anterior chamber, vitreous cavity and subretinal space, can survive well for at least 12 to 17 days and the grafts induce a suppression of DH directed at donor-specific alloantigens [13,14]. It has also been shown that several immunosuppressive cytokines (e.g., TGF-b, VIP, and MSH-a) constituitively exist in the aqueous humor [15,16]. These cytokines are secreted by the resident cells of the eye, and in the retina the RPE has been identified as one of the major sources of TGF-b [17].

In the degenerative retina of either rd mice or RCS rats, the photoreceptor cells are progressively lost after birth. Although the primary cause of degeneration may be different in these two rodents, pathological changes in the retina occur in a similar fashion [9]. It has been reported that this degenerative process is also accompanied by abnormalities of the blood vessels [10]. Therefore, the microenvironment in the eye of retinal degenerative rodents may be altered by degeneration of the retina, and whether these pathological changes in the retina can also alter the immunological properties of the eye remains unknown. Our study suggests that allogeneic retinal grafts implanted into the eye of rd mice survive for a comparable period of time as grafts implanted into a normal mouse eye. This indicates that

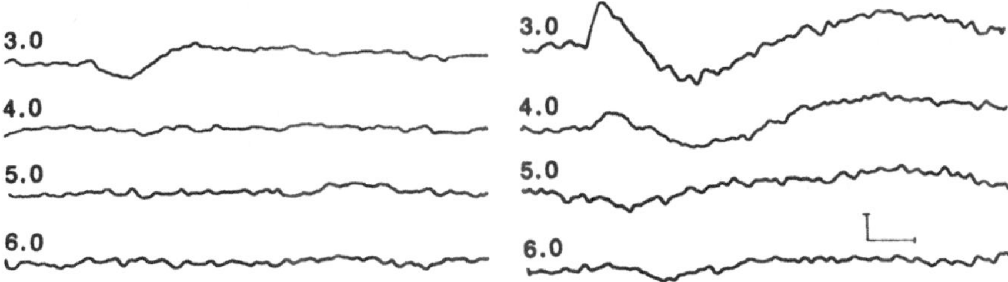

Figure 9. Corneal ERGs recorded from the right eye (left column) and left eye (right column) of a 17 week old RCS rat which received RPE allografts into the SR of the left eye at 20 days of age. The animal received a sham injection of HBSS into the subcutaneous space at age 60 days and served as a non-challenged control. The number to the left of each ERG represents the value of the ND filter used to reduce stimulus intensity. The full intensity stimulus was 5.431 log cd/m^2. Calibration: 12.5 μV; 200 msec.

the property of immune privilege can be sustained in the retinal degenerative eye. Since the allogeneic retinal graft can induce suppression of donor-specific DH at 12 days, it appears that photoreceptor cells do not play a significant role in the induction of this suppression. Interestingly, neonatal neural retinal allografts implanted into either the anterior chamber or subretinal space of rd mice survive well for 12 days; however, histological examination reveals that the photoreceptor layer is significantly reduced in grafts implanted in the subretinal space but not in the anterior chamber. Thus, under similar immunologically privileged conditions, grafts implanted into the anterior chamber appear healthier than grafts in the subretinal space, suggesting that in addition to an immunological influence, there may also be another influence, possibly neurobiological, on graft survival. A previous retinal transplantation study using rd mice suggests that retinal graft survival is dependent on both the intrinsic properties of the graft and the intraocular microenvironment [18]. It is possible that the neurobiological microenvironments of the anterior chamber and subretinal space are different. Further experiments are necessary to confirm this speculation.

We have learned from our previous studies that immune privilege to retinal allografts implanted into the eye is not permanent. When allografts mismatched at both major and minor histocompatibility loci are implanted into the host eye, a suppression of donor-specific DH is induced at 12 days. However, this suppression is eventually overcome as DH emerges, and by day 35 the graft has deteriorated [8]. It is known that histocompatibility does play an important role in the emergence of DH and systemic donor-specific alloimmunity can accelerate rejection of the retinal allograft [8,14].

In our study using the RCS rat model, significant rescue of corneal ERG was detected following implantation of normal RPE when compared to the ERG of the eye not receiving the graft. Although the corneal ERG rescued by RPE grafts was able to be sustained for several months, its amplitude declined remarkably during this period. After the host was immunized with spleen cells bearing donor-specific transplantation alloantigens, the previously rescued ERG was depressed in an accelerated fashion. These results imply that a potential risk of rejection exists for retinal allografts. Even if the graft is immunologically privileged, the emergence of systemic immunity is capable of overcoming this privilege, thereby causing graft rejection. The emergence of systemic immunity is dependent on the disparity between the donor and host [8]. The emergence of systemic immunity can also occur when donor cells accidentally escape from the eye into a conventional site (e.g., subconjunctival space), thereby immunizing the host. In addition, either infection or injury can alter the immunosuppressive intraocular microenvironment, therefore the previously privileged graft may immunize the host and induce a conventional donor-specific systemic immunity. Subsequently, the primary intraocular graft can be rejected. Therefore, systemic donor-specific alloimmunity is an immunological obstacle for functional RPE allografts.

Understanding the immunological aspects of the retinal degenerative eye is critical for achieving successful therapeutic retinal transplantation. It is encouraging to learn that properties of immune privilege exist in the eyes of two different retinal degenerative rodent models. However, immune privilege is not absolute and the alloimmunity induced by donor antigens prejudice graft survival. Searching for an effective immunological strategy for overcoming alloimmunity and preventing graft rejection is a key step for the ultimate goal of successful retinal transplantation.

ACKNOWLEDGMENT

Supported by a grant from the USPHS NEI EY-09595 and a grant from the National Retinitis Pigmentosa Foundation.

REFERENCES

1. Li, X.L., and Turner, J.E., 1988, Inherited retinal dystrophy into the RCS rat: Prevention of photoreceptor degeneration by pigment epithelial cell transplantation. *Exp. Eye. Res.* 47:911-947.
2. Gouras, P.G., and Lopez, P., 1989, Transplantation of retinal epithelial cells. *Invest. Ophthalmol. Vis. Sci.* 30:1681-1683.
3. Del Cerro, M., Notter, M.F., Wiegand, S.J., Jiang, L.Q., and Del Cerro, C., 1988, Intraretinal transplantation of fluorescently labeled retinal cell suspensions. *Neurosci. Lett.* 92:21-26.
4. Ehinger, B., Bergstrom, A., Seiler, M., Aramant, R.B., Zucker, C.I., Gustavii, B., and Adolph, A.R., 1991, Ultrastructure of human retinal cell transplants with long survival retinal times in rats. *Exp. Eye Res.* 53:447-460.
5. Silverman, M.S., and Hughes, S.E., 1989, Photoreceptor transplantation in inherited and environmentally induced retinal degeneration: Anatomy, immunohistochemistry and function. In "Inherited and Environmentally Induced Retinal Degenerations", (Ed. LaVail, M.). Alan R. Liss, Inc., p. 687-704.
6. Jiang, L.Q., and Hamasaki, D., 1994, Corneal electroretinographic function rescued by normal retinal pigment epithelial grafts in retinal degenerative Royal College of Surgeons rats. *Invest. Ophthalmol. Vis. Sci.* 35:4300-4309
7. LaVail, M.M., Li, L., Turner, J.E., and Yasumura, D., 1992, Retinal pigment epithelial cell transplantation in RCS rats. Normal metabolism in rescued photoreceptors. *Exp. Eye Res..* 55:555-562.
8. Jiang, L.Q., Jorquera, M., Streilein, J.W., and Ishioka, M., 1995, Unconventional rejection of neural retinal allografts implanted into the immunologically privileged site of the eye. *Transplantation.* 59, in press.
9. LaVail, M.M., 1981, Analysis of neurological mutants with inherited retinal degenerations: Friedenwald Lecture. *Invest. Ophthalmol. Vis. Sci.* 21:638-657.
10. Blank, J.C., and Johnson, L.V., 1986, Vascular atrophy in the retinal degenerative rd mouse. *J. Comp. Neurol.* 254:543-553.
11. Kunz, H.W., Dixon-McCarthy, B., Lepre, M.A., Hansen, C.T., and Gill, T.J., 1991, RT1.A (MHC), RT2, and RT3 (Blood Group) specificities of 44 inbred and congenic rat strains from the NIH General Resource. *ILAR News.* 33:44-47.
12. Chang, C.W., Roque, R.S., Defoe, D.M., and Caldwell, R.B., 1991, An improved method for isolation and culture of pigment epithelial cells from rat retina. *Curr. Eye Res.* 10:1081-1086.
13. Jiang, L.Q., Jorquera, M., and Streilein, J.W., 1993, Subretinal space and vitreous cavity as immunologically privileged sites for retinal allografts. *Invest. Ophthalmol. Vis. Sci.* 34:3347-3354.
14. Jiang, L.Q., and Streilein, J.W., 1990/1991, Immunologic privilege evoked by histoincompatible intracameral retinal transplants. *Reg. Immunol.* 3:121-130.
15. Cousins, S.W., McCabe, M.M., Danielpour, D., and Streilein, J.W., 1991, Identification of transforming growth factor-beta as an immunosuppressive factor in aqueous humor. *Invest. Ophthalmol. Vis. Sci.* 32:2201-2211.
16. Taylor, A.W., Streilein, J.W., and Cousins, S.W., 1993, Neuropeptides contribute to the immunosuppressive activity of aqueous humor. *Invest. Ophthalmol. Vis. Sci. (Suppl)* 34:903.
17. Tanihara, H., Yoshida, M., Matsumoto, M., and Yoshimura, N., 1993, Identification of transforming growth factor-b expressed in cultured human retinal pigment epithelial cells. *Invest. Ophthalmol. Vis. Sci.* 34:413-419.
18. Jiang, L.Q., and Del Cerro, M., 1992, Reciprocal retinal transplantation: A tool for the study of an inherited retinal degeneration. *Exp. Neurobiol.* 115:325-334.

20

IRON AND HEREDITARY DEGENERATION OF THE RETINA

R. N. Etingof, N. D. Shushakova, and M. G. Yefimova

I.M.Sechenov Institute of Evolutionary Physiology and Biochemistry
Russian Academy of Sciences
M. Torez pr. 44, 194223 St. Petersburg, Russia

INTRODUCTION

Because of its involvement in haem and nonhaem iron containing enzymes iron is required by all cells of organism. The particular role of this ion was shown for the nervous tissue and especially for brain (Connor et al.,1990). It is known, that iron status of organism has an influence on the iron level in brain. The iron status is estimated by the content of nonhaem iron (NHI) in the complex with transferrin (TF) in blood plasma and by the quantity of NHI in hepatic cells ferritin (F) (Taylor et al,1991;). The decreased iron status of the organism (in contrast to the increased one) causes the most significant effects: a decrease of NHI content in brain and different biochemical and functional abnormalities in the latter (Youdim et al,1989; Oloyede et al.,1992). Recent studies have accumulated interesting data showing that disturbances of brain iron homeostasis have a crucial role in the development of some neurodegenerative diseases, such as Alzheimer's, Parkinson's, Wilson's diseases etc. (Connor et al.,1992; Youdim & Riederer,1993; Hafkemeyer et al., 1994; Saija et al., 1994).

The neural retina is a peripheral extension of the forebrain with which it has common embryological origins. The presence of iron in eye tissues seems to be very important because of its involvement in many processes, such as rod outer segment growth, photoreceptor membrane biogenesis, neurotransmitter and neuromodulator functions of iron ions in the complex with apoTF in developing retina (Cho & Hyndmann,1991; Davis,1993; Hunt & Davis,1992; Hyndmann et al,1991). Guanylate-cyclase, the enzyme which maintains the necessary level of cyclic GMP is also an iron containing protein . The presence of receptor complexes binding TF was shown in plasma membranes of photoreceptors and retinal pigment epithelium (RPE) cells (Cho & Hyndmann, 1991); the ferritin-like protein also was found there (Hunt & Davis,1992). At the same time we could not find any data about changes of iron containing components in eye tissues in the different forms of retinal and RPE degeneration, as it usually occurs at some neurodegenerative processes in brain. Only two reports are available at present, that the artificial decrease of iron level in eye tissues as well as its increase may provoke some degenerative disturbances in the retina (Anderson et al., 1984; Lakhanpal, 1984).

Degenerative Diseases of the Retina, Edited by Robert E. Anderson et al.
Plenum Press, New York, 1995

This chapter reviews the results of our recent studies and some unpublished data concerning investigation of some iron depending processes and iron containing components in the different tissues in rats with hereditary degeneration of the retina (HDR) as well as evaluation of iron status in these animals (see in details: Yefimova et al., 1987; Yefimova et al.,1988; Yefimova & Etingof, 1992; Shushakova et al., 1993; Shushakova & Yefimova,1994; Yefimova et al., 1994).

MATERIALS AND METHODS

Animals

Male rats of Wistar strain as control and male rats of Campbell strain (a branch of RCS rats) of different ages, affected with HDR, were used in this study.

Methods

Subcellular fractions of the liver and brain cortex tissues were isolated by the differential centrifugation methods (Karuzina et al, 1979) and (Vijayalakshmi et al,1990; Dambinova & Gorodinsky, 1984), respectively.

The lipid peroxidation (LP) was induced by Fe(II)-ascorbic acid system in accordance with the classical method (Vladimirov & Archakov,1972).The rate of lipid peroxidation was estimated by malondialdehyde (MDA) accumulation, dependent on the time of incubation (Buege & Aust, 1978).

The content of cytochrome P-450 was determined in the microsomal fraction of liver tissue as described (Omura and Sato,1964) and in the brain cortex microsomal fraction (Nabeshima et al,1981).

Partial purification of liver F was carried out by the method of Arosio et al. (1978). ApoTF from rat blood plasma were partially purified as described (Welch and Skinner,1989) with some modifications. To study the apoTF ability to be saturated with iron ions, the complex of iron with nitrilotriacetic acid (FENTA) was used (Bates and Schlach,1973).

Determination of the NHI content in microsomal fractions of the liver and brain cortex, in F preparations and blood plasma was made by the method of Brumby and Massey (1967) using the o-phenanthroline test following the iron extraction by 5% trichloroacetic acid or by 11 Mol nitric acid.

Nondenaturating polyacrylamide gel electrophoresis of F was performed (Catsimpoolas et al,1975) followed by protein staining with Coomassie Blue R-250. Simultaneously the specific staining for determination of Fe was carried out in these gels using the reaction with $K_4Fe(CN)_6$ (Linder-Horowitz et al,1970).The gels were scanned after staining (see below).

Nondenaturating polyacrylamide gel electrophoresis of rat blood plasma proteins was performed (Welch and Skinner,1989) followed by staining with Coomassie Blue R-250. Combined blood plasma of 8 rats of each studied age was used. After staining the gels were cut along the strips corresponding to each of the blood plasma samples and these strips were scanned using a special device for a spectrophotometer Specord M-40, wavelength 680 nm. The TF content was estimated as a ratio of the peak area to the total area of all peaks corresponding to the quantity of blood plasma proteins. Rat blood plasma TF (Sigma) was used as a marker. The degree of TF saturation with iron ions was calculated on the basis of NHI content in the samples. Calculation was performed taking into account that 1 mol of apoTF is capable of binding no more than 2 mol of iron. Two series of experiments were carried out.

The hemoglobin content in blood was determined (Derviz,1973). The quantity of erythrocytes was counted (Predtechenskiy et al, 1950).

A protein concentration was routinely determined (Marwell et al, 1987), using bovine serum albumin as a standard.

The results of experiments were submitted to statistical analysis using Student's criterion.

RESULTS AND DISCUSSION

NHI in Microsomes in Rats with HDR

In our previous studies it has been shown that changes in LP induced by iron-ascorbate system occur in the retina, RPE and brain cortex in rats with HDR in comparison with a control at early stages of postnatal ontogenesis (20th day of life - a period before an appearance of some detectable disturbances of visual function). The phenomena revealed took place only in the tissues mentioned above but not in liver and lungs. These shifts disappeared by 40-90 days of life (a period of blindness). It was unexpected that similar differences occurred both in brain and eye tissues of diseased animals. The phenomena revealed in brain tissue seem to be very interesting, especially because of the lack of reports about similar biochemical shifts in eye and brain tissues at HDR. So we tried to clarify the brain cortex subcellular fractions where these changes were most pronounced. The data are summarized in Table 1.

It is obvious that the enhanced rate of LP takes place only in the microsomal fraction of brain cortex. One of the reasons that the phenomenon revealed may be believed takes into account the theory that the stimulating effect of exogenously added iron on LP rate in membranes depends on a certain quantity of free functional groups being able to bind the added iron ions. These iron ions bound to the functional groups of some membrane components are then involved in LP (Vladimirov,1991). Recently one kind of membrane iron binding component (microsomal protein 66 000 Da) has been purified and the direct evidence of its involvement in LP has been obtained (Minotti, 1992). Another important factor determining the LP rate is Fe(II)/Fe(III) ratio (Minotti & Aust,1987). On the basis of these data, we suggest, that the content of NHI in brain cortex microsomes of Campbell rats may be less than in those of healthy ones. Hence, some additional functional groups may

Table 1. LP induced by iron-ascorbic acid system in different subcellular fractions of brain cortex in Wistar and Campbell strain rats at the 20th day of age

Subcellular fractions	Strain	MDA accumulation (nmol/mg of protein)	Time of induction (minutes)
Brain cortex homogenate	Wistar	5.17 ± 0.64 (6)	6
	Campbell	7.29 ± 0.31* (6)	6
Brain cortex: nuclear fraction	Wistar	1.75 ± 0.32 (6)	3
	Campbell	1.30 ± 0.10 (6)	3
Brain cortex: mytochondrial fraction	Wistar	6.60 ± 0.65 (6)	6
	Campbell	6.73 ± 0.65 (6)	6
Brain cortex: microsomal fraction	Wistar	5.54 ± 0.49 (10)	6
	Campbell	8.94 ± 0.73* (10)	6

Notes: in brackets - the number of experiments; * - values differ significantly from the control ($P \leq 0,05$).

Table 2. NHI content in brain cortex and liver microsomal fraction of Wistar and Campbell strain rats at the 20th day of age

Tissue	Strain	NHI content (μg/mg of protein) Total	Fe (II)	Fe (III)	Fe (II)/Fe (III) ratio
Brain cortex	Wistar	0.591 ± 0.061 (12)	0.199 ± 0.009 (12)	0.392 ± 0.029 (12)	0.51
	Campbell	0.385 ± 0.058*(12)	0.323 ± 0.031* (12)	0.062 ± 0.008* (12)	5.21
Liver	Wistar	2.30 ± 0.21 (6)	1.62 ± 0.32 (6)	0.68 ± 0.11 (6)	2.38
	Campbell	0.89 ± 0.14 * (6)	0.61 ± 0.12 * (6)	0.28 ± 0.10 * (6)	2.19

Notes: in brackets - the number of experiments; * - values differ significantly from the control ($P \leq 0,05$).

remain free and available for binding of exogenously added iron ions. If that is so, the exogenous iron would provide more intensive LP in brain cortex microsomes of rats with HDR. On the other hand, the Fe(II)/Fe(III) ratio also may be changed there. The results of the following series of experiments confirming this hypothesis are presented in Table 2.

It was found that both factors mentioned above occur in the brain cortex of Campbell rats at early stages of postnatal ontogenesis. The Fe(II)/Fe(III) ratio was increased 10-fold and NHI content significantly decreased in Capmbell rats in comparison with control. As for the liver tissue, only a change of NHI content but not of Fe(II)/Fe(III) ratio was found. All these shifts disappeared in both tissues by the 90th day of age (data not shown). Thus, it seems that the interaction of both of these factors (decreased NHI content and changed Fe(II)/Fe(III) ratio) is more important for the increase of induced LP rate in Campbell rats, than the individual action of each of them.

The similarity of changes of induced LP in brain cortex and eye tissues leads us to suggest that the enhanced rate of induced LP in the retina and RPE of diseased animals may be caused by the same reasons as in brain cortex.

Haem Iron Containing Components in Rats with HDR

It seemed to be of interest to investigate haem iron containing component cytochrome P-450 in microsomes of rats with HDR. The results of this study are presented in Table 3.

We did not succeed in determining the content of cytochrome P-450 in the microsomal fraction of the brain cortex at the 20th day of animal life, which is in agreement with the reports about the formation of this protein at relatively late stages of ontogenesis (Mishin and Lyachovitsh,1985). Neither was this cytochrome present in the retina, which coincides with findings of other authors (Shichi, 1976). By the 45th and especially 90th day of life there was a significant increase (2.3 and 4.3 times, respectively) of the level of haemoprotein in the microsomal fraction of the brain cortex in diseased rats as compared to controls. As for the liver microsomes of the diseased animals, the content of cytochrome P-450 did not differ from those of normal rats at all ages studied. It is noteworthy, that genetically mediated aromatic hydrocarbon inducible hydroxylase activity of cytochrome P-450 in the RPE has been suggested to be one of the contributing factors to retinal degeneration in some mice with HDR (Shichi, 1976). However only the data on basal activity, which was equal in diseased and healthy mice but not on the content of cytochrome P-450 were presented in this report. The causes underlying increase of the cytochrome P-450 content in the brain microsomes of the diseased rats have not been found yet. Nevertheless the data obtained in our experiments suggest that metabolic disorders in the peripheral part of visual oyotem (photoreceptors) are accompanied by essential metabolic shifts in its central part, i.e. in the brain cortex.

Table 3. The content of cytochrome P-450 in microsomal fractions of brain cortex and liver of Wistar and Campbell strain rats of different ages

Tissue	Age (days of age)	Strain	Content of cytochrome P-450 (pmol/mg of protein)
Brain cortex	20	Wistar	Not detectable
	20	Campbell	Not detectable
	45	Wistar	15.8 ± 0.2 (20)
	45	Campbell	37.3 ± 1.0* (20)
	90	Wistar	17.5 ± 0.1 (10)
	90	Campbell	81.3 ± 1.6* (20)
Liver	20	Wistar	590.0 ± 40.0 (6)
	20	Campbell	540.0 ± 60.0 (6)
	45	Wistar	1040.0 ± 80.0 (6)
	45	Campbell	1160.0 ± 70.0 (6)
	90	Wistar	1200.0 ± 100.0 (6)
	90	Campbell	1250.0 ± 90.0 (6)

Notes: in brackets - the number of experiments; * - values differ significantly from the control ($P \leq 0{,}05$).

Results of experiments with soluble haem iron containing protein hemoglobin show that its content as well as the quantity in erythrocytes (data not shown) did not differ in Campbell and Wistar rats at all ages studied. All these data allow the assumption that in the diseased rats NHI containing proteins are the first to undergo damage at the early stages of postnatal life.

Iron Status in Rats with HDR

Recent studies on iron homeostasis disturbances in different brain areas in some forms of neurodegenerative diseases and our own data mentioned above enable us to suppose some disorders in iron status in rats with HDR. To clarify this hypothesis we investigated the content and properties of liver F and blood plasma TF in Campbell rats in comparison to controls.

Fig. 1 shows densitograms of electrophoregrams of partially purified preparations of F from the liver of control and diseased animals at the 20th day of postnatal life. As is seen, the content of this protein in liver tissue of the affected rats is much lower compared to control.

Since F has generally a rather higher content of iron (Chasteen et al,1985), one of the ways to detect this protein may be the determination of iron ions in it. On the densitogram of electrophoregram presented in Fig. 2 a marked decrease in coloring characteristic of the presence of iron in F preparations from the diseased rats can be interpreted as another evidence of a significant lowering of the F and iron content in the liver tissue of HDR-affected animals.

In the course of postnatal development a difference in the F content of the normal and the diseased rats revealed at the 20th day of postnatal life became less evident and practically vanished by day 90 (Figs.1, 2).

In the next series of experiments the contents of blood serum TF, another NHI containing protein excreted into blood from liver and the content of NHI in blood plasma were determined. The obtained results are summarized in Table 4.

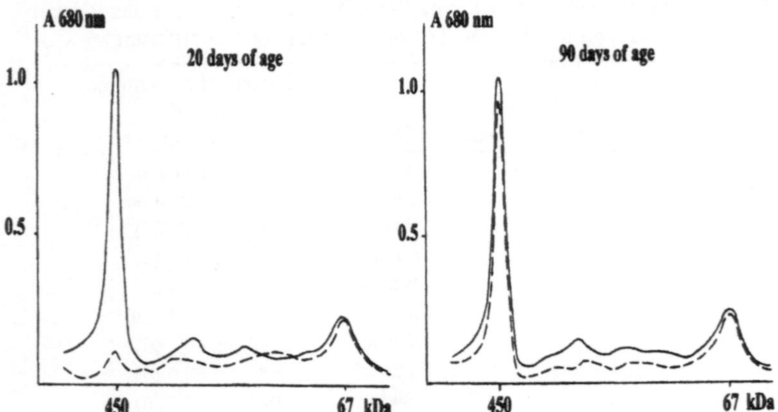

Figure 1. Densitometrical analysis of electrophoregrams of ferritin prerarations from liver of Wistar (solid line) and Campbell (dashed line) rats. Determination of protein content. The gels were stained with Coomassie Blue R-250. Abscissa - molecular weight of rat liver ferritin (450 kDa) and bovine serum albumin (67 kDa). Ordinate - optical density.

As is seen, by the 20th day of life the content of TF in blood plasma of the diseased animals was significantly (4,3 times) lower, than in control. The difference disappeared by the 40th day of postnatal development. The content of NHI was also lower (2 times) at the 20th day of Campbell rats' life. By the 40th day there was not any difference in NHI content in diseased and control rats.

The above data concerning the contents of NHI and TF in blood plasma made it possible to calculate the degree of protein saturation with iron, since the main pool of blood plasma NHI is, as a rule, bound with TF. As it follows from Table 4, in rats with HDR the degree of TF iron saturation at the 20th day of life was 2.3 fold that of control. By the 40th day the difference vanished. To find what causes such changes of TF iron saturation in the diseased rats, the ability of apoTf to bind iron ions was studied in the diseased and normal

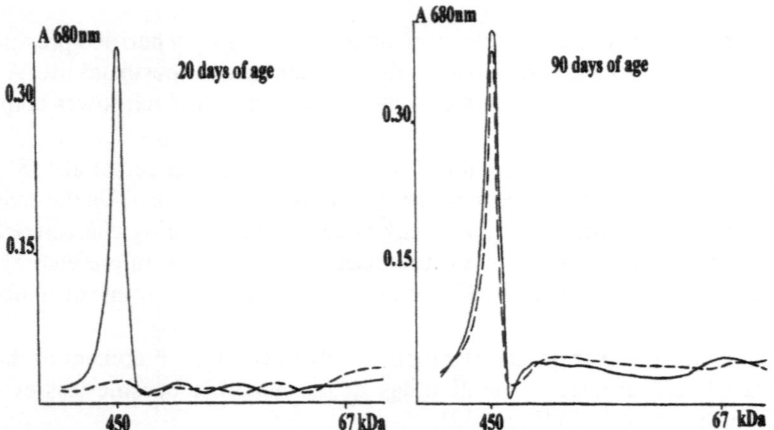

Figure 2. Densitometrical analysis of electrophoregrams of ferritin preparations from liver of Wistar (solid line) and Campbell (dashed line) rats. The gels were stained with $K_4Fe(CN)_6$ for determination of iron in the proteins. Axes - the same as in Figure 1.

Table 4. Transferrin and NHI content in blood plasma of Wistar and Campbell strain rats of different ages

Age (days of age)	Strain	NHI content (μg/ml of blood plasma)	Transferrin (mg/ml of blood plasma)	Saturation degree of transferrin with iron (%)
20 20	Wistar	1.74 ± 0.04 (16)	4.30 ± 0.40 (16)	29.0
	Campbell	0.93 ± 0.08* (16)	1.00 ± 0.08* (16)	66.4
40 40	Wistar	1.48 ± 0.01 (16)	4.20 ± 0.50 (16)	25.0
	Campbell	1.42 ± 0.06 (16)	4.80 ± 0.60 (16)	22.0

Notes: in brackets - the number of experiments; * - values differ significantly from the control ($P \leq 0{,}05$).

rats. The results obtained show this ability to be practically the same in both animal groups (Fig. 3).

The data obtained confirm our hypothesis about a certain discrepancy of iron status in rats with HDR at early stages of disease development. However, it should be pointed out, that the disturbances revealed differ from those taking place when iron deficiency is induced by iron depleted diets. In the latter case the content of TF protein increases compensatory and its saturation with iron markedly decreases (Taylor et al., 1991).

Some Speculation about Possible Mechanisms of Iron Ions Involvement in Pathogenesis of HDR

Cells maintain iron levels in part by regulating uptake of iron into the cell via the transferrin receptor (TFR) and by regulating the sequestration of iron into F. The syntheses of F, TF and TFR are regulated by iron, but in opposite directions. The synthesis of F is increased by iron; conversely, TF and TFR syntheses are decreased by iron and increased by

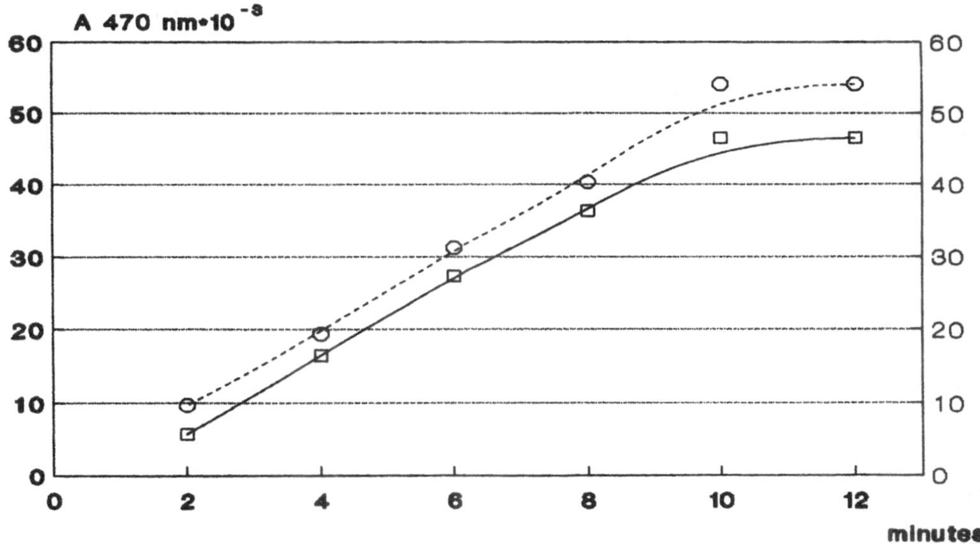

Figure 3. Kinetic dependences of blood serum transferrin saturation with iron in Wistar (solid line) and Campbell (dashed line) rats at 20 days of age.Absciss axis - time (minutes). Ordinate axis - optical density at wavelength 470 nm (* 10-3).

iron deprivation. The post-transcriptional regulation of F, TF and TFR expression by iron is controlled by a cytosolic protein termed as an iron responsive element binding protein (IRE-BP).

IRE-BP was found in the different tissues: liver, spleen, lungs, marrow (Yu et al.,1992; Cox & Adrian, 1993). However, we did not find reports about the presence of this protein in eye tissues (retina, RPE). It is widely recognized, that IRE-BP level is regulated in cells of different tissues. The mechanism(s) by which cells establish tissue-specific levels of IRE-BP appear to be complex. Finally, the functional activity of IRE-BP determines the content of basic proteins (TF, TFR, F) involved in iron metabolism in cells.

Taking into account our data, we suppose the disturbance of the regulation of iron metabolizing proteins (TF, TFR, F) synthesis by IRE-BP occurs in the different tissues of rats with HDR at early stage of postnatal life. The discrepancy between the shifts of the values of blood plasma TF and its saturation with iron on the one hand and the values of the liver F and NHI content in the latter on the other hand supports this assumption. We believe, that the disturbance of iron metabolizing proteins synthesis mechanisms is caused by a change of the functional activity of IRE-BP.

As for eye tissues of Campbell rats, the disturbance of the IRE-BP functional activity may provoke a decrease of TFR amount on the basolateral membrane of RPE cells in diseased animals. It is generally believed, that iron ions are put into the retina from RPE cells. The lack of these ions evoked by the reduced ability of the RPE cells to accept these ions from the blood plasma may be a reason for a discrepancy of the retina's growing processes and one of the contributing factors to the retinal degeneration. Moreover, it is widely recognized that the iron ions are necessary for phagocytosis (it was shown for macrophages (Reit & Simmons, 1990)), hence it is possible to believe, that the decreased content of iron may also provoke the discrepancy of phagocytic ability in RPE cells.

Our hypothesis coincides with studies on apoptotic cell death at HDR (Portera-Caillian et al., 1994; Chang et al., 1993), because it has been shown recently that iron deprivation may result in apoptosis in some cells (Fukuchi et al.,1994).

As mentioned above, there is a compensation of shifts concerning the proteins involved in the iron metabolism in Campbell rats by 40-45 days of postnatal life, probably, due to the involvement of some additional mechanisms of regulation, which are unknown up to now.

Apparently, in the early stage of postnatal life the changes in the iron status of organism are not dramatic for tissues, which are already formed on the whole by the animal's birth. On the contrary, these changes may play a crucial role for the developing eye tissues at the early stage of the postnatal life of Campbell rats.

It seems to be very important to clarify whether the phenomena revealed as the basis for our hypothesis also occur in other forms of HDRs, especially taking into account the genetic heterogeneity of this pathology. A positive answer to this question will be a good stimulus for a study of the common mechanism in different forms of HDR.

REFERENCES

Anderson, R.E., Rapp, L.M., and Wiegand, R.D., 1984, Lipid peroxidation and retinal degeneration, Curr. Eye Res., 3: 223-227.

Arosio, P., Adelman, T.G., and Drysdale, J.W., 1978, On ferritin heterogeneity. Further evidence for heteropolymers, J.of Biol. Chem., 253(12): 4451-4458.

Bates, G.W., and Schlach, M.R., 1973, The reaction of ferric salts with transferrin. J. of Biol. Chem., 248: 3228-3232.

Brumby, P.E., and Massey, V., 1967, Determination of nonheme iron, total iron and copper, Methods in Enzymology, 10:463-474.

Buege, J.A., and Aust, S.D., 1978, Microsomal lipid peroxidation, Methods in Enzymol., 52: 302-310.

Catsimpoolas, L.N., Griffith, A., Linder, M., and Munro, H.N., 1975, Size and charge heterogenity of rat tissue ferritins, Biochimica et Biophysica Acta, 412: 148-156.

Chang G.-G., Hao, J., and Wong, F., 1993, Apoptosis: final common pathway of photoreceptor death in rd, rds and rhodopsin mutant mice, Neuron, 11(4): 595-605.

Chasteen, N.D., Antanaitis, B.C., and Aisen, P., 1985, Iron deposition in apoferritin. Evidence for the formation of a mixed valence binuclear iron complex, J. of Biol. Chem., 260: 2926-2929.

Cho, S.S., and Hyndmann, A.G., 1991, The ontogenesis of transferrin receptors in the embryonic chick retina, Brain Res, 549(2): 327-331.

Connor, J.R., Menzies, S.L., Martin, S.M.St, and Mufson, E.J., 1990, Cellular distribution of transferrin, ferritin and iron in normal and aged human brains, J.Neurosci.Res., 27: 595-611.

Connor, J.R., Snyder, B.S., Beard, J.L., Fine, R.E., and Mufson, E.J., 1992, Regional distribution of iron and iron-regulatory protein in the brain in aging and Alzheimer's disease, J.Neurosci.Res., 31(2):327-335.

Cox, L.A., and Adrian, G.S., 1993, Posttranscriptional regulation of chimeric human transferrin genes by iron, Biochemistry, 32: 4738-4745.

Dambinova, S.A., and Gorodinsky, A.I., 1984, The molecular organisation of glutamate - sensitive chemoexited membranes in neurons. Binding of L-/3H/glutamate with synaptic membranes in rat brain cortex, Biockimiya, 49(1): 67-75. In Russian.

Davis, A.A., 1993, Transferrin is made and bound by photoreceptor cells, J. Cell. Physiol., 156(2): 280-285.

Derviz, G.V., 1973, Application of the photoelectrocolorimeters for the hemoglobin determination in blood by the cyan-methemoglobin method, Laboratornoye Delo, (2): 67-72. In Russian.

Fukuchi, K., Tomoyasu, S., Tsuruoka, N., and Gomi, K., 1994, Iron deprivation induced apoptosis in HL-60 cells, FEBS lett, 350: 139-142.

Hafkemeyer, P., Schupp, M., Storch, M., Gerok, W., and Haussinger, D., 1994, Excessive iron storage in a patient with Wilson's disease, Clinical Invest., 72(2): 134-136.

Hunt, R.C., and Davis, A., 1992, Release of iron by human retinal pigment epithelium, J. of Cell. Physiol., 152: 102-110.

Hyndmann, A.G., Hockberger, P.E., Zeevalk, G.D., and Connor, J.A., 1991, Transferrin can alter physiological properties of retinal neurons, Brain Res., 561(2): 318-323.

Karuzina, I.I., Bachmanova, G.I., Mengazetdinov, D.E., Myasoyedova, K.N., Zhyhareva, V.O., Kuznetsova, G.P., and Artchakov A.I., 1979, Purification and properties of cytochrome P-450 rabbit liver microsomes, Biockimiya, 44: 1049-1057. In Russian.

Lakhanpal, V., Schacket, S., and Rouben, J., 1984, Desferrioxamine (Desferal) - induced toxin retinal pigmentary degeneration and presumed optic neuropathy, Ophtalmology, 91: 443-451.

Linder-Horowitz, M., Ruettinger, R.T., and Munro, H.M., 1970, Iron induction of electrophoretically different ferritin in rat liver, heart and kidney, Biochimica et Biophysica Acta, 200:442-446.

Marwell, M.A.K., Haas, S.H., Beiler, L.L., and Tolbert, N.E., 1978, A modification of Lowry procedure to simplify protein determination in membrane and lipoprotein samples, Analyt. Biochem., 87: 206-210.

Minotti, G., 1992, The role of an endogeneous nonhene iron in microsomal redox reactions, Arch. Biochem. Biophys., 297(2):189-198.

Minotti, G., and Aust, S.D., 1987, The requirement for iron (III) in the initiation of lipid peroxidation by iron (II) and hydrogen peroxide, J. of Biol. Chem., 262:1098-1104.

Mishin, V.M., and Lyakhovitch, V.V., 1985, The multiple forms of cytochrome P-450. Novosibirsk, Nauka, 142 p. In Russian.

Nabeshima, T., Fontenot, J., and Ing, K.Ho., 1981, Effect of chronic administration of phenobarbital or morphine on the brain microsomal cytochrome P-450 system, Biochem. Pharmacol., 30: 1142-1144.

Oloyede, O.B., Folayan, A.T., and Odutuga, A.A., 1992, Effects of low-iron status and deficiency of essential fatty acids on some biochemical constituents of rat brain, Biochem. Internat., 27(5): 913-922.

Omura, T., and Sato, R., 1964, The carbon monoxide-binding pigment of liver microsome, J. of Biol. Chem., 239:2379-2385.

Predtechenskiy, V.E., Borovskaya, V.M., and Margolina, L.T., 1950, The laboratory methods of the studies, Moscow:Medgiz. In Russian.

Portera-Caillian, C., Sung, C.-H., Nathans, J., and Adler, R., 1994, Apoptotic photoreceptor cell death in mouse models of retinitis pigmentosa, Proc. Nat. Amer. Sci., 91(3): 970-978.

Reit, D.W., and Simmons, R.D., 1990, Nitric oxide mediates iron release from ferritin, Arch Biochem. Biophys.,283:537-541.

Saija, A., Princi, P., Casuscelli, F., Lanza, M., Scalese, M., Trombetta, D., Costa, G., and De Sarro, G., 1994, Genetically epilepsy -prone rodents show some changes of iron levels in the brain, Brain Res. Bul., 33(1): 1-6.

Shichi, H., Tsunematsu, Y., and Nebert, D.W., 1976, Aryl hydrocarbon hydroxlase induction in retinal pigment epithelium: possible association of genetic differences in a drug-metabolizing enzyme system with retinal degeneration, Exp. Eye Res., 23:165-176.

Shushakova, N.D., Rychkova, M.P., and Yefimova, M.G., 1993, The activities of NADPH-cytochrome C reductase and aldose reductase in the retina, brain cortex and liver in Wistar rats and in Campbell rats with hereditary degeneration of the retina in early stages of postnatal life, J. of Evol. Biochem.& Physiol., 29: 146-154.

Shushakova, N.D., and Yefimova, M.G., 1994, About the changes of cytochrome P-450 and NADPH-cytochrome C reductase at hereditary degeneration of the retina in rats, Bull.of Exper.Biol.& Med.,117(3):259-261.

Taylor, E.M., Crowe, A., and Morgan, E.H., 1991, Transferrin and iron uptake by the brain: effects of altered iron status, J.of Neurochemistry, 57: 1584-1592.

Vijayalakshmi, R., Anadatheerthavarada, H.K., and Shankar, S.K., 1990, NADPH-cytochrome P-450 reductase in rat, mouse and human brain, Biochem. Pharmacol., 39: 1013-1018.

Vladimirov, J.A., and Artchakov, A.I., 1972, Lipid peroxidation in biological membranes, Moscow, Nauka, 252 p. In Russian.

Vladimirov, J.A., and Azizova, O.A., 1991, The free radicals in vital systems, Moscow, Nauka, 341 p. In Russian.

Welch, S., and Skinner, A., 1989, A comparison of the structure and properties of human, rat and rabbit serum transferrin, Comparative Biochemistry and Physiology, 93B: 417-424.

Yefimova, M.G., Nikolayeva, J.A., and Shushakova, N.D., 1994, The changes of the serum transferrin content and its Fe-saturation at hereditary degeneration of the retina in rats, Bull. of Exper. Biol.& Med., 118(7): 24-26.

Yefimova, M.G., Ostapenko, I.A., and Etingof, R.N., 1987, Peculiarities of lipid peroxidation process in retina and brain tissues in rats with hereditary degeneration of the retina, Neirockimia, 6: 406-412. In Russian.

Yefimova, M.G., Ostapenko, I.A., Brailovskaya, I.V., and Etingof, R.N., 1988, About the changes of lipid peroxidation during hereditary retinal degeneration, Neirockimia, 7(4): 574-582. In Russian.

Yefimova, M.G., and Etingof, R.N., 1992, The location and the reasons of lipid peroxidation disturbances in rat brain cortex at early stages of hereditary degeneration of the retina, Ukrainian Biochem. J., 64(2):66-71. In Russian.

Youdim, M.B.H., Ben-Schachar, D., and Yehuda, S., 1989, Putative biological mechanisms of the effect of iron deficiency on brain biochemistry and behaviour, Amer. J. Clin. Nutr., 50: 607-617.

Youdim, M.B.H., and Riederer, P., 1993, The role of iron in senescence of dopaminergic neurons in Parkinson's disease, J. of Neural Transmission. General Section, Suppl.40: 57-68.

Yu, Y., Radisky, E., and Leibold, E.A., 1992, The iron-responsive element binding protein. Purification, cloning and regulation in rat liver, J.Biol.Chem., 267: 19005-19010.

21

RECEPTOR DEGENERATION IS A NORMAL PART OF RETINAL DEVELOPMENT

Juliani Maslim,[1] Rupert Egensperger,[2] Horstmar Holländer,[2]
Martin Humphrey,[3] and Jonathan Stone[1]

[1] Department of Anatomy and Histology, University of Sydney
Sydney, Australia
[2] Department of Neuromorphology, Max Planck Institute for Psychiatry
Munich, Germany
[3] Lions Eye Institute
Perth, Western Australia, Australia

INTRODUCTION

The degeneration of photoreceptors is regarded as pathological, for the tangible reason that it causes blindness. The causes of such degeneration include genetic defects specific to photoreceptors, many specific to the rhodopsin molecule, and genetic defects in the retinal pigment epithelium (reviewed in ref. 5 and this volume). The death of affected photoreceptors involves DNA fragmentation (2, 7, 8,18) characteristic of apoptosis or programmed cell death (3). This paper presents evidence from the rat and rabbit that DNA fragmentation occurs in committed photoreceptors in normally developing retina; that the affected cells undergo death; that this physiological death of photoreceptors occurs during a discrete period in retinal development, which coincides with the growth of inner and outer segments; and that, in the rcs strain of rat in which photoreceptors degenerate pathologically, the onset of their degeneration coincides with the onset of physiological degeneration.

Death of cells in the outer retinal layers during normal development was described some years ago (4,14). The present description builds on these earlier reports in two ways. First, we followed the time course of cell death beyond postnatal life. The division of the neuroblast layer into the outer and inner nuclear layers marks the end of the major period of retinal neurogenesis (11,12). By tracing cell death beyond time of this division, we were able to examine whether retinal neurones die after they are committed, postmitotic neurones. Second, we identified dying cells with the TUNEL technique (3), which detects cells undergoing death at an earlier stage than the detection of pyknotic bodies, on which earlier studies relied. Our observations were made on the sections of the retina of albino and hooded rats, of two strains of rabbits (chinchilla and New Zealand), of a strain of rat (the Royal College of Surgeons (rcs) rat) in which the receptors undergo total degeneration, and of a congenic control (rdy) strain.

Degenerative Diseases of the Retina, Edited by Robert E. Anderson et al.
Plenum Press, New York, 1995

MATERIALS AND METHODS

Eyes from albino and hooded rats, from rcs and rdy rats and from albino rabbits were fixed by immersion in 4% buffered formaldehyde. Fixation time was not critical and was usually overnight. Whole eyes were frozen sectioned at 10μm. The rat eyes were sectioned horizontally, and the rabbit eyes vertically. Sections were mounted onto the slides pretreated with poly-L lysine (Sigma).

TUNEL labelling was performed following Gavrieli et al. (3). Each slide was immersed in 70% alcohol for 30 minutes, then in distilled water for 10 minutes. Endogenous peroxidase was inactivated with 3% H_2O_2 for 5 min at room temperature. The sections were then rinsed in distilled water (2 x 5 minutes). After this preparation, the elongation of exposed DNA fragments was performed in the following steps. Each slide was washed in TdT buffer (30 mM Trizma base, 140 mM sodium cacodylate, 1mM cobalt chloride) for 5 minutes. Then each slide was covered with the reaction solution (100μl pipetted onto the section) and incubated in a humid chamber at 37°C for 1h. The reaction solution was a mixture of the TdT enzyme (terminal deoxytransferase, Boehringer-Mannheim, final concentration 0.3eu/μl) and biotinylated dUTP (final concentration 40μM) in TdT buffer. The reaction was stopped by immersion in SSC (300mM NaCl, 30mM sodium citrate) for 15 minutes at room temperature. Non-specific binding was blocked by incubating the section in 1% human or bovine serum albumin in PBS at room temperature for 20 minutes. The sections were then incubated in avidin-peroxidase, and the peroxide was visualised with the diaminobenzidine reaction.

RESULTS

Morphological Differentiation of Photoreceptors

Two steps in the differentiation of photoreceptors were recognised morphologically. One was the formation of the outer nuclear layer (onl), which becomes distinct when the outer plexiform layer (opl) forms, dividing the neuroblast layer into inner and outer parts, which become the inner nuclear layer (inl) and the onl (Figures 1A,B,C). This occurs between E17 and P10 in the rat (1); between E31 and P6 in the rabbit (15). The neuroblast layer is the major site of retinal neurogenesis and, at its division, neurogenesis in the layer ceases (11,12). The cells of the onl are postmitotic, committed to become photoreceptors. The second morphologically distinct step in photoreceptor differentiation is the formation of the inner and outer segments, which occurs in the rat by Pcheck (1) and in the rabbit by Pcheck (13).

DNA Fragmentation Occurs in All Layers of Retinal Neurones

At birth in both rat and rabbit, $TUNEL^+$ cells were prominent throughout the thickness of the neuroblast layer (Figure 1A) and in the ganglion cell layer. The labelling in the ganglion cell layer of the rat, in which blood vessels develop, may occur among endothelial cells as well as among neurones. Once the neuroblast layer had divided $TUNEL^+$ cells were detected in the inl (Figure 1B) and onl (Figure 1C).

TUNEL-Labelled Cells Undergo Death

Several lines of evidence suggest that the $TUNEL^+$ cells undergo death. First, the patterns of $TUNEL^+$ cells described here match descriptions of cell death based on identifi-

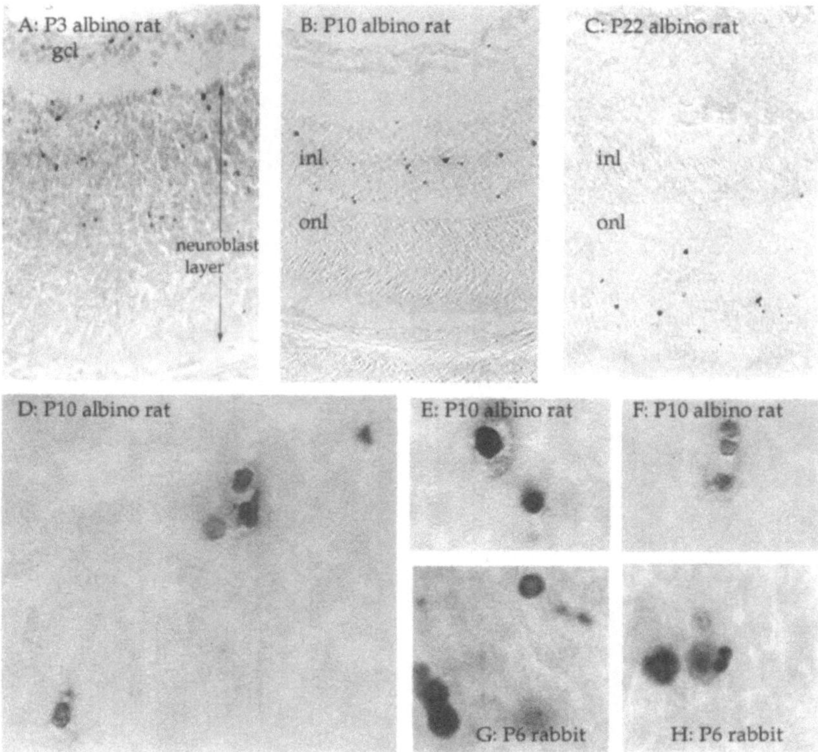

Figure 1. A: At P(postnatal day)3 in the rat the neurones of the ganglion cell layer (gcl) have separated from the neuroblast layer, which is still mitotically active. TUNEL$^+$ (dark) nuclei are apparent throughout the neuroblast layer. B: At P10, TUNEL$^+$ cells are confined to the inner nuclear layer (inl). C: At P22,. TUNEL$^+$ cell are confined to the outer nuclear layer (onl). D-H: At high magnification, the TUNEL$^+$ structures include clumped nuclei, and often appear to be surrounded by a non-labelled cell. In some nuclei the label is dark around the edge of the nucleus (upper G), suggesting that DNA fragmentation occurs first at the edge of the nucleus. In other cases the labelled structures resemble pyknotic bodies (H). Occasionally the label appeared in the cytoplasm of a cell adjacent to a darkly labelled nucleus (H).

cation of pyknotic bodies (4,14.16,19). Second, although some of the TUNEL$^+$ structures resembled nuclei which might still be in situ in their cells (e.g. the upper labelled structure in Figure 1E), others showed evidence of condensation and crowding suggestive of degeneration, and some appeared surround by a phagocytotic cell which was TUNEL$^-$ (Figures 1D-H), and others resembled pyknotic bodies (Figure 1G, right). Further, some TUNEL$^+$ structures were apposed by cells in which the cytoplasm was lightly TUNEL$^+$ (example next to the pyknotic bodies in Figure 1G). This cytoplasmic labelling is presumably occurring in microglia or neurones which have ingested the fragmenting DNA of a dying neighbour. Finally, similar patterns of TUNEL labelling are seen in the rcs rat (below) in which photoreceptors do degenerate.

DNA Fragmentation Occurred Successively in the inl and onl

After the division of the neuroblast layer into the inl and onl, TUNEL labelling was prominent in the inl in both rat (from P5) and rabbit (from P3). During this period TUNEL$^+$

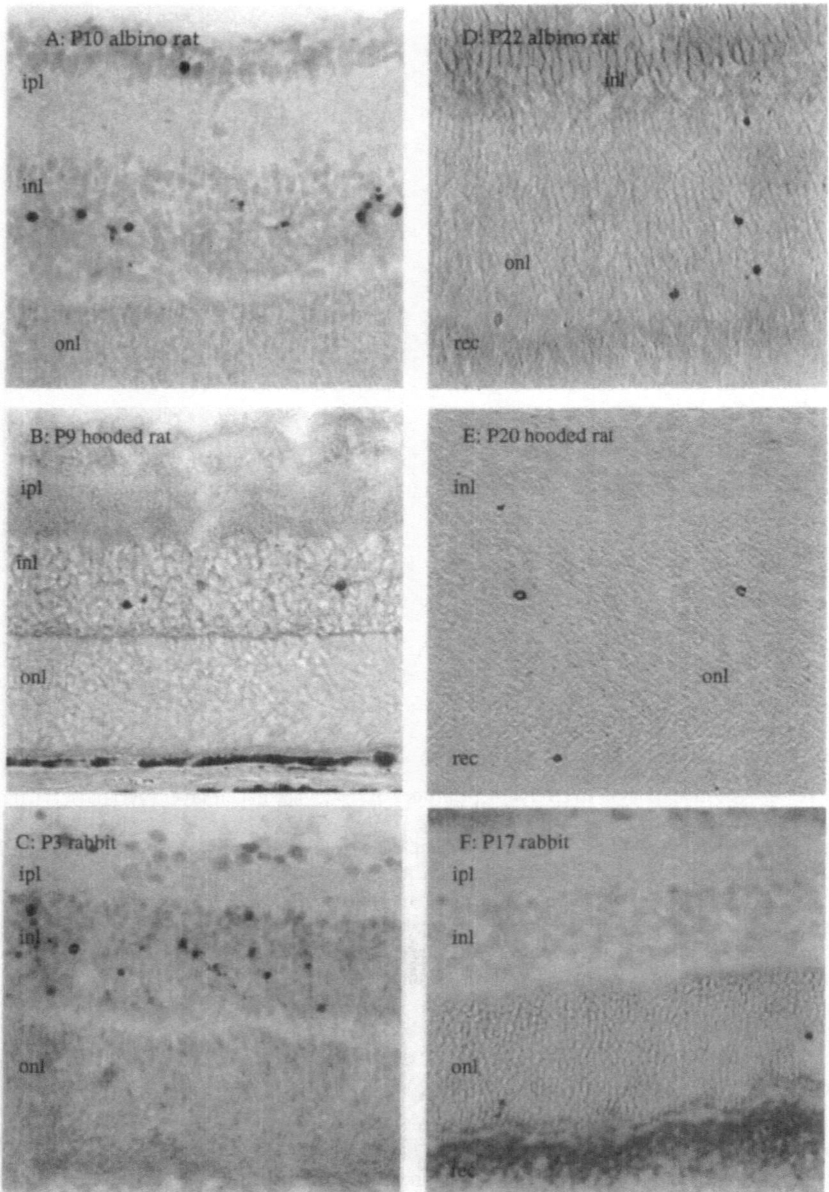

Figure 2. After formation of the inl and onl, labelling concentrated first in the inl, then in the onl. Labelling in the inl was strong in the albino rat at P10 (A), in the hooded rat at P9 (B) and in the rabbit at P3 (C). Labelling in the onl was strong in the albino rat at P22 (D), in the hooded rat at P20 (E) and in the rabbit at P17 (F). At these ages, the receptor layer (rec) is well formed.

nuclei were found throughout the thickness of the inl, but were absent from the onl (Figures 2A,B,C). DNA fragmentation in the inl persisted until approximately P13 in the rat and P10 in the rabbit.

As TUNEL$^+$ structures became less frequent in the inl, TUNEL labelling became apparent throughout the thickness of the onl (Figures 2D, E,F). TUNEL$^+$ structures in the onl were most numerous in the albino rat, in which they were most prominent at age P22. When we searched this age period in the hooded rat, TUNEL$^+$ cells were again detected in the onl, shown for age P20 in Figure 2D. In the two rabbit strains, TUNEL$^+$ cells were detected in the onl from P9 -P13. At later ages TUNEL$^+$ nuclei were not prominent in the onl.

The rcs Rat

In the developing rcs rat, TUNEL$^+$ cells appeared in the inl in numbers and patterns similar to those seen in the albino and hooded strains, and in an rdy reference strain (not shown). Figure 3A shows the appearance of TUNEL$^+$ cells in the inl at P15. The onl was normal in its morphology and, as in the control strains, was free of TUNEL$^+$ cells until approximately P18. Thereafter TUNEL$^+$ cells were rare in the inl but occurred at abnormally high frequency in the onl (Figure 3B). By P20 labelling was prominent not only in nuclei of the inl, but also in the cytoplasm of cells which spread horizontally in the opl or extended radially across the opl, and were occasionally seen in the receptor layer (Figure 3C). At P40, TUNEL$^+$ labelling was still prominent in the onl (Figure 3D), and the onl was relatively thin; at this age TUNEL-labelling was no longer detectable in albino, hooded or rdy strains.

DISCUSSION

The TUNEL technique polymerises labelled UTP to the 3′ OH ends of DNA, and dense labelling is evidence of the presence of many such ends, exposed by the action of endogenous nucleases (3). In our material the procedure labelled cells in all cell layers of the developing retina. The labelled structures included nuclei apparently in situ, but also pyknotic bodies, bodies enclosed by microglia, and the cytoplasm of neighbouring cells, confirming that the labelled DNA is from cells undergoing death. Recent interpretations (e.g. 3,9, 10) suggest that such cells are undergoing apoptotic or programmed cell death.

The TUNEL$^+$ cells occur in distinct spatial and temporal patterns. In the youngest rat retinas examined (P0), TUNEL$^+$ cells were common in the neuroblast layer, indicating that some potential neurones die during the process of neurogenesis. TUNEL$^+$ cells were also

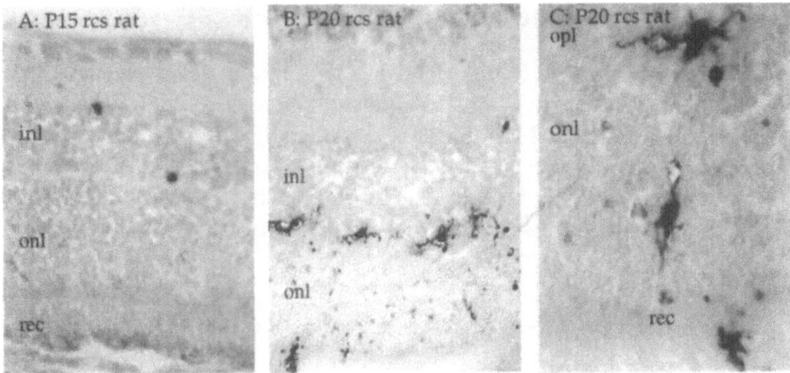

Figure 3. In the rcs rat, as in the other rat strains and the rabbit, TUNEL labelling appeared in the inl at about P9, and at P15 was still restricted to the inl (A). By P20, labelling in the inl had subsided, but a massive labelling of cells in the onl was observed (B). At this age, the label appeared in a population of cells which lined up along the outer plexiform layer (B) but also were found in and even external to the onl (C).

prominent among postmitotic cells. During the late stages of neurogenesis, TUNEL$^+$ cells were prominent in the ganglion cell layer, which contains ganglion and amacrine cells which have left the mitotic cycle. As neurogenesis ended and the neuroblast layer separated into the inl and onl, TUNEL$^+$ cells became prominent first in the inl, then in the onl. These patterns suggest that waves of cell death occur in each of the three layers of postmitotic retinal neurones, first in the ganglion cell layer, then in the inl, then in the onl.

The Normal Death of Photoreceptors

The present results suggest that, in species free from genetic abnormality, a minority of photoreceptors undergo degeneration and that their death is part of an episode of cell death which affects all classes of postmitotic retinal neurones. Death among ganglion cells has been widely reported as part normal development (4,14.16,17,19). Death among cells of the inl has also been previously reported (4.14), but the present report is the first to observe a distinct late period of receptor degeneration. This is partly because earlier studies relied on the detection of pyknotic bodies in Nissl stained material, which may be a less sensitive indicator of cell death, and partly because those studies did not follow cell death after the formation of the onl. One recent study (7) used the TUNEL technique to characterize cell death in mice with genetic defects affecting photoreceptors, and reported a low level of cell death among photoreceptors in control mice. They did not, however, note a discrete period of raised frequency of receptor death.

Comparative Observations

Two comparative observations suggest interesting variants in photoreceptor degeneration between presumed normal species and strains. Within the rat, the numbers of dying cells were consistently more numerous in the albino than in the hooded rat. Whether the higher death rate in the albino is related to the gene which causes its lack of pigmetnation, or due to higher light levels in the eye resulting from the lesser pigmentation of the iris and choroid, deserves investigation. Comparing rat and rabbit, dying cells were particularly prominent in the inl of the rabbit, being several times more frequent than in any layer of the rat retina. Whether the high rate of cell death in the rabbit relates to the larger size of the eye in this species, or to the lack of vasculature in the retina, also deserves investigation.

Cytoplasmic Labelling by the TUNEL Procedure

We were initially surprised to see TUNEL-labelling in the cytoplasm of cells. Such labelling may occur in (at least) two ways. The endonucleases which attack nucleic DNA may attack the periphery of the nuclear chromatin, creating the annular labelling of a nucleus seen in Figure 1G (S. Ben-Sasson, personal communication). From the periphery of the nucleus DNA fragments may then spread into the cytoplasm of the affected cell. Alternatively, the DNA fragments may disperse in the cytoplasm of cells which ingest the dying cell.

Receptor Differentiation May Be a Cause of Receptor Death

Two features of the developmental timing of photoreceptor degeneration deserve note. First, the degeneration does not occur until the receptors have developed their inner and outer segments. Second, the pathological degeneration of photoreceptors in the rcs rat begins at approximately the same stage of photoreceptor development. These times of degeneration suggest that the onset of receptor function is a factor in the induction of receptor death in both normal development and in the rcs rat. The role of receptor function in the induction of neuronal death remains to be elucidated.

ACKNOWLEDGMENTS

This work was supported by grants from the Australian Retinitis Pigmentosa Association, the National Health and Medical Research Council of Australia and the Sir Zelman Cowen Universities Fund.

REFERENCES

1. Braekevelt, C. R., and Hollenberg, M. J., 1970, The development of the retina of the albino rat, *Am. J. Anat.* 127: 281-302.
2. Chang, G. -Q, Hao, Y., and Wong, F., 1993, Apoptosis: final common pathway of photoreceptor death in rd, rds and rhodopsin mutant mice, *Neuron* 11: 595-605.
3. Gavrieli, Y., Sherman, Y., and Ben-Sasson, S. A., 1992, Identification of programmed cell death in situ via specific labeling of nuclear DNA fragmentation, *J. Cell Biol.* 119: 493-501.
4. Harman, A. M., Snell, L. L., and Beazley, L. D., 1989, Cell death in the inner and outer nuclear layers of the developing retina in the wallaby *Seytonix brachyurus* (quokka), *Journal of Comparative Neurology* 289: 1-10.
5. Holleyfield, J., Anderson, D., and Lavail, M. M., 1993, Retinal degeneration, *Plenum Press, New York*
6. Ilschner, S. U., and Aring, P., 1992, Fragmentation of DNA in the retina of chicken embryos coincides with retinal ganglion cell death, *Biochemical and Biophysical Research Communications* 183: 1056-.
7. Lolley, R. N., Rong, H., and Craft, C. M., 1994, Linkage of photoreceptor degeneration by apoptosis with inherited defect in phototransduction, *Investigative Ophthalmology and Visual Science* 35: 358-362.
8. Portera-Cailliau, C., Sung, C. -H, Nathans, J., and Adler R, 1993, Apoptotic cell death in mouse models of retinitis pigmentosa, *Proc Natl Acad Sci USA* 91: 974-978.
9. Raff, M. C., 1992, Social controls on cell survival and death, *Nature* 356: 397-40.
10. Raff, M. C., Barres, B. A., Burne, J. F., Coles, H. S., Ishizaki, Y., and Jacobson, M. D., 1993, Programmed cell death and the control of cell survival: lessons from the nervous system, *Science* 262: 695-.
11. Rapaport, D. H., and Stone, J., 1984, The area centralis of mammalian retina: focal point for function and development of the visual system, *Neuroscience* 11: 289-301.
12. Rapaport, D. H., and Stone, J., 1983, The topography of cytogenesis in the retina of the cat, *J. Neuroscience* 3: 1824-1834.
13. Reichenbach, A., Schnitzer, J., Friedrich, A., Ziegert, W., Brückner, G., and Schober, W., 1991, Development of the rabbit retina I. Size of eye and retina, and postnatal cell proliferation, *Anat. Embryol.* 183: 287-297.
14. Robinson, S. R., 1988, Cell death in the inner and outer nuclear layers of the developing cat retina, *Journal of Comparative Neurology* 267: 506-515.
15. Stone, J., Egan, M., and Rapaport, D. H., 1985, The site of commencement of retinal maturation in the rabbit, *Vision Res.* 25: 309-317.
16. Stone, J. and Rapaport, D.H., 1986 The role of cell death in shaping the ganglion cell population of the adult cat retina. In: Festschrift for P.O. Bishop, Cambridge University Press.
17. Thanos, S., 1991, The relationship of microglial cells to dying neurons during natural neuronal cell death and axotomy-induced degeneration of the rat retina, *European Journal of Neuroscience* 3: 1188-1207.
18. Tso, M. O. M., Zhang, C., Abler, A. S., Chang, C. J., Wong, F., Chang, G. Q., and Lam.T.T, 1994, Apoptosis leads to photoreceptor degeneration in inherited retinal dystrophy of rcs rats, *Investigative Ophthalmology and Visual Science* 35: 2693-2699.
19. Wong, R. O. L., and Hughes, A., 1987, Role of cell death in the topogenesis of neuronal distribution in the developing cat retinal ganglion cell layer, *Journal of Comparative Neurology 262: 496-511.*

22

DEVELOPMENT OF OPSIN AND SYNAPSES IN MONKEY PHOTORECEPTORS

A. Hendrickson, E. Dorn, K. Bumsted, and A. Szel

Departments of Biological Structure and Ophthalmology
University of Washington
Seattle Washington 98195
Semmelweise University
Budapest

INTRODUCTION

Primate vision is characterized by high acuity and color perception, each of which originate in cone photoreceptors. Visual sensitivity resides in rod photoreceptors. Adult *Macaca* monkey retina contains an average of 2.9-3.5 million cones and rods average 60.1 million with considerable individual variation[1]. *Macaca* retinal topography is centered on the fovea, a specialized region which contains only cones and has a depression in the inner retina. The monkey fovea has a cone density around 200,000/mm^2 which drops rapidly into the periphery. Beginning on the foveal edge, rods rapidly increase in number and rod density peaks at the eccentricity of the optic disc in a circular ring that has a density close to that of foveal cones[1-3]. This distinctive topography results in a cone-dominated fovea and a rod-dominated periphery whose developmental patterns can be studied separately.

Wavelength specificity is mediated by opsin which is the major membrane protein of cone and rod outer segment (OS)[4,5]. Monkey cones are divided into red or long wave-length specific (L), green or medium wave-length specific (M), and blue or short wave-length specific (S) subtypes. Immunocytochemical staining in adult human retina using an antibody specific for human S opsin has demonstrated that S cones are absent in the center of the fovea [6]. In more peripheral retina of both humans and monkeys, immunocytochemical staining with antibodies to S opsin[6-8], morphological characterization[9] or vital dye marking of OS[10] finds that S cones comprise 6-10% of the total population. Rods pack the spaces between cones at all eccentricities except the fovea. Rods are absent from the foveal center in monkeys[1] and from a larger foveal central region in humans[2]. Rods are well characterized morphologically, show relatively little change with eccentricity[1,2] and contain a single photopigment called rhodopsin which also can be labeled immunocytochemically with well characterized antisera[5].

Degenerative Diseases of the Retina, Edited by Robert E. Anderson et al.
Plenum Press, New York, 1995

The macaque monkey retina develops over a long period with birth occuring on fetal day (Fd) 165-170. Each developmental event involving photoreceptors such as cessation of mitosis, morphological differentiation, synapse formation or specific protein expression occurs first in the fovea at Fd 45-60. Each of these events then follows a central to peripheral sequence[11-13] which reaches the retinal edge between Fd125 and birth. At a given retinal eccentricity, rods are generated 10-14 days after cones[11]. These developmental patterns make three predictions: 1) foveal cones should be the first to express opsin, 2) S opsin should be expressed after L/M opsin because S cones are lacking in the fovea and 3) rod opsin should lag both L/M and S opsin expression. Another measure of photoreceptor differentiation is the formation of synapses which are necessary for phototransduction information to be carried centrally. An additional question covered in this paper asks what is the sequence of synapse maturation compared to opsin protein expression.

METHODS

A series of fetal infant and adult retinal frozen sections containing the entire horizontal meridian were stained using standard immunocytochemical techniques for single and double fluorescent labeling or for visualization with diaminobenzidine. Adult monkey retina sections were used to determine whether different antibody sets were recognizing the same cone types. Antibodies used to identify opsin phenotypes included 1) two monoclonal antibodies generated to chicken opsins which recognize L/M (COS-1) and S (OS-2) opsin[8] 2) two polyclonal antibodies generated to known peptide sequences of human L/M and S cone opsins[14] and 3) a monoclonal antibody (4D2) to the N-terminus of rhodopsin[5]. Synapses were labeled with a monoclonal antibody to the synaptic vesicle protein SV2 which has been shown to be present in both rod and cone synaptic vesicles[13].

RESULTS

Double-label immunofluorescent staining in adult retina comparing mouse monoclonal anti-L/M opsin (L/Mmab) to rabbit polyclonal anti-L/M opsin (L/Mpoly) showed completed double labeling of all L/M cone OS (Fig.1 A,B), as did the comparable pair of S mab and S poly (Fig.1 C,D). Rod OS are not labeled by either cone antibody set. In adult retina, cone cell bodies are not labeled. When Smab and L/Mpoly antibodies were used for double labels, each OS was stained by only one antibody. Rod mab labeling was confined to rod OS which were stained very heavily while cell bodies were lightly stained, in contrast to cones (Fig.3 D). Labeling patterns were markedly different for the antibody sets in fetal retina. L/Mmab was only detected when presumptive OS appeared (Fig. 2B), whereas the L/Mpoly, Spoly, Smab and rod mab labeled the entire fetal cell membrane in addition to the developing OS (Fig. 2 A,C, D; Fig.3B). In all fetal retinas, Smab stained much more intensely than Spoly (Fig. 2C,D), each photoreceptor labeled for only one rod or cone opsin, and there was no cross-reactivity between antibody sets.

An opsin comparison strategy was applied to serial frozen sections throughout monkey retinal development from Fd55-postnatal (P) 1year. Adjacent sections were stained with Smab, L/Mpoly or rod mab to determine the temporal and spatial sequences for opsin expression. Graphic summaries of the overall expression pattern of S, L/M and rod opsin expression are shown in Figs. 4 and 5. At Fd55-65 no opsin expression was detected. The first opsin detected was rod opsin which was present at Fd66 in rods on the foveal edge and scattered throughout the fovea. By Fd70, the most central foveal cones were labeled for L/M opsin, but no Smab immunoreactive cones were detected. At Fd75-80, rod, S and L/M opsin

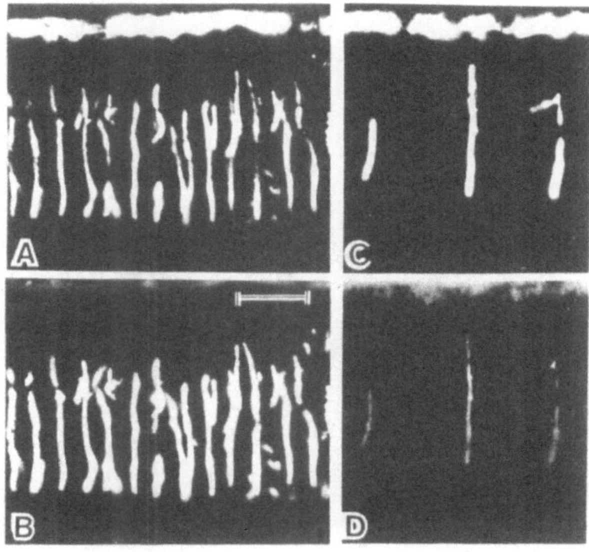

Figure 1.

stained individual perifoveal photoreceptors, and in and around the fovea it appeared that all photoreceptors were expressing their opsin.

All cell types showed the same general sequence of maturation. The most prominent staining at early ages in both rods and cones was in the presumptive developing OS embedded in the apex of the enlarging inner segment (Fig.3 A,B). The entire photoreceptor cell

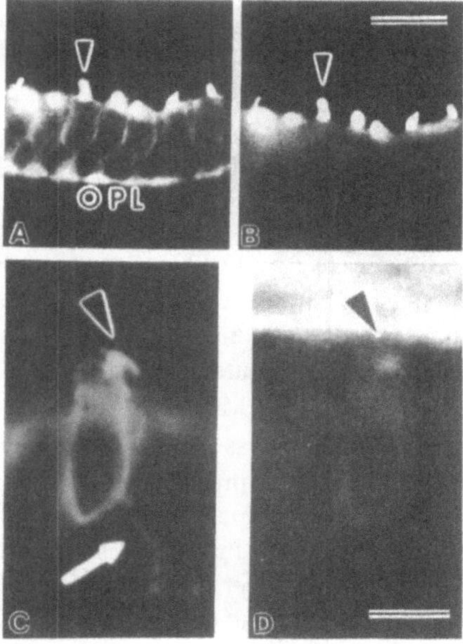

Figure 2.

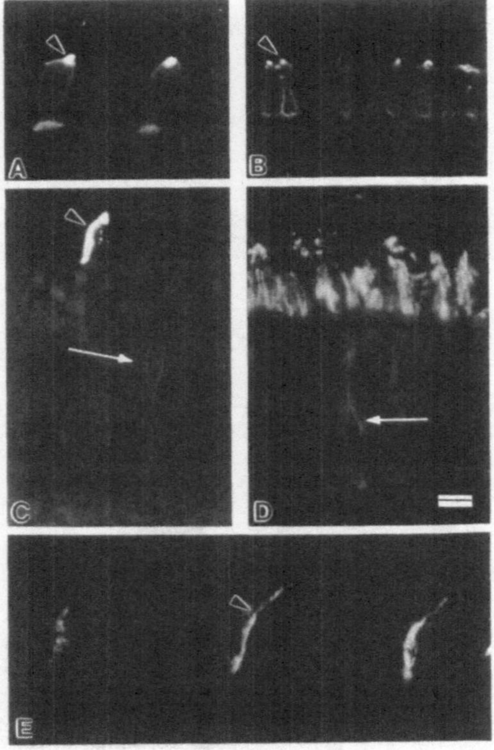

Figure 3.

membrane was more lightly labeled including the developing synaptic pedicle. As the OS lengthens, cell membrane labeling decreases. Cone membrane labeling disappears 30-40 days after initial opsin expression but rods continue to show very light cell body labeling into adulthood.

Once rod and S opsin began to be expressed in photoreceptors outside the fovea, labeled rods and S cones always were present at more peripheral eccentricities than labeled L/M cones (Fig. 4 and 5). Rods typically were labeled slightly more peripherally than S cones so that rods reach the retinal edge by Fd132, S cones by Fd140, and L/M cones at Fd145. A major difference between photoreceptor types is that the most peripheral and presumably youngest S cones at all ages have fine processes resembling telodendria extending from the synaptic pedicle into the outer plexiform, inner nuclear and outer nuclear layer (Fig. 2C). These were not present on L/M cones or rods.

Previous work from our laboratory has shown that immunoreactivity for SV2 and ribbon synapses identified by EM are both present in foveal cones at Fd60 and in cone pedicles at the retinal edge by Fd125[13]. This means that synapses form about a week before cone opsin expression in central retina, and this delay is closer to three weeks in the far periphery. Perifoveal rods label with SV2 later than cones, and also form ribbon synapses later than adjacent cones, but the exact timing of rod opsin expression and synaptic formation still is uncertain for most retinal eccentricities. These data are summarized graphically in Fig.5.

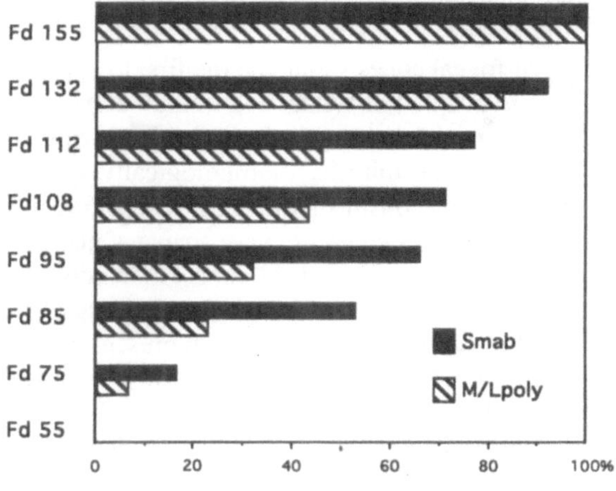

Figure 4.

DISCUSSION

Based on staining with these immunocytochemical markers, a complex pattern of primate photoreceptor differentiation emerges. In the fovea, cone synapses are generated about a week before L/M cone opsin is expressed. In rod-dominated peripheral retina, cone synaptic pedicle labeling for SV2 reaches the retinal edge prior to either S or L/M opsin, indicating that the same pattern of synapses>opsin continues across the retina. There is less hard data on synaptic formation in rods, so the exact timing of rod opsin expression and rod synapse formation is incomplete. In the perifoveal retina opsin expression and synapse

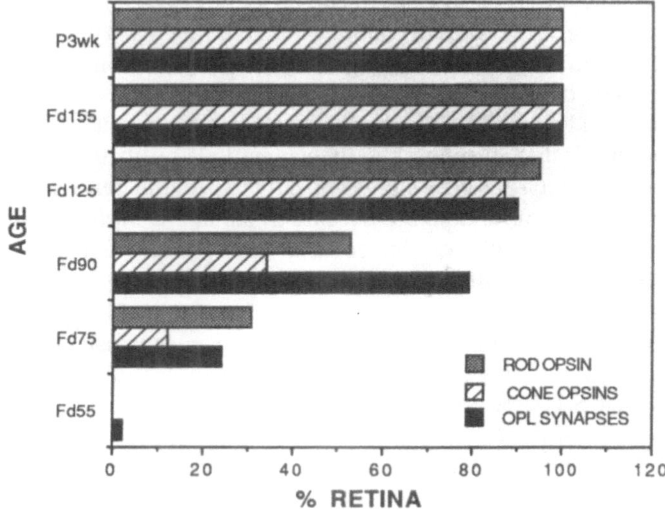

Figure 5.

formation in rods occur very close together[13, 15] so it is possible that they form simultaneously.

Our first prediction that foveal cones should be the first to express opsin is substantiated with these data. This result and that of Wikler and Rakic[16] adds cone opsin expression to previous studies which have shown that the fovea is the focus of differentiation in the primate retina where cells first cease mitosis, morphologically differentiate, and form synapses. Our second prediction that S opsin should lag L/M opsin expression in primates was supported only in the fovea where S cones are very sparse. Outside the fovea, S cones are labeled farther in the periphery compared to L/M cones, indicating that proportionately "younger" S cones express opsin. An earlier retinal wholemount study using L/Mpoly and Spoly antisera in the fetal monkey retina[16] found L/M opsin present in foveal cones at Fd80, but S opsin did not appear until Fd100-110. This is opposite to opsin expression sequences described for other vertebrate retinas. In Zebrafish[17] and mouse[18], S opsin appears before L/M opsin, similar to our results in peripheral primate retina. Both primate studies agree that the first detectible cone opsin is L/M and that this is found in the foveal cones. Given that few rods or S cones are present in the fovea, this region might have different developmental conditions from the rest of the retina. Peripheral primate retina appears to have a pattern more consistent across species because we find the same S>L/M sequence as in the rod-dominated nonfoveate retina of mice[18] and the cone-dominated retina of Zebrafish[17]. We ascribe the difference in sequences between the two primate studies to the relatively weak immunoreactivity shown by the Spoly antisera in fetal retina which could influence the ability to recognize early S cones in retinal wholemounts. In preliminary studies we have confirmed that Smab detects cones consistently more peripherally than Spoly.

Our third prediction that rods will express opsin after cones was not supported by our results. This means that birthdate order for photoreceptors is not necessarily a predictor of the order in which specific proteins are expressed. In addition, it appears that even though rods become postmitotic at least 2 weeks after cones, they express their opsin on a faster time schedule. For instance on the foveal edge rods are born at Fd45-50[11] and express their opsin by Fd66-70[15]. Because rods express only one opsin protein the need for a secondary decision of phenotype choice is unneccessary, unlike cones which must "decide" whether to be L/M or S. Thus, shortly after a rod becomes postmitotic, the cell begins a differentiation program to express the requisite rod proteins.

There is a long "dark period" between cone generation and opsin expression; for instance, foveal cones are generated at Fd36-45[11] but we find that opsin is not expressed in these cells until Fd70. In the case of cones, does the cone "know" which opsin it will express when it is generated, or is this information subsequently impressed onto the cone by epigenetic factors? Wikler and Rakic[16] proposed two possible mechanisms for cone mosaic organization. The first is that all cones initially express L/M opsin and then a small number later switch to S opsin which has been shown recently to occur in gerbil retina[19]. We find no evidence for this mechanism in monkeys because Smab and L/Mpoly double-label immunocytochemistry found no double-labeled cones at any age. The second is that a subset of "precocious" peripheral cones directly organize the subsequent photoreceptor mosaic via unspecified intercellular interactions. They proposed that scattered L/M cones play this role. We did not find this population at the leading edge of opsin expression in our study, rather L/M cones formed a solid front which was consistently more central than the single S cones. Instead, we find that S cones, are 10% of the population even at young fetal ages, and consistently are the most peripheral cones which express opsin. These peripheral S cones have fine basal processes resembling telodendria which extend 6-10 μm from the pedicle into the surrounding outer retina (Fig. 2 C). As proposed by Wikler and Rakic[16] these S cones could be the organizers who use their telodendria to set up a scaffold of cell-cell contacts to inhibit surrounding cones from expressing S opsin, but allow them to randomly express either L or M opsin. Random choice could yield the final adult Old World monkey cone ratio of 1M:1L[20].

Another factor in determining opsin expression could involve information transfer from deeper retinal layers, with the signal to express opsin induced by or dependent on the formation of synapses. Before there is any evidence of opsin expression, foveal cones form synapses with their bipolar cells which in turn already have formed synapses onto ganglion cell dendrites[13, 21]. Transgenic mice carrying the human S opsin gene express S opsin in S cones and a bipolar cell subset[22,23]. Although we have no evidence that any fetal bipolar cells express opsin in primate retina, these early cone/bipolar synaptic contacts could have other functions. Current EM studies are underway to obtain more exact chronological data on whether rods also form synapses before they express opsin. If they do not, the role of the inner retina in opsin phenotypic choice would be less likely.

ACKNOWLEDGMENTS

We thank K. Lerea for the gift of Spoly and L/Mpoly which were originally made by the late C. Lerea, and A. Szel for the gift of Smab and L/M mab. This work was supported by EY01208, EY04536, and in part by VISION CORE EY01730 and RR00166. KB is a predoctoral trainee supported by EY07031 and AH is a Senior Scholar of Research to Prevent Blindness, Inc..

REFERENCES

1. Packer, O., Hendrickson, A. and Curcio, C., Photoreceptor topography of the adult pigtail macaque (*M. nemestrina*). *J.Comp.Neurol.* 288, 165, 1989.
2. Curcio, C.A., Sloan, K.R., Kalina, R.E., and Hendrickson, A.E., Human photoreceptor topography. *J. Comp. Neurol.* 292, 497, 1990.
3. Wikler, K.C., Williams, R.W., and Rakic, P., Photoreceptor mosaic: number and distribution of rods and cones in the rhesus monkey retina. *J. Comp. Neurol.* 297, 499, 1990.
4. Nathans, J., Thomas, D., and Hogness, D.S., Molecular genetics of human color vision: the genes encoding blue, green and red pigments. *Science.* 232,193, 1986.
5. Molday, R., Monoclonal antibodies to rhodopsin and other membrane proteins of rod outer segments. *Progr. Ret. Res.* 8, 173 1988.
6. Curcio, C.A., Allen, K.A., Sloan, K.R., Lerea, C.L, Hurley, J.B., Klock, I.B. and Milam, A.H., Distribution and morphology of human cone photoreceptors stained with antiblue opsin. *J. Comp. Neurol.* 312, 610, 1991.
7. Wikler, K. and Rakic, P., Distributiion of photoreceptor subtypes in the retina or diurnal and nocturnal primates. *J.Neurosci.* 10,3390, 1990.
8. Szel, A., Diamantstein, T., and Rohlich, P., Identification of the blue sensitive cones in the mammalian retina by antivisual pigment antibody. *J. Comp. Neurol.* 273, 593 1988.
9. Anhelt, P.K., Kolb, H. and Pflug, R., Identification of a subtype of cone photoreceptor, likely to be blue sensitive, in the human retina. *J.Comp.Neurol.* 255, 18, 1987.
10. de Monasterio, F. M., McCrane, E. P. Newlander, J. K., and Schein,S. J., Density profile of blue sensitive cones along the horizontal meridian of macaque retina. *Invest. Ophathalmol. Vis. Sci.* 26, 289, 1985.
11. La Vail, M. M., Rapaport, D. H. and Rakic, P., Cytogenesis in the monkey retina. *J. Comp. Neurol.* 309, 86 (1991).
12. Hendrickson, A.E., Development of the primate retina , *Early Visual Development, Normal and Abnormal*, K.Simons ed., Oxford, New York, 1993, 287.
13. Okada, M., Erickson, A., and Hendrickson, A. E., Light and electron microscopic analysis of synaptic development in Macaca monkey retina as detected by immunocytochemical labeling for the synaptic vesicle protein, SV2. *J. Comp. Neurol.* 339, 535, 1994.
14. Lerea, C.L., Milam, A.H. and Hurley, J.B., a transducin is present in blue, green and red sensitive cone photoreceptors in the human retina. *Neuron* 3, 367, 1989.

15. Fulton, A., Hansen, R., Dorn, E. and Hendrickson, A., Development of primate rod structure and function, *Infant Vision,* F. Vital-Durand, ed., Oxford, New York, 1995, in press.
16. Wikler, K.C. and Rakic, P., Relation of an array of early differentiating cones to the photoreceptor mosaic in the primate retina. *Nature* 351, 397, 1991.
17. Matesic, L.E., Robinson, J., and Dowling, J.E., Opsin gene expression in zebrafish (*Brachydanio rerio*): developmental time course.*Invest.Ophthal.Vis.Sci.(Supp)* 35,1728, 1994.
18. Szel, A., Röhlich, P., Meiziewska, K., Aguirre, G. and van Veen, T., Spatial and temporal differences between the expression of short and middle wave sensitive cone pigments in the mouse retina: a developmental study. *J. Comp. Neurol.* 331, 564 1993.
19. Szel, A., van Veen, T., and Röhlich, P., Retinal cone differentiation. *Nature* 390, 336, 1994.
20. Mollen, J.D., and Bowmaker, J.K., The spatial arrangement of cones in the primate fovea. *Nature* 360, 677, 1992.
21. Nishamura, Y. and Rakic, P., Development of the rhesus monkey retina. I. Emergence of the inner plexiform layer and its synapses. *J. Comp. Neurol.* 241, 420, 1985.
22. Chui, M.I. and Nathans, J., Blue cones and cone bipolar cells share transcriptional specificity as determined by expression of human blue visual pigment transcribed transgenes. *J.Neurosci.* 14, 3426, 1994.
23. Chen, J., Tucker, C. L., Woodford, B., Szel, A., Lem, J., Gianella-Borradori, A., Simon, M.I. and Bogenmann., E., The human blue opsin promoter directs transgene expression in short wave cones and bipolar cells in the mouse retina. *Proc. Natl. Acad. Sci. USA.* 91, 2611, 1994.

23

THE NATURE OF NEWLY FORMED CAPILLARIES IN EXPERIMENTAL NAPHTHALENE-INDUCED RETINAL DEGENERATION IN RABBIT

Nicola Orzalesi, Luca Migliavacca, and Stefano Miglior

Eye Clinic
University of Milan
Institute of Biomedical Sciences
S. Paolo Hospital
via di Rudini' 8
20142 Milan, Italy

INTRODUCTION

Papers relating to previous experimental and clinicopathologic studies have reported contradictory results concerning the "maturity" of subretinal new vessels, and the development and evolution of new morphological features related to changes in the permeability of subretinal membranes.

The literature included descriptions of both "immature" and "mature" features; the former characterized by a substantial lack or a decrease in fenestrations (Gehrs et al, 1992; Ishibashi et al, 1987; Okhuma et al, 1983), the latter as having a well fenestrated appearance (El Dirini et al, 1991; Van der Zypen et al, 1985; Miller et al, 1986).

In previous papers (Orzalesi et al, 1993; Orzalesi et al, 1994), we reported the experimental retinal degeneration and subretinal neovascularization (SRN) induced by naphthalene in the rabbit, which was characterized by initial photoreceptor damage but without any lesion of Bruch's membrane or of the retinal pigment epithelium (RPE).

In order to clarify the nature of these newly-formed vessels, we here report the results of a morphometric study of the different types of capillaries found during the course of this process. We investigated a series of parameters which are considered as "standard references" for the evaluation of the "maturity" and the proliferative state of the choriocapillaries during the course of the development of experimental SRNs: wall thickness, the lumen of the vessels, the number of fenestrations and cytoplasmic buds, the interruptions and duplications of the basement membrane and the presence of pericytes.

Degenerative Diseases of the Retina, Edited by Robert E. Anderson et al.
Plenum Press, New York, 1995

MATERIALS AND METHODS

A 10% (wt/vol) solution of naphthalene dissolved in paraffin oil was administered to 20 adult rabbits by gavage on alternate days for 5 weeks. The intoxicated rabbits were killed 30, 90 and 180 days after the beginning of treatment and their eyes were processed for light and electron microscopy. All of the rabbits sacrificed at 90 and 180 days had developed subretinal new vessels. Four rabbits were used as controls (for further details see Orzalesi et al, 1993 and Orzalesi et al, 1994).

Six different samples were considered:

1. the choriocapillaries of control eyes,
2. the choriocapillaries 30 days from the beginning of treatment in the areas facing the foci of initial retinal degeneration
3. the choriocapillaries 90 and 180 days from the beginning of treatment, around the subretinal new vessels and below the intraretinal "neovascular complex",
4. the newly-formed subretinal vessels,
5. the vessels of the late intraretinal "neovascular complex",
6. the choriocapillaries in areas showing no retinal or pigment epithelium damage 90 and 180 days from the beginning of treatment.

The vessels were photographed by means of transmission electron microscopy (TEM) at a magnification of 13,000x for montaging. The measurements were made on photographic enlargements (final magnification 32,500X) of the TEM photographs of the vessels of the six samples. The calibration of the electron microscope and the enlarger were carefully controlled.

Forty complete vascular profiles were randomly selected from each sample and analysed by means of a MOP Videoplan (Kontron, Munich, Germany) image analysis system.

For each vessel, the following parameters were measured: the perimeter (μm) and area of the lumen of the vessel (μm^2), the outer perimeter of the endothelial cells lining the vessel (μm), the total area of the vessel (lumen plus endothelial cells) (μm^2), and the number of cytoplasmic buds (buds/10 μm).

The wall thickness (i.e.the average endothelial cell thickness) was calculated using the method of Wallow et al.

The number of fenestrations was referred to the perimeter of the capillary (number/μm), normalized to the standard of 10 μm.

The pericyte coverage was calculated by measuring the right-angle projection of these cells on to the endothelial cells. This measurement was referred as the percentage of the perimeter of the vessel. The endothelial cell basement membrane interruptions and duplications were calculated in a similar way.

The measurements were statistically analyzed by means of independent Student T-test.

RESULTS

The complete results are reported in Tables 1 and 2. No statistically significant differences were found between sample 1 and 2 and between sample 1 and 6.

In comparison with controls, the vessels of sample 3 showed a reduced area of the lumen (37.2 μm^2 vs. 60.6 μm^2), and lumen/vessel area (48.4% vs. 77.1%), a reduced number of fenestrations (1.66/10 μm vs. 2.6/10 μm), an increased wall thickness (0.27 μm vs. 0.16

Table 1.

	Samples					
	1	2	3	4	5	6
Vessel area (μm^2)	78.7 (± 24)	77.8 (± 21)	74.9 (± 23)	74 (± 17)	75.2 (± 23)	81.4 (± 22)
Lumen area (μm^2)	60.6 (± 19)	58.6 (± 14)	37.2 (± 13)‡	42.9 (± 10)‡	54.9 (± 21)	60.5 (± 22)
% lumen	77.1 (± 5)	76.3 (± 5)	48.4 (± 8)‡	59 (± 11)‡	71.2 (± 11)†	74.4 (± 4.2)
Pericytes*	3.17 (± 5)	3.88 (± 4)	12 (± 6.6)‡	14.5 (± 7)‡	14.7 (± 5)‡	2.01 (± 3.5)
Buds/10 µm	0.01 (± 0.02)	0.02 (± 0.05)	0.15 (± 0.13)‡	0.16 (± 0.1)‡	0.12 (± 0.11)‡	0.02 (± 0.0)
Membr. interr.*	0.2 (± 1)	1.77 (± 3.9)	10.1 (± 6.3)‡	10.4 (± 4)‡	8.79 (± 6)‡	0.2 (± 0.9)
Membr. duplic.*	0.61 (± 3.2)	1.3 (± 3.6)	10.8 (± 6)‡	8 (± 4.2)‡	15.4 (± 8)‡	0.1 (± 0.8)

*Percent of the vessel perimeter covered by pericytes or interruptions or duplications of the basal membrane.
†$p < 0.05$.
‡$p < 0.01$.

µm) and pericytes coverage, and a greater number of buds, interruptions and duplications of the basement membrane. These differences were statistically significant ($p< 0.0001$).

In comparison with controls, the newly-formed subretinal vessels (sample 4) and the neovascular complex (sample 5) showed a reduced area of the lumen (42.9 μm^2 and 54.9 μm^2 vs. 60.6 μm^2) and lumen/vessel area (59% and 71.2 % vs. 77.1%), a reduced number of fenestrations (2.16/10 µm and 2.54 /10 µm vs. 2.6/10 µm), an increased vessel wall thickness (0.20 µm and 0.18 µm vs. 0.16 µm) and pericyte coverage, and a greater number of buds, interruptions and duplications of the basement membrane.

For sample 4, these differences were all statistically significant; for sample 5, only the differences in lumen/vessel area, pericyte coverage, buds, basal membrane duplications and interruptions were statistically significant.

Between samples 4-5 and sample 3 there were statistically significant differences in lumen/vessel area (59% and 71% vs. 48%), the number of fenestrations (2.16/10 µm and 2.51/10 µm vs. 1.66/10 µm) and wall thickness (0.20 µm and 0.17 µm vs. 0.27 µm).

The wall thickness and the number of fenestrations of sample 4 were further divided in two sub-samples: the part of the vessel below (sub-sample 4A) and above Bruch's membrane (sub-sample 4B). These two sub-samples differed substantially, as sub-sample 4B was similar to the controls, whereas sub-sample 4A showed statistically significant differences (the wall thickness was 0.18 µm in 4B vs. 0.21 µm in 4A and 0.16 µm in controls;

Table 2.

Sample	Fenestrations*	Vessel wall (µm)
1	2.6 (± 0.4)	0.16 (± 0.04)
2	2.65 (± 0.3)	0.16 (± 0.04)
3	1.66 (± 0.3)†	0.27 (± 0.07)†
4a	2.16 (± 0.2)†	0.22 (± 0.07)†
4b	2.56 (± 0.3)	0.18 (± 0.05)
5	2.51 (± 0.3)	0.17 (± 0.05)
6	2.66 (± 0.2)	0.17 (± 0.04)

*n° of fenestrations/10µm.
†p 0.01.

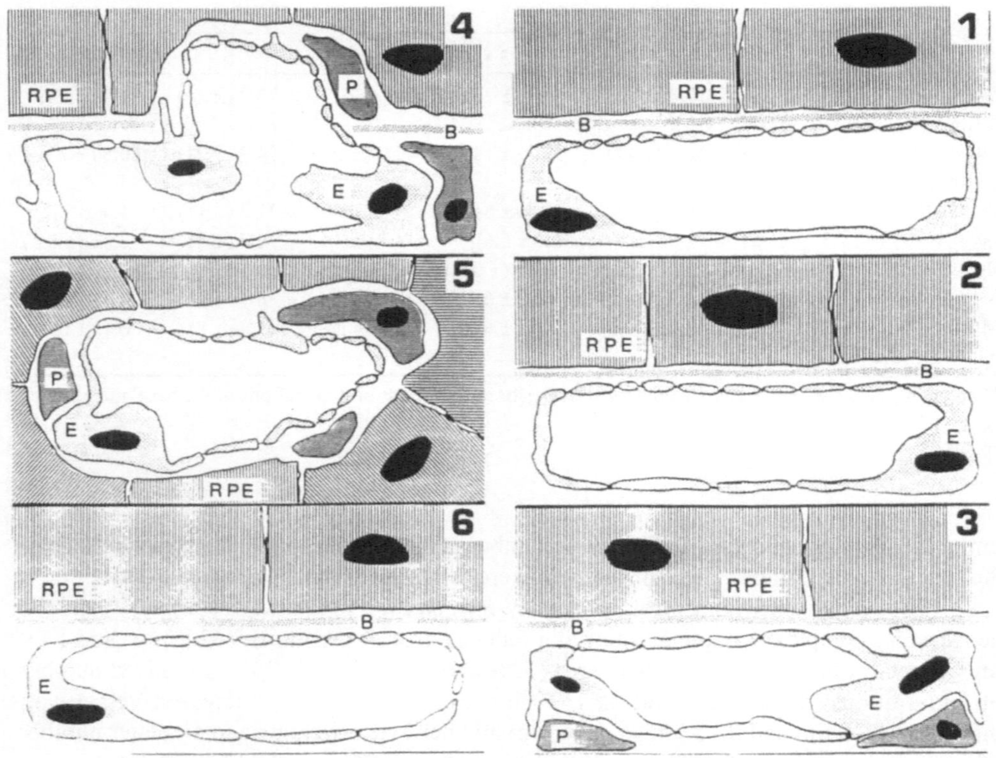

Figure 1. **(A)** Sketches of the samples 1, 2, 3. **(B)** Sketches of the samples 4, 5, 6. RPE: retinal pigment epithelium; B: Bruch's membrane; E: endothelial cell; P: pericyte. (See text for explanation.)

the number of fenestrations was 2.56/10 μm in 4B vs. 2.16/10 μm in 4A and 2.6/10 μm in controls) (Fig. 1A, 1B).

DISCUSSION

The above results show that there are remarkable differences in a series of vascular features of the choriocapillaries which develop during the course of subretinal neovascular proliferation after naphthalene poisoning of the retina.

In sample 1, the choriocapillaries showed their mature features, with thin-walled endothelial cells rich in fenestrations, and lined by a continuous monolayered basement membrane, and rare pericytes.

The choriocapillaries in sample 2, taken at the level of the early retinal lesion, did not show any significant changes from those of the control sample, thus confirming that SRN in this experimental model follows the degeneration of photoreceptors and the reaction of RPE.

In sample 3, the choriocapillaries below the foci of advanced retinal degeneration have immature features and are lined by an increased number of pericytes that suggests the proliferating activity of both types of cells.

Of particular interest is the appearance of the subretinal newvessels analyzed in sample 4, which show relatively "mature" features mainly represented by thin-walled and fenestrated endotelial cells (4B) at the tip of the vascular sprout, which is ensheathed by RPE cells, as well as an "immature" thick-walled, non-fenestrated endothelium on the choroidal side of the vessel (4A).

In sample 5, the newly-formed capillaries, which give rise to "neovascular complexes" completely surrounded by polymorphus RPE cells, show fully established mature features with a loss of polarization.

The normal appearance of the choriocapillaries in sample 6, taken from areas of the retina spared by the degenerative and neovascular changes, further demonstrate the focal nature of the retinal damage induced by naphthalene.

The reconstruction of dynamic processes like SRN on the basis of morphological data suffers from the discontinous analysis of the event. Nevertheless the difficulty to investigate SRN by other means and the increasing importance of this condition as a devastating retinal disease prompts us to draw some conclusions about the data supplied by the naphthalene model of SRN.

In this model, which from the morphological point of view closely resembles SRN found in age related macular degeneration, the newly-formed vessels seem to arise from the normal choriocapillaries in the area where the poison has induced damage to the photoreceptors and RPE reaction.

Damage to the choriocapillaries which characterizes other models of vascular reaction such as iodate poisoning (Korte G, 1989) was not seen in this case, but vascular proliferation was accompanied by dedifferentiation of the endothelium with the appearance of "immature" choriocapillary characteristics. These changes were limited to the choroidal side of the vessel, but on the retinal side, "mature" features were generally seen. This is in agreement with the autoradiographic study in the rat model of SRN following Krypton laser photocoagulation, where the tips of the neovascular tufts, once they had penetrated Bruch's membrane did not proliferate, and the endothelial cells further back along the vascular stalk divided (Zhang NL et al, 1993). According to previous studies (Mancini et al, 1986), it is also possible that RPE modulates the vascular tufts, once they have proceeded to the retinal side, thus inducing the appearance of "mature" features. These features characterize the "neovascular complexes", seen in the advanced stages, which are completely surrounded by RPE cells.

Another significant aspect is the dramatic increase in the number of pericytes covering the newly formed vessels in all stages of their proliferation. The role of these cells in SRN is controversial as they have been found to increase in some cases (Archer DB and Gardiner TA, 1981) but be reduced in number or absent in others (Van der Zypen et al, 1985; Ishibashi et al, 1987; Ishibashi et al, 1994).

REFERENCES

Archer D.B. et Gardiner T.A., 1981, Electron microscopic features on experimental choroidal neovascularization, Am. J. Opthalmol. 91:433-457.

El Dirini A., Ogdent E., Ryan S.J., 1991, Subretinal endophotocoagulation. A new model of subretinal neovascularization in the rabbit, Retina 11:244-249.

Gehrs K.M., Heriot W.J., De Juan E., 1992, Study of a subretinal neovascular membrane due to age-related macular degeneration, Arch. Ophthalmol. 110:833-837.

Ishibashi T., Miller H., Orr G., Sorgente N., Ryan S.J., 1987, Morphologic observations on experimental subretinal neovascularization in the monkey, Invest. Ophthalmol. Vis. Sci. 28:1116-1130.

Ishibashi T., Inomata H., Ryan S.J., 1994, Pericytes of newly-formed vessels in experimental subretinal neovascularization, Invest. Ophthalmol. Vis. Sci. 35(Suppl.):1505.

Korte G., E., Choriocapillaris regeneration in the rabbit, 1989, Invest. Ophthalmol. Vis. Sci. 30:1938-1950.

Mancini M.E., Frank R.N., Keirn R.J., Kennedy A., Khoury J.K.,1986, Does the retinal pigment epithelium polarize the choriocapillaris?, Invest. Opthalmol. Vis. Sci. 27:336-345.

Miller H., Miller B., Ryan S. J., 1986, Newly-formed subretinal vessels. Fine structure and fluorescein leakage, Invest. Ophthalmol. Vis. Sci. 27:204-213.

Ohkuma H., Ryan S.J., 1983, Experimental subretinal neovascularization in the monkey, Arch. Ophthalmol. 101:1102-1110.

Orzalesi N., Migliavacca L., Miglior S., 1993, The effect of naphthalene on the retina of the rabbit, in "Retinal Degeneration", J.G. Hollyfield, R.E. Anderson, M.M. LaVail editors, Plenum Press, New York, London, pp. 343-353.

Orzalesi N., Migliavacca L., Miglior S., 1994, Subretinal neovascularization after naphthalene damage to the rabbit retina, Invest. Ophthalmol. Vis. Sci. 35:696-705.

Van der Zypen E., Fankhauser F., Raes K., 1985, Choroidal reaction and vascular repair after chorioretinal photocoagulation with the free-running Neodymium-Yag laser, Arch. Ophthalmol. 103:580-588.

Zhang N.L., Samadani E.E., Frank R.N., Mitogenesis and retinal pigment epithelial cell antigen expression in the rat after kripton laser photocoagulation, Invest. Ophthalmol. Vis. Sci. 34: 2412-2424.

24

RETINAL CELL RESPONSES TO ARGON LASER PHOTOCOAGULATION

Martin F. Humphrey,[1,3] Yi Chu,[1] Claudia Sharp[1], Krishna Mann, and Piroska Rakoczy[2]

[1] WARP Research Centre
[2] Molecular Biology Unit
Lions Eye Institute
Perth, Western Australia
[3] Medical Psychology Institute
University of Magdeburg
Magdeburg, Germany

INTRODUCTION

Argon laser photocoagulation is a technique which is routinely used in clinical practise for a wide variety of retinal problems. Early studies largely done prior to and shortly after the introduction of laser photocoagulation examined the mechanisms of photocoagulation and the histological consequences for the retina at the coagulation site (1). These studies showed that the light energy was largely absorbed by the pigment epithelial cells and that damage to the overlying photoreceptors and inner retina was predominantly due to thermal conduction from this area of absorption, and sometimes also due to mechanical shock wave effects especially when a Q-switched laser was used. However, many of the effects of laser photocoagulation are not readily explained by the known changes in retinal structure at the coagulation site. For example, the dramatic effect of pan-retinal photocoagulation in reversing the invasion of the macula by newly formed blood vessels which occurs in diabetic retinopathy appears to be due to action at a distance because the photocoagulation is done in the periphery (2). Similarly a grid photocoagulation pattern is as effective in containing sub-retinal vascular growth as coagulation aimed specifically at the growing vessels. In certain animal models of the inherited retinal degeneration retinitis pigmentosa it had been noted that mechanical injury to the retina could paradoxically slow the photoreceptor loss (3). This prompted us to examine whether controlled photocoagulation could have similar effect and we found that there was indeed increased survival of photoreceptors on the flanks of the coagulation lesions (4). All of these cases suggest that there may be alterations to the retina at some distance from photocoagulation lesions which have profound effects on the maintenance of retinal cells and on blood vessel stability. Therefore we have examined various cellular markers to examine whether there are changes in the retina distant from

Degenerative Diseases of the Retina, Edited by Robert E. Anderson et al.
Plenum Press, New York, 1995

photocoagulation lesions and to better understand what happens following photocoagulation. Photocoagulation also provides a uniquely localised injury which is useful for basic studies of the injury response and how it can be modulated. All of the results presented here are from studies of dystrophic or congenic control RCS rats with argon green laser lesions performed in the superior retina as previously described (4).

MÜLLER CELL RESPONSE

Glial fibrillary acidic protein (GFAP) is an intermediate filament cytoskeletal protein which is constitutively expressed in retinal and brain astrocytes (5). The Müller glial cells do not normally express GFAP but will do so after a variety of injuries (6). Therefore GFAP expression is a good marker that the Müller cell is responding to an injury stimulus. Many of the injury stimuli that have been shown to induce Müller cell GFAP expression are themselves of long duration however following photocoagulation healing is quite rapid. Immunocytochemical labelling for GFAP demonstrated that in both dystrophic and control retinas the GFAP is first weakly detectable by 12hrs and at 24hrs, prior to any changes in retinal thickness due to degeneration there is expression in Müller cells surrounding the coagulated region although not in the core itself (Fig 1a). As phagocytosis of the lesion core proceded the Müller cells sprouted to form a glial scar in the core which extended into the choroid in heavy lesions where Bruch's membrane was penetrated (Fig 1b). The core scarring was well advanced by 7 days and appeared settled by 14 days but GFAP expression in surrounding Müller cells was present for at least 1.5 months. GFAP is a cytoskeletal protein and may therefore have a relatively long half-life. Therefore we examined the expression of GFAP mRNA following pan-retinal photocoagulation in congenic control RCS rats using a semi-quantitative PCR technique (7).

The results showed that the mRNA levels were normally not detectable after one round of 25 cycles of PCR amplification but that at 1, 2 and 3 days after photocoagulation the expression of GFAP was readily detectable. After 3 days the levels dropped back to normal despite the continued GFAP immunoreactivity of the Müller cells for over a month after this (Fig 2). Therefore it would seem that the stimulus for increased GFAP production is actually present for only a few days after photocoagulation and that the GFAP has a slow turnover rate. This pattern of a transient elevation in mRNA levels is similar to that found in the brain after spreading cortical depression, a phenomenon which is reversible and not thought to produce neural destruction (8,9). When neuronal destruction does occur mRNA levels stay high for a much more extended period (10). It is not known whether the GFAP expression significantly alters the function of Müller cells however blockade of GFAP expression in astrocytes by using antisense DNA prevents hypertrophy and division in response to injury (11).

bFGF LOCALIZATION

Elevated bFGF in brain lesions is thought to be a stimulus for glial reactions and it is known that Müller cells can also respond to bFGF (12, 13) and that exogenous bFGF can prolong the survival of photoreceptors in the RCS rat retina (14). Therefore we also examined bFGF localization after photocoagulation. Immunolocalization initially showed increased bFGF levels in the coagulated outer segment region for the first few days but otherwise very little change. However at later stages (from 2 to 21 days) there was an elevation of bFGF in the endothelial cells of the retinal blood vessels in the outer retina

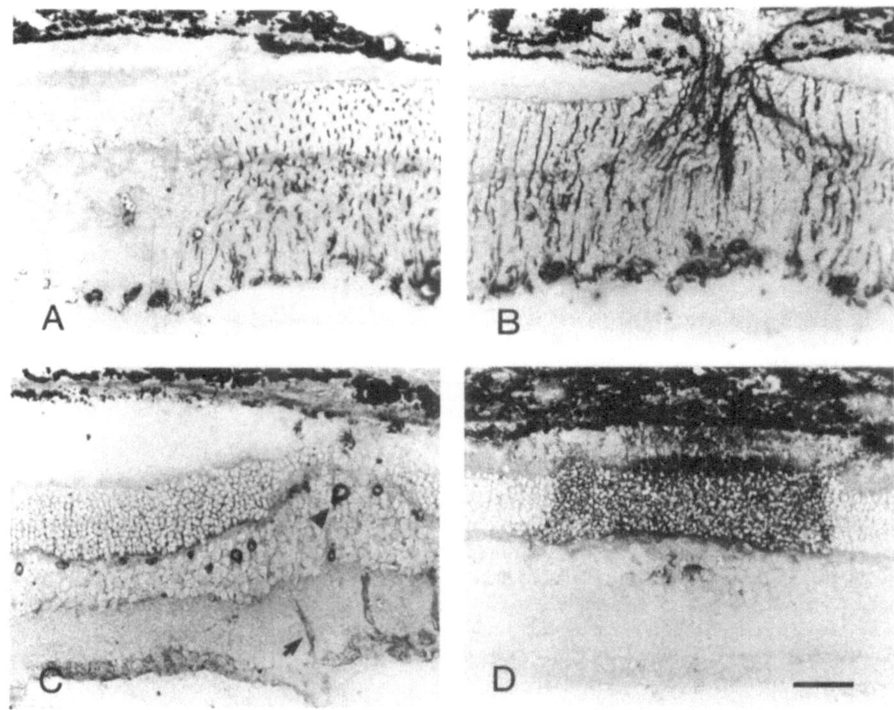

Figure 1. Macroglial, bFGF and IgG changes after photocoagulation. A: GFAP immunoreactivity 1 day after coagulation. In the core of the lesion there was no reaction but in the flanking region the Müller cells expressed GFAP along their entire length. B: At 7 days the coagulated photoreceptors had been removed and GFAP positive Müller cell processes formed a dense scar which sometimes (as here) extended into the choroid. C: At 14 days after coagulation the intense bFGF immunoreactivity in the outer blood vessels was particularly evident (arrow head), the inner vessels were also labeled but less strongly (arrow). The photoreceptors on the flanks also showed an elevation of bFGF immunoreactivity. D: At 1 day after coagulation IgG was localised to the outer segment region and outer plexiform layer. Occasionally cells in the inner nuclear layer were labelled and the inner segment/outer limiting membrane region was labelled. Scale bar = 50 microns and applies to all micrographs.

(Fig 1c). This elevation was not localized exclusively to the lesion core but extended for several hundred microns. The inner vessels were also labelled but much less strongly than the outer vessels. At 7-14 days there was also a slight but consistent elevation in the outer nuclear layer on the flanks of the lesion core. Thus the bFGF changes surrounding the coagulation core tended to occur at late stages but spread considerably and were focussed on the blood vessels and photoreceptor cells, the two cell types for which we have the most evidence for responding to photocoagulation. Therefore bFGF is likely to be involved in the effects on these cell types but the precise mechanisms may be difficult to work out. Many retinal cells contain bFGF and therefore subtle changes in localization or action may be difficult to detect also the form of the bFGF may be important, for example whether bound to heparan sulphate proteoglycans or not. In our PCR studies it was paradoxically found that bFGF mRNA levels dropped dramatically from 3 days until beyond 7 days (Fig 2). The significance of this is obscure although it may be a down regulation in response to the shifts in bFGF.

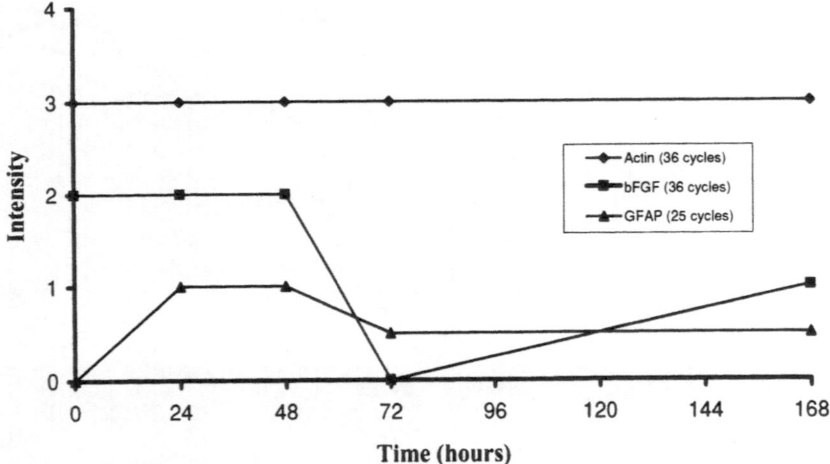

Figure 2. mRNA after laser photocoagulation. Following pan-retinal photocoagulation RCS-rdy+ retinas were removed at various times, the RNA extracted, converted to cDNA and then the beta-actin, GFAP and bFGF amplified by PCR. The samples were then run on a gel and the relative intensities of the bands graded qualitatively on a scale of 0-4. The constancy of the beta-actin levels are controls for the quality of the RNA extraction and amplification for each sample. The GFAP mRNA was elevated at 24 and 48 hrs but was already reduced by 72 hrs. bFGF mRNA was unaltered by 48hrs but at 72 hrs was undetectable and had not recovered to normal levels by 7 days.

BLOOD-RETINAL-BARRIER BREAKDOWN

It is known that the blood-retinal-barrier (BRB) breaks down following photocoagulation (15). This breakdown allows serum molecules to enter the retina and this may be important in the generation of cellular responses, particularly at a distance from the lesions as it has been reported that some molecules may enter the vitreous. Müller cells, in particular, have been shown to respond to thrombin in a dose dependent manner by dividing (16). In order to examine the entrance of serum molecules into the retina in the context of the other changes we have been examining we used immunolabelling for the rats own immunoglobulins. The IgG molecule is relatively large and therefore indicates where the breakdown was extensive. At no stage was there any evidence for IgG leakage from the retinal vessels. However, for the first three days after photocoagulation there was IgG present in the retina consistent with a breakdown of the barrier between the retina and choroid formed by the pigment epithelial cells. The IgG which entered the retina had a very characteristic distribution which varied according to the retinal layer. At 1-2 days the coagulated outer segment region contained IgG, as would be expected after a breakdown of the pigment epithlial barrier. Strong immunoreactivity was also located in the outer nuclear layer but this was confined to the coagulated region with a very sharp boundary to the flanking regions (Fig 1d). By contrast label in the region of the outer limiting membrane/ inner segments spread for a considerable distance from the lesion site. For example, after a grid of lesions superior to the optic nerve head, the inner segment region was labelled inferior to the optic nerve. In the first 24 hours there were occasional cells or regions labelled in the inner nuclear layer but otherwise there was very little entry of IgG to the inner retina. Thus, at least for IgG molecules there is extensive spread at the level of the outer limiting membrane but otherwise the label is confined to the outer retina within the coagulated lesion core.

PHAGOCYTIC CELLS

In recent years there has been a growing awareness that interactions between the immune and inflammatory system and the nervous system can be very important for the response of neural tissue to injury. In photoreceptor dystrophies phagocytic cells migrate from the inner retina to remove the dying cells (17). However, in the case of ganglion cell death after optic nerve section there is evidence that the phagocytic cells activily destroy cells which would normally survive for longer (18). The response of phagocytic cells to photocoagulation is not well understood largely because of the complexity of the situation. Thus, there are retina-resident microglial cells which can act as phagocytic cells, as well as blood borne cells such as monocytes which can enter the injured region and also pigment epithelial cells which bud from Bruch's membrane and migrate into the retina. In order to get a better understanding of which phagocytic cells are primarily involved in the response to photocoagulation we immunolabelled adjacent sections with Ox-42 antibodies, which identify phagocytic cells by C3b-complement receptor expression (18), ED1 antibodies which identify monocyte derived cells, ED2 antibodies which identify tissue resident macrophages (20), and Ox-6 antibodies which identify the MHC type II molecule and therefore cells capable of antigen presentation (21). Ox-42 normally labels the microglial population which is located in the inner retina (up to and including the outer plexiform layer). In the first 6-12 hours elongated Ox-42 positive cells extended into the outer nuclear layer and into the outer segment region in the coagulated core. This location and the morphology of the cells identified them as microglia. Alhough actively migrating these cells did not express either of the ED antigens nor MHC II. However, at the same time there was an aggregation of Ox-42 positive globular cells within the choroid underlying the coagulated core. Many of these cells were ED1 positive, indicating that they are of monocyte origin, and many were also MHC II positive indicating that they are potential antigen presenting cells. Only a few of the cells were ED2 positive so there was less of an aggregation by tissue resident macrophages. For the first 7-14 days there was a dense aggregation of Ox-42 positive cells in the coagulated core where phagocytosis was most active (Fig 3a) and also where Müller cells were most actively sprouting (Fig 1b). Many of these cells were ED1 positive but at no stage were ED2 or MHC II labelled cells found within the retina (Fig 3b,c,d). At 7 days, comparatively late in the sequence of phagocytosis, some of the inner retinal microglia which had migrated into the injury area expressed ED1 antigen patchily on their somas and processes. Thus microglia are able to express ED1 antigen but they only do so quite late in the sequence. Our data is therefore consistent with a very early migration of adjacent microglia into the lesion core, as we have previously demonstrated in the rabbit (6), and a clustering of monocytes and some tissue-resident macrophages in the choroid and sclera underlying the lesion zone. A dense aggregation of phagocytic cells is formed in the lesion core and these must be partly made up of migrated microglia and monocytes. The absence of ED2 expression suggests that the choroidal tissue-resident macrophages do not cross Bruch's membrane. None of these cells expressed MHC II antigen despite expression by many of those cells aggregated in the choroid. In the regions flanking the coagulated zone there was little change apart from a replacement of the cells which migrated into the injured zone. Thus it would appear that the phagocytic and inflammatory response is relatively mild and contained, possibly due to the very rapid response of the microglia reducing the stimulus to non-retinal phagocytic cells. The absence of any MHC II expression

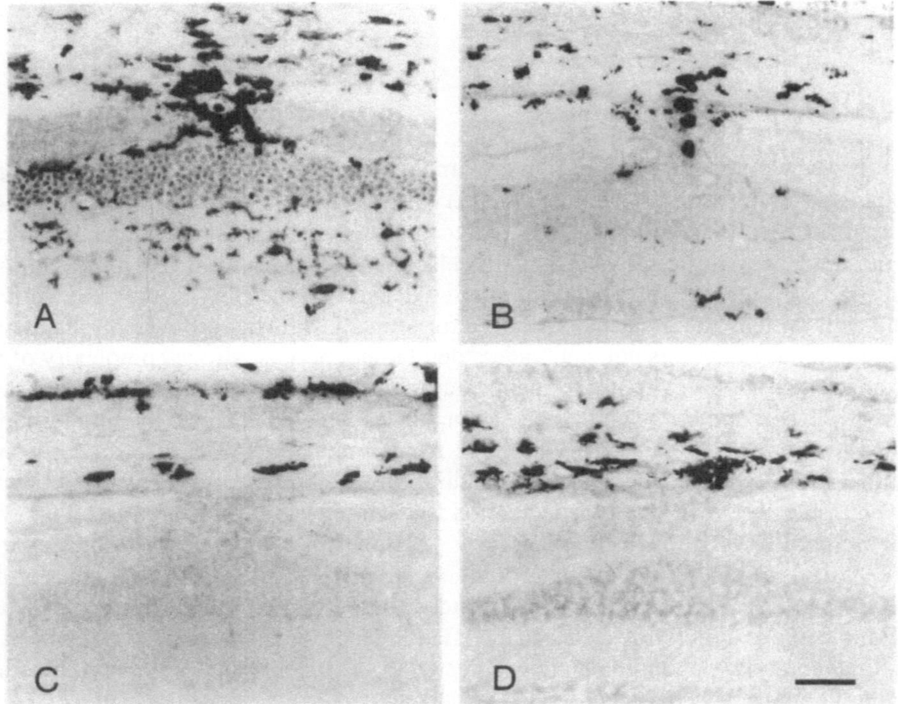

Figure 3. Phagocytic cells at 7 days after coagulation. Parallel cryostat sections through a coagulation lesion at 7 days, scale bar = 50 microns. A: Ox-42 immunoreactivity labels microglial cells in the inner retina and also a dense aggregation of cells in the lesion core and in the choroid and sclera. B: ED1 label for monocyte origin cells labels some cells in the core, choroid and sclera and, at this stage also labels some microglial processes. C: ED2 label for resident macrophages only labels cells in the choroid and sclera. D: Ox-6 label for MHC II expression only labelled cells in the choroid and sclera.

actually within the retina is consistent with the reduced immune responses reported in the sub-retinal space.

SUMMARY

In summary, argon laser photocoagulation produces an intense but rapidly resolved cellular reaction in the coagulated core where there is an elevated level of serum molecules and bFGF for the first 3 days. Rapid phagocytosis occurs through the action of migrated microglia and monocytes which do not, however, express MHC II markers at any stage. This phagocytosis is accompanied by dense Müller cell scarring which can extend into the choroid when Bruch's membrane is breached. In the retina surrounding the coagulated zones there are also changes. The most dramatic is the expression of GFAP by the Müller cells which persists for 1-1.5 months. However, our PCR studies indicate that elevated synthesis only occurs for 3-4 days after the coagulation and therefore that the stimulus for this is only transient. The extent to which GFAP expression alters Müller cell function is a relatively unexplored area. Accompanying these distant Müller cell reactions we have found that even large serum molecules such as IgG can spread over wide distances at the level of the photoreceptor inner segments/outer limiting membrane. The outer retinal blood vessels also become bFGF immunoreactive over a considerable distance while the outer nuclear layer has a similar increase but in a more restricted area. The phagocytic cell response by contrast was more focussed on the primary injury area. Thus the longer distance interactions, at least with the markers we have used, emphasise the Müller cells and the blood vessels. There is evidence that the Müller cells are responsible for blood-retinal-barrier formation (22) and so this is a natural link. When we understand the Müller cell-blood vessel interaction more fully we may better understand the mechanisms of the laser effect in diabetic retinopathy and thus be in a better position to prevent or control this condition.

REFERENCES

1. Marshall, J. and Mellerio, J., 1970, Laser irradiation of retinal tissue. *Br. Med. Bull.* 26: 156-161.
2. Meyer-Schwickerath, G. and Fried M., 1981, Treatment of diabetic retinopathy with photocoagulation. How many coagulations have to be performed in the individual case? *Dev. Ophthal.* 2: 265-273.
3. Silverman, M.S. and Hughes, S.E. (1990) Photoreceptor rescue in the RCS rat without pigment epithelium transplantation. Curr. Eye Res., 9, 183-191.
4. Humphrey, M.F., Parker, C., Chu, Y. and Constable, I.J. (1993) Transient preservation of photoreceptors on the flanks of argon laser lesions in the RCS rat. Curr. Eye Res., 12, 367-372.
5. Bignami, A. and Dahl, D., 1976, Astroglial response to stabbing. Immunofluorescence studies with antibodies to astrocyte-specific protein (GFA) in mammalian and submammalian vertebrates. *Neuropath. Appl. Neurobiol.*, 2: 99-110.
6. Humphrey, M.F., Constable, I.J., Chu, Y. and Wiffen, S., 1993, A quantitative study of the lateral spread of Müller cell responses to retinal lesions in the rabbit. *J. Comp. Neurol.*, 334: 545-558.
7. Rakoczy, P.E., Humphrey, M.F., Cavaney, D.M., Chu, Y. and Constable, I.J., 1993, Expression of basic fibroblast growth factor and its receptor in the retina of Royal College of Surgeons rats. A comparative study. *Invest. Ophthalmol. Vis. Sci.*, 34: 1845-1852.
8. Bonthius, D.J. and Steward, O., 1993, Induction of cortical spreading depression with potassium chloride upregulates levels of mRNA for glial fibrillary acidic protein in cortex and hippocampus: inhibition by MK-801. *Brain Res.*, 618: 83-94.
9. Nedergaard, M. and Hansen, A.J., 1988, Spreading depression is not associated with neuronal injury in the normal brain. *Brain Res.*, 449: 395-398.
10. Cancilla, P.A., Bready, J., Berliner, J., Sharifi-Nia, H., Toga, A.W., Santori, E.M., Scully, S. and DeVellis, J., 1992, Expression of mRNA for glial fibrillary acidic protein after experimental cerebral injury. *J. Neuropath. Exp. Neurol.*, 51: 560-565.
11. Ghirnikar, R.S. Yu, A.C.H. and Eng, L.F., 1994, Astrogliosis in culture: III. Effect of recombinant retrovirus expressing antisense glial fibrillary acidic protein RNA. *J. Neurosci. Res.*, 38: 376-385.
12. Finkelstein, S.P., Apostolides, P.J., Caday, C.G., Prosser, J., Philips, M.F. and Klagsbrun, M., 1988, Increased basic fibroblast growth factor (bFGF) immunoreactivity at the site of focal brain wounds. *Brain Res.*, 460: 253-259.
13. Puro, D.G. and Mano, T., 1991 Modulation of calcium channels in human retinal glial cells by basic fibroblast growth factor: A possible role in retinal pathology. *J. Neurosci.*, 11: 1873-1880.
14. Faktorovich, E.G., Steinberg, R.H., Yasumura, D., Matthes, M.T., and LaVail, M.M., 1990, Photoreceptor degeneration in inherited retinal dystrophy delayed by basic fibroblast growth factor. *Nature*, 347: 83-86.
15. Peyman, G.A. and Bok, D., 1972, Peroxidase diffusion in the normal and laser-coagulated primate retina. *Invest. Ophthalmol. Vis. Sci.*, 11: 35-45.
16. Puro, D.G., Mano, T., Chan, C-C., Fukuda, M. and Shimada, H., 1990, Thrombin stimulates the proliferation of human retinal glial cells. *Graefe's Arch. Clin Exp. Ophthalmol.*, 228: 169-173.
17. Thanos, S., 1992, Sick photoreceptors attract activated microglia from the ganglion cell layer: a model to study the inflammatory cascades in rats with inherited retinal dystrophy. *Brain Res.*, 588: 21-28.
18. Thanos, S., Mey, J. and Wild, M., 1993, Treatment of the retina with microglia-suppressing factors retards axotomy-induced neural degredation and enhances axonal regeneration in vivo and in vitro. *J. Neurosci.*, 13: 455-466.
19. Robinson, A.P., White, T.M. and Mason, D.W., 1986, Macrophage heterogeneity in the rat as delineated by two monoclonal antibodies MRC Ox-41 and MRC Ox-42, the latter recognising the complement receptor type 3. *Immunol.*, 57: 239-247.

20. Dijkstra, C.D., Döpp, E.A., Joling, P. and Kraal, G., 1985, The heterogeity of mononuclear phagocytes in lymphoid organs: distinct macrophage subpopulations in the rat recognized by monoclonal antibodies ED1, ED2 and ED3. *Immunol.*, 54: 589-599.
21. Gehrmann, J., Banati, R.B. and Kreutzberg, G.W., 1993, Microglia in the immune surveillance of the brain: Human microglia constitutively express HLA-DR molecules. *J. Neuroimmunol.*, 48: 189-198.
22. Tout, S., Chan-Ling, T., Holländer, H. and Stone, J., 1993, The role of Müller cells in the formation of the blood-retinal barrier. Neurosci., 55: 291-301.

25

DROSOPHILA AS A MODEL FOR PHOTORECEPTOR DYSTROPHIES AND CELL DEATH

William S. Stark, David Hunnius, Jennifer Mertz, and De-Mao Chen

Department of Biology
Saint Louis University
St. Louis, Missouri 63103-2010

INTRODUCTION

It is the silver anniversary of the isolation of Drosophila retinal degeneration mutants. The genes are still of substantial interest. Progress in neurobiology, developmental genetics and signal transduction has intensified as new molecular biology methods are applied. Retinal degeneration mutants display substantial cell death, informative by its specificity and its relationship with the visual excitation cascade. Other visual mutants best known for defects in visual pigment or phototransduction have various dystrophies. Light treatments and carotenoid deprivation also result in well-defined photoreceptor cell abnormalities. Since retinal degeneration in Drosophila was recently reviewed (1, 2, 3), the purpose of this paper is limited to presenting an overview and our recent data.

Historical Perspective

Pioneering mutant isolations (4, 5, 6, 7, 8, 9, 10, 11) ushered in our increasing interest in Drosophila vision. Electrophysiology, behavioral analyses and histology resulted in initial descriptions. From the outset, mutants of two genes, receptor degeneration I and II cistrons (6), were noteworthy: gross cell death devastated the retina. The genes were renamed rdgA and rdgB respectively in a paper devoted to Drosophila retinal degeneration (12). Also, rdgB was used with other mutants in a "genetic dissection" of the Drosophila retina (13): mutants with vs. without specific receptors functional were used to simplify the retina so that crude techniques like the electroretinogram [ERG] could be used to make inferences about specific receptor types (14).

Parallels between Drosophila and Vertebrate Studies

By this time, work was under way in vertebrates with retinal degeneration. There are striking similarities in these two fields. Genetic mosaic analyses suggested autonomy to

Degenerative Diseases of the Retina, Edited by Robert E. Anderson et al.
Plenum Press, New York, 1995

visual receptors in rdgB Drosophila (12) and rd mice (15) in contrast with the RCS rat with its defect in retinal pigment epithelium (16). Receptor degeneration uncovered function of minor receptor types in Drosophila (13) and rat (17). Vitamin A deprivation protects against light-induced retinal degeneration in rdgB (12) and intense ultraviolet [UV] and blue light treated Drosophila (18) and in the rat (19, 20). Transduction defects were suggested in rdgB Drosophila (12) and in rd mice (21). Intense UV and blue lights cause degeneration in Drosophila (18, 22). Intensities and wavelengths are remarkably similar to those in the aphakic monkey (23, 24); they are brighter than the blue light causing blue cone loss in work that Sperling and co-workers (25) presented in a journal from an early retinal degeneration symposium. White-eyed but otherwise non-mutant Drosophila (26) and albino rodents (17, 19, 20) have light damage from moderate, photopic light intensities.

Present Interest in Drosophila rdg Genes

Interest in rdg genes in Drosophila has flourished as molecular defects are elaborated. For example, rdgA codes for diacylglycerol [DAG] kinase (27). This enzyme is relevant to the visual cascade since it phosphorylates and recycles DAG, a product of phospholipase C [PLC], to phosphadidic acid; PLC is coded by the transduction mutant norpA [no receptor potential A] (28, 29). RdgB mutants are of particular interest since R1-6, the most numerous and sensitive compound eye receptors, thus analogous to rods, function until stimulation converting substantial rhodopsin to metarhodopsin (12). RdgB was isolated by Hyde and O'Tousa, contributors to this symposium (30, 31). It codes for a phosphatidylinositol transfer protein (32) located near rhabdomere bases (33, 34), pertinent to phosphatidylinositol [PI] signaling. Supplementing molecular advances, progress using cell biology came from Minke, another participant in this symposium (35), and his co-workers; pharmacological manipulations related to photic excitation can replace light to induce degeneration (36). Also, calcium, implicated in the IP3 [inositol trisphosphate] pathway, accumulates in photoreceptors of degeneration mutants (35). Another gene with light induced retinal degeneration, rdgC, encodes a serine/threonine protein phosphatase (37); the kinase vs. phosphatase cycle is obviously important in visual excitation.

Cell Death vs. Dystrophy of the Photoreceptive Rhabdomere in Drosophila

Mutants of several genes associated with phototransduction defects also cause cell death, judging from the ultrastructure, trp [transient receptor potential] being an example (3). Also, gross devastation is seen after diverse light treatments as mentioned above. For example, intense UV and blue stimulation causes cell death (18). By contrast mutants of the aforementioned norpA gene cause visual loss (38, 39) without appreciable cell death (2, 26, 40). Also, vitamin A deprivation causes visual decreases without cell death from which the animal can recover. Vitamin A deprivation and replacement in the fly have been the subject of thorough recent treatments elsewhere (41, 42, 43, 44); parallels in visual loss and recovery upon deprivation and replacement in the rat are also documented elsewhere (45, 46).

Ora and Other Mutants of ninaE as Retinal Dystrophy Models

Drosophila ninaE [neither inactivation nor afterpotential] (47, 48) codes for Rh1, the major compound eye rhodopsin, in R1-6 receptors (49, 50). As ninaE was mapped with respect to specific mutations (51, 52, 53, 54, 55), this laboratory's long-standing interest in the ora allele (13, 56, 57) was rekindled. Outer rhabdomeres absent [ora] was the first characterized ninaE mutant (11, 13). Ora is a nonsense mutant at codon 251 (51). The name "outer rhabdomeres absent" over-states the diminution of R1-6 rhabdomeres, called "outer" because of their position just outside the central ommatidial axis (56). Small rhabdomeres in newly-emerged animals diminish with age (52, 57). Accumulation of dense membranes with a zipper-like structure, some apparently intracellular [and possibly derived from rhabdomeres] (56) and some associated with abnormal junctions between photoreceptors (57) is a hallmark. Photoreceptors are remarkably resistant to cell death (57). Since cells die in human opsin mutants with autosomal dominant retinitis pigmentosa [ADRP], this contrast was considered noteworthy (2). Now, this discrepency appears to be resolved with the recently discovered dominant opsin mutants in Drosophila (31, 58) [O'Tousa, this symposium].

A transgenic fly with an engineered ninaE gene lacking opsin's glycosylation site (53) in a background with a ninaE deletion [oI17 allele] (49) also has rhabdomere diminution without cell death (59). Our recent data lead us to retract our data on the negative control [oI17] used in this study by Brown et al. (59). Although the stock we used lacked Rh1, genetic crosses suggest that it did not behave strictly as an allele of ninaE. A cross between ora females and oI17 males yielded sons and daughters with R1-6 functional, as judged by the ERG. A cross between oI17 females and ora males gave sons without R1-6 function and daughters with R1-6 function. The now suspect Brown et al. (59) oI17 data would only be of interest should the nature of our faulty stock be discovered. This finding eliminates a contradiction since other research (55) suggested that oI17 has more severe defects than Brown et al. reported, important since oI17 is widely used (60, 61).

METHODS

The Introduction reviews Drosophila retinal degeneration historically and with relation to work on vertebrates. New research lines are also presented. Microspectrophotometric [MSP], electron microscopic [EM] and morphometric studies of flies reared under different photic conditions were presented at ICER (62). Sensitivity and morphometric analyses of ora heterozygotes were presented at ARVO (63). Previously published techniques are overviewed only in sufficient detail to direct the reader to the appropriate literature.

Animals

Drosophila melanogaster were reared on our standard food. The most relevant feature is yellow cornmeal, a source of carotenoids, and a supplement of beta-carotene (41), to insure adequate vitamin A (64, 65). Strains were white-eyed otherwise wild-type flies [w = white] and white-eyed ninaE mutants [ora allele]. For light treatments, vials of flies were transferred from a 25 degree C incubator with a 12 hr on / 12 hr off cycle of fluorescent light and allowed to mature in a room temperature chamber near a fluorescent desk lamp with these conditions: [1] dark; [2] white light; [3] green light [behind a green acetate sheet, 0.25 log unit absorbance at peak transmission, 525 nm, 2.5 at 420 nm, peak absorbance, half peak cutoff at 480 nm]; and [4] red-green light [with a broad band red acetate increasing the short wavelength cutoff of the green acetate, peak transmission at 510 - 540 nm, half peak cutoff at 480 nm]. Intensities, calibrated with a portable radiometer / photometer [Ealing, Holliston, MA 01746, model 27-5479], were: [a] incubator, 49 lux; [b] light,1330 lux; [c] green, 190 lux; and [d] red-green 24 lux.

Microspectrophotometry [MSP] , Electron Microscopy [EM] and Electroretinography [ERG]

This laboratory developed (65) and refined (59) visual pigment measurement from R1-6 rhabdomeres of live, white-eyed Drosophila to assay receptor demise in mutants (66) or after light treatments (22); it can be averaged from many flies (67). Heads were fixed for transmission electron microscopy [TEM] using our long-standing protocols (59, 68). For morphometry, sections with [Fig. 4] and without [Fig. 2] xylene expansion, were picked up on grids (42). Electroretinograms [ERGs] were first used in this laboratory's research in 1972 (69); methodology has been refined in the interim (59). Careful wavelength and intensity controls and accurate calibrations allow sensitivity measurements. A recent innovation involves signal amplification and display using a MacLab/2e [Millford, MA] with a Macintosh LC II computer.

RESULTS AND DISCUSSION

Heterozygotes of Ora, the Nonsense Opsin Mutant

Our renewed interest in ora [Introduction] was motivated in part by studies on a null mutation at codon 249 in the human rod opsin gene which causes autosomal retinitis pigmentosa which is recessive [ARRP] (70). We determined ERG sensitivity of heterozygotes as a function of days post-eclosion and compared responsivity with homozygous mutant and wild-type flies [Fig. 1]. Sensitivity was 0.5 log units lower for heterozygotes than for positive controls at 1, 3, 5, 7, and 9 days. ERG waveforms, specifically the presence of on- and off-transients and the inducibility and reversibility of the PDA [prolonged depolarizing afterpotential] by intense short and long wavelength stimulation respectively, indicated sustained R1-6 function. We also measured cross sections of R1-6 rhabdomeres in heterozygotes using EM morphometry [Fig. 2]. Rhabdomeres were the same size in heterozygous specimens as in wild-type controls when newly-eclosed. Interestingly, R1-6 rhabdomeres diminished to half their newly-eclosed size after one week. Since diminished sensitivity suggests lower visual pigment, it is no surprise that ora heterozygotes have small rhabdomeres. What was surprising was the normal size of rhabdomeres in newly-emerged flies. Apparently, diminution of the photoreceptive organelle is an indirect result of decreased visual pigment. It would be interesting to determine whether this diminution is light dependent. The lower sensitivity is like that obtained from heterozygous relatives of the blind patient with homozygous ARRP (70).

The Green Light Experiment

The photointerconvertible R-480 - M-570 of Drosophila was considered to be non-bleaching (71, 72). Later, green light was found to decrease Calliphora visual pigment (73). Green light maintains moderate metarhodopsin, more labile than rhodopsin. It limits short wavelengths, needed to photoisomerize the chromophore from trans to cis. This phenomenon has been utilized in chromophore deprivation in Calliphora (74, 75, 76, 77). We wanted to replicate this experiment on Drosophila. MSP showed that green light decreased visual pigment in Drosophila, but only by about 1/4 [Fig. 3]. This decrease occurred within the first 24 hr of light exposure and remained the same for 2, 5 and 8 days. Additional filtering [rg], which decreased intensity and blocked blue light more effectively, gave the same result. We confirm Schwemer's Calliphora finding and extend it to Drosophila.

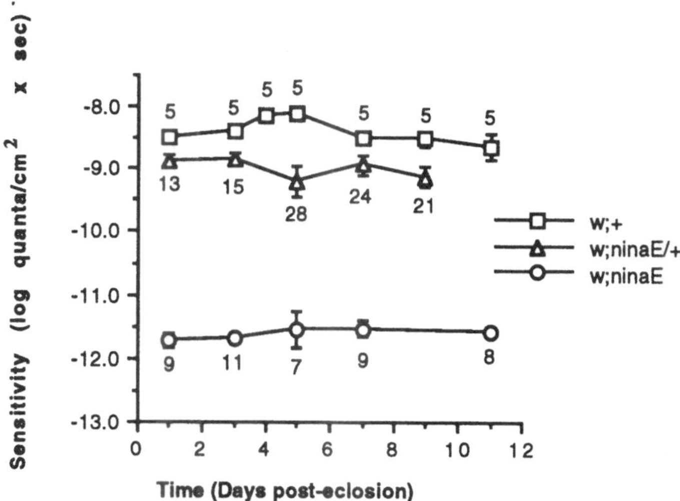

Figure 1. ERG sensitivity, high in white-eyed otherwise wild-type flies [w;+]; low in white-eyed flies homozygous for the ora mutant allele of ninaE [w;ninaE]; and intermediate in white-eyed heterozygotes of ora [w;ninaE/+]. Sensitivity was determined for a 2 mV peak-to-peak ERG at 470 nm and averaged [with SE] from the numbers of flies shown. Data were collected for flies one to 11 days post-eclosion as shown. The very low sensitivity of the ora homozygotes reflects the well documented absence of R1-6 function, uncovering the residual R7/8 contribution to the ERG. Importantly, ora heterozygotes remain half a log unit less sensitive than the positive wild-type control for over one week.

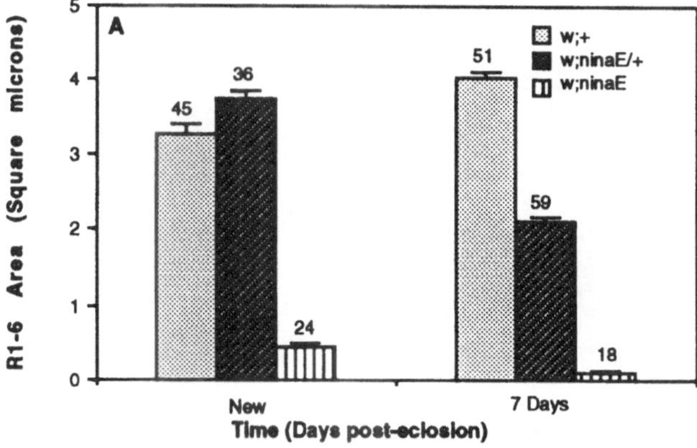

Figure 2. EM morphometry of R1-6 in the same fly types as in Fig. 1. Importantly, even though R1-6 in ora start out in newly-emerged flies the same size as in wild type, they are half the size after one week.

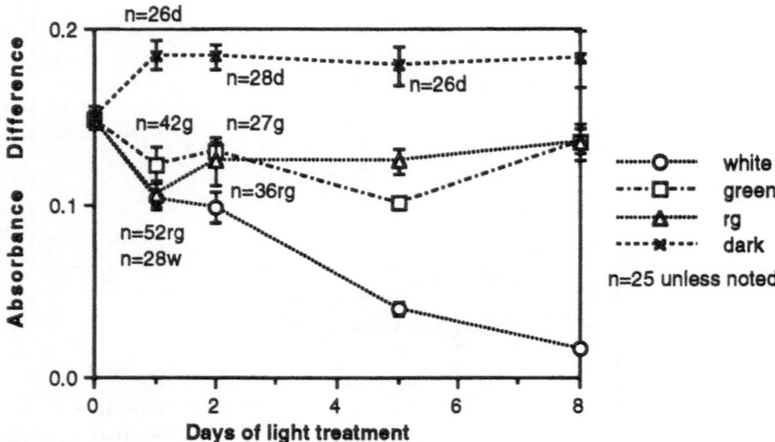

Figure 3. MSP of green and white light induced changes to visual pigment in white-eyed Drosophila. Absorbance difference was measured for newly-emerged flies out of incubator and flies at 1, 2, 5 and 8 days post-eclosion. Data are averaged from 25 or more preparations as indicated, and the SE is shown. When the pseudopupil was indistinct, which was the case especially after 5 and 8 days of white light treatment, measurement was from the area 150 microns beneath the most vertical corneal hairs, the strategy used to measure from degeneration mutants [see (66)].

However, even in pilot experiments (62), our reduction in visual pigment in Drosophila was not as striking as in Calliphora. This could be because Drosophila is different than Calliphora. Alternatively, high carotenoids in our food [Methods] may make optical chromophore deprivation marginal. Finally, since light is used in preparation and measurement, regeneration may have occurred, though we worked quickly with small numbers of flies per experiment to minimize this problem. EM showed that [1] green light did not damage receptors; [2] membrane turnover, as judged by multivesicular bodies, was normal; and [3] there may be a slight diminution of rhabdomere size by 2 days, extending to 8 days, as determined by morphometry [Fig 4]. Earlier, it was reported that constant moderate white

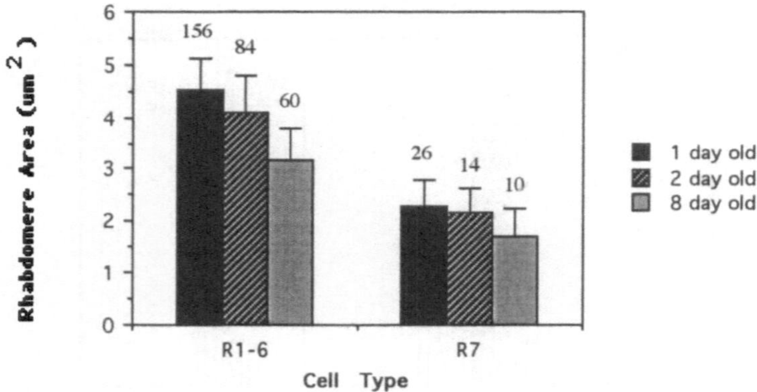

Figure 4. Morphometric analysis of flies reared under the green light condition. The numbers of rhabdomeres averaged and the standard deviations of the measurements are shown. Rhabdomeres under all conditions are larger than in other studies since measurements were made from sections expanded with xylene [see Methods].

light reduced visual pigment and caused structural degeneration (26). In the Fig. 3 experiment, the green light effect can be directly compared with the more profound losses induced by white light.

ACKNOWLEDGMENTS

Supported by NIH grant RO1 EY07192. We thank Gary Brown, Randall Sapp and Ronnie Lee for help in EM and morphology.

REFERENCES

1. Pak, W. L., 1994, Retinal degeneration mutants of Drosophila, in "Molecular genetics of inherited eye disorders," A. Wright and B. Jay, eds., Reading, Berkshire, UK, Harwood Academic Publishers, pp. 29-52.
2. Stark, W. S., J. S. Christianson, L. Maier, and D.-M. Chen, 1991, Inherited and environmentally induced retinal degenerations in Drosophila, in "Retinal Degenerations," R. E. Anderson, J. G. Hollyfield and M. M. LaVail, eds., New York, CRC Press, Inc., pp. 61-75.
3. Stark, W. S., and R. J. Sapp, 1989, Retinal degeneration and photoreceptor maintenance in Drosophila: rdgB and its interaction with other mutants, in "Inherited and Environmentally Induced Retinal Degenerations," M. M. LaVail, R. E. Anderson and J. G. Hollyfield, eds., New York, Liss, pp. 467-489.
4. Benzer, S., 1967, Behavioral mutants of Drosophila isolated by countercurrent distribution, Proc. Nat. Acad. Sci. (USA) 58:1112-1119.
5. Hotta, Y., and S. Benzer, 1969, Abnormal electroretinograms in visual mutants of Drosophila, Nature (Lond.) 222:354-356.
6. Hotta, Y., and S. Benzer, 1970, Genetic dissection of the Drosophila nervous system by means of mosaics, Proc. Nat. Acad. Sci. (USA) 67:1156-1163.
7. Pak, W. L., J. Grossfield, and N. V. White, 1969, Nonphototactic mutants in a study of vision of Drosophila, Nature (Lond.) 222:351-354.
8. Pak, W. L., J. Grossfield, and K. Arnold, 1970, Mutants of the visual pathway of Drosophila melanogaster, Nature 222:518-520.
9. Heisenberg, M., 1971, Isolation of mutants lacking the optomotor response, Dros. Inf. Serv. 46:68.
10. Cosens, D., and A. Manning, 1969, Abnormal electroretinogram from a Drosophila mutant, Nature (Lond.) 224:285-287.
11. Koenig, J. H., and J. R. Merriam, 1977, Autosomal ERG mutants, Drosoph. Inform. Serv. 52:50-51.
12. Harris, W. A., and W. S. Stark, 1977, Hereditary retinal degeneration in Drosophila melanogaster: a mutant defect associated with the phototransduction process, J. Gen. Physiol. 69:261-291.
13. Harris, W. A., W. S. Stark, and J. A. Walker, 1976, Genetic dissection of the photoreceptor system in the compound eye of Drosophila melanogaster, J. Physiol. (Lond.) 256:415-439.
14. Stark, W. S., A. M. Ivanyshyn, and K. G. Hu, 1976, Spectral sensitivities and photopigments in adaptation of fly visual receptors, Naturwissen. 63:513-518.
15. LaVail, M. M., and R. J. Mullen, 1976, Role of pigment epithelium in inherited retinal degeneration analysed with experimental mouse chimeras, Exp. Eye Res. 23:227-245.
16. Mullen, R. J., and M. M. LaVail, 1976, Inherited retinal dystrophy: Primary defect in pigment epithelium determined with experimental rat chimeras, Science 192:799-801.
17. Cicerone, C. M., 1976, Cones survive rods in the light-damaged eye of the albino rat, Science 194:1183-1185.
18. Stark, W. S., and S. D. Carlson, 1984, Blue and ultraviolet light induced damage to the Drosophila retina: ultrastructure., Curr. Eye Res. 3:1441-1454.
19. Noell, W. K., M. C. Delmelle, and R. Albrecht, 1971, Vitamin A deficiency effect on the retina: dependence on light, Science 172:72-76.
20. Noell, W. K., and R. Albrecht, 1971, Irreversible effects of visible light on the retina: Role of vitamin A, Science 172:76-80.
21. Farber, D. B., and R. N. Lolley, 1973, Proteins in the degenerative retina of C3H mice: Deficiency of a cyclic nucleotide phosphodiesterase and opsin, J. Neurochem. 21:817-828.
22. Stark, W. S., K. D. Walker, and J. M. Eidel, 1985, Ultraviolet and blue light induced damage to the Drosophila retina: Microspectrophotometry and electrophysiology, Curr. Eye Res. 4:1059-1075.

23. Ham, W. T. Jr., H. A. Mueller, J. J. Jr. Ruffolo, and A. M. Clarke, 1979, Sensitivity of the retina to radiation damage as a function of wavelength, Photochem. Photobiol. 29:735-743.
24. Ham, W. T. Jr., H. A. Mueller, J. J. Jr. Ruffolo, J. E. Millen, S. F. Cleary, R. K. Guerry, and D. I. Guerry, 1984, Basic mechanisms underlying the production of photochemical lesions in the mammalian retina, Curr. Eye Res. 3:165-174.
25. Sperling, H. G., R. S. Johnson, and R. S. Harwerth, 1980, Differential spectral photic damage to primate cones, Vision Res. 20:1117-1125.
26. Zinkl, G., L. Maier, K. Studer, R. Sapp, D. M. Chen, and W. S. Stark, 1990, Microphotometric, ultrastructural and electrophysiological analyses of light dependent processes on visual receptors in white-eyed wild-type and norpA (no receptor potential) mutant Drosophila., Vis. Neurosc. 5:429-439.
27. Masai, I., A. Okazaki, T. Hosoya, and Y. Hotta, 1993, Drosophila retinal degeneration A gene encodes an eye-specific diacylglycerol kinase with cysteine-rich zinc-finger motifs and ankyrin repeats, Proc. Natl. Acad. Sci. USA 90:11157-11161.
28. Yoshioka, T., H. Inoue, and Y. Hotta, 1985, Absence of phosphotidylinositol phosphodiesterase in the head of a Drosophila visual mutant, norpA (no receptor potential A), J. Biochem. 97:1251-1254.
29. Bloomquist, B. T., R. D. Shortridge, S. Schnewly, M. Perdew, C. Montell, H. Steller, G. Rubin, and W. L. Pak, 1988, Isolation of a putative phospholipase C gene of Drosophila, norpA, and its role in phototransduction, Cell 54:723-739.
30. Hyde, D. R., S. Milligan, D. Paetkau, and T. S. Vihtelic, 1994, The role of the retinal degeneration B protein in the Drosophila visual system, Int. Symp. Ret. Degen. 6:18.
31. Kurada, P., and J. E. O'Tousa, 1994B, The role of dominant rhodopsin mutations in Drosophila retinal degeneration, Int. Symp. Ret. Degen. 6:17.
32. Vihtelic, T. S., D. R. Hyde, and J. E. O'Tousa, 1991, Isolation and characterization of the Drosophila retinal degeneration B (rdgB) gene, Genetics 127:761-768.
33. Vihtelic, T. S., M. Goebl, S. Milligan, J. E. O'Tousa, and D. R. Hyde, 1993, Localization of Drosophila retinal degeneration B, a membrane-associated phosphatidylinositol transfer protein, J. Cell Biol. 122:1013-1022.
34. Suzuki, E., and K. Hirosawa, 1994, Immunolocalization of a Drosophila phosphatidylinositol transfer protein (rdgB) in normal and rdgA mutant photoreceptor cells with special reference to the subrhabdomeric cisternae, J. Electron Microsc. 43:183-189.
35. Sahly, I., W. H. Schroder, K. Zierold, and B. Minke, 1994, Accumulation of calcium in degenerating photoreceptors of several Drosophila mutants, Vis. Neurosci. 11:763-772.
36. Rubenstein, C. T., S. Bar-Nachum, Z. Selinger, and B. Minke, 1989A, Chemically induced retinal degeneration in the rdgB (retinal degeneration B) mutant of Drosophila, Vis. Neurosci. 2:541-551.
37. Steele, F. R., T. Washburn, R. Rieger, and J. E. O'Tousa, 1992, Drosophila retinal degeneration C (rdgC) encodes a novel serine/threonine protein phosphatase, Cell 69:669-676.
38. Ostroy, S. E., 1978, Characteristics of Drosophila rhodopsin in wild-type and norpA vision transduction mutants, J. Gen. Physiol. 72:717-732.
39. Meyertholen, E. P., P. J. Stein, M. A. Williams, and S. E. Ostroy, 1987, Studies of the Drosophila norpA phototransduction mutant, J. Comp. Phys. 161:793-798.
40. Stark, W. S., R. J. Sapp, and S. D. Carlson, 1989A, Photoreceptor maintenance and degeneration in the norpA (no receptor potential-A) mutant of Drosophila melanogaster, J. Neurogenet. 5:49-59.
41. Chen, D.-M., and W. S. Stark, 1992, Electrophysiological sensitivity of carotenoid deficient and replaced Drosophila, Vis. Neurosci. 9:461-469.
42. Sapp, R. J., J. S. Christianson, L. Maier, K. Studer, and W. S. Stark, 1991A, Carotenoid replacement therapy in Drosophila : recovery of membrane, opsin and rhodopsin., Exp. Eye Res. 53:71-79.
43. Lee, R. D., C. F. Thomas, R. G. Marietta, and W. S. Stark, 1995, Vitamin A, visual pigments and visual receptors in Drosophila, Micros. Res. Tech. in:press.
44. Sun, D., D.-M. Chen, A. Harrelson, and W. S. Stark, 1993, Increased expression of chloramphenicol acetyltransferase by carotenoid and retinoid replacement in Drosophila opsin promoter fusion stocks, Exp. Eye Res. 57:177-187.
45. Katz, M. L., C. L. Gao, M. Kutryb, N. Norberg, R. H. White, and W. S. Stark, 1991, Maintenance of opsin density in photoreceptor outer segments of retinoid-deprived rats, Invest. Ophthalmol. Vis. Sci. 32:1968-1980.
46. Katz, M. L., D.-M. Chen, H. J. Stientjes, and W. S. Stark, 1993, Photoreceptor recovery in retinoid-deprived rats after vitamin A replenishment, Exp. Eye Res 56:671-682.
47. Schinz, R. H., M. V. C. Lo, D. C. Larrivee, and W. L. Pak, 1982, Freeze-fracture study of the Drosophila photoreceptor membrane: Mutations affecting membrane particle density, J. Cell Biol. 93:961-969.

48. Scavarda, N. J., J. O'Tousa, and W. L. Pak, 1983, Drosophila locus with gene-dosage effects on rhodopsin, Proc. Nat. Acad. Sci. (USA) 80:4441-4445.
49. O'Tousa, J. E., W. Baehr, R. L. Martin, J. Hirsh, W. L. Pak, and M. L. Applebury, 1985, The Drosophila ninaE gene encodes an opsin, Cell 40:839-850.
50. Zuker, C. S., A. F. Cowman, and G. M. Rubin, 1985, Isolation and structure of a rhodopsin gene from D. melanogaster, Cell 40:851-858.
51. Washburn, T., and J. E. O'Tousa, 1989, Molecular defects in Drosophila rhodopsin mutants, J. Biol. Chem. 15464-15466.
52. O'Tousa, J. E., D. S. Leonard, and W. L. Pak, 1989, Morphological defects in oraJK84 photoreceptors caused by mutation in R1-6 opsin gene in Drosophila, J. Neurogenet. 6:41-52.
53. O'Tousa, J. E., 1992, Requirement of N-linked glycosylation site in Drosophila rhodopsin, Vis. Neurosci. 8:385-390.
54. Washburn, T., and J. E. O'Tousa, 1992, Nonsense suppression of the major rhodopsin gene in Drosophila, Genetics 130:585-595.
55. Leonard, D. S., V. D. Bowman, D. F. Ready, and W. L. Pak, 1992, Degeneration of photoreceptors in rhodopsin mutants of Drosophila, J. Neurobiol. 23:605-626.
56. Stark, W. S., and S. D. Carlson, 1983, Ultrastructure of the compound eye and first optic neuropile of the photoreceptor mutant oraJK84 of Drosophila, Cell Tiss. Res. 233:305-317.
57. Stark, W. S., and R. J. Sapp, 1987, Ultrastructure of the retina of Drosophila melanogaster: The mutant ora (outer rhabdomeres absent) and its inhibition of degeneration in rdgB (retinal degeneration-B), J. Neurogenet. 4:227-240.
58. Kurada, P., and O'Tousa, 1994A, The role of ninaE dominant mutations in retinal degeneration, Dros. Res. Conf. 35:11.
59. Brown, G., D.-M. Chen, J. S. Christianson, R. Lee, and W. S. Stark, 1994, Receptor demise from alteration of glycosylation site in Drosophila opsin: Electrophysiology, microspectrophotometry, and electron microscopy, Vis. Neurosci. 11:619-628.
60. Britt, S. G., R. Feiler, K. Kirschfeld, and C. S. Zuker, 1993, Spectral tuning of rhodopsin and metarhodopsin in vivo, 11:29-39.
61. Colley, N. J., E. K. Baker, M. A. Stamnes, and C. S. Zuker, 1991, The cyclophilin homolog ninaA is required in the secretory pathway., Cell 67:255-263.
62. Stark, W. S., 1994, R1-6 visual pigment is reduced by green light in Drosophila, Exp. Eye Res. 59:S.33.
63. Chen, D.-M., G. Brown, and W. S. Stark, 1993, Sensitivity and rhabdomere decreases in heterozygotes of ora, a nonsense mutant allele of ninaE, the Drosophila rh1 opsin gene, Invest. Ophthalmol. Vis. Sci. 34:808.
64. Stark, W. S., A. M. Ivanyshyn, and R. M. Greenberg, 1977, Sensitivity and photopigments of R1-6, a two-peaked photoreceptor, in Drosophila, Calliphora and Musca, J. Comp. Physiol. 121:289-305.
65. Stark, W. S., and M. A. Johnson, 1980, Microspectrophotometry of Drosophila visual pigments: Determinations of conversion efficiency in R1-6 receptors, J. Comp. Physiol. 140:275-286.
66. Stark, W. S., D.-M. Chen, M. A. Johnson, and K. L. Frayer, 1983, The rdgB gene in Drosophila: Retinal degeneration in different mutant alleles and inhibition of degeneration by norpA, J. Insect Physiol. 29:123-131.
67. Chen, D.-M., J. S. Christianson, R. J. Sapp, and W. S. Stark, 1992, Visual receptor cycle in normal and period mutant Drosophila: Microspectrophotometry, electrophysiology, and ultrastructural morphometry, Vis. Neurosci. 9:125-135.
68. Stark, W. S., and A. W. Clark, 1973, Visual synaptic structure in normal and blind Drosophila, Drosoph. Inform. Serv. 50:105-106.
69. Stark, W. S., and G. S. Wasserman, 1972, Transient and receptor potentials in the electroretinogram of Drosophila, Vision Res. 12:1771-1775.
70. Rosenfeld, P. J., G. S. Cowley, T. L. McGee, M. A. Sandberg, E. L. Berson, and T. P. Dryja, 1992, A Null mutation in the rhodopsin gene causes rod photoreceptor dysfunction and autosomal recessive retinitis pigmentosa, Nature Genetics 1:209-213.
71. Ostroy, S. E., M. Wilson, and W. L. Pak, 1974, Drosophila rhodopsin: photochemistry, extraction and differences in the norpAP12 phototransduction mutant, Biochem. Biophys. Res. Comm. 59:960-966.
72. Pak, W. L., and K. L. Lidington, 1974, Fast electrical potential from a long-lived, long-wavelength photoproduct of fly visual pigment, J. Gen. Physiol. 63:740-756.
73. Schwemer, J., 1984, Renewal of visual pigment in photoreceptors of the blowfly, J. Comp. Physiol. 154:535-547.
74. Smakman, J. G. J., and D. G. Stavenga, 1986, Spectral sensitivity of blowfly visual receptors: dependence on waveguide effects and pigment concentration, Vision Res. 26:1019-1025.

75. Hamdorf, K., P. Hochstrate, G. Hoglund, M. Moser, S. Sperber, and P. Schlecht, 1992, Ultra-violet sensitizing pigment in blowfly photoreceptors R1-6; probable nature and binding sites, J. Comp. Physiol. 171:601-615.
76. Schwemer, J., and F. Spengler, 1992, Opsin synthesis in blowfly photoreceptors is controlled by an 11-cis retinoid, in "Structures and Functions of Retinal Proteins," J. L. Rigaud, eds., Montrouge, France, John Libbey Eurotext, pp. 277-280.
77. Huber, A., U. Wolfrum, and R. Paulsen, 1994, Opsin maturation and targeting to rhabdomeral photoreceptor membranes requires the retinal chromophore, Eur. J. Cell Biol. 63:219-229.

26

ABNORMAL CA^{2+} MOBILIZATION AND EXCESSIVE PHOTOPIGMENT PHOSPHORYLATION LEAD TO PHOTORECEPTOR DEGENERATION IN *DROSOPHILA* MUTANTS

Baruch Minke[1] and Zvi Selinger[2]

[1] Kuhne Minerva Center for Studies of Visual Transduction and Department of Physiology
[2] Kuhne Minerva Center for Studies of Visual Transduction and Department of Biological Chemistry
The Hebrew University
Jerusalem, Israel 91120

INTRODUCTION

A great deal is known today about the identity of several gene products which are targets for mutations that induce retinal degeneration both in vertebrates and invertebrates (1-15). These mutant genes lead to various forms of retinal degeneration. However, the molecular mechanisms underlying the sequence of events which bring about retinal degeneration is still obscure. A common denominator of these mutant gene products is that most of them are proteins important for phototransduction. A clue to a molecular mechanism which initiates the degeneration process came from recent studies on retinal degeneration in *Drosophila* mutant photoreceptors in which the degeneration process is light-dependent, namely, the photoreceptors do not degenerate if the fly is raised in the dark.

It has been well established that photoreceptors are efficient photon counters. To achieve this function, each step in the excitatory cascade needs an efficient turn off mechanism. To account for the conditional (light-dependent) phenotype of photoreceptor degeneration, it was suggested that the normal counterpart of the retinal degeneration gene product in wild type fly counteracts one of the steps in the phototransduction cascade to turn its activity off (4,10,16). When this gene is mutated and either becomes non-functional or completely absent, it leads to abnormal activity of the phototransduction step with which the retinal degeneration gene product normally interacts and this, in turn, gives rise to light-dependent retinal degeneration.

Degenerative Diseases of the Retina, Edited by Robert E. Anderson et al.
Plenum Press, New York, 1995

An example for the above mechanism is the light-induced degeneration of the photoreceptors in the *retinal degeneration C (rdgC)* mutant (7,8,10).

Genetic analysis of the *Drosophila* mutant *rdgC* revealed that retinal degeneration is dependent on high levels of activated rhodopsin and placed the site of action of the *rdgC* gene product, before phospholipase C (7). More recently, molecular cloning of the *rdgC* gene showed sequence similarity with mammalian serine/threonine phosphatase and identified an appended domain containing putative direct Ca^{2+} binding sites (8).

PHOSPHORYLATION-DEPHOSPHORYLATION REACTIONS OF THE PHOTOPIGMENT AND THEIR RELATION TO DEGENERATION IN THE *rdgC* MUTANT

A characteristic of fly photoreceptors is that their photopigment is thermostable and photoreversible. Blue light (<490 nm) converts rhodopsin (R) to metarhodopsin (M) (80%), and orange light (>580 nm) regenerates it to rhodopsin (100%). *Drosophila* eye membranes, preilluminated with blue light, revealed substantial phosphorylation of metarhodopsin. This phosphorylation was strictly dependent on conversion of rhodopsin to metarhodopsin, (Fig. 1, lanes 0,0′) as no phosphorylation of the photopigment took place in membranes that had been preilluminated with orange light (10).The highest extent of metarhodopsin phosphorylation was obtained in the presence of the Ca^{2+} chelator EGTA, whereas Ca^{2+} considerably reduced the extent of metarhodopsin phosphorylation. This effect was not dependent on calmodulin, as an efficient peptide inhibitor of calmodulin, M5, did not change the extent of metarhodopsin phosphorylation (10).

Ca^{2+} was required for the dephosphorylation reaction. The rate of dephosphorylation was much faster in membranes in which phosphorylated metarhodopsin (p-M) had been first

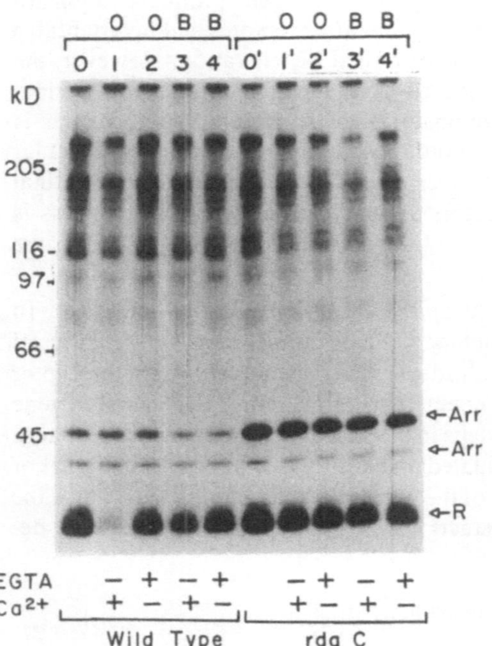

Figure 1. Dephosphorylation of proteins in wild-type and *rdgC* eye membranes. Membranes were derived from blue-illuminated eyes and phosphorylated in the presence of 10μM EGTA, followed by addition of unlabelled ATP to 0.5 mM and centrifugation. The membrane pellets were suspended in homogenization buffer and illuminated with orange (O) or blue (B) lights. Dephosphorylation was carried out for 5 min at 25°C, with the indicated additions. Lanes in radiogram: 0 and 0′, phosphoproteins before dephosphorylation; 1, 1′, 3, and 3′, dephosphorylation in the presence of 0.1 mM Ca^{2+}; 2, 2′,4 and 4′, dephosphorylation in the presence of 1mM EGTA. Arr, arrestin, R, rhodopsin (From Ref. 10).

converted to phosphorylated rhodopsin (p-R) by illumination with orange light and then subjected to dephosphorylation in the presence of Ca^{2+} (Fig. 1, lanes 1, 2 and 4).

In the experiments illustrated in Fig. 1, parallel assays were conducted on *Drosophila* eye preparations of wild type and the *rdgC* mutant. Optimal dephosphorylation conditions resulted in removal of 95% of the phosphate from wild-type rhodopsin following conversion of p-M to p-R, whereas little if any dephosphorylation of the *rdgC* rhodopsin was observed under identical conditions (Fig. 1 lanes 1 and 1′). These results indicate that phosphorylated rhodopsin is a major substrate for the *rdgC* protein phosphatase, which requires Ca^{2+} for activation.

The *rdgC* mutant has thus demonstrated that excessive light-dependent phosphorylation of the photopigment unbalanced by the normal dephosphorylation by *rdgC* phosphatase leads to photoreceptor degeneration.

INHIBITION OF CA^{2+} MOBILIZATION BY MUTATIONS LEADS TO RETINAL DEGENERATION

In wild type fly, the smooth operation of the photopigment cycle requires light-induced elevation of cytosolic Ca^{2+} during illumination to ensure dephosphorylation of p-R (Fig. 1). It has been recently shown that illumination is accompanied by a large increase in cellular Ca^{2+} (17-19) in *Drosophila* photoreceptors and that this increase depends on activation of the inositol-lipid cascade (17). Figure 2 shows fluorescence measurements of intracellular Ca^{2+} during intense illumination in dialyzed photoreceptors during whole cell recordings. Figure 2 shows that in wild type photoreceptors the light-induced current (LIC) is accompanied by a simultaneous increase of cellular Ca^{2+}. Figure 2 also shows that under identical recording conditions in the no receptor potential A (*norpA*) mutant which lacks light-dependent phospholipase C (PLC; 5), neither the LIC nor the increase in cellular Ca^{2+} could be observed. Accordingly, elimination of the light-activated phospholipase C (PLC), by mutation in the *norpA* gene (5) blocks the light-induced current and the increase in cellular Ca^{2+} (Fig. 2).

The dependence of the *rdgC* phosphatase on Ca^{+2}, which is one of the end products of the inositol lipid phototransduction cascade is interesting as it bears on the unexplained observation that several phototransduction mutants like the *norpA*, and the transient receptor potential *(trp)*, which codes for a putative light-sensitive channel with high permeability for Ca^{+2} (19), both undergo age and light dependent retinal degeneration (11,20). Retinal degeneration of the *norpA* mutant which was studied more extensively showed that strong *norpA* alleles which have no electrical response to light demonstrated more prominent retinal degeneration than weak alleles that have some light dependent electrical activity. Surprisingly, in *norpA* alleles, which do not respond electrically to light, retinal degeneration was still light dependent (11). We have suggested (10,16) that the mechanism of retinal degeneration in the strong *norpA* alleles is similar to the mechanism of retinal degeneration in the *rdgC* mutant. In both mutants, retinal degeneration is initiated by light dependent phosphorylation of the photopigment which is not adequately followed by dephosphorylation. In the *rdgC* mutant, this is due to deficiency in rhodopsin phosphatase, while in the *norpA* mutant deficient dephosphorylation of rhodopsin is due to the inability to generate the Ca^{+2} signal which is required for activation of rhodopsin phosphatase (10). This mechanism can account for light dependent retinal degeneration in other transduction mutants which block the increase of cellular Ca^{+2}. We have corroborated this putative mechanism of light dependent retinal degeneration by phosphorylation experiments in the intact fly *in-vivo*. In these experiments rhodopsin was found to be hyperphosphorylated both in the *rdgC* and in the

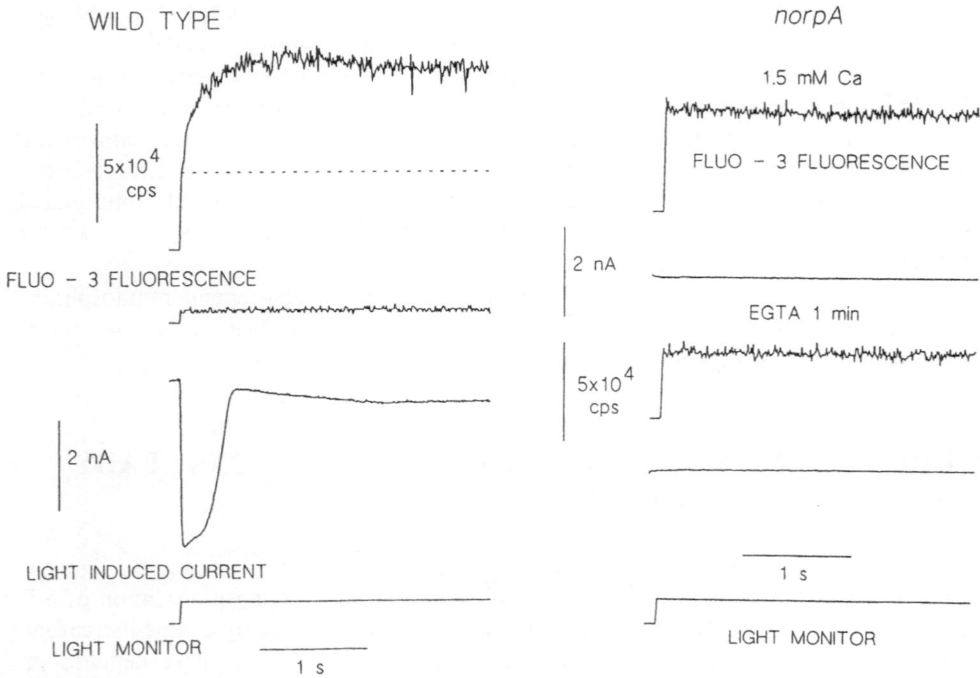

Figure 2. The light-evoked inward current and the Ca^{2+} signal of Wild-Type are abolished in the *norpA* mutant. Simultaneous measurement of fluo-3 fluorescence (upper traces) and LIC (middle traces) at -50mV in the photoreceptors of Wild-Type (WT) (left column) and the no receptor potential A (*norpA*^{P24}) mutant of *Drosophila* (right column). The horizontal dotted line indicates the fluorescence level at the time of full opening of the shutter that controlled the exciting light. The background fluorescence (measured after removal of the ommatidium) is indicated below the upper trace (left). The lower trace shows the output of a photodiode that monitored the exciting light. The current traces showed very pronounced light-induced current (LIC) in WT, but no LIC could be observed in the mutant. Also the fluorescence signal of the mutant showed only a fast initial rise that accompanies the opening of the shutter which mainly reflects the resting Ca^{2+}. Removing external Ca for 1 min by exposure to 0-Ca^{2+} EGTA Ringer reduced the resting Ca^{2+} level (Modified from ref. 17).

norpA mutant, while it was minimally phosphorylated in the wild type fly (unpublished experiments).

A strong independent support for this mechanism of degeneration in *rdgC* and *norpA* came from recent studies by O'Tousa and colleagues (this volume) which screened for suppressor mutants of *rdgC* mutant degeneration. They have isolated 2 classes of dominant suppressors of *rdgC* degeneration and found that these suppressors are defective in the rhodopsin molecules leading to reduced pigment content. They furthermore found that the dominant rhodopsin alleles act to suppress both *rdgC* and *norpA* induced degeneration and conclude that the *norpA*$^{+}$ activity is required for *rdgC*$^{+}$ activity. All the above data indicate that degeneration in *rdgC* and *norpA* photoreceptors is due to excessive phosphorylation of the photopigment due to deficient *rdgC* phosphatase *activity* and when the pigment level is reduced, degeneration is suppressed even in the absence of the *rdgC* phosphatase activity.

Figure 3 summarizes the cascade of molecular steps and interactions of fly photopigment and illustrates the tight control of phosphorylation and dephosphorylation reactions of the photopigment on the photopigment cycle: Photoconversion of rhodopsin (R) to metarhodopsin (M) results in rapid phosphorylation of M by rhodopsin kinase to give phosphorylated M (p-M) followed by binding of 49 kDal arrestin. Phosphorylation of M decreases its ability

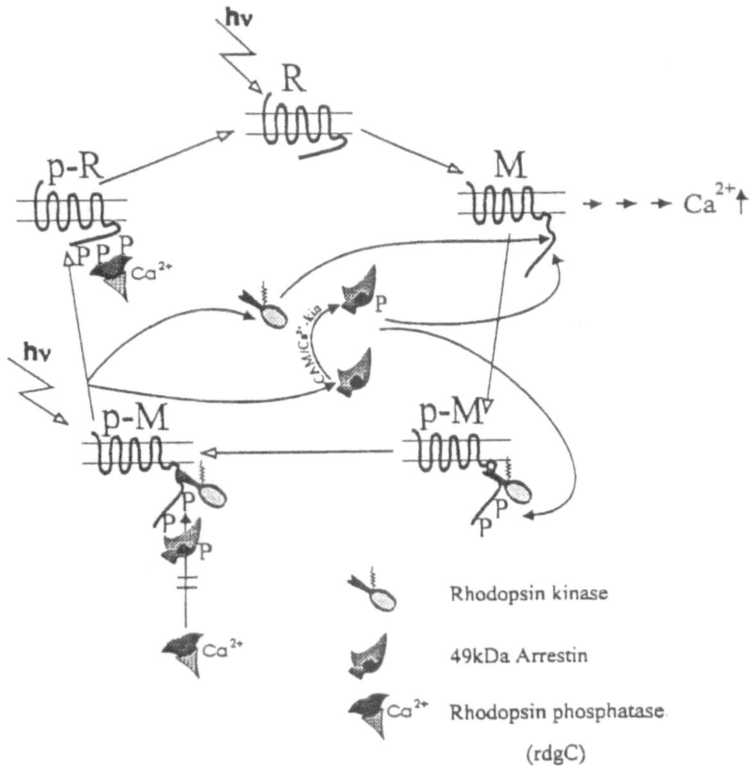

Figure 3. The cascade of molecular steps of fly photopigment (see text).

to activate the G protein. Binding of arrestin further quenches the ability of M to activate G protein and protect it from phosphatase activity. Absorption of a photon by the p-M/arrestin complex regenerates it to phosphorylated R (p-R) with concomitant release of arrestin. The p-R then becomes a substrate for *rdgC* rhodopsin phosphatase that reintroduces it to the excitable rhodopsin pool without reinitiation of phototransduction in the dark. Binding and release of arrestin to p-M and p-R, respectively, confines the rhodopsin phosphatase activity only to p-R, thereby securing the fidelity of phototransduction (see ref. 10 for experimental details).

ACCUMULATION OF CALCIUM IN DEGENERATING PHOTORECEPTORS OF *DROSOPHILA* MUTANTS

How excessive phosphorylation of the photopigment leads to photoreceptor degeneration is still unclear. Nevertheless, a step in the degeneration process involves a large increase in cellular Ca^{2+} as evidenced by accumulation of Ca^{2+} in subcellular organelles (21).

Total calcium level was measured in photoreceptors of mutant *Drosophila* which show light-dependent degeneration: the *rdgB* (4,9), *rdgC* (7), *norpA* (5,11) and *ninaC* (13) mutants raised in light or darkness (21). A unique and powerful technique of energy dispersive x-ray analysis (e.d.x) of sections from quick frozen retinae was used to examine if a large increase in cellular calcium is a phenomenon which accompanies retinal degeneration even when induced by different mutations. Figure 4 shows representative original e.d.x. spectra without any data processing, measured in sections from the photoreceptor cell body of a shock frozen dark (A) and *light-raised* retinae of *rdgC* mutant (B,C). For comparison Fig. 4 shows spectra of cryosubstituted material (Fig. 4B) and of cryosections (Fig. 4C). The location of the energy line for the various elements in the sample are indicated. K_α energy lines for potassium and calcium are of particular interest. The small peak of Fig. 4A which is indicated by the arrow ($Ca_{K\alpha}$+$K_{K\beta}$) contains both the overlapping $Ca_{K\alpha}$ peak and the $K_{K\beta}$ peak. Peak deconvolution of the overlapping potassium and calcium signals reveals, in these samples, a calcium peak to background (P/B) ratios near the detection limit of calcium under these conditions which is about 1 - 2 mmole/l calcium for both e.d.x. analysis methods used. This detection limit did not allow analyses of calcium changes induced by light in WT photoreceptors or in mutants raised in the dark or at very early stages of degeneration (21). A systematic and detailed scanning of different areas of photoreceptors from sections of dark-raised mutant flies (e.g. *rdgC, ninaC and norpA*) showed spectra very similar to the light-raised WT. Very small levels of calcium were observed in samples of shock frozen cryosubstituted retinae from the above mutants *raised in the dark* which did not show any

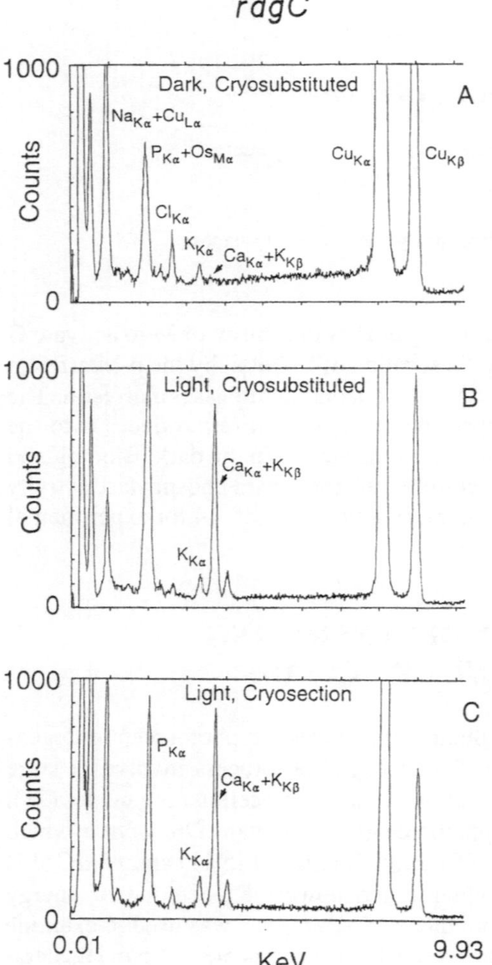

Figure 4. E.d.x. spectra from cross sections of dark raised (A) and light-raised (B,C) *rdgC* eyes (5 day old, at 24°C). The $K_{K\alpha}$ and $Ca_{K\alpha}$+$K_{K\beta}$ peaks are indicated in addition to the peaks of the other elements. A,B) E.d.x. spectra of shock frozen cryosubstituted plastic embedded material of dark-raised *rdgC* mutant (A) and light-raised *rdgC* fly (B). The measurements were carried out during 457 s in A. Measurement was taken from the globular bodies during 100 s in B. C) E.d.x. spectrum from cryosection of shock frozen *rdgC* retina measured during 103 s. In addition to the large $Ca_{K\alpha}$+$K_{K\beta}$ peak, a prominent peak of phosphate is observed ($P_{K\alpha}$) which arises from granule of calcium phosphate precipitate (from Ref. 21).

degeneration. In contrast, globular bodies appeared in the degenerating photoreceptors of all mutants (e.g. *rdgC, norpA, rdgB, ninaC*) with a dramatic calcium content as reflected in the large $Ca_{K\alpha}$+$K_{K\beta}$ peak to background ratio (P/B) (see ref. 21), while the $K_{K\alpha}$ P/B did not change significantly. Since $K_{K\alpha}$ reflects the major potassium content of the tissue, the increase in $Ca_{K\alpha}$+$K_{K\beta}$ in the degenerating photoreceptors indicated that the increase in $Ca_{K\alpha}$+$K_{K\beta}$ peak is due to an increase in the calcium content. The examples of spectra in Fig. 4A,B are representative of the analysis of large number of flies including all the mutants which showed very similar spectra.

As in *rdgC* flies, the large calcium peak observed in *ninaC, rdgB, and norpA* flies was found either in the globular bodies or in the smaller mass dense structures - (possibly lysosomes). In many cases of cryosections (in *rdgC*), spectra of the calcium-containing bodies showed only two major bands in the relevant region, those of phosphorus and calcium (Fig. 4C).

The fact that the high calcium level of the degenerating photoreceptors was found in subcellular organelles, which may possibly be MVB or lysosomes, suggest that active processes leading to calcium accumulation against a large concentration gradient exist in the degenerating cells. The observations that the photoreceptors which contain sequestered calcium in membrane bound organelles (i.e the globular bodies) still have rhabdomeres and surface membrane indicate that their high calcium content does not simply reflect deteriorated cells with perforated plasma membrane that allow free calcium movement into the cells. Furthermore, the careful measurements of the calcium content in the cytosol clearly showed that the calcium content of the cytosol is much smaller than that observed in the organelles. Accordingly, the extremely high level of calcium that was found in the organelles, (e.g. 135 mmole/l calcium, on the average, in light raised *rdgC* retinae) which is about two orders of magnitude above the calcium content usually found in normal cells, reflects an active process of calcium accumulation.

The finding that the high level of calcium is localized exclusively in degenerating photoreceptors suggests that this elevated calcium accompanies retinal degeneration as found in other cells and tissues (22). However, the finding that very different mutations (affecting very different gene products) lead to the same phenomenon of calcium accumulation suggests that this accumulation of calcium is a secondary rather than the primary factor which causes the degeneration.

ACKNOWLEDGMENTS

This work was supported by NIH Grant EY03529 The Israel Science Foundation and the US-Israel Binational Science Foundation (BSF).

REFERENCES

1. Dryja, T.P., McGee, T.L., Reichel, E., Hahn, L.B., Cowley, G.S., Yandell, D.W., Sandberg, M.A., and Berson, E.L., 1990, A point mutation of the rhodopsin gene in one form of retinitis pigmentosa, *Nature* 343:364-366.
2. Bowes, C., Li, T., Danciger, M., Baxter, L.C., Applebury, M.L., and Farber, D.B., 1990, Retinal degeneration in the *rd* mouse is caused by a defect in the beta subunit of rod cGMP-phosphodiesterase, *Nature* 347:677-680.
3. Minke, B., and Selinger, Z., 1991, Inositol lipid pathway in fly photoreceptors, excitation, calcium mobilization and retinal degeneration, in *Progress in Retinal Research*, Vol. 11, ed. Osborne, N.N. and Chader, G.J., pp. 99-124, Oxford: Pergamon Press.

4. Harris, W.A., and Stark, W.L. 1977, Hereditary retinal degeneration in *Drosophila melanogaster*. A mutant defect associated with the phototransduction process, *J. Gen. Physiol.* 69:261-291.
5. Bloomquist, B.T., Shortridge, R.D., Schneuwly, S., Perdew, M., Montell, C., Steller, H., Rubin, G., and Pak, W.L., 1974, Isolation of a putative phospholipase C gene of *Drosophila*, *norpA*, and its role in phototransduction, *Cell*, 54:723-733.
6. O'Tousa, J.E., Leonard, D.S., and Pak, W.L., 1989, Morphological defects in *oraJK84* photoreceptors caused by mutation in R1-6 opsin gene of *Drosophila*, *J. Neurogenet.*, 6:41-52.
7. Steele, F., and O'Tousa, J.E., 1990, Rhodopsin activation causes retinal degeneration in *Drosophila rdgC* mutant, *Neuron*, 4:883-890.
8. Steele, F.R., Washburn, T., Rieger, R., and O'Tousa, J.E., 1992, Drosophila retina degeneration C (rdgC) encodes a novel serine/threonine protein phosphatase, *Cell*, 69:669-676.
9. Vihtelic, T.S., Goebl, M., Milligan, S., O'Tousa, J.E., and Hyde, D.R., 1993, Localization of *Drosophila* retinal degeneration B, a membrane-associated phosphatidylinositol transfer protein, *J. Cell Biol.* 122(5):1013-1022.
10. Byk, T., Bar-Yaacov, M., Doza, Y.N., Minke, B., and Selinger, Z., 1993, Regulatory arrestin cycle secures the fidelity and maintenance of the fly photoreceptor cell, *Proc. Natl. Acad. Sci. U.S.A.* 90(5):1907-1911.
11. Meyertholen, E.P., Stein, P.J., Williams, M.A., and Ostroy, S.E., 1987, Studies of the *Drosophila norpA* phototransduction mutant, *J. Comp. Physiol. A.* 161:793-798.
12. Dolph, P.J., Ranganathan, R., Colley, N.J., Hardy, R.W., Socolich, M., and Zuker, C.S., 1993, Arrestin function in inactivation of G protein-coupled receptor rhodopsin in vivo, *Science* 260:1910-1916.
13. Porter, J.A., and Montell, C., 1993, Distinct roles of the *Drosophila ninaC* kinase and myosin domains revealed by systematic mutagenesis, *J. Cell. Biol.* 122:601-612.
14. Wu, L., Niemeyer, B., Colley, N., Socolich, M., and Zuker, C.S., 1995, Regulation of PLC-mediated signalling in vivo by CDP-diacylglycerol synthase, *Nature* 373:216-222.
15. Inoue, H., Yoshioka, T., and Hotta, Y., 1989, Diacylglycerol kinase defect in a *Drosophila* retinal degeneration mutant *rdgA*, J. Biol. Chem. 264:5996-6000.
16. Selinger, Z., Doza, Y.N., and Minke, B., 1993, Mechanisms and genetics of photoreceptors desensitization in *Drosophila* flies. Biochem. Biophys. Acta. 1179:283-299.
17. Peretz, A., Suss-Toby, E., Rom-Glass, A., Arnon, A., Payne, R., and Minke, B., 1994, The light response of Drosophila photoreceptors is accompanied by an increase in cellular calcium: effects of specific mutations, *Neuron* 12:1257-1267.
18. Ranganathan, R., Bacskai, B.J., Tsien, R.Y., and Zuker, C.S., 1994, Cytosolic calcium transients: Spatial localization and role in Drosophila photoreceptors cell function, *Neuron*, 13:837-848.
19. Hardie, R.C., and Minke, B., 1992, The *trp* gene is essential for a light-activated Ca^{2+} channel in Drosophila photoreceptors, *Neuron*, 8:643-651.
20. Minke, B., 1982, Light-induced reduction in excitation efficiency in the *trp* mutant of *Drosophila*, *J. Gen. Physiol.* 79:361-385.
21. Sahly, I., Schroder, W.H., Zierold, K., and Minke, B., 1993, Accumulation of calcium in degenerating photoreceptors of several *Drosophila* mutants, *Vis. Neurosci.* 11:763-772.
22. Farber, L., 1981, The role of calcium in cell death, *Life Sciences* 29:1289-1295.

27

THE ROLE OF DOMINANT RHODOPSIN MUTATIONS IN *DROSOPHILA* RETINAL DEGENERATION

Phani Kurada, Timothy D. Tonini, and Joseph E. O'Tousa

Department of Biological Sciences
University of Notre Dame
Notre Dame, Indiana 46556-0369

INTRODUCTION

Mutations within the human rod opsin gene are responsible for approximately 25% of the autosomal dominant forms of retinitis pigmentosa (ADRP) afflicting human populations (1).There are over 60 different mutations known in the rhodopsin gene that cause ADRP. To account for the mechanisms by which these mutations can cause retinal degeneration, three general types of models have been considered. First, improper folding mutant protein may prevent its proper maturation (2). A second model is that the mutated forms may produce rhodopsins that are constitutively active, leading to increased metabolic activity (3). A third possibility is that the mutant proteins show improper cellular localization (4). All three views are consistent with a scenario by which the dominant rhodopsin mutation reduces the health of the photoreceptor, eventually causing the cell to degenerate. Understanding the molecular basis of rhodopsin-based forms of ADRP represents a major advance, but it is likely that successful strategies for alleviating the disease will require additional knowledge on the mechanisms by which these rhodopsin mutants trigger retinal degeneration.

The invertebrate, *Drosophila melanogaster*, has been developed during the last twenty five years as a genetic model for vision. Among the first genes identified in this organism that affected vision were several that caused retinal degeneration (5, 6, 7). More recently, my laboratory has produced and analyzed dominant mutations in the rhodopsin gene to gain insights into mechanisms by which rhodopsin mutations may trigger retinal degeneration. In this report, I will describe work in the *Drosophila* field that is relevant to studies of rhodopsin-based retinal degeneration, emphasizing our recent studies on dominant rhodopsin mutations.

THE MAJOR RHODOPSIN GENE IS ENCODED BY *ninaE*

The *Drosophila ninaE* (neither inactivation nor afterpotential) gene is named for a characteristic electroretinogram that lacks the inactivation and prolonged depolarizing

Degenerative Diseases of the Retina, Edited by Robert E. Anderson et al.
Plenum Press, New York, 1995

afterpotential seen in wild type flies after stimulation of the retina with a bright blue light (8). This phenotype can be produced by mutations in many genes (8), as well as in wild type flies that have been deprived for vitamin A (9). However, *ninaE* mutants have two additional properties that suggested the gene encoded the visual pigment expressed in the R1-6 class of photoreceptors. First, only the rhodopsin in the R1-6 class of the photoreceptors was affected (10), and the gene exhibited a gene dosage effect on rhodopsin content (11). Molecular analysis of the *ninaE* gene (12, 13) first showed that visual pigment genes of invertebrates and vertebrates share a common ancestry, and encode proteins possessing many common features. This conclusion has been confirmed by subsequent cloning of additional vertebrate and invertebrate visual pigments genes.

PHOTORECEPTORS OF *ninaE* MUTANTS LACK RHABDOMERES

In addition to the physiological defects, *ninaE* mutants show structural defects of the photoreceptor. In historical context, it was the discovery of the ora^{JK84} (ora=outer rhabdomeres absent) mutation by Koenig and Merriam in 1977 (14) that first indicated the effect of rhodopsin mutations on photoreceptor structure. Subsequent to the cloning of the *ninaE* locus, ora^{JK84} was shown to have a mutant allele at the *ninaE* locus (15). The actual molecular defect is a stop codon resulting in a truncated protein lacking the sixth and seventh transmembrane domains (16).

The widely used null allele of the *ninaE* gene is the $ninaE^{I17}$ allele. This irradiation induced allele lacks about 1.6 kb of the gene, including the transcriptional and translational start sites. and fails to make any detectable transcript of the gene. (12). Morphological studies on ora^{JK84} and $ninaE^{I17}$ show that both mutations affect the structure of the photoreceptor in similar ways, as described below.

The rhabdomere is a microvillal extension of the plasma membrane in invertebrate photoreceptors (17). Rhodopsin and other proteins required for phototransduction are sequestered within the rhabdomere. Ultrastructuctural studies on oraJK84, as well as the $ninaE^{I17}$ allele (15, 18, 19) showed that photoreceptors initially build a rhabdomere even in the absence of rhodopsin synthesis, but the rhabdomere decays and is eventually lost as the animal ages. By 5-7 days of age, most photoreceptors lacking any rhodopsin protein have completely lost the rhabdomeric membranes (15, 18, 19).

The severity of the mutant alleles of *ninaE* have been classified in two different ways. One approach carried out by Johnson and Pak (20) used intracellular recordings to assess the amount of rhodopsin remaining in the R1-6 photoreceptor cells. A different approach assesses the integrity of the rhabdomere as a function of the age of the fly (21). Data collected from these two approaches suggested that the rhabdomeric membranes are maintained for longer periods in the *ninaE* mutants that produce higher levels of functional rhodopsin. Nonetheless, the data of Leonard et al (21) suggested that all *ninaE* mutants will eventually lose the rhabdomeric membranes as the fly ages.

The finding that *Drosophila ninaE* mutants show an age dependent degeneration of the rhabdomere suggests that rhodopsin plays a structural role in the maintenance of this structure. However, this phenomenon is distinct from the retinal degeneration seen in the human ADRP disease for several reasons. First, the mutations are causing loss of the rhabdomeric membranes, not degeneration of the entire photoreceptor cell. In fact, most studies referred to above note that the cell body of these photoreceptors appear to maintain normal structure. However, only one study (21) evaluated retinas aged for extended periods. This study suggested that photoreceptor cell bodies are eventually affected in severe *ninaE* mutants. Second, and perhaps more important, the *Drosophila* mutants are genetically recessive with respect to all structural changes to the photoreceptor. In the case of human

ADRP, rhodopsin mutations are genetically dominant with respect to their ability to induce retinal degeneration. Only recently have dominant mutations of the *Drosophila* rhodopsin gene been recognized. These mutations were independently discovered in our laboratory (22) as well as in the laboratory of Charles Zuker (23).

IDENTIFICATION AND CHARACTERIZATION OF DOMINANT *ninaE* MUTANTS

In our laboratory, dominant *ninaE* mutants were identified in a genetic screen designed to select for suppressors of the retinal degeneration caused by the *rdgC* (retinal degeneration C) mutant. The *rdgC* gene encodes a novel serine/threonine protein phosphatase (24). Figure 1 shows the type of data generated for *ninaE*^{D1}, the first dominant allele identified. The first row shows genotypes at the *ninaE* locus. The wild type allele is designated as "+", the other two are the null *ninaE*^{I17} allele and the *ninaE*^{D1} dominant allele. The second row shows that some of these genotypes were also homozygous for the *rdgC* mutant (m), some were *rdgC*$^{+}$ (+). The third row provides the data on *rdgC*-based degeneration in the various *ninaE* genotypes. The key result is that degeneration is seen in the *ninaE*^{I17}/+ heterozygote, but not in the *ninaE*^{D1}/+ heterozygote. The identification of *ninaE*^{D1} as a dominant negative mutation is provided by assays for rhodopsin levels in these flies that is summarized in the fourth row. Rhodopsin levels are conveniently measured by use of a polyclonal antibody generated in my laboratory in dot blot or western blot assays. One physiological consequence of low rhodopsin, failure to generate a prolonged depolarizing afterpotential in the electroretinogram, was also monitored to confirm this result (22).

These data suggested that the dominant *ninaE* mutations prevent *rdgC*-triggered degeneration because they lower rhodopsin expression in heterozygotes significantly below the expected decrease of 50% in *ninaE* heterozygotes. Byk et al. (25) have proposed that the rdgC phosphatase is responsible for the desphosphorylation of rhodopsin based on their studies of rhodopsin dephosphorylation. They suggest that the excess phosphorylated rhodopsin seen in *rdgC* mutants is toxic to the photoreceptor and therefore leads to degeneration. With this model, it is easy to accommodate the effect of the dominant rhodopsin mutants: rhodopsin fails to accumulate to the levels required for toxicity, and therefore the rdgC phosphatase activity is not needed to prevent degeneration.

The proposal by Byk et al (25) also explains the observation that *norpA* (no receptor potential A) mutations degenerate with similar time course as *rdgC* (26, 27). In the absence of norpA phospholipase C activity, intracellular Ca^{2+} levels do not rise above basal levels

ninaE genotype:	+/+	+/+	D1/D1	I17/I17	I17/+	D1/+	I17/+	D1/+	D1/I17
rdgC genotype:	+	m	m	m	+	+	m	m	m
retinal degeneration:	--	yes	no	no	--	--	yes	no	no
% wild type rhodopsin:	100	100	<1	<1	~50	~5	~50	~5	<1

Figure 1. Effects of dominant *ninaE* mutants on rhodopsin content and *rdgC*-induced retinal degeneration. The *ninaE* alleles are designated as follows: wild type (+), *ninaE*^{D1} (D1), *ninaE*^{D2} (D2), *ninaE*^{I17} (I17). Retinal degeneration is assessed by histological methods. % wild-type rhodopsin is measured from western blot analysis of retinas from the indicated genotypes.

and hence the Ca^{2+} dependent rdgC phosphatase will not be activated. As a result, rhodopsin is not dephosphorylated, and would trigger retinal degeneration. Results obtained with the dominant rhodopsin mutants support this proposal, as similar levels of rhodopsin are required to induce degeneration in both *rdgC* and *norpA* flies (22).

DOMINANT *ninaE* MUTANTS CAUSE RETINAL DEGENERATION

The *Drosophila* dominant mutants were isolated based on their ability to prevent the rapid onset of retinal degeneration caused by the *rdgC* mutations. To determine if the mutations might mimic the human ADRP mutations, we assayed the ability of the mutations to cause retinal degeneration in an $rdgC^{+}$ background. Figure 2 shows an experiment comparing retinal structure in $ninaE^{D1}$/+, $ninaE^{D2}$+ and $ninaE^{I17}$/+ heterozygotes. In these experiments, the flies were reared in three different light regimes: constant light, 12 hour light/12 hour dark cycle and in constant darkness. Figure 2 shows that $ninaE^{D1}$/+, $ninaE^{D2}$/+ flies reared in constant light show significant retinal degeneration by twenty days of age. The $ninaE^{I17}$/+ fly serves as a control, showing minimal damage to the retina by 20 days of age. These results suggested that the dominant alleles were responsible for retinal degeneration, and in this fashion mimicked the effects seen in human ADRP disease.

The importance of the light treatment was evident from the results of rearing flies in the reduced light conditions. Longer periods of time were required to induce marked degeneration in $ninaE^{D1}$/+ and $ninaE^{D2}$/+ flies when reared on the 12 hour light/12 hour dark cycle. Flies reared in constant darkness gave no clear indication of retinal degeneration.

The ultrastructure of the retina was examined in both young and old $ninaE^{D1}$/+, $ninaE^{D2}$/+ and $ninaE^{I17}$/+ heterozygotes. No differences were noted in the young heterozygotes of these genotypes, but at older ages, the dominant heterozygotes showed much greater evidence of retinal degeneration. The R1-6 photoreceptor class, the cells expressing the *ninaE* gene product, were the only affected cell type in this study.

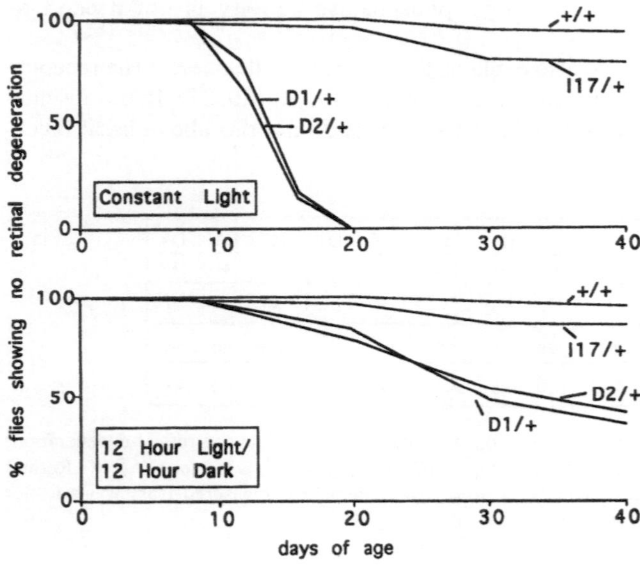

Figure 2. Time course of retinal degeneration in dominant rhodopsin mutants. Flies were aged either in 24 hour light of approximately 3.7 x 10^2 LUX (top graph) or on a 12 hour light/12 hour dark cycle (bottom graph). Retinal degeneration in individuals flies was assessed by use of the deep pseudopupil.

DOMINANT *ninaE* MUTANTS AFFECT RHODOPSIN MATURATION

Our analysis of the dominant *ninaE* alleles suggest that all act to inhibit the maturation of the wild type rhodopsin. In principle, this could occur at transcriptional, translational, or posttranslational steps. We discounted the possibility that there is transcriptional control because the steady state levels of rhodopsin mRNA is not affected by the mutations. The initial translation of the protein also appears normal, as shown by the behavior of the mutant protein in the presence of mutant *ninaA* protein. The ninaA protein a cyclophilin required for rhodopsin exit from the endoplasmic reticulum (ER) (28). In flies producing rhodopsin, mutant flies show a build up of rough ER membranes. It is clear that this phenotype is due to accumulation of rhodopsin in the endoplasmic reticulum (ER) as *ninaA*; *ninaE*^{I17} flies lack excess ER membranes (28). However, *ninaA* mutants either homozygous or heterozygous for a dominant *ninaE* allele still show build up of ER membranes, despite low overall rhodopsin levels in the cell (22). Therefore, it appears that the mutant rhodopsin is translated and stable through the posttranslational processing event in the ER that requires *ninaA* function.

ACTION OF DOMINANT RHODOPSIN MUTANTS

The data presented above, as well as data described by Colley et al (23), suggest that the dominant rhodopsin mutants reduce rhodopsin content in photoreceptors by interfering with the post-translational maturation process. Several steps of this process have been defined, as summarized in Figure 3. Initial glycosylation of rhodopsin is the first discernible posttranslational step. Flies raised on media lacking vitamin A also fail to mature rhodopsin (29), and this rhodopsin remains in the endoplasmic reticulum with the 40 kD preprocessed form of glycosylation (30). Colley et al (28) showed that the 40 kD premature form also accumulates in *ninaA* mutants, suggesting that the next events include both the addition of retinal and the requirement for *ninaA* function.

The 40kD rhodopsin is then modified into several intermediate forms (38-36 kD, likely due to processing of the oligosaccharide) then is deglycosylated to its mature form of

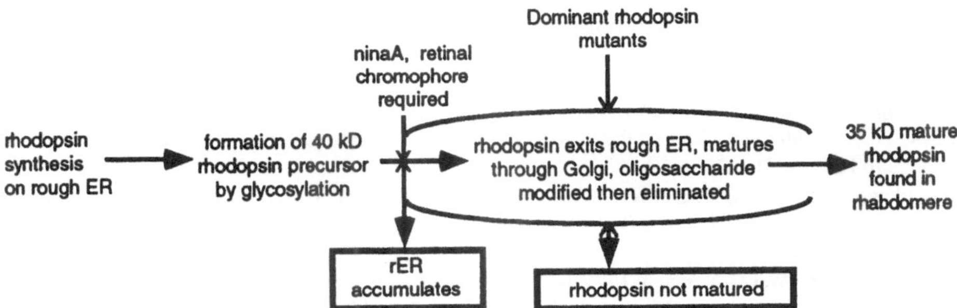

Figure 3. Experimental perturbations of the *Drosophila* rhodopsin maturation pathway. The first event in maturation is the addition of a oligosaccharide side chain to generate 40 kD rhodopsin precursor. This precursor requires both *ninaA* function and association with retinal to exit from the ER. Without these events, the rhodopsin remains in the ER, causing build up of ER membranes. Dominant rhodopsin mutants are thought to effect rhodopsin production at a subsequent stage as they do not elicit build up of ER membranes. Later events in rhodopsin maturation include the further modification and finally loss of the oligosaccharide side chain.

35 kD before or at the time it is deposited in rhabdomere membranes (30,31). Although the mature rhodopsin has no or little oligosaccharide, the glycosylation appears important in the maturation process because rhodopsin protein lacking a glycosylation site fails to produce mature rhodopsin (32). The dominant rhodopsins could interfere with rhodopsin maturation at any step of the maturation pathway, as indicated in Figure 3. Our recent unpublished data shows that the initial glycosylation of the mutant rhodopsin is normal, placing the defective step at or after the time of the *ninaA* requirement.

Better definition of the steps and the required components for the rhodopsin maturation is an obvious direction for future research. This work will generate a clearer picture of how these dominant mutants interfere with rhodopsin maturation. The work could have potential benefits for the study of human ADRP. The finding that these dominant mutants do trigger degeneration suggests that defects in intracellular trafficking are capable of causing retinal degeneration. Such a model has already been advanced for rhodopsins responsible for ADRP (2, 33). We expect that the study of *Drosophila* rhodopsin mutants will provide insights into these mechanisms, thereby contributing to strategies for the design of therapeutic agents to treat this disease.

ACKNOWLEDGMENTS

This work was supported by NIH grant R01 EYO6808. We also acknowledge a graduate student summer fellowship funded in honor of Neil and Donna Weisman by the Fight for Sight research division of Prevent Blindness America awarded to P.K.

REFERENCES

1. Dryja, T. P., T. L. Mcgee, L. B. Hahn, G. S. Cowley, J. E. Olsson, E. Reichel, M. A. Sandberg and E. L. Berson, 1990, Mutations within the rhodopsin gene in patients with autosomal dominant retinitis-pigmentosa, *New England J. Medicine*. 323: 1302-1307.
2. Sung, C.-H., C. M. Davenport and J. Nathans, 1993, Rhodopsin mutants responsible for autosomal dominant retinitis pigmentosa, *J. Biol. Chem.* 268: 26645-26649.
3. Robinson, P. R., G. B. Cohen, E. A. Zhukovsky and D. D. Oprian, 1992, Constitutively active mutants of rhodopsin, *Neuron*. 9: 719-725.
4. Sung, C.-H., C. Makino, D. A. Baylor and J. Nathans, 1994, A rhodopsin gene mutation responsible for autosomal dominant retinitis pigmentosa result in a protein that is defective in localization to the photoreceptor outer segment, *J. Neurosci.* 14: 5818-5833.
5. Hotta, Y. and S. Benzer, 1969, Abnormal electroretinogram in visual mutants in *Drosophila*, *Nature*. 222: 354-356.
6. Pak, W. L., J. J. Grossfield and N. V. N.V. White, 1969, Nonphototactic mutants in a study of vision of *Drosophila*, *Nature*. 222: 351-354.
7. Pak, W. L. "Mutations affecting the vision of *Drosophila melanogaster*." Handbook of Genetics. King ed. 1976 Plenum. New York.
8. Stephenson, R. S., J. O'Tousa, N. J. Scavarda, L. L. Randall and W. L. Pak. "*Drosophila* mutants with reduced rhodopsin content." The Biology of Photoreception. Cosens and Vince-Price ed. 1983 Cambridge University Press. Cambridge.
9. Stark, W. S. and W. G. Zitzmann, 1976, Isolation of adaptation mechanisms and photopigment spectra by vitamin A deprivation in *Drosophila*, *J. Comp. Physiol.* 105: 15-27.
10. Schinz, R. H., M.-V. C. Lo, D. C. Larrivee and W. L. Pak, 1982, Freeze-fracture study of the *Drosophila* photoreceptor membrane: mutations affecting membrane particle density, *J. Cell Biol.* 93: 961-969.
11. Scavarda, N. J., J. O'Tousa and W. L. Pak, 1983, *Drosophila* locus with gene dosage effect on rhodopsin., *Proc. Natl. Acad. Sci.* 80: 4441-4445.
12. O'Tousa, J. E., W. Baehr, R. L. Martin, J. Hirsh, W. L. Pak and M. L. Applebury, 1985, The *Drosophila* *ninaE* gene encodes an opsin., *Cell*. 40: 839-850.

13. Zuker, C. S., A. F. Cowman and G. M. Rubin, 1985, Isolation and structure of a rhodopsin gene from D. melanogaster, *Cell*. 40: 851-858.
14. Koenig, J. and J. Merriam, 1977, Autosomal ERG mutants, *Drosophila Information Service*. 52: 50-51.
15. O'Tousa, J. E., D. S. Leonard and W. L. Pak, 1989, Morphological defects in *ora*^{JK84} photoreceptors caused by mutation in R1-6 opsin gene of *Drosophila.*, *J. Neurogenetics*. 6: 41-52.
16. Washburn, T. and J. E. O'Tousa, 1989, Molecular defects in *Drosophila* rhodopsin mutants, *J. Biol. Chem.* 264: 15464-15466.
17. Blest, A. D., S. Stowe and W. Eddey, 1982, A labile, Ca^{+2}-dependent cytoskeleton in rhabdomeral microvilli of blowflies, *Cell Tissue Res.* 223: 553-573.
18. Stark, W. S. and S. D. Carlson, 1983, Ultrastructure of the compound eye and the first optic neuropile of the photoreceptor mutant *ora*^{JK84} of *Drosophila*, *Cell Tiss. Res.* 233: 305-317.
19. Stark, W. S. and R. Sapp, 1987, Ultrastructure of the retina of *Drosophila melanogaster*: the mutant *ora* (outer rhabdomeres absent) and its inhibition of degeneration in *rdgB* (retinal degeneration-B), *J. Neurogenet.* 4: 227-240.
20. Johnson, E. C. and W. L. Pak, 1986, Electrophysiological study of *Drosophila* rhodopsin mutants, *J. Gen. Physiol.* 88: 651-673.
21. Leonard, D. S., V. D. Bowman, D. F. Ready and W. L. Pak, 1992, Photoreceptor degeneration associated with mutations in presumptive opsin structural gene of *Drosophila*, *J. Neurobiology*. 23: 605-626.
22. Kurada, P. and J. E. O'Tousa, 1995, Retinal degeneration caused by dominant rhodopsin mutants in *Drosophila.*, *Neuron*. 14 (in press).
23. Colley, N. J., J. A. Casill, E. K. Baker and C. S. Zuker, 1995, Defective intracellular transport is the molecular basis of rhodopsin-dependent dominant retinal degeneration, *Proc. Natl. Acad. Sci. USA*. 92: 3070-3074.
24. Steele, F., W. T., R. Rieger and J. E. O'Tousa, 1992, *Drosophila rdgC* encodes a novel protein phosphatase, *Cell*. 69: 669-676.
25. Byk, T., M. Bar-Yaacov, Y. N. Doza, B. Minke and Z. Selinger, 1993, Regulatory arrestin cycle secures the fidelity and maintenance of the fly photoreceptor cell, *Proc. Natl. Acad. Sci. USA*. 90: 1907-1911.
26. Meyertholen, E. P., P. J. Stein, M. A. Williams and S. E. Ostroy, 1987, Studies of the *Drosophila norpA* phototransduction mutant II. Photoreceptor degeneration and rhodopsin maintenance, *J. Comp. Physiol.* 161: 793-798.
27. Stark, W. S., R. Sapp and S. D. Carlson, 1989, Photoreceptor maintenance and degeneration in the *norpA* (no receptor potential-A) mutant of *Drosophila melanogaster*, *J. Neurogenetics*. 5: 49-59.
28. Colley, N. J., E. K. Baker, M. A. Stamnes and C. S. Zuker, 1991, The cyclophilin homolog *ninaA* is required in the secretory pathway, *Cell*. 67: 255-263.
29. Stark, W. S., R. Sapp and D. Schilly, 1988, Rhabdomere turnover and rhodopsin cycle: maintenance of retinula cells in *Drosophila melanogaster*, *J. Neurocyto.* 17: 499-509.
30. Ozaki, K., H. Nagatani, M. Ozaki and F. Tokunaga, 1993, Maturation of the major *Drosophila* rhodopsin, ninaE, requires chromophore 3-hydroxylretinal, *Neuron*. 10: 1113-1119.
31. Huber, A., U. Wolfrum and R. Paulsen, 1994, Opsin maturation and targeting to rhabdomeral photoreceptor membranes requires the retinal chromophore, *Eur. J. Cell Biol.* 63: 219-229.
32. O'Tousa, J. E., 1991, Requirement of N-linked glycosylation site in *Drosophila* rhodopsin, *Visual Neuroscience*. 8: 385-390.
33. Sung, C.-H., B. G. Schneider, N. Agarwal, D. S. Papermaster and J. Nathans, 1991, Functional heterogeneity of mutant rhodopsins responsible for automsomal retinitis pigmentosa, *Proc. Natl. Acad. Sci. USA*. 88: 8840-8844.

28

THE ROLE OF THE *RETINAL DEGENERATION B* PROTEIN IN THE *DROSOPHILA* VISUAL SYSTEM

Function of *Drosophila* rdgB Protein in Photoreceptors

D. R. Hyde,[1] S. Milligan,[1] and T. S. Vihtelic[2]

[1] Department of Biological Sciences
University of Notre Dame
Notre Dame, Indiana 46556
[2] Massachusetts Eye and Ear Infirmary
Department of Ophthalmology
Howe Laboratory, 243 Charles Street, Boston, Massachusetts 02114

INTRODUCTION

The *Drosophila melanogaster* visual system has been extensively studied using molecular, genetic and biochemical approaches. These analyses have uncovered several key components that are required for phototransduction. At least four different rhodopsin molecules are utilized to absorb light [1]. The rhodopsins have different spectral sensitivities and are expressed in mutually exclusive photoreceptor cells of the larval photoorgan, the adult ocelli, and the compound eye. The *ninaE* gene encodes the opsin that is expressed in photoreceptors R1-6 in the compound eye [2, 3]. The *Drosophila* phototransduction cascade has an absolute requirement for the *norpA*$^+$ gene [1, 4, 5], which encodes a phosphatidylinositol-specific phospholipase C-β (PLC) protein [6, 7]. A retinal-specific heterotrimeric G-protein links the photoactivated metarhodopsin and the *norpA*-encoded PLC. The G protein's alpha subunit (DGqα), which is a member of the Gqα subfamily, is encoded by the *dgq* gene [8]. Biochemical and genetic data demonstrate that the DGqα protein responds to light-activated metarhodopsin and in turn, stimulates the *norpA*-encoded PLC [9]. A mutation in the retinal-specific Gβ subunit, gbe, produces an abnormal electrophysiological response to light [10]. The *inaC*-encoded, retinal-specific protein kinase C [11] is thought to be stimulated by diacyl glycerol, which is generated from the PLC-mediated hydrolysis of PIP_2. Even though this *inaC*-encoded protein kinase C is activated by the products of PLC hydrolysis, it is not required for activation of the light channels. Rather, this protein kinase C plays a role in the deactivation of the light-response [11, 12, 13, 14, 15]. While the ligands of the invertebrate light-ac-

Degenerative Diseases of the Retina, Edited by Robert E. Anderson et al.
Plenum Press, New York, 1995

tivated channels are not known, it appears that the *transient receptor potential* (*trp*) gene encodes one type of these channels [16, 17, 18].

The *Drosophila retinal degeneration B* (*rdgB*) mutation was originally identified as exhibiting light-enhanced photoreceptor cell degeneration [4]. The initial light-response of *rdgB* photoreceptors is drastically reduced relative to wild-type and it deteriorates further with light exposure [19]. Genetic evidence suggests the rdgB protein acts within the light-initiated phosphoinositide cascade. Mutations in either the *ninaE* or the *norpA* genes suppress *rdgB* degeneration [19, 20]. The allele-specific suppression of $rdgB^{KS222}$ degeneration by $norpA^{suII}$ suggests that the rdgB protein may interact directly with the *norpA*-encoded PLC [19] or that these proteins share a common intermediate. Constitutively activated dominant *dgq* mutations stimulate degeneration of *rdgB* mutant photoreceptors in the absence of light [9]. This *dgq*-stimulation of *rdgB* degeneration is $norpA^{+}$-dependent, which suggests that DGqα, PLC and rdgB all function in the visual transduction cascade. Both genetic and biochemical experiments are consistent with *rdgB* functioning subsequent to protein kinase C [11, 21]. The rdgB protein also appears to function in the *Drosophila* olfactory system. An *rdgB* allele was recovered in a screen for *Drosophila* olfaction mutants and some previously isolated *rdgB* alleles were subsequently demonstrated to be defective in olfaction [22]. Study of the *rdgB* gene and protein should identify the biochemical defect responsible for retinal degeneration and specify its roles in the visual phosphoinositide cascade and the olfactory system.

While many differences exist between the vertebrate and *Drosophila* visual transduction systems, recent work demonstrated that analogous mutations in both cascades lead to the common phenotype of photoreceptor degeneration. The most obvious example is the rhodopsin mutations in both *Drosophila* and humans. One form of autosomal dominant retinitis pigmentosa is due to various mutations in the human opsin gene [23]. Similarly, dominant *ninaE* mutations also cause degeneration of *Drosophila* photoreceptor cells [24]. Strong *ninaE* recessive mutations can also cause rhabdomere loss and will show sporadic photoreceptor degeneration over prolonged periods [25, 26]. Additionally, mutations that affect the heterotrimeric G protein's effector molecule in both systems leads to degeneration [27, 28, 29]. However, the *Drosophila norpA* degeneration appears to be due to the buildup of phosphorylated metarhodopsin [30], while the *rd* degeneration is via an apoptotic pathway [31]. While the mechanisms may differ, the loss or aberrant activity of the phototransduction effector molecule is intolerable to the photoreceptor. Proteins involved in the development and maintenance of the membrane-rich light gathering organelles are also critical for the photoreceptor. Both the *Drosophila* chaoptin and the vertebrate peripherin (*rds* gene) proteins function to align microvilli and discs during the development of the rhabdomere and outer segment, respectively [32, 33]. Adult *chaoptic* mutants exhibit photoreceptors that are devoid of rhabdomeres, whereas *rds* mutant mice possess rods and cones that lack outer segments and exhibit degeneration over the course of a year. Mutations in the human peripherin gene can lead to either retinitis pigmentosa or macular degeneration [34, 35].

It is possible that the highly specialized photoreceptor cells are so sensitive that even slight perturbations in their ability to biochemically respond to light leads to degeneration. Because analogous mutations in vertebrates and *Drosophila* exhibit a common retinal degeneration phenotype, the *Drosophila* system provides an excellent system to identify and examine the mechanisms of retinal degeneration. For this reason, we are characterizing the *Drosophila rdgB* mutation at the molecular, biochemical and genetic levels. Additionally, we began to examine if a vertebrate homolog of the rdgB protein exists. Isolation of such a molecule will possibly identify a candidate for vertebrate retinal degeneration.

RESULTS

rdgB Is a Membrane-Associated Protein Possessing Phosphatidylinositol Transfer Activity

The *rdgB* gene encodes a putative protein of 1054 amino acids that has six hydrophobic regions that may function as transmembrane domains [36]. The rdgB protein segregates with the membrane fraction of homogenized heads and is not removed by alkaline washes, which demonstrates that rdgB is likely an integral membrane protein [36]. The absence of a cleavable signal peptide in the rdgB sequence prior to the first hydrophobic domain [37] suggests that the amino-terminal 498 amino acids are within the cytoplasm [38]. Within this amino-terminal domain, the first 281 amino acids possess 42% identical residues and 11% conserved substitutions when compared to the entire rat brain phosphatidylinositol transfer protein (PI-TP; [39]). While the amino acid homology extends the length of the PI-TP sequence, the first 65 amino acids show 75% identity or conserved substitutions. This high degree of homology is surprising because the PI-TP molecules are soluble proteins of approximately 24 to 36 kilodaltons [40], while rdgB is an integral membrane protein of 160 kilodaltons [36].

To determine if rdgB's PI-TP domain has phosphatidylinositol (PI) transfer activity, we assayed if the truncated amino terminus of the *rdgB* protein could transfer PI *in vitro*. A soluble, truncated form of the rdgB protein, which contained amino acids 1-296, was expressed in *E. coli* under the transcriptional control of the T7 RNA polymerase promoter. We partially purified the truncated rdgB protein and isolated the equivalent fraction from E. coli that did not express the rdgB construct [36]. We assayed PI transfer activity by following the movement of L-3-phosphatidyl [2-^{3}H]inositol from unilamellar vesicles to mitochondrial membranes (Figure 1A). As a control for the specificity of the transfer, we incorporated cholesteryl [1-^{14}C] oleate in the vesicles, which is not transferred by a phosphatidylinositol transfer protein. We determined that the protein fraction containing the soluble truncated rdgB protein transferred 80-fold more PI than the control fraction (Figure 1B). The observed rates of PI transfer using the truncated rdgB protein were similar to those reported for the rat brain PI-TP.

Identification of an rdgB Protein that Restores the Wild-Type Phenotype

The *rdgB* gene encodes at least 5 different sized mRNAs [37]. We screened several different *Drosophila* head cDNA libraries to identify clones that correspond to the different mRNAs and to examine if those mRNAs encode the same rdgB protein. We identified 4 different rdgB cDNAs that are alternatively spliced at two different locations within the open reading frame. The first alternatively spliced exon is within the acidic amino acid region, which is thought to be involved in Ca^{2+}-binding [36]. This splice variant introduces 13 amino acids, without affecting the subsequent amino acid sequence. The second splice variant is within the lumenal loop portion of the protein, between the fifth and sixth hydrophobic domains. This variant introduces 13 amino acids without affecting the subsequent amino acid sequence.

We examined the potential functional differences between these protein variants by germline transformation and expression of the cDNAs in *rdgB* mutant flies. In this manner, the only rdgB protein expressed in these flies is encoded by the cDNA. We germline transformed the cDNA that lacked the first alternatively spliced exon, but possessed the second alternatively spliced exon. Based on immunohistochemical detection, the protein was properly expressed in the retina. We assayed the electrophysiological light response of two

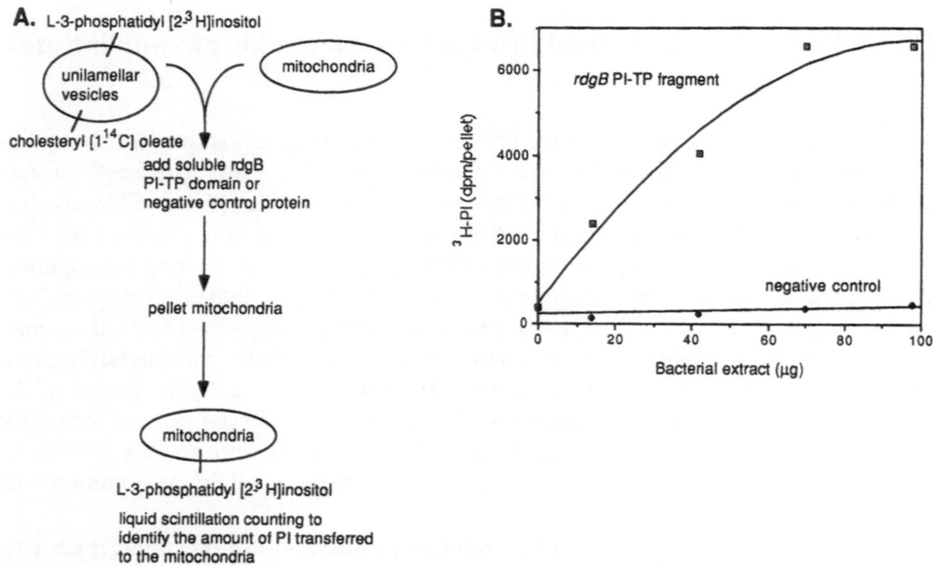

Figure 1. The rdgB protein possesses a PI transfer activity. A. We prepared small unilamellar vesicles containing L-3-phosphatidyl [2-^{3}H]inositol, phosphatidylinositol, phosphatidylcholine, and cholesteryl [1-^{14}C] oleate [36]. Fresh bovine heart muscle mitochondria were used as the recipient membrane. Various concentrations of either the truncated rdgB protein or an equivalent protein mixture from bacteria that did not express the rdgB protein were incubated with the two membrane preparations at 37°C. The reactions were terminated and centrifuged. The mitochondrial pellet was washed and resuspended in buffer and Liquiscint. The transfer of ^{3}H-phosphatidylinositol to mitochondria was determined relative to ^{14}C-cholesteryl oleate, a non-transferable marker. B. The results from the PI transfer assay are plotted. The amount of L-3-phosphatidyl [2-^{3}H]inositol that was transferred to the mitochondrial pellet (removing vesicle contamination of the pellet as determined by measuring ^{14}C in the pellet) was plotted as dpm in the pellet versus the amount of bacterial protein added to the reactions. The reactions were carried out for 60 minutes.

day-old flies using the electroretinogram (ERG) and the degree of degeneration using light microscopy of fixed tissue sections. The ERG and histology of a wild-type fly are shown (Figure 2, Panels A and B). The ERG has a large amplitude that is maintained for the duration of the light stimulation and the retinal section shows large rhabdomeres and a lack of holes in the tissue. The ERG of a *rdgB* mutant fly has very little, if any, amplitude (Figure 2C) and the *rdgB* mutant rhabdomeres are small and holes are present throughout the section where cells have either died or are dying (Figure 2D). The germline transformed rdgB cDNA restored the wild-type ERG light response (Figure 2E) and prevented the light-enhanced degeneration associated with the *rdgB* mutation (Figure 2F). This confirms that we correctly cloned the *rdgB* gene and also demonstrates that not all the rdgB isoforms are required to prevent the two retinal mutant phenotypes.

We also expressed the truncated soluble rdgB PI-TP domain in *rdgB* mutant flies. These flies expressed this truncated protein in the retina based on tissue staining (data not shown). While this protein possesses PI-TP activity *in vitro* (Figure 1B), it failed to prevent the light-enhanced degeneration and did not restore the ERG response. This suggests that the rdgB protein does not function simply as a PI-TP molecule. It is possible that the subcellular localization of rdgB must be tightly restricted within the photoreceptor cell for proper function.

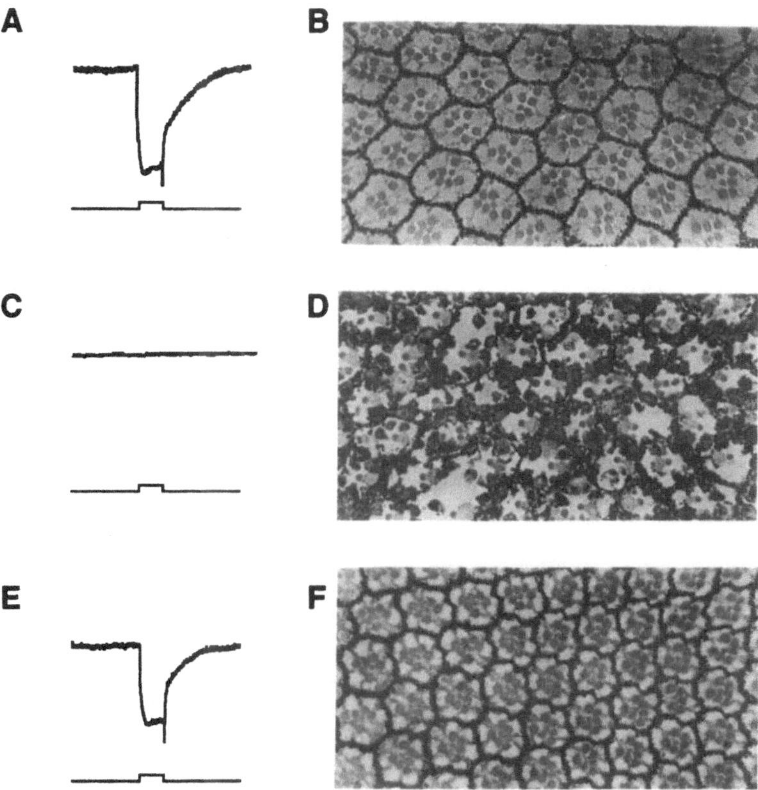

Figure 2. Germline transformation of a rdgB cDNA to restore the wild-type phenotypes. A. The electroretinogram (ERG) of a two day-old wild-type *Drosophila*. The ERG measures the sum electrical potential of the retina in response to a light stimulus. The duration of a 1 second light response is shown at the bottom of the trace. B. A retinal section of a 11 day-old wild-type *Drosophila* does not exhibit signs of degeneration. The rhabdomeres are large and round, there are no holes in the section, and the ommatidia are properly arranged. C. The ERG of a two day-old *rdgB*^{KS222} mutant fly shows no light response. D. A retinal section of a nine day-old *rdgB*^{KS222} mutant fly at the light microscope level. The small rhabdomeres, holes in the section, and the irregular arrangement of the ommatidia are characteristic signs of degeneration. E. The ERG of a two day-old *rdgB*^{KS222} mutant fly that expresses a germline transformed copy of the rdgB cDNA. The amplitude and kinetics of the light-response are essentially wild-type. F. A retinal section of a ten day-old *rdgB*^{KS222} mutant fly that expresses a germline transformed copy of the rdgB cDNA. No signs of light-enhanced degeneration are apparent.

Immunolocalization of the rdgB Protein

A rdgB fusion protein, containing 30% of the proposed rdgB protein was expressed in the IPTG-inducible pMal-cRI expression vector [36] and used to generate a monoclonal antibody. This monoclonal antibody yields results that are identical to the data we previously reported for the rdgB polyclonal antiserum [36]. We determined the size and expression of the rdgB protein in wild-type and *rdgB* mutant flies using immunoblots. The monoclonal antibody detected a 160 kDa protein in both wild-type and *eyes absent* (*eya*; [41]) heads (Figure 3, Lanes 1 and 3, respectively). The rdgB protein is expressed at much lower, but detectable, levels in bodies (Figure 3, Lane 2). Previously, we detected *rdgB* mRNAs by RNA Northern blots in wild-type and *eya* heads, but not in bodies [37]. The specificity of

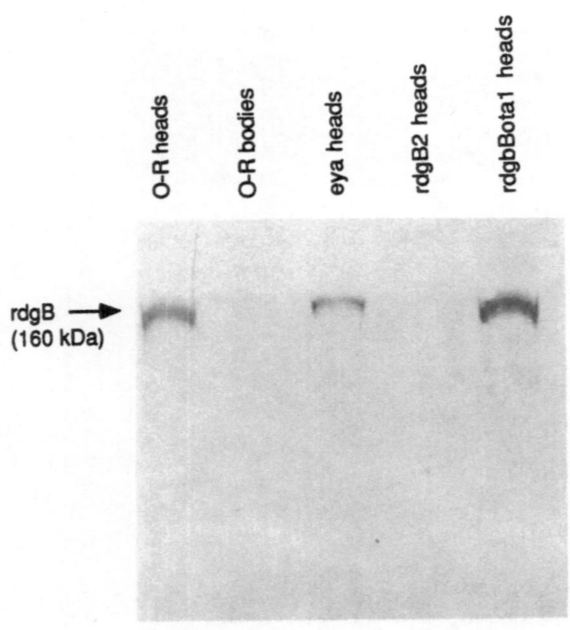

Figure 3. Immunoblot of the rdgB protein in the *Drosophila* head. Oregon-R heads (O-R heads), Oregon-R bodies (O-R bodies), *eyes absent* heads (*eya* heads), and heads of *rdgB* mutants ($rdgB^2$ heads and $rdgB^{ota1}$ heads) were homogenized in a 2% SDS/3% urea extraction buffer [36]. The samples were electrophoresed on a 5% SDS-polyacrylamide gel. The gel was electroblotted overnight. An anti-rdgB monoclonal antibody was used, followed by detection with a goat anti-mouse alkaline phosphatase-conjugated secondary antibody. The 160 kDa rdgB protein is marked with an arrow.

the monoclonal antibody was confirmed by its failure to detect the 160 kD protein in $rdgB^2$ mutant head extracts (Figure 3, Lane 4). This allele lacks the *rdgB* gene region that is expressed in the fusion protein [37]. However, some *rdgB* alleles ($rdgB^{ota1}$, Figure 3, Lane 5) show both wild-type size rdgB protein and wild-type abundance. It is possible that these alleles could be missense mutations. Preliminary DNA sequence data suggests that the $rdgB^{ota1}$ allele possesses a single amino acid change in a critical functional domain of the rdgB protein.

Previously, we immunolocalized the rdgB protein to the retina, the optic lobes, the ocelli, the central brain, and the antennal segments with polyclonal antiserum [36]. This antiserum also gave a low level of general staining throughout the brain. The monoclonal antibody yielded essentially an identical staining pattern. The retina and the underlying optic lobes, the lamina and medulla, stain positive with the monoclonal antibody in wild-type heads, but not in $rdgB^2$ heads (Figure 4A and B, respectively). The expression in the lamina and medulla is likely due to the presence of rdgB in the photoreceptor axons because the medulla shows a regular repeating array of individual units that are consistent with the positions and morphologies of the R7 and R8 axons [42]. The protein's axonal localization is consistent with the observation that *rdgB* degeneration initiates in the laminal receptor terminals [43] and suggests that the rdgB protein is required in the photoreceptor axons. In addition, the mushroom bodies and antennal lobes stain positive with the monoclonal antibody in wild-type heads, but not in $rdgB^2$ heads (data not shown). Therefore, all of these structures must express rdgB protein because of the lack of staining in the $rdgB^2$ null mutant heads. This suggests that the rdgB protein plays a general role in neuronal function instead of a highly specialized role in photoreceptors.

Using transmission electron microscopy, we immunolocalized most of the rdgB signal on membranes closely associated with, but not part of, the base of the rhabdomeric microvilli in all the photoreceptor cells [36]. A lower level of staining may be associated with the base of the rhabdomeric microvilli. This is in stark contrast to the subcellular distribution of the *norpA*-encoded phospholipase C in the photoreceptor rhabdomere [7],

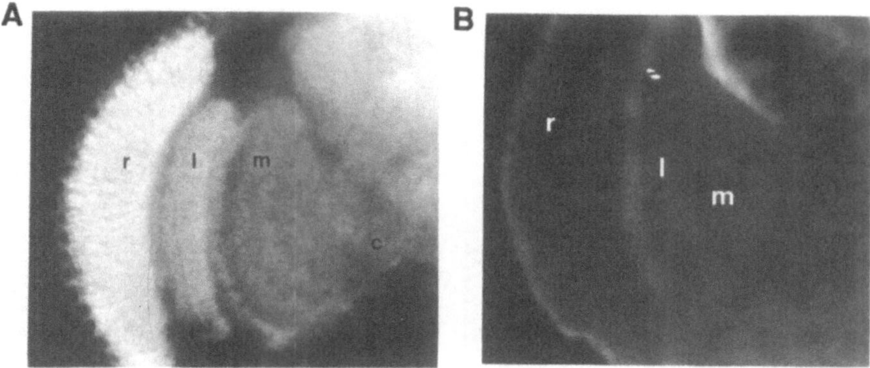

Figure 4. Immunolocalization of the rdgB protein in the *Drosophila* head. A. An eight micron section of a frozen white-eyed (w^{1118}) *Drosophila* head was stained with a rdgB monoclonal antibody and detected with FITC-conjugated goat-anti-mouse IgG as described [44]. Several structures of the head are stained as follows: retina, r; lamina, l; medulla, m; and cortex, c. B. An eight micron section of a frozen white-eyed (w^{1118}), $rdgB^2$ *Drosophila* head was stained with a rdgB monoclonal antibody and detected with FITC-conjugated goat-anti-mouse IgG. There is a lack of staining in the retina, optic lobes, ocelli, and brain.

which argues against the protein-protein interaction suggested by the allele-specific suppression of $rdgB^{KS222}$ by $norpA^{sull}$ [19]. rdgB's subcellular localization coincides with the subrhabdomeric cisternae (SRC), which are elaborate extensions of the endoplasmic reticulum running the length of the photoreceptor [45, 46]. The SRC likely plays a role in rhabdomere maintenance by transporting membrane proteins to the rhabdomeric microvilli [45] and acts as an intracellular Ca^{2+} store [46, 47]. The rdgB protein is not essential for establishing the SRC's structure, because the $rdgB^{EE170}$ allele, which lacks detectable rdgB protein [36], has a normal SRC ultrastructure prior to the onset of degeneration [45]. The exact role of the rdgB protein in the SRC membrane is currently unknown and is a major focus of our future research.

A Crossreacting rdgB Protein Is Expressed in the Mouse Retina

If the rdgB protein has an important neuronal function, then it is possible that a rdgB homolog exists in vertebrates. Therefore, we examined if the polyclonal antiserum detected crossreacting epitopes in the vertebrate retina. Because the rdgB protein exhibits a very limited subcellular localization within the *Drosophila* photoreceptor, we expected that a homolog would exhibit an analogous subcellular distribution. We sectioned and stained nonpigmented mouse retinas with both the preimmune serum and an immunopurified rdgB immunoglobulin, followed by detection with Cy3-conjugated goat anti-mouse secondary antibody. The immunopurified rdgB immunoglobulin detected two primary regions in the mouse retina, the photoreceptor inner segments and the outer plexiform layer (Figure 5A). When mouse preimmune serum was used in the primary incubation, only an irregular staining pattern was detected in the mouse retina (Figure 5B). This signal is also seen in tissue sections that were treated with only the secondary antibody (data not shown). The inner segment staining is analogous to the *Drosophila* photoreceptor localization, in that the signal is not in the rhodopsin-enriched region (rhabdomere and outer segments), but rather in the cell body that supports the formation of this organelle.

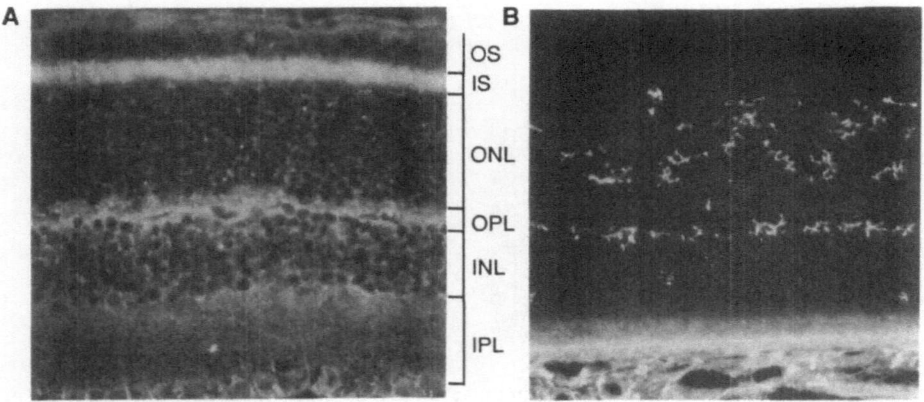

Figure 5. Immunolocalization of a crossreacting protein in the mouse retina. A. A seven micron frozen section of a nonpigmented mouse retina was stained with an immunopurified rdgB immunoglobulin and detected with a Cy3-conjugated goat-anti-mouse IgG. The layers of the retina are labeled as follows: outer segments, OS; inner segments, IS; outer nuclear layer, ONL; outer plexiform layer, OPL; inner nuclear layer, INL; inner plexiform layer, IPL. B. A seven micron frozen section of a nonpigmented mouse retina was stained with the rdgB mouse preimmune serum and detected with a Cy3-conjugated goat-anti-mouse IgG. The irregular signal throughout the retina is also observed when the sections were treated without any primary antiserum (data not shown).

DISCUSSION

The *rdgB* mutation in *Drosophila* provides an unique opportunity to gain insight into the general processes that are required for maintaining the highly specialized photoreceptor cell as well as being a model for studying retinal degeneration mechanisms. Shortly after eclosion, *rdgB* mutants exhibit loss of the electrophysiological light response. Several days later, there are obvious signs of photoreceptor degeneration, such as rhabdomere loss and photoreceptor cell death. The connection between these two phenotypes and the mechanisms that lead to these phenotypes are not known. We anticipate that a combination of molecular, genetic, and biochemical approaches will help to elucidate the role of the rdgB protein in visual transduction and in the prevention of photoreceptor degeneration.

The *rdgB* gene encodes a novel phosphatidylinositol transfer protein. We demonstrated that the amino terminus of rdgB possesses the ability to move phosphatidylinositol between two different membranes. As rdgB is an integral membrane protein of approximately 160 kDa [36], it is unlikely to move between membranes as observed for the soluble PI-TPs, which range from 24 to 36 kDa [40]. However, the distance between the SRC and the rhabdomere is approximately 20 nm [45], which could be spanned by the 498 amino-terminal amino acids of rdgB. It is interesting to note that the rdgB protein possesses six hydrophobic domains. If the purpose of the hydrophobic domains are to embed the protein in a membrane, then one domain would be sufficient. The presence of six domains suggests that this region of the protein plays a critical role in the function of the molecule and that rdgB is not functioning simply as a PI-TP molecule. This is consistent with the soluble rdgB PI-TP truncated protein failing to restore the wild-type light response and prevent degeneration in germline transformed flies. Finally, the rdgB protein binds Ca^{2+} *in vitro*. One of the consequences of invertebrate visual transduction is the IP_3-mediated rise in intracellular Ca^{2+}. Because *rdgB* exhibits a light-enhanced degeneration, it is possible that the inability of rdgB to respond to the rise in intracellular Ca^{2+} leads to the degeneration.

When we put all of the data together for the functional domains of the rdgB protein, we can envision three potential roles for the molecule. The first is based upon the role of PI-TPs in membrane trafficking and the inositol signaling cascade. Reconstitution experiments in permeabilized PC12 cells suggest that PI-TP activity is necessary for regulated secretion, specifically the ATP-dependent Ca^{2+}-activated fusion of secretory granules with the plasma membrane [48]. Constitutive secretion dependence on PI-TP is observed in the *Saccharomyces cerevisiae sec14* mutant, which displays an abnormal accumulation of Golgi membranes [49]. The *sec14* gene encodes a PI-TP that has similar *in vitro* catalytic properties as mammalian PI-TP. *sec14* mutants are defective in transport of secretory glycoproteins from a late Golgi compartment [49, 50]. This suggests that PI-TPs can stimulate membrane transport from the Golgi by elevating the phosphatidylinositol:phosphatidylcholine ratio in the Golgi membrane [51]. In photoreceptor cells, rdgB protein is found in the region where the SRC juxtaposes to the rhabdomeric membranes, suggesting that it is involved in membrane movement from the SRC to the rhabdomere. Many essential phototransduction components are expected to move to the rhabdomere through such a transport process [45]. For example, rhodopsin has been localized in the SRC [52]. Additionally, the vesicles that transport rhodopsin to the rhabdomeres will likely be enriched in PI, which can be phosphorylated to produce PIP_2, the substrate for the *norpA*-encoded phospholipase C. A role for PI-TPs in inositol signaling has been demonstrated by its prerequisite in reconstituting GTPγS-mediated PLC activity in partially differentiated granulocytes [53]. It is possible that rdgB functions to stimulate vesiculation from the SRC as well as provide PI for the phototransduction cascade. This rdgB-dependent membrane transport to the rhabdomere may be regulated by the intracellular Ca^{2+} level, which is increased by the light-activated visual cascade.

A second model supposes that phosphatidylinositol and/or its phosphorylated derivatives may bind as a ligand to activate an unidentified function of the rdgB protein. It was recently shown that SEC14p regulates choline-phosphate cytidylytransferase (CCTase), the rate determining enzyme of the CDP-choline pathway for phosphatidylcholine biosynthesis in *S. cerevisiae*, differentially depending upon the lipid ligand bound to the PI-TP [54]. The transfer of PI or PC appears unrelated to this regulation as the rat brain PI-TP is incapable of mimicking the effect of SEC14p on CCTase. Similarly, PI, phosphorylated PI derivatives, or PC could be the ligand for rdgB which affects another function, possibly Ca^{2+} uptake into the SRC. The Ca^{2+}-binding domain of the rdgB protein lies on the presumed cytosolic face of the SRC membrane, which is adjacent to the base of the rhabdomeric microvilli. The close apposition between the microvilli and the SRC membrane may facilitate loading of the SRC Ca^{2+} stores via an rdgB-mediated process. In the *rdgB* mutant, this activity would be missing and the photoreceptor cell would exhibit a rise in intracellular Ca^{2+} upon light stimulation. Because the resting potential of the cell would have difficulty being restored, further electrophysiological light responses could not occur. Similarly, the increased Ca^{2+} in the cell could cause the photoreceptor to degenerate.

Finally, the rdgB protein could exhibit some novel function that is not clear. The PI-TP domain may be used solely for binding the amino-terminus of the protein in the rhabdomeric microvillar membrane. It is possible that binding of Ca^{2+} could affect the affinity of the amino-terminus for PI and PC, which could alter its binding affinity for the rhabdomere. This would suggest that rdgB could stabilize the SRC compartment adjacent to the rhabdomere based on the PI/PC ratio of the microvillar membranes. In *rdgB*^{EE170} flies, the SRC gradually disintegrate upon light exposure and prior to the disintegration of the rhabdomere [45], which implicates rdgB in SRC maintenance. The light-induced hydrolysis of PIP_2 could have a dramatic effect on the PI content in the rhabdomeric membranes, which could initiate the *rdgB*-dependent degeneration. However, the loss of the ERG light response is observed well before any morphological change in the SRC, which also establishes a role for rdgB in photoreceptor physiology.

While our work established potential activities associated with the rdgB protein, the mechanism for *rdgB*-induced retinal degeneration remains elusive. Our working model is that *rdgB* is required to transport PI or proteins involved in phototransduction from the SRC to the photoresponsive rhabdomeric membranes. Because *rdgB* mutants show an aberrant light response before there is substantial degeneration [19], the SRC and rhabdomeres must not have established proper conditions for phototransduction. The SRC degenerates before the rhabdomeres, which suggests that the rhabdomeres are gradually lost as a consequence of inadequate maintenance by the SRC [45]. Localization of the rdgB protein to the SRC membrane is consistent with the importance of this organelle in *rdgB* degeneration. Thus, we imagine that regulated interactions between the SRC and the rhabdomere normally establish conditions for phototransduction, and the regulation of these are aberrant in *rdgB* mutants. As a consequence, a cascade of events, perhaps involving a rise in intracellular Ca^{2+} via voltage-gated Ca^{2+} channels [55], leads to degeneration.

ACKNOWLEDGMENTS

This work was supported by the National Institutes of Health (EY08058) and The George Gund Foundation and The National Retinitis Pigmentosa Foundation, Inc. (#93-152).

REFERENCES

1. Smith, D. P., Stamnes, M. A. and Zuker, C. S., 1991, Signal Transduction in the Visual System of *Drosophila*, *Annu. Rev. Cell Biol.* 7: 161-190.
2. O'Tousa, J. E., Baehr, W., Martin, R. L., Hirsh, J., Pak, W. L. and Applebury, M. L., 1985, The *Drosophila ninaE* gene encodes an opsin., *Cell.* 40: 839-850.
3. Zuker, C. S., Cowman, A. F. and Rubin, G. M., 1985, Isolation and structure of a rhodopsin gene from *D. melanogaster*, *Cell.* 40: 851-858.
4. Hotta, Y. and Benzer, S., 1970, Genetic dissection of the *Drosophila* nervous system by means of mosaics, *Proc. Natl. Acad. Sci. USA.* 67: 1156-1163.
5. Pak, W. L., Grossfield, J. and Arnold, K., 1970, Mutants in the visual pathway of *Drosophila melanogaster*, *Nature (Lond.).* 227: 518-520.
6. Bloomquist, B. T., Shortridge, R. D., Schneuwly, S., Perdew, M., Montell, C., Stellar, H., Rubin, G. and Pak, W. L., 1988, Isolation of a putative phospholipase C gene of *Drosophila*, *norpA*, and its role in phototransduction, *Cell.* 54: 723-733.
7. Schneuwly, S., Burg, M. G., Lending, C., Perdew, M. H. and Pak, W. L., 1991, Properties of photoreceptor-specific phospholipase C encoded by the *norpA* gene of *Drosophila melanogaster*, *J. Biol. Chem.* 266: 24314-24319.
8. Lee, Y.-G., Dobbs, M. B., Verardi, M. L. and Hyde, D. R., 1990, *dgq*: a *Drosophila* gene encoding a visual system-specific Gα molecule, *Neuron.* 5: 889-898.
9. Lee, Y. J., Shah, S., Suzuki, E., Zars, T., O'Day, P. M. and Hyde, D. R., 1994, The *Drosophila dgq* gene encodes a Gα protein that mediates phototransduction, *Neuron.* 13: 1143-1157.
10. Dolph, P. J., Mansonhing, H., Yarfitz, S., Colley, N. J., Deer, J. R., Spencer, M., Hurley, J. B. and Zuker, C. S., 1994, An eye-specific Gβ subunit essential for termination of the phototransduction cascade, *Nature.* 370: 59-61.
11. Smith, D. P., Ranganathan, R., Hardy, R. W., Marx, J., Tsuchida, T. and Zuker, C. S., 1991, Photoreceptor deactivation and retinal degeneration mediated by a photoreceptor-specific protein kinase C, *Science.* 254: 1478-1484.
12. Ranganathan, R., Harris, G. L., Stevens, C. F. and Zuker, C. S., 1991, A *Drosophila* mutant defective in extracellular calcium-dependent photoreceptor deactivation and rapid desensitization, *Nature.* 354: 230-232.

13. Hardie, R. C., Peretz, A., Susstoby, E., Romglas, A., Bishop, S. A., Selinger, Z. and Minke, B., 1993, Protein kinase-C is required for light adaptation in *Drosophila* photoreceptors, *Nature*. 363: 634-637.
14. Selinger, Z., Doza, Y. N. and Minke, B., 1993, Mechanisms and genetics of photoreceptors desensitization in *Drosophila* flies, *Biochim Biophys Acta*. 1179: 283-299.
15. Hardie, R. C. and Minke, B., 1994, Calcium-dependent inactivation of light-sensitive channels in *Drosophila* photoreceptors, *J Gen Physiol*. 103: 409-427.
16. Montell, C. and Rubin, G. M., 1989, Molecular characterization of the *Drosophila trp* locus: a putative integral membrane protein required for phototransduction, *Neuron*. 2: 1313-1323.
17. Hardie, R. C. and Minke, B., 1992, The *trp* gene is essential for a light-activated Ca^{2+} channel in *Drosophila* photoreceptors, *Neuron*. 8: 643-651.
18. Hardie, R. C. and Minke, B., 1994, Spontaneous activation of light-sensitive channels in *Drosophila* photoreceptors, *J Gen Physiol*. 103: 389-407.
19. Harris, W. A. and Stark, W. S., 1977, Hereditary retinal degeneration in *Drosophila melanogaster*: a mutant defect associated with the phototransduction process, *J. Gen. Physiol*. 69: 261-291.
20. Stark, W. S., Chen, D.-M., Johnson, M. A. and Frayer, K. L., 1983, The *rdgB* gene of *Drosophila*: retinal degeneration in different alleles and inhibition by *norpA*, *J. Insect Physiol*. 29: 123-131.
21. Minke, B., Rubinstein, C. T., Sahly, I., Bar-Nachum, S., Timberg, R. and Selinger, Z., 1990, Phorbol ester induces photoreceptor-specific degeneration in a *Drosophila* mutant, *Proc. Natl. Acad. Sci. USA*. 87: 113-117.
22. Woodard, C., Alcorta, E. and Carlson, J., 1992, The *rdgB* gene of *Drosophila*: a link between vision and olfaction, *J. Neurogenetics*. 8: 17-31.
23. Nathans, J., 1994, In the eye of the beholder: visual pigments and inherited variation in human vision, *Cell*. 78: 357-360.
24. Kurada, P. and O'Tousa, J. E., 1995, Retinal degeneration caused by dominant rhodopsin mutations in *Drosophila*, *Neuron*. in press.
25. O'Tousa, J. E., Leonard, D. S. and Pak, W. L., 1989, Morphological defects in *ora*^{JK84} photoreceptors caused by mutation in R1-6 opsin gene of *Drosophila*., *J. Neurogenetics*. 6: 41-52.
26. Leonard, D. S., Bowman, V. D., Ready, D. F. and Pak, W. L., 1992, Degeneration of photoreceptors in rhodopsin mutants of *Drosophila*, *J Neurobiol*. 23: 605-626.
27. Meyertholen, E. P., Stein, P. J., Williams, M. A. and Ostroy, S. E., 1987, Studies of the *Drosophila norpA* phototransduction mutant II. Photoreceptor degeneration and rhodopsin maintenance, *J. Comp. Physiol*. 161: 793-798.
28. Stark, W. S., Sapp, R. and Carlson, S. D., 1989, Photoreceptor maintenance and degeneration in the *norpA* (*no receptor potential-A*) mutant of *Drosophila melanogaster*, *J. Neurogenetics*. 5: 49-59.
29. Bowes, C., Li, T. S., Danciger, M., Baxter, L. C., Applebury, M. L. and Farber, D. B., 1990, Retinal degeneration in the rd mouse is caused by a defect in the beta-subunit of rod cGMP-phosphodiesterase, *Nature*. 347: 677-680.
30. Byk, T., Baryaacov, M., Doza, Y. N., Minke, B. and Selinger, Z., 1993, Regulatory arrestin cycle secures the fidelity and maintenance of the fly photoreceptor cell, *Proc Natl Acad Sci USA*. 90: 1907-1911.
31. Portera-Cailliau, C., Sung, C.-H., Nathans, J. and Adler, R., 1994, Apoptotic photoreceptor cell death in mouse models of retinitis pigmentosa, *Proc. Natl. Acad. Sci. USA*. 91: 974-978.
32. Van Vactor, D., Jr., Krantz, D. E., Reinke, R. and Zipursky, S. L., 1988, Analysis of mutants in chaoptin, a photoreceptor cell specific glycoprotein in *Drosophila*, reveals its role in cellular morphogenesis, *Cell*. 52: 281-290.
33. Arikawa, K., Molday, L. L., Molday, R. S. and Williams, D. S., 1992, Localization of peripherin/rds in the disk membranes of cone and rod photoreceptors: Relationship to disk membrane morphogenesis and retinal degeneration, *J. Cell Biol*. 116: 659-667.
34. Wells, J., Wroblewski, J., Keen, J., Inglehearn, C., Jubb, C., Eckstein, A., Jay, M., Arden, G., Bhattacharya, S., Fitzke, F. and Bird, A., 1993, Mutations in the human retinal degeneration slow (RDS) gene can cause either Retinitis-Pigmentosa or macular dystrophy, *Nat Genet*. 3: 213-218.
35. Kemp, C. M., Jacobson, S. G., Cideciyan, A. V., Kimura, A. E., Sheffield, V. C. and Stone, E. M., 1994, RDS gene mutations causing retinitis pigmentosa or macular degeneration lead to the same abnormality in photoreceptor function, *Invest Ophthalmol Visual Sci*. 35: 3154-3162.
36. Vihtelic, T. S., Goebl, M., Milligan, S., O'Tousa, J. E. and Hyde, D. R., 1993, Localization of *Drosophila retinal degeneration B*, a membrane-associated phosphatidylinositol transfer protein, *J. Cell Biol*. 122: 1013-1022.
37. Vihtelic, T. S., Hyde, D. R. and O'Tousa, J. E., 1991, Isolation and characterization of the Drosophila *retinal degeneration B* (*rdgB*) gene, *Genetics*. 127: 761-768.

38. Engelman, D. M. and Steitz, T. A., 1981, The spontaneous insertion of proteins into and across membranes: the helical hairpin hypothesis, *Cell.* 23: 411-422.
39. Dickeson, S. K., Lim, C. N., Schuyler, G. T., Dalton, T. P., Helmkamp, G. M. J. and Yarbrough, L. R., 1989, Isolation and sequence of cDNA clones encoding rat phosphatidylinositol transfer protein, *J. Biol. Chem.* 264: 16557-16564.
40. Wirtz, K. W. A., 1991, Phospholipid transfer proteins, *Ann. Rev. Biochem.* 60: 73-99.
41. Sved, J., 1986, *eyes absent (eya), Dros. Inf. Service.* 63: 169.
42. Zipursky, C. S., Venkatesh, T. R., Teplow, D. B. and Benzer, S., 1984, Neuronal development in the *Drosophila* retina: monoclonal antibodies as molecular probes, *Cell.* 36: 15-26.
43. Carlson, S. D., Stark, W. S. and Chi, C., 1985, Rapid light induced degeneration of photoreceptor terminals in *rdgB* mutant of *Drosophila, Invest. Opthal. Vis. Sci. Suppl.* 26: 131.
44. Fujita, S. C., Zipursky, S. L., Benzer, S., Ferrus, A. and Shotwell, S. L., 1982, Monoclonal antibodies against the *Drosophila* nervous system, *Proc. Natl. Acad. Sci. USA.* 79: 7929-7933.
45. Matsumoto-Suzuki, E., Hirosawa, K. and Hotta, Y., 1989, Structure of the subrhabdomeric cisternae in the photoreceptor cells of *D. melanogaster, J. Neurocyto.* 18: 87-93.
46. Baumann, O. and Walz, B., 1989, Topography of Ca^{+2}-sequestering endoplasmic reticulum in photoreceptors and pigmented glial cells in the compound eye of the honeybee drone, *Cell Tissue Res.* 255: 511-522.
47. Payne, R., Walz, B., Levy, S. and Fein, A., 1988, The localization of calcium release by inositol trisphosphate in *Limulus* photoreceptors and its control by negative feedback, *Phil. Trans. R. Soc. London.* 320: 359-379.
48. Hay, J. C. and Martin, T. F. J., 1993, Phosphatidylinositol transfer protein is required for ATP-dependent priming of Ca^{2+}-activated secretion, *Nature.* 366: 572-575.
49. Bankaitis, V. A:, Malehorn, D. E., Emr, S. D. and Greene, R., 1989, The *Saccharomyces cerevisiae SEC14* gene encodes a cytosolic factor that is required for transport of secretory proteins from the yeast Golgi complex, *J. Cell Biol.* 108: 1271-1281.
50. Franzukoff, A. and Schekmann, R., 1989, Functional compartments of the yeast Golgi apparatus are defined by the *sec7* mutation, *EMBO J.* 8: 2695-2702.
51. Cleves, A. E., McGee, T. P., Whitters, E. A., Champion, K. M., Aitken, J. R., Dowhan, W., Goebl, M. and Bankaitis, V., 1991, Mutations in the CDP-choline pathway for phospholipid biosynthesis bypass the requirement for an essential phospholipid transfer protein, *Cell.* 64: 789-800.
52. Suzuki, E. and Hiosawa, K., 1991, Immunoelectron microscopic study of the opsin distribution in the photoreceptor cell of *Drosophila melanogaster, J. Electron Microsc.* 40: 187-192.
53. Thomas, G. M. H., Cunningham, E., Fensome, A., Ball, A., Totty, N. F., Truong, O., Hsuan, J. J. and Cockcroft, S., 1993, An essential role for phosphatidylinositol transfer protein in phospholipase C-mediated inositol lipid signaling, *Cell.* 74: 919-928.
54. Skinner, H. B., McGee, T. P., McMaster, C. R., Fry, M. R., Bell, R. M. and Bankaitis, V. A., 1995, The *Saccharomyces cerevisiae* phosphatidylinositol-transfer protein effects a ligand-dependent inhibition of choline-phosphate cytidylyltransferase activity, *Proc. Natl. Acad. Sci. USA.* 92: 112-116.
55. Sahly, I., Bar-Nachum, S., Suss-Toby, E., Rom, A., Peretz, A., Kleiman, J., Byk, T., Selinger, Z. and Minke, B., 1992, Calcium channel blockers inhibit retinal degeneration in the *retinal degeneration-B* mutant of *Drosophila, Proc. Natl. Acad. Sci. USA.* 89: 435-439.

29

DROSOPHILA VISUAL TRANSDUCTION, A MODEL SYSTEM FOR HUMAN EYE DISEASE?

Retinal Degenerations

Jude Fitzgibbon and David Hunt

Institute of Ophthalmology
Department of Molecular Genetics
Bath Street, London ECIV 9EL, United Kingdom

INTRODUCTION

Animal models have been successfully used for the identification of a number of genes that are mutated in human retinal degenerations. For example, the study of the *rds* (Travis *et al.*, 1989), *rd* (Bowes *et al.*, 1993; Pittler and Baehr, 1991) and *shaker-1* (Gibson *et al.*, 1995) mutants in the mouse has led to the identification of mutations in the genes for peripherin/rds (Farrar *et al.*, 1991; Kajiwara *et al.*, 1991) and rod cGMP phosphodiesterase β subunit (McLaughlin *et al.*, 1993) that give rise to retinitis pigmentosa (RP) in man, and to a mutation in the gene for a myosin type V11 in Usher syndrome type 1B (Weil *et al.*, 1995). An additional source of mutations with effects on the visual process is present in the fruitfly, *Drosophila melanogaster*. As phototransduction in mammals involves a signal transduction pathway that is distinct from that used in invertebrate photoreceptors, the genes mutated in these retinal degenerations (Smith *et al.*, 1991; Wu *et al.*, 1995) have not been considered in detail as a cause of human retinal disease.

In *Drosophila,* phototransduction involves the light stimulated breakdown of phosphatidylinositol-4-5-bisphosphate (PIP_2) into two second messengers, inositol triphosphate (IP_3) and diacylglycerol (DAG) using a phosphoinositide-specific phospholipase C (PI-PLC). These second messengers modulate intracellular signaling through the activation of protein kinase C and the release of intracellular calcium. Several members of this pathway are mutated in *Drosophila* retinal disease (see Table 1) and there is now convincing evidence to support an analogous pathway(s) in the mammalian retina. As a consequence, we are considering whether parallels exist between *Drosophila* and mammalian retinal disease. Initially we have focused on the chromosomal map position of genes which are involved in PI-PLC-mediated signaling and their comparison with known retinal disease causing loci.

Degenerative Diseases of the Retina, Edited by Robert E. Anderson et al.
Plenum Press, New York, 1995

Table 1. Examples of the genes mutated in *Drosophila* retinal degenerations

Drosophila mutant	Reference	Gene	Reference
norpA (no receptor potential A)	Hotta and Benzer, 1970 Pak *et al.*, 1970	PLC	Bloomquist *et al.*, 1988
rdgA (retinal degeneration A)	Hotta and Benzer, 1970	DAGK	Masai *et al.*, 1993
rdgB (retinal degeneration B)	Hotta and Benzer, 1970	MBPITPN	Vihtelic *et al.*, 1993
eye-CDS (CDP-DAG synthase)	Wu *et al.*, 1995	CDP-DAG	Wu *et al.*, 1995
inaC (inactivation no after potential C)	Pak, 1979	PKC	Smith *et al.*, 1991

EVIDENCE FOR A PHOSPHOLIPASE C- MEDIATED LIGHT-ACTIVATED SIGNAL TRANSDUCTION PATHWAY IN THE MAMMALIAN RETINA

Several lines of evidence support the existence of a light-activated PLC-mediated signal transduction pathway in the mammalian retina. The light-activated breakdown of PIP_2 is well established (Das *et al.*, 1987; Millar *et al.*, 1988); furthermore the identification of a PLC that is predominately expressed in the retina, and shares strong homology to the *norpA* PLC identified in *Drosophila* (Bloomquist *et al.*, 1988), suggests that this particular pathway is confined to the retina. Subsequent immunocytochemical studies have localised this PLC specifically to cone outer segments indicating a possible role for this PI-PLC in cone phototransduction (Ferreira and Pak, 1994). The observation that protein kinase C α (PRKCA) is also expressed specifically in cones (Ohki *et al.*, 1994) provides further evidence in favour of a cone-specific PLC-mediated signaling cascade. This does not preclude the existence of a similar pathway in rod photoreceptors since PLC and PKC activities have been observed in the bovine ROS (rod outer segments) (Kelleher and Johnson, 1985; Ghalayini *et al.*, 1987, 1991; Gehm *et al.*, 1990; Wolbring *et al.*, 1991).

The generation of second messengers has been more difficult to confirm. Jung *et al.* (1993) have observed light activated IP_3 release in the rat retina, although the only evidence for its origin in photoreceptors is the predominant localisation of its precursor molecule, PIP_2 in these cells (Das *et al.*, 1987). The identification of IP_3 receptors in photoreceptor cells (Day *et al.*, 1993), although probably confined to the inner segments, also supports this suggestion. Kai *et al.* (1994) have recently identified a DAGK which is expressed predominately in the retina, with a truncated and inactive form transcribed in most other tissues. The cellular localisation in the retina of this DAGK has yet to be determined. As DAGK is thought to modulate the activity of protein kinase C through its regulation of the second messenger, DAG (Azzi *et al.*, 1992), its localisation to photoreceptors might also be expected.

The observation by Newton and Williams (1991, 1993) that protein kinase C phosphorylates rhodopsin in a light-dependent manner suggests that this pathway is involved in desensitising rhodopsin. They propose that a balance may exist between PKC and rhodopsin kinase phosphorylation activities, with each predominating at low and high light levels respectively. Arrestin, which prevents activation of transducin by rhodopsin, has also been identified as an activator of PLC (Ghalayini *et al.*, 1992). It is therefore possible that arrestin is involved in regulating the activity of both phosphorylation pathways. The PKC isozyme responsible for this phosphorylation activity has not been determined. Indeed immunocytochemical studies have failed to localise any of the known PKC isozymes to the rod outer segments (Usada *et al.*, 1991; Osborne *et al.*, 1992; Huwiler *et al.*, 1992; Ghaylayini *et al.*, 1994; Ohki *et al.*, 1994). It would appear that the PKC involved in the phosphorylation of rhodopsin has not as yet been identified.

PI-PLC Mediated Signal Transduction and the Defect in the RCS Rat

The recent observation that stimulation of a PI-PLC signal pathway is absent in the Royal College of Surgeons (RCS) rat (Heth and Marescalchi, 1994) provides the first evidence to implicate an alteration in a light-activated PLC-mediated signal transduction pathway in mammalian retinopathy. In this recessive disorder, the process of phagocytosis and degradation of outer segment membranes is disrupted, leading to the accumulation of photoreceptor debris and retinal degeneration. Heth and Marescalchi (1994) show that binding of the rod outer segments (ROS) to the retinal pigment epithelium (RPE) fails to evoke a transmembrane signal with subsequent decrease in PIP_2 level and generation of the second messenger IP_3. The pathway however can be stimulated by treatment with carbachol, demonstrating that the components of the pathway are all present and functional. Previously it has been shown that there are changes in the phosphorylation state of the RCS RPE compared to normal RPE, after phagocytic challenge. This is consistent with a change in the activation of protein kinases (Heth and Schmidt, 1992) which may underly the absence of the IP_3 pathway in these rats. This is not a phenomenon confined to the RCS rat as changes in the phosphorylation state of rhodopsin have also been observed in other retinal degenerations (Shuster and Farber, 1986).

CHROMOSOMAL LOCALISATION OF MAMMALIAN GENES INVOLVED IN THE PLC MEDIATED SIGNAL TRANSDUCTION PATHWAY

If a PI-PLC signaling pathway plays a part in the causation of retinal disease, then the chromosomal position of genes present in this pathway may coincide with the locations of retinal diseases. Previously, such fortuitous associations have led investigators to identify rhodopsin (McWilliam *et al.*, 1989) and tissue inhibitor of metalloproteinases-3 (Weber, 1994a) as the causative genes for RP (Dryja *et al.*, 1990) and Sorby's macular dystrophy respectively (Weber, 1994b). Table 1 briefly summarises the *Drosophila* phototransduction mutants which result from a defect in some component of the PI-PLC signaling cascade. Mammalian genes which share sequence and functional homology to these genes and their chromosomal positions are described in Table 2.

The *norpA*-like mammalian PLC, initially identified by Ferriera *et al.* (1993) has been mapped to human chromosome 20p (Baehr *et al.*, 1994). Although no retinopathies have so far been localised to this region, PLCβ4 would appear an obvious candidate for future linkage studies of macular disease. It could also be supposed that, given the restricted expression of this gene to cone photoreceptors, a rod specific PLC also exists. In *Drosophila*, mutations in an eye-specific DAGK encodes the *retinal degeneration A* mutant (*rdgA*); degeneration in these flies is thought to arise either from chronic activation of PKC or from a lack of phosphatidic acid (PA). The identification of one form of DAGK, DAGK3 (Kai *et al.*, 1994), that is predominately expressed in the retina, is therefore particularly interesting. Our current mapping studies have assigned this gene to the distal region of human chromosome 3q (Fitzgibbon, unpublished observation). Although there are no retinal disease loci in this region, it will be important to determine the cellular localisation of this gene in the retina and compare it to that of other members of the pathway.

The most recent mutant to be discovered in *Drosophila* is *eye-CDS* (Wu *et al.*, 1995), which encodes a CDP-diacylglycerol synthetase required for the regeneration of PI from PA. No mammalian homologues of this gene have so far been reported. This is also the case for

Table 2. Summary of the chromosomal localisations of human genes sharing sequence and functional homology to genes causing retinal degenerations in *Drosophila.* As PLCβ shares strongest homology to the norpA PLC other PLC isozymes have not been included. The sequence of DAGK4 has been submitted to Genbank, Acc. No. L38707. No homologues of the *Drosophila* eye-CDS have been identified

Gene name	Reference	Human chromosomal position	Reference
Phospholipase C (PLC)			
NorpA			
PLCβ4	Baehr *et al.*, 1994	20p	Baehr *et al.*, 1994
Diacylglycerol kinase (DAGK)			
rdgA			
DAGK1	Schaap *et al.*, 1990	12q13.3	Hart *et al.*, 1994
DAGK3	Kai *et al.*, 1994	3q	Fitzgibbon, unpubl. observ.
DAGK4	Schaap, per. comm..	4 or 12	Pilz *et al.*, 1995
EST D0S2077E	Adams *et al.*, 1992	2 or 11	Pilz *et al.*, 1995
Protein Kinase C (PRKC)			
PRKCA	Knopf *et al.*, 1986	17q22-q24	Leach *et al.*, 1989
PRKCB1	Knopf *et al.*, 1986	16p	Coussins *et al.*, 1986
PRKCD	Mischak *et al.*, 1991	3p	Huppi *et al.*, 1994
PRKCG	Knopf *et al.*, 1991	19q13.4	Saunders *et al.*, 1990
PRKCQ	Baier *et al.*, 1993	10p15	Erdel *et al.*, 1995
Phosphatidylinositol transfer protein (PITPN)			
rdgB			+
PITPN1	Dickeson *et al.*, 1989	17p13.3	Fitzgibbon *et al.*, 1994

the *rdgB* protein, a membrane bound phosphatidylinositol transfer protein (MBPITPN) (Vihtelic *et al.*, 1993) which, as the name implies, is thought to play a role in the transport of phospholipids. Although two mammalian PITPNs have so far been identifed (Dickeson *et al.*, 1989; Tanaka and Hosaka, 1994), unlike *rdgB*, these do not contain any membrane associated domains. We have mapped one of these genes, PITPN1, to human chromosome 17p13.3 (Fitzgibbon *et al.*, 1994), a region which contains a locus for an autosomal dominant RP (Greenberg *et al.*, 1994); however since this gene is expressed in a number of different tissues (Dickeson *et al.*, 1989), there is no compelling evidence to suppose that it is a strong candidate for this disease. The chromosomal position of PRKCG (19q13,3) also co-localises to a region containing an adRP locus (Al-Maghtheh *et al.*,1994). However the absence of its expression in photoreceptors would suggest that an involvment in this form of RP is unlikely (Osborne *et al.*, 1992). It will however be worth refining the chromosomal localisations of both these genes to exclude their involvment.

CONCLUSION

Although there is convincing evidence to support a light-activated PLC-mediated signaling pathway(s) in the mammalian retina, there is no absolute evidence to date to implicate mutations in this pathway as a cause of retinal degeneration. It seems likely that other components of the pathway with a similarly restricted pattern of expression, as shown for both PLCβ4 and DAGK3, will be identified in due course. We will then be in a better position to consider whether this pathway plays any role in the aetiology of retinal disease. A recent report showing that eye morphogenesis is under similar genetic controls in vertebrates and *Drosophila* (Quiring *et al.*, 1994) suggests that this is more likely than previously envisaged.

ACKNOWLEDGMENTS

We would like to thank our collaborators, Lyn Yarbrough, Dick Schaap and Hideo Kanoh for providing the probes used in our mapping studies.

REFERENCES

Adams, M.D., Dubnick, M., Kerlavage, A.R., Moreno, R., Kelley, J.M., Utterback, T.R., Nagle, J.W., Fields, C., and Venter, J.C., 1992, Sequence identification of 2,375 human brain genes, *Nature* 355: 632-634.

Al-Maghtheh, M., Inglehearn, C.F., Keen, T.J., Evans, K., Moore, A.T., Jay, M., Bird, A.C., and Bhattacharya, S.S., 1994, Identification of a sixth locus for autosomal dominant retinitis pigmentosa on chromosome 19, *Hum. Mol. Genet.* 3: 351-354.

Azzi, A., Boscoboinik, D., and Hensey, C., 1992, The protein kinase C family, *Eur. J. Biochem.* 208: 547-557.

Baehr, W., Alvarez, R.A., Xu, P., Hardcastle, A., Bhattacharya, S., and Anderson, R.E., 1994, cDNA sequence of a human retinal phosphoinositide-specific phospholipase Cβ4 and chromosomal localisation of its gene, VI International symposium on retinal degeneration, Abstract.

Baier, G., Telford, D., Giampi, L., Coggeshall, K.M., Baier-Bitterlich, G., Isakov, N., and Altman, A., 1993, Molecular cloning and characterisation of PKC-θ, a novel member of the PKC gene family expressed predominantly in hematopoietic cells, *J Biol. Chem.* 268: 4997-5004.

Bloomquist, B.T., Shortridge, R.D., Schneuwly, S., Perdew, M., Montell, C., Stellar, H., Rubin, G., and Pak, W.L., 1988, Isolation of a putative phospholipase C gene of *Drosophila, norpA*, and its role in phototransduction, *Cell* 54: 723-733.

Bowes, C., Li, T., Frankel, W.N., Danciger, M., Coffin, J.M., Applebury, M.L., and Farber, D.B., 1993, Localisation of a retroviral element within the rd gene coding for the β subunit of cGMP phosphodiesterase, *Proc. Natl. Acad. Sci.* USA 90: 2955-2959.

Coussens, L., Parker, P.J., Rhee, I., Yang-Feng, T.L., Chen, E., Waterfield, M.D., Francke, U., and Ullrich, A., 1986, Multiple distinct forms of bovine and human protein kinase C suggests diversity in cellular signaling pathways, *Science* 233: 859-866.

Das, N.D., Yoshioka, T., Samuelson, D., Cohen, R.J., and Shichi, H., 1987, Immunochemical evidence for the light regulated modulation of phosphatidylinositol 4,5-bisphosphate in rat photoreceptor cells, *Cell Struct. Funct.* 12, 471-481.

Day, N.S., Koutz, C.A., and Anderson, R.E., 1993, Inositol-1,4,5-triphosphate receptors in the vertebrate retina, *Curr. Eye. Res.* 12: 981-992.

Dickeson, S.K, Lim, C.N., Schuyler, G.T., Dalton, T.P., Helkamp, G.M. Jr., and Yarbrough, L.R., 1989, Isolation and sequence of cDNA clones encoding rat phosphatidylinositol transfer protein, *J. Biol. Chem.* 264: 16557-16564.

Dryja, T.P., McGee, T.L., Reichel, E., Hahn, L.B., Cowley, G.S., Yandell, D.W., Sandberg, M.A., and Berson, E.L., 1990, A point mutation in the rhodopsin gene in one form of retinitis pigmentosa. *Nature* 343: 364-365.

Erdel, M., Baier-Bitterlich, G., Duba, C., Isakov, N., Altman, A., Utermann, G., and Baier, G., 1995, Mapping of the human protein kinase C-θ (PRKCQ) gene locus to the short arm of chromosome 10 (10p15) by FISH. *Genomics* 25: 595-597.

Farrar, G.J., Kenna, P., Jordan, S., Kumar-Singh, R., Humphries, M.M., Sharp, E.M., Sheils, D.M., and Humphries, P., 1991, A three-base-pair deletion in the peripherin-RDS gene in one form of retinitis pigmentosa, *Nature* 354: 478-479.

Ferreira, P., Shortridge, R.D., and Pak, W., 1993, Distinctive subtypes of bovine phospholipase C that have preferential expression in the retina and high homology to the *norpA* gene product of *Drosophila*, *Proc Natl. Acad. Sci* USA 90: 6042-6046.

Ferreira, P., and Pak, W., 1994, Bovine phospholipase C highly homologous to the NorpA protein of *Drosophila* is expressed specifically in cones, *J. Biol. Chem.* 269: 3129-3131.

Fitzgibbon, J., Pilz, A., Gayther, S., Appukutan, B., Dulai, K., Delhanty, J.D.A., Helmkamp, G.M., Yarbrough, L.R., and Hunt, D., 1994, Localisation of the gene encoding human phosphatidylinositol transfer protein (PITPN) to 17p13.3: a gene showing homology to the *Drosophila* retinal degeneration B gene (*rdgB*), *Cytogenet. Cell Genet.* 67: 205-207.

Gehm, B.D., and McConnell, D.G., 1990, Phosphatidylinositol 4,5-bisphosphate phospholipase C in bovine rod outer segments, *Biochemistry* 29: 5447-5452.

Ghalayini, A.J., and Anderson, R.E., 1987, Activation of bovine rod outer segment phospholipase C by ATP and GTP, *Neurosci. Res. Commun.* 1: 119-127.

Ghalayini, A.J., Tarver, A.P., Machin, W.M., Koutz, C.A., and Anderson, R.E., 1991, Identification and immunolocalisation of phospholipase C in bovine rod outer segments, *J. Neurochem.* 57: 1405-1412.

Ghalayini, A.J., and Anderson, R.E., 1992, Activation of bovine rod outer segment phospholipase C by arrestin, *J. Biol. Chem.* 267: 17977-17982.

Ghalayini, A.J., Koutz, C.A., Wetsel, W.C., Hannun, Y.A., and Anderson, R.E., 1994, Immunolocalisation of PKCζ in rat photoreceptor inner segments, *Curr. Eye. Res.* 13: 145-150.

Gibson, F., Walsh, J., Mburu, P., Varela, A., Brown, K.A., Antonio, M., Beisel, K.W., Steel, K.P., and Brown, S.D.M., 1995, A type VII myosin encoded by the mouse deafness gene *shaker-1*, *Nature* 374: 62-64.

Greenberg, J., Goliath, R., Beighton, P., and Ramesar, R., 1994, A new locus for autosomal dominant retinitis pigmentosa on short arm of chromosome 17: *Hum. Mol. Genet.* 3: 915-918.

Hart, T.C., Zhou, J., Champagne, C., Van Dyke, T.E., Rao, P.N., and Pettenati, M.J, 1994, Assignment of the human diacylglycerol kinase gene (DAGK) to 12q13.3 using fluorescence *in situ* hybridisation analysis, *Genomics* 22: 246-247.

Heth, C.A., and Marescalchi, P.A., 1994 Inositol triphosphate generation in cultured rat retinal pigment epithelium, *IOVS* 35: 409-416.

Heth, C.A., and Schmidt, S.Y., 1992, Protein phosphorylation in retinal pigment epithelium of Long-Evans and Royal College of Surgeons *IOVS* 33: 2839-2847.

Hotta, Y., and Benzer, S., 1970, Genetic dissection of the Drosophila nervous system by means of mosaics, *Proc Natl. Acad. Sci.* USA 67: 1156-1163.

Huppi, K., Siwarski, D., Goodnight, J., and Mischak., 1994, Assignment of the protein kinase C d polypeptide gene (PRKCD) to human chromosome 3 and mouse chromosome 14, *Genomics* 19: 161-162.

Huwiler, A., Jung, H.H., Pfeilschifter, J., and Reme, C.E., 1992, Protein kinase C in the rat retina: immunocytocharacterisation of calcium-independent δ, ε, ζ isoenzymes, *Molec. Brain Res.* 16: 360-364.

Jung, H.H., Reme, C.E., and Pfeilschifter, J., 1993, Light evoked inositol triphosphate release in the rat retina *in vitro, Curr. Eye. Res.* 12: 727-732.

Kai, M., Sakane, F., Imai, S., Wada, I., and Kanoh, H., 1994, Molecular cloning of a diacylglycerol kinase isozyme predominantly expressed in the human retina with a truncated and inactive enzyme expression in most other cells, *J. Biol. Chem.* 269: 18492-18496.

Kajiwara, K., Hahn, L.B., Mukai, S., Travis, G.H., Berson, E.L., and Dryja, T.P., 1991, Mutations in the human retinal degeneration slow gene in autosomal dominant retinitis pigmentosa, *Nature* 354: 480-483.

Kelleher, D.J., and Johnson, G.L., 1985, Purification of protein kinase C from bovine rod outer segments, *J. Cyclic Nucleot. Prot. Phosphoryl. Res.* 10: 579-591.

Knopf, J.L., Lee, M., Sultzmann, L.A., Kritz, R.W., Loomis, C.R., Hewick, R.M., and Bell, R.M., 1986, Cloning and expression of multiple protein kinase C cDNAs, *Cell* 255: 2273-2276.

Leach, R.J., Thayer, M.J., Schafer, A.J., and Fournier, R.E.K., 1989, Physical mapping of human chromosome 17 using fragment-containing microcell hybrids, *Genomics* 5, 167-176.

Masai, I., Okazaki, A., Hosoya, T., and Hotta, Y., 1993, Drosophila retinal degeneration A gene encodes an eye-specific diacylglycerol kinase with cysteine-rich zinc finger motifs and ankyrin repeats, *Proc Natl Acad. Sci.*, USA 90: 11157-11161.

McLaughlin, M.E., Sandberg, M.A., Berson, E.L., and Dryja, T.P., 1993, Recessive mutations in the gene encoding the β-subunit of rod phosphodiesterase in patients with retinitis pigmentosa, *Nature Genetics* 4: 130-134.

McWilliam, P., Farrar, Kenna, P., Bradley, D., Humphries, M.M., Sharp, E.M., and Humphries, P., 1989, Autosomal dominant retinitis pigmentosa (ADRP): localisation of an ADRP gene on the long arm of chromosome 3, *Genomics* 5: 619-622.

Millar, F.A., Fisher, S.C., Muir, C.A., Edwards, E., and Hawthorne, J.N., 1988, Polyphosphoinositide hydrolysis in response to light stimulation of rat and chick retina and retinal rod outer segments, *Biochem. Biophys. Acta.* 970: 205-211.

Mischak, H., Bodenteich, A., Kolch, W., Goodnight, J., Hofer, F., and Mushinski, J.F., 1991, Mouse protein kinase C-d the major isoform expressed in mouse hemopoietic cells: Sequence of the cDNA, expression patterns and characterisation of the protein, *Biochemistry* 30: 7925-7931.

Newton, A.C., and Williams, D.S., 1991, Involvment of protein kinase C in the phosphorylation of rhodopsin, *J. Biol. Chem.* 266: 17725-17728.

Newton, A.C., and Williams, D.S., 1993, Rhodopsin is the major *in situ* substrate of protein kinase C in rod outer segments of photoreceptors, *J. Biol. Chem.* 268: 18181-18186.

Ohki, K., Yoshida, K., Imaki, J., Harada, T., and Matsuda, H., 1994, The existance of protein kinase C in cone photoreceptors in the rat retina, *Curr. Eye. Res.* 13, 547-550.

Osborne, N.N., Barnett, N.D., Morris, N.J., and Huang, F.L., 1992, The occurrence of three isoenzymes of protein kinase C (α, β and γ) in retinas of different species, *Brain Res.* 570: 161-166.

Pak, W.L., Grossfield, J., and Arnold, K., 1970, Mutants in the visual pathway of *Drosophila melanogaster, Nature* 227: 518-520.

Pak, W.L., 1979, Study of photoreceptor function using Drosophila mutants. In Neurogenetics: Genetic approaches to the nervous system, X.O. Breakfield, eds (New York; Elsevier north-Holland), 67-99.

Pilz, A., Schaap, D., Hunt, D., and Fitzgibbon, J., 1995 Chromosomal localisation of three mouse diacylglycerol kinase (DAGK) genes: genes sharing sequence homology to the *Drosophila retinal degeneration A* (*rdgA*) gene, in press *Genomics*.

Pittler, S.J., and Baehr, W., 1992, Identification of a nonsense mutation in the rod photoreceptor cGMP phosphodiesterase β subunit gene of the *rd* mouse, *Proc Natl Acad. Sci.* USA 88: 8322-8326.

Quiring, R., Walldorf, U., Kloter, U., and Gehring, W.J., 1994, Homology of the eyeless gene of Drosophila to the small eye gene in mice and aniridia in humans, *Science* 265: 785-789.

Saunders, A.M,. and Seldin, M.F., 1990, The syntenic relationship of proximal mouse chromosome 7 and the myotonic dystrophy gene region on human chromosome 19q, *Genomics* 6: 324-332.

Schaap, D., de Widt, J., van der Wal, J., Vanderkerckhove, J., van Damme, J., Gussow, D., Ploegh, H.L., van Blitterswijk, W. J., and van der Bend, R., 1990, Purification, cDNA cloning and expression of human diacylglycerol kinase, *FEBS Lett.* 275, 151-158.

Shuster, T.A., and Farber, D.B., 1986, Rhodopsin phosphorylation in developing normal and degenerative mouse retinas, *IOVS* 27: 264-268.

Smith, D.P., Stamnes, M.A. and Zuker, C.S., 1991, Signal transduction in the visual system of *Drosophila Annu. Rev. Cell. Biol.* 7: 161-190.

Tanaka, S., and Hosaka, K., 1994, Cloning of a cDNA encoding a second phosphatidylinositol transfer protein of rat brain by complementation of the yeast *sec14* mutation, *J. Biochem.* 115: 981-984.

Travis, G.H., Brennan, M.B., Danielson, P.E., Kozak, C.A., and Sutcliffe, J.G.,1989, Identification of a photoreceptor specific mRNA encoded by the gene responsible for retinal degeneration slow (rds), *Nature* 388: 70-73.

Usada, N., Kong, Y., Hagiwara, M., Uchida, C., Terasawa, M., Nagata, T. and Hidaka, H., 1991, Differential localisation of protein kinase C isoenzymes in retinal neurons, *J. Cell Biol.* 112: 1241-1247.

Vihtelic, T.S., Goebl, M., Milligan, S., O'Tousa, J.E. and Hyde, D.R., 1993, Localisation of *Drosophila* retinal degeneration B, a membrane associated phosphatidylinositol transfer protein, *J Cell Biol.* 122: 1013-11022.

Weber, B.H.F., Vogt, G, Wolz, W., Ives, E.J., and Ewing, C.C., 1994a, Sorby's fundus dystrophy is genetically linked to chromosome 22q13-qter, *Nature Genetics* 7: 158-161.

Weber, B.H.F., Vogt, G, Pruett, R.C., Stohr, H., and Felbor, U., 1994b, Mutations in the tissue inhibitor of metalloproteinases-3 (TIMP3) in patients with Sorby's fundus dystrophy, *Nature Genetics* 8: 352-355.

Weil, D., Blanchard, S., Kaplan, J., Guilford, P., Gibson, F., Walsh, J., Mburu, Varela, A., Levilliers, J., Weston, M.D., Kelley, P.M., Kimberling, W.J., Wagenaar, M., Levi-Acobas,F., Larget-Piet, D., Munnich, A., Steel, K.P., Brown, S.D.M., and Petit, C., 1995, Defective myosin V11 a gene responsible for Usher syndrome type 1B. *Nature* 374: 60-61.

Wolbring, G., and Cook, N.J., 1991, 'Rapid purification and characterisation of protein kinase C from bovine retinal rod outer segments, *Eur. J. Biochem.* 210: 601-606.

Wu, L., Niemeyer, B., Colley, N., Socolich, M., and Zuker, C.S., 1995, Regulation of PLC-mediated signaling *in vivo* by CDP-diacylglycerol synthase, *Nature* 373: 216-222.

30

CHARACTERIZATION OF VERTEBRATE HOMOLOGS OF *DROSOPHILA* PHOTORECEPTOR PROTEINS

Paulo A. Ferreira[1] and William L. Pak[2]

[1] Psychiatry and Neuroscience Department
University of Texas Southwestern Medical Center
5323 Harry Hines Blvd., Dallas, Texas 75235-9111.
[2] Department of Biological Sciences
Purdue University
West Lafayette, Indiana 47907

INTRODUCTION

In invertebrates, a large body of evidence about the molecular mechanisms governing the visual cascade has been derived from the molecular genetic dissection of the components of the *Drosophila* photoreceptor machinery (1 - 5). As the best genetically characterized metazoan system, *Drosophila* provides an advantage in some respects over the mammalian system in that the genes can be both molecularly and genetically dissected facilitating the association of a particular gene product with its *in vivo* cellular function. Several of the *Drosophil*a mutants identified with changes in the light evoked responses also lead to photoreceptor degeneration. These are of particular interest, as retinal degeneration is a major cause of inherited blindness in humans. In addition, electrophysiological studies in *Limulus* photoreceptors cells have also provided us with an important insight into the physiology of the invertebrate visual cascade (6).

Vertebrate rod photoreceptor cells are more amenable to biochemical manipulations than their invertebrate counterparts. A great deal of what is known about vertebrate phototransduction has been derived from *in vitro* studies on the biochemistry of several components of the visual cascade in rod outer segments, as well as, from electrophysiological recordings in this cell compartment (7, 8, 9). More recently, advances in mammalian genetics have begun to provide important information about the *in vivo* function of certain components involved in photoreceptor function (10).

PHOTOTRANSDUCTION IN VERTEBRATES AND INVERTEBRATES

In the retina, photoreceptors are the first order neurons that confer photosensitivity through a cascade of complex reactions. The cascade is mediated by a G protein-coupled

Degenerative Diseases of the Retina, Edited by Robert E. Anderson et al.
Plenum Press, New York, 1995

mechanism that is analogous between vertebrates and invertebrates. Phototransduction is initiated with the excitation, by a photon, of a chromoprotein receptor, rhodopsin (11, 12). Rhodopsin consists of a chromophore, 11-cis retinal, covalently linked as a protonated Schiff base to the apoprotein, opsin. Photoexcited rhodopsin in vertebrates and invertebrates leads to the activation of a GTP-binding transducer molecule, a heterotrimeric G-protein, which communicates the stimulus to an effector molecule that modulates the levels of one or more second messengers. These in turn, convey the light-stimulus to cation channels that affect the flow of Na^+ and Ca^{2+} ions across the plasma membrane of the outer segment. The resulting receptor potential modulates neurotransmitter release by the presynaptic terminal of photoreceptors, transmitting the light signal to second order neurons.

The deactivation of the phototransduction cascade in vertebrates and invertebrates also appears to be mediated by some common biochemical mechanisms. Although the inactivation mechanisms have yet to be adequately resolved, one common desensitization mechanism is the phosphorylation of one or more of the phototransductive components. A key substrate is rhodopsin. Phosphorylation of rhodopsin by possibly different types of conserved kinases, such as rhodopsin kinase and protein kinase C, leads to the desensitization of the receptor (13, 14, 15). The ability of the phosphorylated receptor to stimulate the GTPase activity of the G-protein is further quenched by the binding to the receptor of a conserved protein, arrestin, which may compete for a common binding site with the G-protein (16, 17). The level of intracellular Ca^{2+} in photoreceptors also appears to be a key factor in mediating a series of inactivation events by influencing the sensitivity and kinetics of several steps in the phototransduction cascade (18 - 24).

Despite the analogy of the phototransduction mechanism in vertebrates and invertebrates, numerous striking differences appear to exist. These occur mainly at three well characterized levels of the visual cascade. First, a G_q-class photoreceptor-specific protein is used by invertebrates (*Drosophila*) to transduce the extracelullar stimulus to the intracellular machinery while transducin is the photoreceptor-specific G protein used in vertebrates (25, 26). Second, another difference is in the G protein-coupled effector and consequently, the second messengers used to convey the light stimulus from the photopigment to the light-activated channel. In invertebrates, the effector molecule is a phosphatidylinositol phospholipase C (PI-PLC) encoded by the *norpA* (no receptor potential) gene of *Drosophila* (27). A strong mutation in the *norpA* gene completely abolishes the photoreceptor potential of the fly, drastically reduces PI-PLC activity in the eye and causes the photoreceptors to degenerate in a light-dependent manner (2, 28-30). This provides critical evidence that the NorpA protein plays a direct role in *Drosophila* phototransduction. In vertebrates, a cGMP-dependent phosphodiesterase, one subunit of which is encoded by the *rd* gene mediates the light response (31, 32). A mutation in this gene leads to high levels of cGMP in the photoreceptor cell and subsequently, to photoreceptor degeneration (33, 34). Finally, another major difference lies in the identitity of the channels that generate the receptor potential in the photoreceptor cells. In *Drosophila,* the photoreceptors depolarize upon a light stimulus and two channels, *trp* and *trp-l*, appear as candidates to mediate a light-induced ionic current (20, 35, 36). In vertebrates, however, it is well established that a cGMP-gated channel mediates a dark-induced cationic current in the outer segments of the photoreceptors (37).

However, reports from electrophysiological studies in *Limulus* photoreceptors seem, in some instances, to contradict those obtained from *Drosophila*. For example, patch clamping studies have shown the presence of light-dependent and cGMP-gated channels in *Limulus* ventral photoreceptors and that these channels could not be opened by Ca^{2+} (38, 39). These results were unexpected since inositol 1,4,5-triphosphate (IP_3) and Ca^{2+} were shown to be necessary for the excitation of *Limulus* photoreceptors similarly to what is thought to occur in *Drosophila* (40-42). Therefore, it appears that other events may take place between the mobilization of Ca^{2+} from the intracellular stores by IP_3 and the activation of the light-dependent channel/s in invertebrates.

These studies have shown therefore, that phototransduction in vertebrate photoreceptors is based on G-protein (transducin)-mediated activation of cGMP phosphodiesterase and apparently, no direct role for a PI-PLC has been identified to date. Yet, there have been reports of light-activated PI-PLC activity in vertebrate photoreceptors and, of PI-PLC and IP_3 receptor immunoreactivity in rod outer and inner segments (43 - 55). IP_3 is known to be a key mediator in mobilizing Ca^{2+} from intracellular stores which is known to be an important mediator in the excitation and adaptation of photoreceptor cells. It is possible therefore, that the PLC product, IP_3, could play a role in the excitation and/or adaptation of certain vertebrate photoreceptor cells. Furthermore, most of the biochemical studies on phototransduction in vertebrates have been done in rods. Most of the information derived from cone phototransduction has come from the identification, in cone outer segments, of orthologous components of the rod phototransduction cascade (56 - 59). In addition, for the first time since 23 years ago, when Tomita recorded light-activated hyperpolarizing responses from cone photoreceptor cells, an interesting study has reported antagonistic chromatically-dependent hyperpolarization and depolarization in the photoreceptors of the parietal eye of the lizard (60, 61). This raises important questions about the molecular mechanisms underlying this chromatic antagonism. The existence of a cone cell expressing more than one visual pigment is known to occur and could provide a helpful insight into the mechanisms responsible for these chromatically-dependent responses (62). Also, it would be interesting to see, for example, if a phopholipase C signaling cascade is present and responsible for the biochemical basis of this depolarizing response. Such a signalling pathway could work in parallel or alternatively to the already known cGMP-dependent phosphodiesterase cascade in some of the photoreceptor cells.

IDENTIFICATION AND CHARACTERIZATION OF *norpA*-HOMOLOGOUS cDNAs

To investigate the role of inositol lipids in the visual cascade of both rods and cones we first tried to answer the question whether NorpA-like proteins exist in the vertebrate retina. Accordingly, we constructed and screened a bovine retina cDNA library at reduced stringency conditions by using a mixture of two radiolabeled fragments representing conserved portions of the translated region of Boxes X and Y of the *norpA* cDNA. These boxes are conserved among all PI-PLCs to date characterized (63). As a result of this screening, one positively hybridyzing cDNA clone was finally selected for further analysis because Southern blot analysis indicated that it may be a yet uncharacterized PLC clone highly homologous to the *norpA*. In Northern blot analysis of several different bovine tissues, this cDNA recognized a 7.4 kb transcript which was expressed at high levels in the retina and at low levels in the cerebellum but otherwise not in brain, spinal cord, retinal pigment epithelium, ovary, kidney, spleen or skeletal muscle (64-66). Sequence analysis of the initial clone and of other clones subsequently isolated, showed that the cDNAs fall into two major classes, I and II, derived by alternative splicing (Figure 1A) (66). Class II is 4705 nucleotides long and contains a complete, long open reading frame that begins with the methionine start codon. Several stop codons are present in all reading frames upstream of the proposed translation start site. This class represents a truncated form, at the NH_2-terminus, of class I. Class I is 4531 nucleotides long and still contains an incomplete long open reading frame suggesting that it represents a PLC class with an extended NH_2-terminus like the NorpA gene product (Figure 1A). When we compare class II with I, class II has an out of frame deletion of 58 nucletides (P3 domain) and a different 5′-end sequence, both derived probably from an alternative splicing mechanism (Figure 1A).

Interestingly, P1 and P2 domains of classes II and I, respectively, seem to have acceptor splice consensus sequences, $Y_{(n)}$T/CAGN (Y, represents a pyrimidine), at the 3′-boundaries, raising the possibility that these domains are produced by splicing within the exon. In addition, there are two other minor cDNA subclasses, A and B, which differ only by the presence, or absence, of 36 nucleotides, P4 domain, within the open reading frame (Figure 1A). Similarly, P3 domain has a consensus sequence of an acceptor splice site (CAG) at the 3′-boundary leading also to the possibility that this domain is produced through the use of a splice site within an exon. A similar instance of usage of a criptic splice site within an exon has been suggested in the generation of inositol 1,4,5-triphosphate receptor subtypes in the mouse brain (67).

PRIMARY STRUCTURE ANALYSIS of NorpA COGNATE PROTEINS

Conceptual translation of the cDNAs, thus, generates four potential isoforms of the bovine protein: Class II, Forms A and B (907 and 919 amino acids, respectively), and Class I, Forms A and B (1011 and 1023 amino acids, respectively). Unlike any other known PLCs,

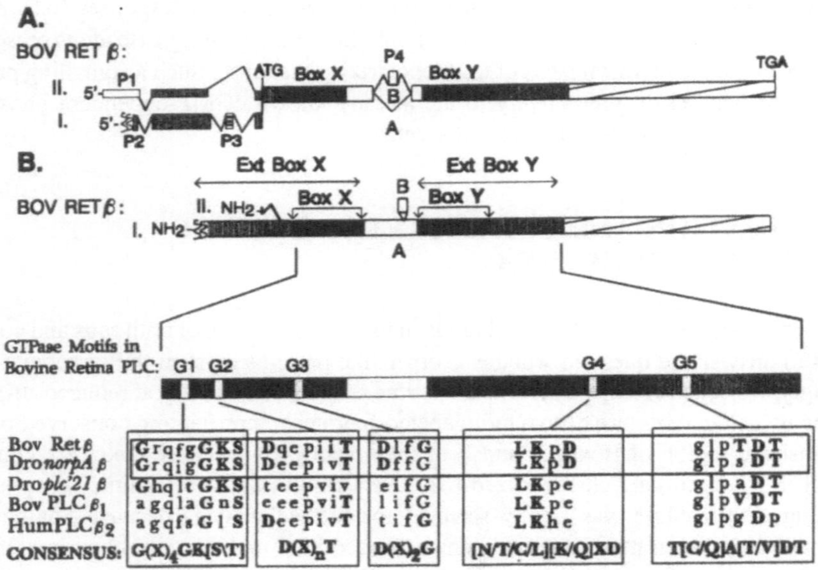

Figure 1. Panel A. Structure of bovine retinal PI-PLC cDNA Classes I and II showing putative, alternatively spliced regions. cDNA segments corresponding to regions that are putatively spliced are labeled P1, P2, P3, and P4. Non-coding regions are shown as thin bars. P1 and P2 are 5′-end sequences unique to Class II and Class I cDNAs, respectively. P3 is present in Class I cDNA but not in Class II cDNA. Because P3 represents a frame-shift deletion for class II cDNA, the same 5′sequences that bare coding in Class I cDNA is non-coding for Class II cDNA. P4 is present in Form B but not in Form A of either Class I or Class II cDNA. Boxes X and Y are regions of highest sequence conservation found in known PI-PLCs. **Panel B.** Structure of protein-coding regions of bovine retinal PI-PLC and identification of conserved GTPase sequence motifs. Ext Box X and Ext Box Y are extended regions of sequence conservation found in known PI-PLCs (Fig. 2). Conserved sequence motifs of the GTPase superfamily, G1, G2, G3, G4, and G5, are compared among five PI-PLCβs at the bottom of the figure. The amino acid residues that match the consensus sequences are capitalized and boldfaced. X is any amino acid residue. Bov Retβ, bovine retinal PI-PLCs (66); Dro *norpA*β, *Drosophila norpA* protein (27); Dro *plc-21*β, *Drosophila plc-21* protein (68); Bov PLCβ1, bovine brain PLCβ (69); Hum PLCβ2, human PLCβ2 (70).

class II retinal PLC lacks most of the N-terminal region, making it a unique PLC. Comparison of the deduced protein sequence with other known PI-PLC sequences showed that (Figure 2): 1) all these new protein isoforms have a high sequence similarity to to PI-PLCs, especially within the two conserved domains X and Y, and their flanking regions, extended boxes X and Y; 2) this bovine retina-specific PI-PLC has the highest sequence similarity to the NorpA protein of any PLCs known to date, 76% and 78% identity in boxes X and Y, respectively; 3) its overall primary structure is that of the beta class containing a long C-terminus sequence and an "acidic" region between boxes X and Y.

When we analyzed the primary structure of all four isoforms of the bovine retinal PLCs for the existence of protein sequence motifs, we observed the presence of all five conserved GTPase sequence motifs, G1, G2, G3, G4, and G5, found in proteins of the GTPase superfamily (73). These motifs are also conserved in *Drosophila* NorpA protein but not in any other PI-PLCs (Figure 1B), suggesting that GTP might play an important role in the function or regulation of these retinal PLCs. In addition, other sequence motifs were present in the bovine NorpA homologue proteins. Most of the beta PI-PLCs cloned to date contain one or more PEST regions suggesting that these proteins are likely to be metabolically very labile (74). PEST proteins are characterized by segments containing negatively charged residues or residues that become so upon phosphorylation. The region between boxes X and Y of this retinal PI-PLC consists of residues that are over 50% negatively charged or which can become so upon phosporylation. Although the C-terminus region of this protein does not exhibit high sequence similarity to that of *norpA*, both have a high number of serines and threonines. The C-terminal region has been implicated in a regulatory role in the activation of these enzymes by G-proteins. The low similarity of this region among PI-PLCs of b class might reflect different interactions with different subclasses of trimeric G-proteins.

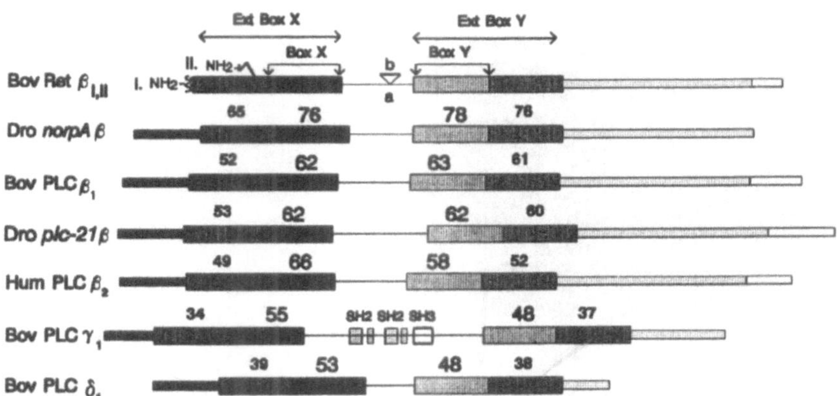

Figure 2. Amino acid conservation between bovine retinal PI-PLC and other PI-PLCs. Percent amino acid identity is compared between bovine retinal PI-PLC and six other representative PI-PLCs within four regions: Box X, Extended Box X, Box Y and Extended Box Y. The bovine retinal PI-PLC structure is shown at the top. Large numbers above the boxes, represent the identity between the retinal PLC and those of other PI-PLCs within boxes X or boxes Y. Small numbers above the boxes, represent the identity between the retinal PLC and those of other PI-PLCs within the entire extended boxes X or boxes Y. The identity is highest between bovine retinal PI-PLC and the *Drosophila norpA* protein. Bov Retβ, bovine retinal PI-PLCs (66); Dro *norpA*β, *Drosophila norpA* protein (27); Dro *plc-21*β, *Drosophila plc-21* protein (68); Bov PLCβ1, bovine brain PLCβ (69); Hum PLCβ2, human PLCβ2 (70); Bov PLCg$_1$, bovine brain PLCg$_1$ (71); Bov PLCd$_1$, bovine brain PLCd$_1$ (72).

TISSUE *IN SITU* AND IMMUNOLOCALIZATION OF THE NorpA HOMOLOGOUS BOVINE PRODUCTS IN RETINA

Localization of the retina PLC transcripts by *in situ* hybridization revealed that the mRNAs were localized in the outer nuclear layer, innner nuclear layer and ganglion cell layers (Figure 3) (66). The staining of the outer nuclear layer was more intense to that region adjacent to the external limiting membrane where over 50 % of cone nuclei are known to lie (75). Therefore, this pattern might reflect a cone-specific distribution. Immunolocalization of the NorpA-homologue PLCs in retina cryosections with antibodies generated against poorly conserved regions of the retinal PLC showed that in fact, the proteins are localized in the inner and outer segments of cones but not of rod photoreceptor cells (76). In addition, the retinal PLCs were also present in the outer and inner plexiform layers as well as horizontal and/or bipolar cells and ganglion cells of the retina.

The NorpA protein of *Drosophila* is the only other known PLC that is expresed predominantly in the retina (27, 77). A recent report however, has shown that the protein is also expressed in the brain and male (but not female) abdomen albeit at a lower level than that in the eye of the fly (78). Northern blot analysis with the same probe used to detect the bovine retinal 7.4 kb transcript detected, in addition, a putative alternatively spliced and lower molecular weight transcript in the mouse testis but not bovine ovary (66). Therefore, the tissue distribution of the bovine retinal PLCs and that of the NorpA *Drosophila* protein appear to parallel each other.

Western blot analysis showed that Class II and I NorpA PLCs are localized, respectively, in the membrane and cytosol fraction of retinal extracts. An antibody against P4 domain predominantly detects an aproximately 100 kD band in the membrane retinal fraction, while an antibody against the C-terminal region detects a 130-140 kD band in the

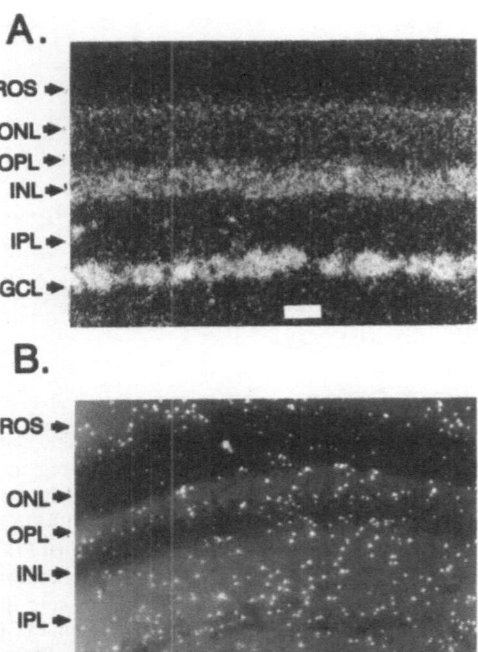

Figure 3. Tissue *in situ* hybridization. Panels A and B show, respectively, light and dark field micrographs of a bovine retinal cryosection probed with antisense cRNA sythesized from the cDNA fragment which comprises Box Y of bovine retinal PI-PLC. Panel C shows a dark field, control micrograph probed with sense cRNA generated from the same cDNA fragment. ROS, rod outer segment; OPL, outer plexiform layer; ONL, outer nuclear layer; IPL, inner plexiform layer; INL, inner nuclear layer; GCL, ganglion cell layer. Scale bar: 30um.

cytosolic fraction. These results are in accordance with the predicted molecular weight of 104 kD for class II PLC and with the biochemical purification from mouse cerebellum of a 97 kD PLC with peptide sequence identical to the deduced protein sequence of class II cDNA (79). Subsequently, Rhee laboratory have purified the 130 kD PLC from bovine retina with identical peptide sequence to class I PLC (80).

CONCLUSION

Taken together, these results show: 1) that the bovine retinal PLCs are closely related to the NorpA protein in structure and probably function, and 2) that cone but not rod photoreceptors outer and inner segments (as well as other retinal neurons) contain a PI-PLC transduction pathway that is mediated by a norpA-like PLC signalling pathway that is probably light activated as suggested by previous biochemical reports. Cone outer segments are the structures specialized to carry out photoreception and phototransduction. The presence of NorpA-homologous PLCs in cone outer segments suggests that one or more of these new retinal PLC isoforms could be involved in phototransduction and that the light transduction pathway in cones is not identical to that in rods as it had been assumed to date. For example, cones or a certain subclass of cone photoreceptor cells could employ a PI-PLC pathway in parallel with a cGMP phosphodiesterase light-activated cascade similar to that of rods. Alternatively, these retinal PLCs could be involved in some other signaling cascade. Finally, since *norpA* mutations cause retinal degenerations in *Drosophila* in a light-dependent manner, a defect in this PLC could lead to specific degeneration of cone photoreceptor cells. Degeneration of the macula, the cone dominated central patch of the retina, does occur and leads to severe visual impairment in humans.

PROCESSING COMPONENTS OF THE PHOTOTRANSDUCTION MACHINERY

In order for the phototransduction machinery of invertebrates and vertebrates to work, all of their protein components need to be correctly processed and sorted in photoreceptor cells. This requires the existence of a processing machinery, in which a defect could seriously impair the function of the visual cascade. One might ask for example, how the vast amounts of opsin in vertebrate and invertebrate photoreceptors are properly processed and sorted to the outer segment subcellular compartment.

Once again, *Drosophila* has proved to be an excellent system to dissect out genetically this processing machinery. William Pak and coworkers have isolated several mutations, named *nina* (*n*either *i*nactivation *n*or *a*fterpotential) based on their ERG phenotype (3, 81). These fall into several complementations groups (*ninaA-H*), all of which cause severe reductions in the amount of rhodopsin in the *Drosophila* compound eye and consequently, a reduced prolonged depolarizing afterpotential (PDA). However, mutations in two of these complementation groups, *ninaA* and *ninaE,* only affected the rhodopsin content in a subset of photoreceptor cells, R1-6 (82). *ninaA gene* was shown to encode for a photoreceptor-specific and ER resident small transmembrane protein with strong homology to the cyclophilin proteins (83-85). The cyclophilin domain of NinaA is thought to lie inside the lumen of the ER while a short stretch of amino acids at the C-terminal region face the cytoplasm. Discovery of this gene came as a big surprise as over a century of research on vertebrate rhodopsin gave no clue that a cyclophilin-like protein would play a role in the photoreceptor cells.

The *in vivo* function of the NinaA protein is not yet well understood, but reports that the enzyme, peptidyl prolyl cis-trans isomerase (PPIase), which is identical to cyclophilins, accelerates protein folding *in vitro*, led to the proposal that NinaA might be involved in the folding of opsin (83-87). However, mutations in the *ninaA* gene affecting regions lying outside the PPIase domain, for example, in the C-terminal cytoplasmic domain still lead to reduced levels of R1-6 opsins (83, 88). It is therefore, conceivable that the PPIase activity of cyclophilins observed *in vitro* might not reflect the functional role of the NinaA protein (and other cyclophilins) *in vivo*. Another puzzling observation is that although *ninaA* is expressed in all subclasses of photoreceptor cells, *ninaA* mutations only affect R1-6 but not R7 and R8 photoreceptor opsins (82, 85). This protein could have an additional and/or alternative function to the proposed PPIase activity. Recently, Zucker and coworkers showed that this protein forms a stable complex with R1-6 rhodopsin *in vivo* suggesting therefore, that the NinaA protein might work as a chaperone for R1-6 opsin (89). Thus, to date, the NinaA protein is the only molecularly characterized member of the large cyclophilin family for which the specific substrate and a probable function *in vivo* have been identified.

If *Drosophila* rhodopsin requires a cyclophilin for its synthesis, transport, or maturation, one might ask whether mammalian opsin also require cyclophilins in a similar role. However, no evidence has ever been reported that suggests the need for a cyclophilin-like protein in the processing of vertebrate opsins.

PRELIMINARY IDENTIFICATION AND CHARACTERIZATION OF *ninaA*-LIKE cDNAs

To investigate the role of cyclophilins in the processing and sorting of vertebrate opsins, we first asked whether NinaA-like proteins exist in the vertebrate retina. We screened a bovine retina cDNA library at reduced stringency conditions by using *ninaA* cDNA of *Drosophila* as a probe. Sequence analysis of the *ninaA*-hybridizing bovine cDNA clones showed that they represent two isoforms, Type I and II, of a novel *ninaA*-related bovine protein that are generated by alternative RNA splicing (90). In Northern analysis, Type I and II cDNA probes detected, respectively, 10 kb and 3.1 kb transcripts that are present in the retina but not in other bovine tissues tested. Longer exposures times detect the larger transcript also in the brain. *In situ* hybridizations of antisense RNA probes to bovine retinal sections showed that both Type I and II RNAs are found in all nuclear layers including the photoreceptor nuclear layer. Immunocytochemical results suggest that the proteins are localized in the cone but not rods inner and outer segments of photoreceptor, outer and inner plexiform layer and some ganglion cells.

The deduced sequences of both Type I and Type II isoforms of the new bovine proteins have near their C-terminus a conserved "cyclophilin domain," which is, on the average, 60% identical to previously characterized cyclophilins. Otherwise, they are quite unlike any other cyclophilin characterized to date. They are characterized by the presence of extended N- terminal sequences, which are over 900 and 100 amino acids long for Type I and II, respectively. The very long N-terminal region of the Type I protein contains at least two sets of tandem repeat domains and another cyclophilin-like domain (CLD), where most of the structural residues present in the C-terminal cyclophilin domain are conserved. Biochemical analysis of these cyclophilins expressed in *E. coli* showed that 1) all exhibit low PPIase and CsA-binding activities; 2) type I cyclophilin contained a conserved domain responsible for interacting with a 24 kDa GTP-binding protein; 3) they are capable of binding red/green but not the blue cone or rod opsins in a GTP-independent manner in the presence of crude retinal extracts.

CONCLUSION

The results, thus, identify a novel class of mammalian cyclophilins with unique primary structures and retina-specific tissue localization. The cone-specific localization among the photoreceptor cells suggests that these cyclophilins are involved in retina-specific functions. One of these functions could be specific interactions with opsin during its synthesis or maturation, as in the case of the NinaA protein. It has been suggested that the NinaA protein acts as a folding catalyst for rhodopsin/opsin during its synthesis or maturation. However, the NinaA protein, as well as the bovine retinal proteins identified in this study, the only other retina-specific cyclophilins known, are neither very abundant nor have high PPIase activity. Thus, the notion that they act as folding catalysts for rhodopsin may require revision. For example, interactions with the long-wavelength cone opsin may involve some other specialized cell function during the synthesis or maturation of opsin, such as subcellular targeting and/or chaperoning of opsin.

ACKNOWLEDGMENTS

The authors thank G. Travis, S. Azarian, D. Horvath and W. Kedzierski for helpful commments with the preparation of the manuscript. This work was supported by a grant from the National Eye Institute (EY00033). P.A.F. was supported in part by a North Atlantic Treaty Organization (NATO) postdoctoral fellowship.

REFERENCES

1. Pak, W. L., Grossfield, J., and White, N.V. (1969). Nonphototactic mutants in a study of vision of *Drosophila*. Nature (London) *222*, 351-354.
2. Pak, W. L., Grossfield, J., and Arnold, K. (1970). Mutants of the visual pathway of *Drosophila melanogaster*. Nature (London) *227*, 518-520.
3. Pak, W. L. (1979) Study of photoreceptor function using *Drosophila* mutants. In Breakefield XO (ed): "Neurogenetics: Genetic Approaches to the Nervous System," New York: Elsevier North Holland, pp 67-99.
4. Pak, W. L. (1991) Molecular genetic studies of photoreceptor function using *Drosophila* mutants. In D. Farber and G. Chader (ed): "Molecular Biology of the Retina: Basic Clinically Relevant Studies," New York: Wiley-Liss, Inc. pp 1-32.
5. Ranganathan, R., Harris, W. A., and Zucker, C.S. (1991). The molecular genetics of invertebrate phototransduction. Trends in Neuroscience *14*, 486-493.
6. Payne, R. (1986). Phototransduction by microvillar photoreceptors of invertebrates: mediation of a visual cascade by inositol triphosphate. Photobiochem. Photobiophys. *13*, 373-397.
7. Pugh Jr., E. N., and Lamb, T. D. (1990). Cyclic GMP and calcium: the internal messengers of excitation and adaptation in vertebrate photoreceptors. Vision Res. *30*, 1923-1948.
8. Stryer, L. (1991). Visual excitation and recovery. J. Biol. Chem. *266*, 10711-10714.
9. Kaupp, U. B., and Koch, K.-W. (1992). Role of cGMP and Ca^{2+} in vertebrate photoreceptor excitation and adaptation. Annu. Rev. Physiol. *54*, 153-175.
10. Wright, A. F. (1992). New insights into genetic eye disease. Trends in Genetics *8*, 85-91.
11. Hargrave, P. A. and McDowell, J. H. (1992). Rhodopsin and phototransduction: a model system for G-protein-linked receptors. The FASEB J. *6*, 2323-2331.
12. O'Tousa, J. E. and Pak, W. L. (1988). Molecular analysis of visual pigments genes. Photobiochem. Photobiol. *47*, 877-882.
13. Palczewski, K., Buczylko, J., Lebioda, L., Crabb, J. and Polans, A. (1993). Identification of the N-terminal region in rhodopsin kinase involved in its interaction with rhodopsin. J. Biol. Chem. *268*, 6004-6013.
14. Kelleher, D. J. and Jonhson, G.L. (1986). Phosphorylation of rhodopsin by protein kinase C in vitro. J. Biol. Chem. *261*, 4749-4757

15. Doza, Y.N., Minke, B.,Chorev, M. and Selinger, Z. (1992). Characterization of fly rhodopsin kinase. Eur. J. Biochem. *209*, 1035-1040.
16. Dolf, P. J., Ranaganathan, R., Colley, N. J., Hardy, R. W., Socolich, M. and Zucker, C. S. (1993). Arrestin function in inactivation of G protein-coupled receptor rhodopsin *in vivo*. Science *260*, 1910-1916.
17. Gurevich, V. V., and Benovic, J. L. (1993). Visual arrestin interaction with rhodopsin. J. Biol. Chem. *268*, 11628-11638.
18. Payne, R., Flores, T. M. and Fein, A. (1990) Feedback inhibition by calcium limits the release of calcium by inositol triphosphate in Limulus ventral photoreceptors. Neuron *4*, 547-555.
19. Sandler, C. and Kirschfeld, K. (1991). Light-induced extracellular calcium and sodium concentration changes in the retina of Calliphora: involvement in the mechanism of light adaptation. J. Comp. Physiol. *169*, 229-311.
20. Hardie, R. C. and Minke, B. (1993). Novel Ca^{2+} channels underlying transduction in drosophila photoreceptors: implications for phosphoinositide-mediated Ca^{2+} mobilization. Trends in Neuroscience *16*, 371-376.
21. Ranganathan, R., Bacskai, B., Tsien, R. and Zucker, C.S. (1994). Cytosolic calcium transients: spatial localization and role in *Drosophila* photoreceptor cell function. Neuron 13, 837-848.
22. Hardie, R. C. (1995). Photolysis of caged Ca^{2+} facilitates and inactivates but does not directly excite light-sensitive channels in *Drosophila* photoreceptors. J. Neuroscience 15, 889-902.
23. Fain, G. L. and Matthews, H. R. (1990). Calcium and the mechanism of light adaptation in vertebrate photoreceptors. Trends in Neuroscience *13*, 378-384.
24. Gray-Keller, M. P. and Detwiler, P. B. (1994). The calcium feedback signal in the phototransduction cascade of vertebrate rods. Neuron *13*, 849-861.
25. Lee, Y.-J., Shah, S., Suzuki, E., Zars, T., O'Day, P. and Hyde, D. R. (1994) The *Drosophila dgq* gene encodes a Ga protein that mediates phototransduction. Neuron *13*, 1143-1157.
26. Fung, B. K.-K. (1987) Transducin:structure, function, and role in phototransduction. In Osborne N. N., Chader G. J., eds. "Progress in Retinal Research," Oxford: Pergamon Press, *6*, 151-177.
27. Bloomquist, B. T., Shortridge, R. D., Schneuwly, S., Perdew, M., Montell, C., Steller, H., Rubin, G., and Pak, W. L. (1988). Isolation of a putative phospholipase C gene of *Drosophila*, *norpA*, and its role in phototransduction. Cell *54*, 723-733.
28. Hotta, Y., and Benzer, S. (1970). Genetic dissection of the *Drosophila* nervous system by means of mosaics. Proc. Natl. Acad. Sci. USA *67*, 1156-1163.
29. Inoue, H. Yoshioka, T. and Hotta, Y. (1985). A genetic study of inositol triphosphate involvement in phototransduction using *Drosophila* mutants. Biochem. Biophys. Res. Commun. *132*, 513-519.
30. Meyertholen, E. P., Stein, P. J., Williams, M. A., and Ostroy, S. E. (1987). Studies of the *Drosophila norpA* phototransduction mutant, II. Photoreceptor degeneration and rhodopsin maintenance. J. Comp. Physiol. *161*, 793-798.
31. Bowes, C., Li, T. Danciger, M., Baxter, L., Applebury, M., and Farber, D. (1990) Retinal degeneration in the *rd* mouse is caused by a defect in the b subunit of the rod cGMP-phosphodiesterase. Nature *347*, 677-680.
32. Lem, J., Flannery, J., Li, T., Applebury, M., Farber, D. and Simon, M. (1992) Retinal degeneration is rescued in transgenic *rd* mice by expression of the cGMP phosphodiesterase b subunit. Proc. Natl. Acad. Sci. USA *89*, 4422-4426.
33. Farber, D. B. and Lolley, R. N. (1974) Cyclic guanosine monophosphate: elevations in degenerating photoreceptor cells of the C3H mouse retina. Science *186*, 449-451.
34. Farber, D. B. and Lolley, R. N. (1976) Enzymatic basis for cyclic GMP accumulation in degenerative photoreceptor cells of mouse retina. J. Cyclic Nucleotide Res. *2*, 139-148.
35. Hardie, R. and Minke, B. (1992). The *trp* gene is essential for a light-activated Ca2+ channle in Drosophila photoreceptors. Neuron *8*, 643-651.
36. Phillips, A. M., Bull, A. and Kelly, L. E. (1992) Identification of a Drosophila gene encoding a calmodulin-binding protein with homology to the *trp* phototransduction gene. Neuron *8*, 631-642.
37. Yau, K.-W. (1994). Phototransduction mechanism in retinal rods and cones. Invest. Ophtalmol. Vis. Sci. *36*, 263-275.
38. Johnson, E. C., Robinson, P. R., and Lisman, J. E. (1986). Cyclic GMP is involved in the excitation of invertebrate photoreceptors. Nature *324*, 468-470.
39. Bacigalupo, J., Johnson, E., Vergara, C., and Lisman, J. (1991). Light-dependent channels from excised patches of *Limulus* ventral photoreceptors are open by cGMP. Proc. Natl. Acad. Sci. USA *88*, 7938-7942.
40. Fein, A., Payne, R., Corson, D. W., Berridge, M. J. and Irvine, R. (1984) Photoreceptor excitation and adaptation by inositol 1,4,5-triphosphate. Nature *311*, 157-160.

41. Brown, J., Rubin, L., Ghalayini, A., Tarver, A., Irvine, R., Berridge, M. and Anderson, R. E. (1984) Myo-inositol polyphosphate may be a messenger for visual excitation in *Limulus* phootreceptors. Nature *311*, 160-162.
42. Shin, J., Richard, E. and Lisman, J. E. (1993). Ca^{2+} is an obligatory intermediate in the excitation cascade of Limulus photoreceptors. Neuron, *11*, 845-855.
43. Anderson, R. E., and Brown, J. E. (1988). Phosphoinositides in the retina. In Osborne N. N. and Chader G. J., eds. "Progress in Retinal Research," Oxford: Pergamon Press, *9*, 211-228.
44. Das, N. D., Yoshioka, T., Samuelson, D., and Shichi, H. (1986). Immunocytochemical localization of phosphatidyl-4,5-biphosphate in dark- and light-adapted rat retinas. Cell Struc. Funct. *11*, 53-63.
45. Gehm, B. D., and Mc Connel, D. G. (1990). Phosphatidylinositol-4,5-biphosphate phospholipase C in bovine rod outer segments. Biochemistry *29*, 5447-5452.
46. Ghalayini, A., and Anderson, R. E. (1984). Phosphatidylinositol 4,5-bisphosphate: light-mediated breakdown in the vertebrate retina. Biochem. Biophys. Res. Comm. *124*, 503-506.
47. Hayashi, F., and Amakawa, T. (1985). Light-mediated breakdown of phosphatidylinositol-4,5-biphosphate in isolated rod outer segments of frog photoreceptor. Biochem. Biophy. Res. Comm. *128*, 954-959.
48. Jelsema, C. L. (1989). Regulation of phospholipase A2 and phospholipase C in rod outer segments of bovine retina involves a common GTP-binding protein but different mechanisms of action. Ann. N. Y. Acad. Sci. *559* , 158-177.
49. Jelsema, C. L., and Axelrod, J. (1987). In Sensory Transduction (Discussions in Neurosciences), Hudspeth, A., Macleish, P., Margolis, F. eds. (Foundations for the Study of the Nervous System, Geneva), Vol. 4, 79-84.
50. Millar, F., and Hawthorne, J. (1985). Polyphosphoinositide metabolism in response to light stimulation of retinal rod outer segments. Biochem. Soc. Trans. *13*, 984-985.
51. Millar, F. A., Fisher, S. C., Muir, C. A., Edwards, E., and N., H. J. (1988). Polyphosphoinositide hydrolysis in response to light stimulation of rat and chick retina and retinal rod outer segments. Biochem. Biophys. Acta *970*, 205-211.
52. Yoshioka, T., and Inoue, H. (1987). Inositol phospholipid in visual excitation. Neurosc. Res. Suppl. *6*, S15-S24.
53. Ghalayini, A., Tarver, A., Mackin, W.M., Koutz, C.A. and Anderson, R. E. (1991) Identification and immunolocalization of phospholipase C in bovine rod outer segments. J. Neurochem. *57*, 1405-1412.
54. Peng, Y.-W., Sharp, A. H., Snyder, S. H., and Yau, K.-W. (1991). Localization of the inositol 1,4,5-triphosphate receptor in synaptic terminals in the vertebrate retina. Neuron *6*, 525-531.
55. Day, N. S., Koutz, C. A. and Anderson, R. E. (1993). Inositol-1,4,5-trisphosphate receptors in the vertebrate retina. Current Eye Res. *12*, 981-992.
56. Bonigk, W., Altenhofen, W., Muller, F., Dose, A., Illing, M., Molday, R. S., and Kaupp, U. B. (1993). Rod and cone photoreceptor cells express distinct genes for cGMP-gated channels. Neuron *10*, 865-877.
57. Haynes, L. W., and Yau, K.-W. (1990). Single channel measurement from the cGMP-activated conductance of catfish retinal cones. J. Physiol. *429*, 451-481.
58. Lerea, C. L., Bunt-Milam, A. H., and Hurley, J. B. (1989). a-transducin is present in blue-, green-,and red-sensitive cone photoreceptors in the human retina. Neuron *3*, 367-376.
59. Li, T., Volpp, K., and Applebury, M. L. (1990). Bovine cone photoreceptor cGMP phosphodiesterase structure deduced from a cDNA clone. Proc. Natl. Acad. Sci. USA *87*, 293-297.
60. Tomita, T. (1965). Electrophysiological study of the mechanisms subserving color coding in the fish retina. Cold Spring Harb. Symp. Quant. Biol. *30*, 559-566.
61. Solessio, E., and Engbretson, G. A. (1993). Antagonistic chromatic mechanisms in photoreceptors of the parietal eye of lizards. Nature *364*, 442-445.
62. Rohlich, P., van Veen, Th. and Szel, A. (1994). Two different visual pigments in one retinal cone cell. Neuron *13*, 1159-1166.
63. Rhee, S.-G., Suh, P.-G., Ryu, S.-H., and Lee, S. (1989). Studies of inositol phospholipid-specific phospholipase C. Science *244*, 546-550.
64. Ferreira, P., Shortridge, R. and Pak, W. (1991). Identification and characterization of *norpA*-like retina-specific bovine cDNA". Molecular Neurobiology of *Drosophila*, pg. 51, Cold Spring Harbor, New York.
65. Ferreira, P. and Pak, W. (1992). Retina-specific bovine cDNAs encoding homologs of the *norpA* and *ninaA* proteins of *Drosophila*. Society for Neuroscience, vol. 18, pg.137.
66. Ferreira, P., Shortridge, R., and Pak, W. (1993). Distinctive subtypes of bovine phospholipase C that have preferential expression in the retina and high homology to the *norpA* gene product of *Drosophila*. Proc. Natl. Acad. Sci. USA *90*, 6042-6046.

67. Nakagawa, T., Okano, H., Furuichi, T., Aruga, J., and Mikoshiba, K. (1991). The subtypes of the mouse inositol 1,4,5-triphosphate receptor are expressed in a tissue-specific and developmentally specific manner. Proc. Natl. Acad. Sci. USA *88*, 6244-6248.
68. Shortridge, R., Yoon, J., Lending, C., Bloomquist, B., Perdew, M., and Pak, W. (1991). A *Drosophila* phospholipase C gene that is expressed in the central nervous system. J. Biol. Chem. *266*, 12474-12480.
69. Katan, M., Kriz, R., Totty, N., Philip, R., Moldrum, E., Aldape, R., Knopf, J., and Parker, P. (1988). Determination of the primary structure of PLC-154 demonstrates diversity of phosphoinositide-specific phospholipase C activities. Cell *54*, 171-177.
70. Park, D., Jhon, D.-Y., Kriz, R., Knopf, J. and Rhee, S. G. (1992). Cloning, sequencing, expression, and G_q-independent activation of phospholipase C-b2. J. Biol. Chem. *267*, 16048-16055.
71. 71. Stahl M.L., Ferenz, C.R., Kelleher, K.L., Kriz, R.W., Knopf, J.L. (1988). Sequence similarity of phospholipase C with the non-catalytic region of src. Nature *332*, 269-272.
72. Shu, P.-G., Ryu, S., Moon, K.H., Suh, H.W. and Rhee, S. G. (1988). Cloning and sequencing of multiple forms of phospholipase C. Cell *54*, 161-169.
73. Bourne, H. R., Sanders, D. A., and McCormick, F. (1991). The GTPase superfamily: conserved structure and molecular mechanism. Nature (London) *349*, 117-127.
74. Rechsteiner, M. (1990). PEST sequences are signals for rapid intracellular proteolysis. Sem. Cell Biol. *1*, 433-440.
75. Carter-Dawson, L. and LaVail, M. (1979). Rods and cones in the mouse retina. J. Comp. Neur. 188, 245-262.
76. Ferreira, P. and Pak, W. (1994). Bovine phospholipase C highly homologous to the *norpA* protein of *Drosophila* expressed specifically in cones. J. Biol. Chem., 269, 3129-3131.
77. Schneuwly, S., Burg, M., Lending, C., Perdew, M., and Pak, W. L. (1991). Proprieties of photoreceptor-specific phospholipase C encoded by the *norpA* gene of *Drosophila melanogaster*. J. Biol. Chem. *266*, 24314-24319.
78. Zhu, L., McKay, R., and Shortridge, R. (1993). Tissue-specific expression of phopholipase C encoded by the *norpA* gene of *Drosophila melanogaster*. J. Biol. Chem. *268*, 15994-16001.
79. Min, D. S., Kim, D., Lee, Y., Seo, J., Suh, P.-G. and Ryu, S. H. (1993) Purification of a novel phospholipase C isozyme from bovine cerebellum. J. Biol. Chem. *268*, 12207-12212.
80. Lee, C.-W., Park, D. J., Lee, K.-H., Kim, C. G., and Rhee, S. G. (1993). Purification, molecular cloning, and sequencing of phospholipase C-b4. J. Biol. Chem., *268*, 21318-21327.
81. Stephenson, R. S., O'Tousa, J., Scavarda, N. J., Randall, L. L., and Pak, W. L. (1983). *Drosophila* mutants with reduced opsin content. In D. J. In Cosens, & D. Vince-Price (Ed.), The Biology of Photoreception (pp. 447-501). Soc. Exp. Biol., Cambridge, U.K.: Cambridge University Press.
82. Larrivee, D. C., Conrad, S. K., Stephenson, R. S., and Pak, W. L. (1981). Mutation that selectively affect rhodopsin concentration in the peripheral photoreceptors of *Drosophila*. J. Gen. Physiol. *78*, 521-545.
83. Schneuwly, S., Shortridge, R., Larrivee, D., Ono, T., Ozaki, M., and Pak, W. (1989). *Drosophila ninaA* gene encodes an eye-specific cyclophilin (cyclosporine A binding protein). Proc. Natl. Acad. Sci. USA *86*, 5390-5394.
84. Shieh, B.-H., Stamnes, M. A., Seavello, S., Harris, G. L., and Zuker, C. S. (1989). The *ninaA* gene required for visual transduction in *Drosophila* encodes a homologue of cyclosporin A-binding protein. Nature *338*, 67-70.
85. Stamnes, A. M., Shieh, B.-H., Chuman, L., Harris, L. G., and Zuker, C. S. (1991). The cyclophilin homolog *ninaA* is a tissue-specific integral membrane protein required for the proper synthesis of a subset of *Drosophila* rhodopsins. Cell *65*, 219-227.
86. Fisher, G., Wittmann-Liebold, B., Lang, K., Kiefhaber, T., and Schmid, F. X. (1989). Cyclophilin and peptidyl-prolyl cis-trans isomerase are probably identical proteins. Nature *337*, 476-478.
87. Takahashi, N., Hayano, T., and Suzuki, M. (1989). Peptidyl-prolyl cis-trans isomerase is the cyclosporin A-binding protein cyclophilin. Nature *337*, 473-475.
88. Ondek, B., Hardy, R., Baker, E., Stamnes, M., Shieh, B.-H., and Zuker, C. (1992). Genetic dissection of cyclophilin function. J. Biol. Chem. *267*, 16460-16466.
89. Baker, E.K., N.C. Colley, & Zuker, C.S. (1994). The cyclophilin homolog NinaA functions as a chaperone, forming a stable complex in vivo with its protein target rhodopsin. EMBO J. *13*, 4886-4895.
90. Ferreira, P. and Pak, W. (1993). Retina-specific bovine homologs of *ninaA*. Molecular Neurobiology of *Drosophila*, pg. 158, Cold Spring Harbor, New York.

31

RETINAL PATHOLOGY IN RETINITIS PIGMENTOSA

Considerations for Therapy

Ann H. Milam and Zong-Yi Li

Department of Ophthalmology RJ-10
University of Washington
Seattle, Washington 98195

INTRODUCTION

Retinitis pigmentosa (RP) causes primary degeneration of photoreceptors, followed by reactive changes in the retinal pigment epithelium (RPE) and Müller glia, death of inner retinal neurons, and atrophy of the retinal vasculature. Current strategies for retinal therapy include transplantation of normal photoreceptors, delivery of corrective genes to diseased photoreceptors, and electrical stimulation of inner retinal neurons [1-5]. This chapter will review the histopathology of the retina in RP, including the characteristic changes found in the photoreceptors, subretinal space, RPE and choriocapillaris, Müller glia, inner retinal neurons and blood vessels. Changes in these retinal components secondary to death of photoreceptors are important considerations in developing new therapies for RP.

PATHOLOGIC CHANGES IN RP RETINAS

Rods and Cones

Although the list of gene defects associated with RP is rapidly expanding, it is not known how any of the mutations lead to photoreceptor dysfunction and death. Typically the RP patient experiences nyctalopia as a young adult, followed by progressive loss of both rod and cone vision over the ensuing years to decades. Pathologic changes in the rods include shortening of their outer segments, as revealed by immunocytochemistry using anti-rhodopsin (Figure 1). Death of rods usually begins in the mid peripheral retina and progresses with time to involve the macula and far periphery. Rod cell death is accompanied by changes in the adjoining cones, including shortening of their outer segments, and ultimately, cone cell death. Cone degeneration occurs in RP caused by mutations in rod-specific genes, e.g. rhodopsin, which are presumably not expressed in cones (although cones as well as rods

Degenerative Diseases of the Retina, Edited by Robert E. Anderson et al.
Plenum Press, New York, 1995

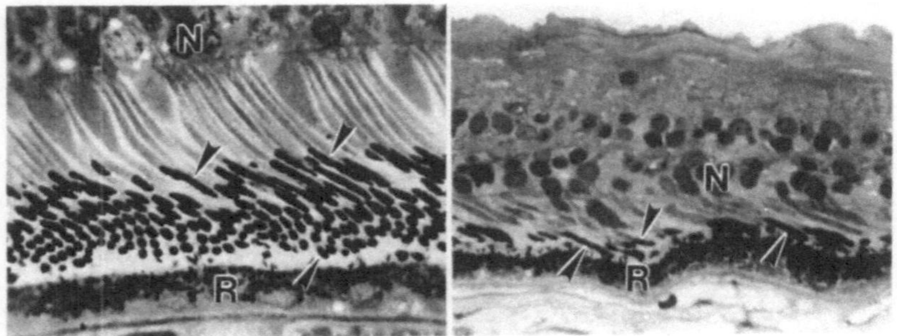

Figure 1. Localization of rhodopsin by immunogold cytochemistry in a normal retina (left) and the retina of a man with X-linked RP (FFB#114) [7] (right). Rods in the normal retina have rhodopsin immunoreactivity restricted to the long, thin outer segments (arrowheads). Rods in the RP retina have labeling for rhodopsin in the short outer segments (arrowheads); R, RPE; N, photoreceptor nuclei. X 500.

show expression of the β-gal gene when driven by the rhodopsin promoter in transgenic mice [6]). The maculas of RP patients who had diminished visual acuity prior to death show considerable loss of both rods and cones (Figure 2), and macular photoreceptors are often reduced to a monolayer of cone somata with very short outer segments (Figure 3).

We recently found [9] that rods in the peripheral regions of RP retinas give rise to long, rhodopsin-positive neurites (Figure 4) that are closely associated with hypertrophied Müller processes. The rod neurites bypass their normal postsynaptic targets, the horizontal and rod bipolar cells, and often reach the inner limiting membrane. Our electron microscopic studies indicate that the rod neurites contain numerous synaptophysin and SV-2 positive synaptic vesicles but do not form true synapses with any other cells [9]. We have documented rod neurite sprouting in fifteen retinas from donors with different genetic forms of RP [9]. Cones do not sprout neurites, although cone axons are abnormally long, reaching the inner plexiform layer (Figure 5). Fundus reflectometry has demonstrated that rhodopsin levels in certain RP patients are higher than suggested by their reduced rod outer segment function [10; 11]. Transmission from rods and cones to neurons of the inner retina can also be abnormal in RP [12-16]. It is unknown if the prominent rhodopsin-positive rod neurites and abnormal cone axons contribute to these functional abnormalities in RP.

To our knowledge, this is the first demonstration of rod neurite sprouting in an intact retina. The neurites are formed by rods in the peripheral regions of RP retinas that have survived death of neighboring rods, but they are not found in the maculas, which have also undergone significant photoreceptor death. Rods sprout similar neurites when cultured on Müller glia or NCAM [17-20]. The close association between rod neurites and Müller processes in the RP retinas suggests that they are stimulated by alterations in Müller surface molecules during the process of reactive gliosis. Neurite formation by rods may also reflect loss of inhibitory factor(s) present in normal retinas but lost *in vitro* and in the diseased RP retinas. Finally, rod neurite formation may represent a rod response to upregulation of one or more growth factors, as occurs in dystrophic animal retinas [21]. Retinal rods in the rodent models for RP show synaptic plasticity [22] but they do not sprout neurites. We have been unable to find rod neurites in older *rds* mice using the same immunolabeling methods that reveal rod neurites in RP retinas [9]. It appears that the phenomenon of rod neurite sprouting is peculiar to human rods or due to to changes in the microenvironment unique to RP retinas. Such changes in the RP retinal milieu are also important considerations for photoreceptor transplantation [2; 3; 23-25], which is based on the assumption that transplanted normal photoreceptors can form synapses on surviving inner retinal neurons and reestablish a

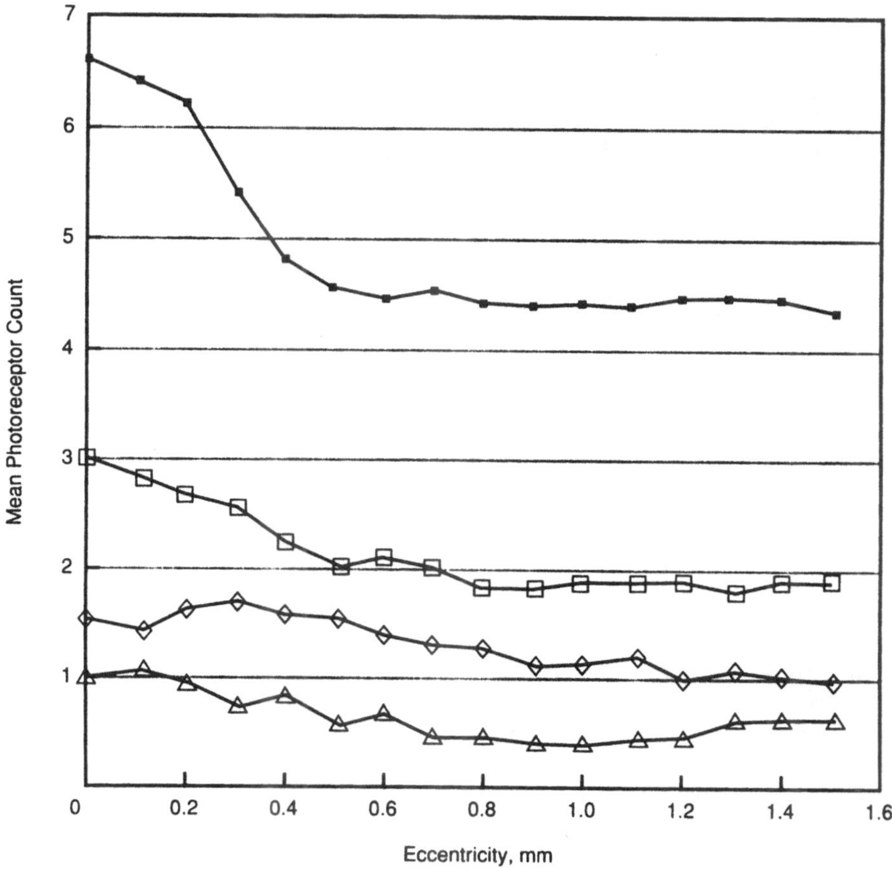

Figure 2. Mean photoreceptor counts at different eccentricities in the maculas of normal donors and patients with RP. Zero eccentricity corresponds to the center of the foveola. Photoreceptor counts from retinas of normal donors (closed squares, n = 20) are significantly higher (P < .05) than those from patients with different forms of RP, including autosomal dominant (open squares, n = 10), simplex (diamonds, n = 22), and X-linked (triangles, n = 9). Reproduced with permission [8].

functional visual pathway. However, changes in the retinal microenvironment that stimulate robust neurite sprouting by rods *in situ* may have the same effect on introduced photoreceptors. A putative increase in growth factor(s) in the diseased retina may serve to prolong survival of the transplanted photoreceptors but may also lead to rod neurite sprouting and cone axonal abnormalities as found in RP. The rod neurites bypass the normal rod axon target cells in the RP retinas, and the rods transplanted into such retinas may not receive the signals for axon termination and synapse formation necessary for their functional integration. Clearly, more complete characterization of the RP retinas is needed, because the dystrophic animal retinas do not show rod neurite sprouting and are thus not completely representative of the retinal pathology found in the human patients [9].

Subretinal Space

The subretinal space is bordered externally by RPE cells linked by *zonulae occludentes* and internally by photoreceptor and Müller cells linked by *zonulae adherentes* that constitute the

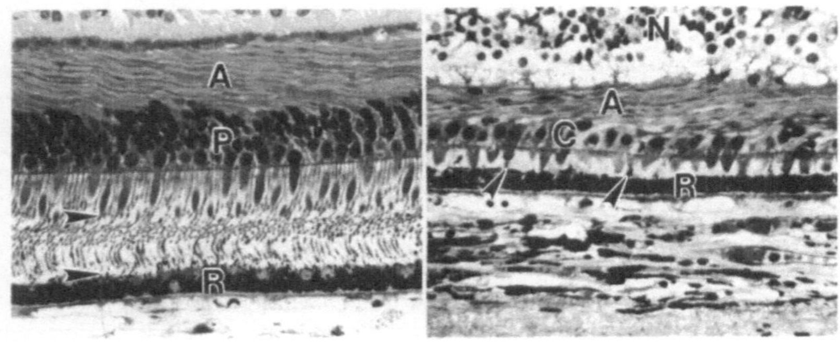

Figure 3. Macula of normal retina (left) contains four to six rows of photoreceptors with long, thin outer segments (between arrowheads); R, RPE; P, photoreceptor nuclei; A, Henle's layer of photoreceptor axons. The right panel shows the macula of a man with X-linked RP (FFB # 114) [7] whose visual acuity was 6/120. The photoreceptors are reduced to a monolayer of cone somata with tiny outer segments (arrowheads); R, RPE; C, cone nuclei; A, Henle's layer; N, inner nuclear layer; R, RPE. X 500.

external limiting membrane (ELM) [26]. In addition to rod and cone outer and inner segments, the subretinal space contains the interphotoreceptor matrix, a complex mixture of glycoproteins and glycosaminoglycans secreted by the bordering cells. Following death of all photoreceptors, the subretinal space is lost, due to RPE migration to the inner retina and extension of the Müller processes as far as Bruch's membrane. Earlier in the RP disease process, the volume of the subretinal space shrinks as the photoreceptor outer segments shorten. This change in the volume of the space may complicate efforts to inject living photoreceptors [23-25] and normal genes in viral vectors [4; 27; 28]. Ongoing photoreceptor death may also lead to alterations in the subretinal microenvironment that are unfavorable for survival of transplanted photoreceptors. Such changes due to death of rods may also be linked to degenerative changes in the cones, including shortening of their outer segments and ultimately, cone cell death. These effects may be due to release of toxic materials from dying rods [29] or to loss of factors secreted primarily by the rods, including interphotoreceptor retinoid binding protein (IRBP) [30]. IRBP is a major component of the

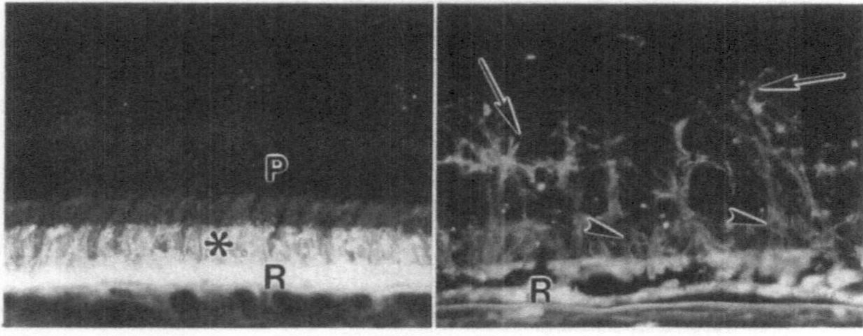

Figure 4. Immunofluorescence demonstration of rhodopsin in a normal retina (left) and the peripheral retina of an RP patient (FFB#424) with autosomal dominant RP due to the Q-64-ter rhodopsin mutation. In the normal retina, rhodopsin immunolabeling is restricted to the long rod outer segments (*); P, photoreceptor somata; R, RPE. In the RP retina, the rods are reduced to two rows of somata (arrowheads). Rhodopsin is delocalized to the surface membranes of the rod inner segments and somata, as well as long, beaded neurites (arrows) that extend into the inner retina; R, RPE. X 230.

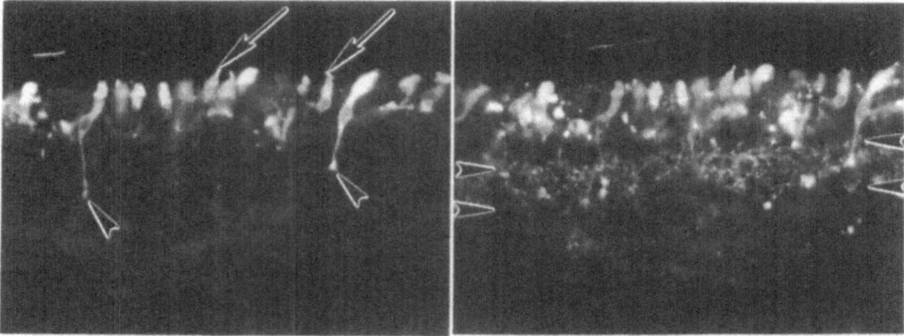

Figure 5. A section from the RP retina shown in Figure 4 has been double labeled with anti-cone transducin-alpha (left panel) and anti-synaptophysin (right panel). The cone outer segments (arrows) are very short and the cone axons (arrowheads) are abnormally long, terminating in the inner plexiform layer (right panel, between arrowheads), which is labeled with anti-synaptophysin. X 230.

interphotoreceptor matrix [31] that is thought to transport retinoids between the RPE and photoreceptors in the visual cycle. A reduction in rod-derived IRBP and RPE-recycled products of rod outer segment shedding/degradation could have significant effects on cones, simulating vitamin A deficiency and causing their outer segments to shorten. It is also possible that rods have direct trophic influences on cones and even on other rods [3; 32; 33] but such trophic factors have not been identified. These cellular interactions and the composition of the interphotoreceptor matrix are probably important for the survival of transplanted photoreceptors in RP retinas.

RPE and Choriocapillaris

Following death of rods and cones, the RPE cells often detach from Bruch's membrane and relocate to perivascular sites in the inner retina, forming so-called bone spicule pigment (BSP). The choriocapillaris is invariably lost from regions containing BSP. Because of the close interdependence of the choriocapillaris, RPE and photoreceptor cells, it is not clear if loss of this capillary bed is secondary to photoreceptor death or to absence of the RPE cells. The RPE cells that have migrated into the inner retina form a continuous epithelial monolayer around a blood vessel, most often a thin walled venule or capillary (Figure 6). The stimulus for RPE migration is unknown, but may include loss of metabolic products from photoreceptors or atrophy of the choriocapillaris following death of the photoreceptors. The phenomenon of RPE migration is highly significant for therapeutic protocols involving transplantation, because RPE cells as well as photoreceptors must be restored to RP retinal regions that contain BSP.

A layer of extracellular matrix is usually deposited between the RPE and the vascular endothelial cells (Figure 6). The deposits of extracellular matrix resemble Bruch's membrane *in situ*, having its characteristic five layers with a middle layer of elastin, which is normally absent from retinal blood vessels. The perivascular layer of extracellular matrix is often thick and contains prominent lipid and calcium deposits typical of the aging Bruch's membrane. In some cases, the ELM deposits occlude the vessel lumen (Figure 7). Thickening of the walls of the vessels and occlusion of their lumina correlates with the sclerosis and atrophy of the retinal vasculature observed funduscopically in RP. The thick perivascular deposits, particularly those rich in hydrophobic lipid and elastin, may compromise the blood supply to neurons of the inner retina and cause their death, an important consideration for design of a therapy based on direct electrical stimulation of neurons of the inner retina [5].

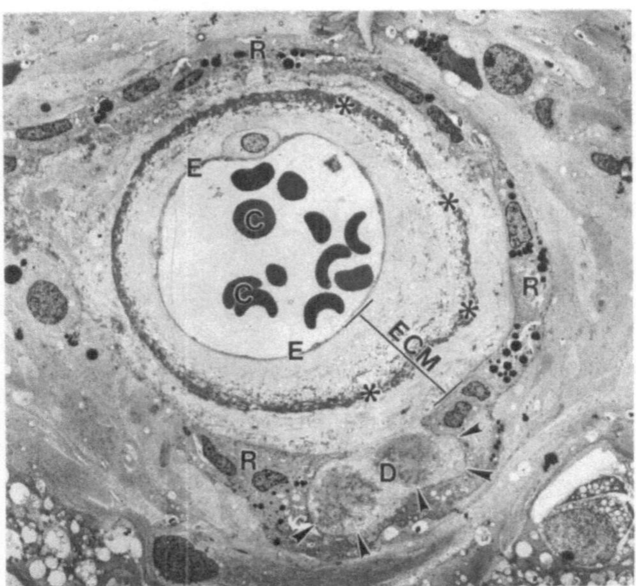

Figure 6. Electron micrograph of BSP, illustrating monolayer of RPE cells (R) encircling a retinal venule. Note thick deposit of extracellular matrix (ECM, bracket) between the RPE and endothelial cells (E), including a middle layer of elastin (*). C, erythrocytes; D and arrowheads, accumulated extracellular matrix resembling a druse. X1,384.

Müller Glia

Müller cells undergo reactive gliosis in response to photoreceptor damage in a number of diseases [34-37]. The Müller gliotic changes include cellular proliferation and hypertrophy, nuclear enlargment and migration, and increased expression of glial fibrillary acid protein. A similar process of reactive gliosis occurs in response to photoreceptor death in RP [7; 38] and the photoreceptor and outer plexiform layers are replaced by a glial scar formed by thick Müller cell processes (see [39]). This scar may constitute a mechanical barrier to synapse formation between transplanted photoreceptors and inner nuclear layer

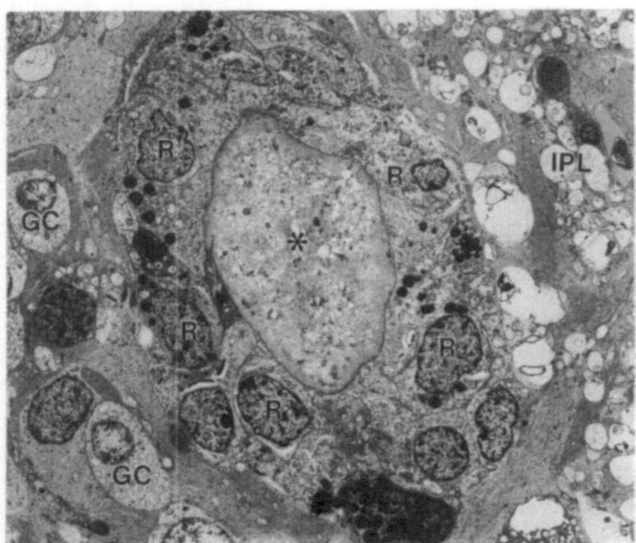

Figure 7. Electron micrograph of BSP, illustrating occlusion of the lumen of a retinal blood vessel by accumulated extracellular matrix (*) associated with the encircling layer of RPE cells (R). IPL, inner plexiform layer; GC, ganglion cells. X 2,284.

neurons, although there is evidence that axonal processes of transplanted photoreceptors can penetrate such a glial barrier [25]. After photoreceptor death, the ELM junctions between photoreceptors and Müller processes are lost, along with the subretinal space they define, and ELM-like junctions are formed between Müller processes opposite the migrated RPE cells in zones of BSP [40]. The Müller cells, along with migrated RPE cells and astrocytes, also contribute to formation of the preretinal membranes that are common in RP [41].

Inner Retinal Neurons

The only morphometric study of inner retinal neurons in RP [8] was limited to ganglion cells in the macula, which were significantly decreased in each form of RP examined (Figure 8). In general, the loss of ganglion cells paralleled the loss of photoreceptors, and was more severe in X-linked and autosomal dominant RP than in simplex RP [8]. The decreased ganglion cell counts were thought to reflect transneuronal degeneration, which may be more severe in X-linked RP because photoreceptor death usually occurs earlier in life. A more recently recognized cause of neuronal death is migration of RPE cells to

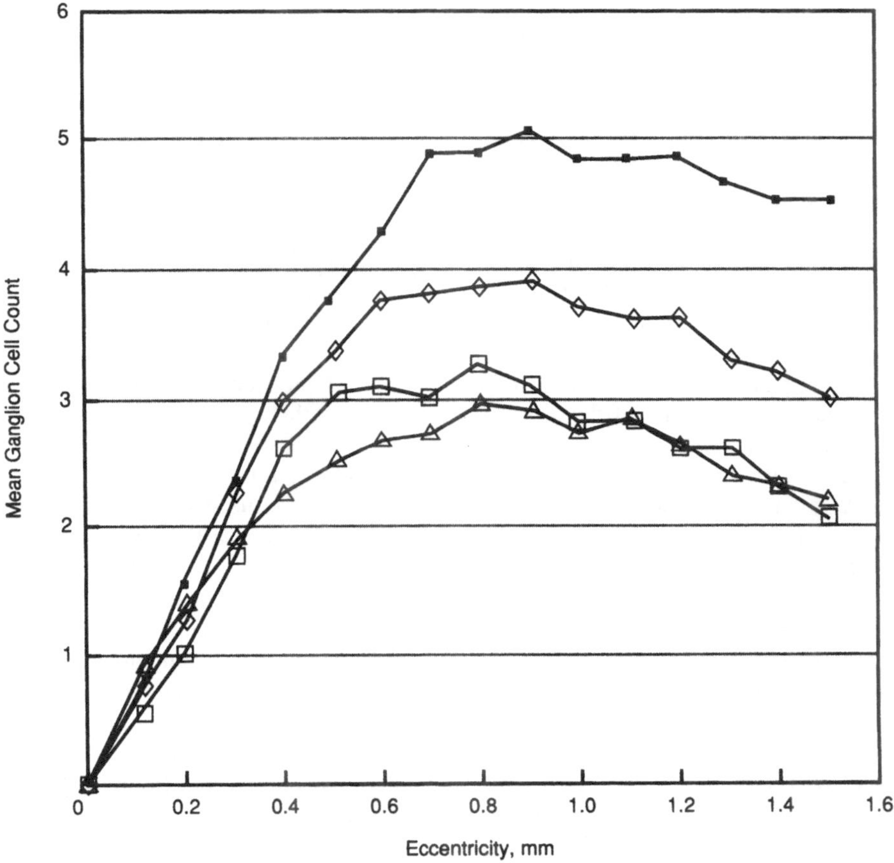

Figure 8. Mean ganglion cell counts at different eccentricities within the maculas of normal donors and patients with RP. Zero eccentricity corresponds to the center of the foveola. Ganglion cell counts from retinas of normal donors (closed squares, n = 20) were significantly higher ($P < .05$) than counts from patients with simplex RP (diamonds, n = 22), autosomal dominant RP (open squares, n = 10), and X-linked RP (triangles, n = 9). Reproduced with permission from [8].

perivascular sites in the inner retina (see above). It is likely that some death of inner retinal neurons results from a compromised blood supply due to perivascular deposits of extracellular matrix [40]. These inner retinal abnormalities are important considerations for any RP therapy based on direct stimulation of surviving ganglion cells.

Retinal Blood Vessels

Attenuation of retinal blood vessels is a hallmark of RP and historically was thought to result from diminished ganglion cell metabolism secondary to photoreceptor degeneration. Our studies indicate that the presence of RPE cells in the inner retina leads to remarkable alterations in the retinal vasculature. In addition to accumulations of perivascular extracellular matrix (see above), the vascular endothelial cells respond to the adjacent RPE cells by developing fenestrations, which leak serum proteins [40], correlating with the leakage of fluorescein observed clinically in RP [42; 43]. The significance of the vascular abnormalities to neurons of the inner retina is discussed above.

CONCLUSIONS

The retinal changes secondary to death of photoreceptors are considerations for certain therapies being developed for RP. With regard to photoreceptor transplantation, a major consideration is the patient's immune response to foreign cells, which may be exacerbated by the RPE-induced leakiness of the inner retinal vessels. An immune response to the donor cells or to tissues damaged by the transplant surgery may also destroy photoreceptors and ocular tissues in the opposite, non-injected eye, as occurs in sympathetic ophthalmia. The milieu of the degenerate RP retina may be altered such that transplanted rods fail to survive or respond abnormally, for example by sprouting neurites. Finally, transplantation into zones of BSP will be complicated by relocation of the RPE cells to the inner retina, loss of the subretinal space, the robust Müller gliotic response, and changes in the inner retinal blood vessels and neurons. Transplantation of normal genes to surviving photoreceptors would appear to be more straightforward and less affected by these retinal changes, although the surgical proceedure will be complicated by shrinkage of the subretinal space. Gene therapy is not likely to correct autosomal dominant diseases, and it is unclear if this form of somatic cell therapy will be permanent. A major problem with the ganglion cell stimulation strategy is the progressive loss of these cells due to transneuronal degeneration and a compromised blood supply. Other therapies on the horizon may be less affected by the secondary disease processes in the RP retinas. These include intravitreally injected growth/survival factors [33; 44; 45] and oral vitamin A treatment [46]. Certain antioxidants and calcium overload blockers have also afforded protection against light-induced photoreceptor damage [47-49]. The mechanism of photoreceptor death in RP, as in several animal models, appears to involve apoptosis (see [50]). It is possible that agents developed to block apoptosis in other cell types will prove to be efficacious for photoreceptors in RP. Success of these experimental therapies will depend on a better understanding of the mechanisms of photoreceptor death and secondary degeneration in the RP retinas. The potential benefits from these treatments are enormous, offering RP patients the possibility of restoration or prolongation of useful vision.

ACKNOWLEDGMENTS

This work was supported by NIH Grants EY0-1311 and -1730, by The Foundation Fighting Blindness, Inc., and by Research to Prevent Blindness, Inc. The authors thank Mr.

D. Possin, Ms. J. Chang, Ms. I. Klock, Mr. C. Stephens, Mr. R. Jones and Ms. J. Huber for assistance.

REFERENCES

1. Bok, D., 1993, Retinal transplantation and gene therapy. Present realities and future possibilities, *Invest. Ophthalmol. Vis. Sci.* 34:473-476.
2. Bok, D., Hageman, G. S. and Steinberg, R. H., 1993, Repair and replacement to restore sight. Report from the panel on photoreceptor/retinal pigment epithelium, *Arch. Ophthalmol.* 111:463-471.
3. Milam, A. H., 1993, Strategies for rescue of retinal photoreceptor cells, *Curr. Opin. Neurobiol.* 3:797-804.
4. Zack, D. J., 1993, Ocular gene therapy. From fantasy to foreseeable reality, *Arch. Ophthalmol.* 111:1477-1479.
5. Humayun, M., Propst, R., de Juan Jr, E., McCormick, K. and Hickingbotham, D., 1994, Bipolar surface electrical stimulation of the vertebrate retina, *Arch. Ophthalmol.* 112:110-116.
6. Woodford, B. J., Chen, J. and Simon, M. I., 1994, Expression of rhodopsin promotor transgene product in both rods and cones, *Exp. Eye Res.* 58:631-635.
7. Milam, A. H., Fliesler, S. J., Chaitin, M. H. and Jacobson, S. G., 1991, Histopathology and immunocytochemistry of human retinal dystrophies, In: Retinal Degenerations, Anderson, R. E., Hollyfield, J. G. and LaVail, M. M., eds., Boca Raton, FL: CRC Press, Inc, pp 361-368.
8. Stone, J. L., Barlow, W. E., Humayun, M. S., de Juan Jr., E. and Milam, A. H., 1992, Morphometric analysis of macular photoreceptors and ganglion cells in retinas with retinitis pigmentosa, *Arch. Ophthalmol.* 110:1634-1639.
9. Li, Z.-Y., Kljavin, I. J. and Milam, A. H., 1995, Rod photoreceptor neurite sprouting in retinitis pigmentosa, *J. Neurosci.* (Submitted):
10. Perlman, I. and Auerbach, E., 1981, The relationship between visual sensitivity and rhodopsin density in retinitis pigmentosa, *Invest. Ophthalmol. Vis. Sci.* 20:758-765.
11. Kemp, C. M., Jacobson, S. G. and Faulkner, D. J., 1988, Two types of visual dysfunction in autosomal dominant retinitis pigmentosa, *Invest. Ophthalmol. Vis. Sci.* 29:1235-1241.
12. Berson, E. L., Gouras, P. and Gunkel, R. D., 1968, Rod responses in retinitis pigmentosa, dominantly inherited, *Arch. Ophthalmol.* 80:58-67.
13. Hood, D. C. and Greenstein, V., 1990, Models of the normal and abnormal rod system, *Vis. Res.* 30:51-68.
14. Greenstein, V. C. and Hood, D. C., 1992, The effects of light adaptation on L-cone sensitivity in retinal disease, *Clin. Vis. Sci.* 7:1-7.
15. Cideciyan, A. V. and Jacobson, S. G., 1993, Negative electroretinograms in retinitis pigmentosa, *Invest. Ophthalmol. Vis. Sci.* 34:3253-3263.
16. Falsini, B., Iarossi, G., Porciatti, V., Merendino, E., Fadda, A., Cermola, S. and Buzzonetti, L., 1994, Postreceptoral contribution to macular dysfunction in retinitis pigmentosa, *Invest. Ophthalmol. Vis. Sci.* 35:4282-4299.
17. Kljavin, I. J. and Reh, T. A., 1991, Müller cells are a preferred substrate for *in vitro* neurite extension by rod photoreceptor cells, *J. Neurosci.* 11:2985-2994.
18. Mandell, J. W., MacLeish, P. R. and Townes-Anderson, E., 1993, Process outgrowth and synaptic varicosity formation by adult photoreceptors *in vitro*, *J. Neurosci.* 13:3533-3548.
19. Kljavin, I. J., Lagenaur, C., Bixby, J. L. and Reh, T. A., 1994, Cell adhesion molecules regulating neurite growth from amacrine and rod photoreceptor cells, *J. Neurosci.* 14:5035-5049.
20. Hicks, D., Forster, V., Dreyfus, H. and Sahel, J., 1994, Survival and regeneration of adult human photoreceptors *in vitro*, *Brain Res.* 643:302-305.
21. Gao, H. and Hollyfield, J. G., 1995, Basic fibroblast growth factor in retinal development: differential levels of bFGF expression and content in normal and retinal degeneration (*rd*) mutant mice, *Devel. Biol.* (In press):
22. Sanyal, S., 1993, Synaptic growth in the rod terminals after partial photoreceptor loss., In: Progress in Retinal Research, Osborne, N. N. and Chader, G. J., eds., Oxford, England: Pergamon Press, pp 247-270.
23. Blair, J. R., Gaur, V., Laedtke, T. W., Li, L., Liu, Y., Sheedlo, H., Yamaguchi, K., Yamaguchi, K. and Turner, J. E., 1991, *In oculo* transplantation studies involving the neural retina and its pigment epithelium, In: Progress in Retinal Research, Osborne, N. N. and Chader, G. J., eds., Oxford, England: Pergamon Press, pp 69-88.
24. Schuschereba, S. T. and Silverman, M. S., 1992, Retinal cell and photoreceptor transplantation between adult New Zealand red rabbit retinas, *Exp. Neurol.* 115:95-99.

25. Gouras, P., Du, J., Kjeldbye, H., Yamamoto, S. and Zack, D. J., 1994, Long-term photoreceptor transplants in dystrophic and normal mouse retina, *Invest. Ophthalmol. Vis. Sci.* 35:3145-3153.
26. Bunt-Milam, A. H., Saari, J. C., Klock, I. B. and Garwin, G. G., 1985, Zonulae adherentes pore size in the external limiting membrane of the rabbit retina, *Invest. Ophthalmol. Vis. Sci.* 26:1377-1380.
27. Li, T., Adamian, M., Roof, D. J., Berson, E. L., Dryja, T. P., Roessler, B. J. and Davidson, B. L., 1994, *In vivo* transfer of a reporter gene to the retina mediated by an adenoviral vector, *Invest. Ophthalmol. Vis. Sci.* 35:2543-2549.
28. Bennett, J., Wilson, J., Sun, D., Forbes, B. and Maguire, A., 1994, Adenovirus vector-mediated *in vivo* gene transfer into adult murine retina, *Invest. Ophthalmol. Vis. Sci.* 35:2535-2542.
29. Bird, A. C., 1992, Investigation of disease mechanisms in retinitis pigmentosa, *Oph. Paed. Genet.* 13:57-66.
30. Hollyfield, J. G., Fliesler, S. J., Rayborn, M. E., Fong, S.-L., Landers, R. A. and Bridges, C. D., 1985, Synthesis and secretion of interstitial retinol-binding protein by the human retina, *Invest. Ophthalmol. Vis. Sci.* 26:58-67.
31. Bunt-Milam, A. H. and Saari, J. C., 1983, Immunocytochemical localization of two retinoid-binding proteins in vertebrate retinas, *J. Cell Biol.* 97:703-712.
32. Huang, P., Gaitan, A., Y., H., M., P. R. and Wong, F., 1993, Cellular interactions implicated in the mechanism of photoreceptor degeneration in transgenic mice expressing a mutant rhodopsin gene, *Proc. Nat. Acad. Sci. USA.* 90:8484-8488.
33. Steinberg, R. H., 1994, Survival factors in retinal degenerations, *Curr. Opin. Neurobiol.* 4:515-524.
34. Eisenfeld, A. J., Bunt-Milam, A. H. and Sarthy, P. V., 1984, Müller cell expression of glial fibrillary acidic protein after genetic and experimental photoreceptor degeneration in the rat retina, *Invest. Ophthalmol. Vis. Sci.* 25:1321-1328.
35. Erickson, P. A., Fisher, S. K., Guerin, C. J., Anderson, D. H. and Kaska, D. D., 1987, Glial fibrillary acidic protein in Müller cells after retinal detachment, *Exp. Eye Res.* 44:37-48.
36. Lewis, G. P., Erickson, P. A., Guerin, C. J., Anderson, D. H. and Fisher, S. K., 1989, Changes in the expression of specific Müller cell proteins during long-term retinal detachment, *Exp. Eye Res.* 49:93-111.
37. Guerin, C. J., Anderson, D. H. and Fisher, S. K., 1990, Changes in intermediate filament immunolabeling occur in response to retinal detachment and reattachment in primates, *Invest. Ophthalmol. Vis. Sci.* 31:1474-1482.
38. Milam, A. H. and Jacobson, S. G., 1990, Photoreceptor rosettes with blue cone opsin immunoreactivity in retinitis pigmentosa, *Ophthalmol.* 97:1620-1631.
39. Korte, G. E., Hageman, G. S., Pratt, D. V., Glusman, S., Marko, M. and Ophir, A., 1992, Changes in Müller cell plasma membrane specializations during subretinal scar formation in the rabbit, *Exp. Eye Res.* 55:155-162.
40. Li, Z.-Y., Possin, D. E. and Milam, A. H., 1995, Histopathology of bone spicule pigmentation in retinitis pigmentosa, *Ophthalmol.* (In press).
41. Szamier, R. B., 1981, Ultrastructure of the preretinal membrane in retinitis pigmentosa, *Invest. Ophthalmol. Vis. Sci.* 21:227-236.
42. Abraham, F. A., Ivry, M. and Tsvieli, R., 1976, Sector retinitis pigmentosa. A fluorescein angiographic study, *Ophthalmol., Basel.* 172:287-297.
43. Krill, A. E., Archer, D. and Newell, F. W., 1970, Fluorescein angiography in retinitis pigmentosa, *Am J Ophthalmol.* 69:826-835.
44. Faktorovich, E. G., Steinberg, R. H., Yasumura, D., Matthes, M. T. and LaVail, M. M., 1990, Photoreceptor degeneration in inherited retinal dystrophy delayed by basic fibroblast growth factor, *Nature.* 347:83-86.
45. Faktorovich, E. G., Steinberg, R. H., Yasumura, D., Matthes, M. T. and LaVail, M. M., 1992, Basic fibroblast growth factor and local injury protect photoreceptors from light damage in the rat, *J. Neurosci.* 12:3554-3567.
46. Berson, E. L., Rosner, B., Sandberg, M. A., Hayes, K. C., Nicholson, B. W., Weigel-DiFranco, C. and Willett, W., 1993, A randomized trial of vitamin A and vitamin E supplementation for retinitis pigmentosa, *Arch. Ophthalmol.* 111:761-772.
47. Rosner, M., Lam, T. T., Fu, J. and Tso, M. O. M., 1992, Methylprednisolone ameliorates retinal photic injury in rats, *Arch. Ophthalmol.* 110:857-861.
48. Li, J., Edward, D. P., Lam, T. T. and Tso, M. O. M., 1993, Amelioration of retinal photic injury by a combination of flunarizine and dimethylthiourea, *Exp. Eye Res.* 56:71-78.
49. Edward, D. P., Lam, T. T., Shahinfar, S., Li, J. P. and Tso, M. O. M., 1991, Amelioration of light-induced retinal degeneration by a calcium overload blocker, *Arch. Ophthalmol.* 109:554-562.
50. Li, Z.-Y. and Milam, A. H., 1995, Apoptosis in Retinitis Pigmentosa, In: Retinal Degeneration II, Anderson, R. E., Hollyfield, J. G. and LaVail, M. M., eds., New York: Plenum, In press.

32

GENOTYPE-PHENOTYPE CORRELATION IN AUTOSOMAL DOMINANT RETINAL DEGENERATION WITH MUTATIONS IN THE PERIPHERIN/*RDS* GENE

M. Nakazawa, E. Kikawa, Y. Chida, K. Kamio, Y. Wada, T. Shiono, and M. Tamai

Department of Ophthalmology
Tohoku University School of Medicine
1-1 Seiryo-machi, Aoba-ku, Sendai 980-77, Japan

INTRODUCTION

Peripherin/*RDS*, a photoreceptor-specific glycoprotein, located in the peripheral portion of the disc membranes of photoreceptor outer segments of both rods and cones (1-4). Although the precise function of this protein is still uncertain, it has been postulated that peripherin/*RDS* plays an important role in maintaining the unique structure of the disc membrane. Since a mutation in the peripherin/*RDS* gene was found to be responsible for retinal degeneration in the *rds* (retinal degeneration slow) mouse strain (3,4), mutations in the human peripherin/*RDS* gene have been identified in patients with autosomal dominant retinitis pigmentosa (adRP) (5-10). In addition to typical retinitis pigmentosa, increased numbers of mutations have been identified in the peripherin/*RDS* gene in families with several kinds of macular dystrophy (8,11), retinitis punctata albescens (12), butterfly-shaped pigment dystrophy of the fovea (13), and fundus flavimaculatus (14). Because peripherin/*RDS* commonly expressed in both rods and cones, mutations in the peripherin/*RDS* gene produce either rod- or cone-predominant retinal degeneration, depending on the kinds and locations of mutations. These genotype-phenotype correlations provide important clues not only for better understanding the mechanisms of photoreceptor degeneration produced by mutations in the peripherin/*RDS* gene, but also for proper genetic counseling based on the accumulated findings regarding clinical courses of patients with known mutations.

We previously have identified two kinds of transversion mutations in codon 244 of the peripherin/*RDS* gene that result in amino acid substitutions of lysine residue and histidine residue for asparagine residue (designated as Asn244Lys and Asn244His, respectively) (15,16). Although these mutations occured in the same codon of the peripherin/*RDS* gene, the phenotypic features of patients associated with these mutations showed marked contrasts.

Degenerative Diseases of the Retina, Edited by Robert E. Anderson et al.
Plenum Press, New York, 1995

Characteristic features produced by the Asn244Lys mutation were rod-cone dystrophy whereas those by the Asn244His mutation were cone-rod dystrophy although the final outcomes of rod-cone dystrophy and cone-rod dystrophy are similar and sometimes indistinguishable, i.e. degeneration of both cone system and rod system. However, the clinical course of each type of retinal dystrophy demonstrates different kinds of visual impairmant on affected patients through-out their lives. In this study, therefore, we characterized clinical features of affected patients with Asn244Lys and Asn244His to provide insights regarding the mechanisms of photoreceptor degeneration at the molecular level.

In addition, we describe the method of molecular genetic screening that we have employed to search for mutations in the peripherin/*RDS* gene or others.

PATIENTS AND METHODS

We screened a total of 58 genomic DNA samples isolated from unrelated patients with adRP or allied retinal degeneration with autosomal dominant inheritance to search for mutations in the peripherin/*RDS* gene using polymerase chain reaction (PCR) followed by nonradioisotopic single-strand conformation polymorphism (SSCP) (17) with a modification previously described (15).

Amplification of the Peripherin/*RDS* Gene by Polymerase Chain Reaction (PCR)

With the use of a thermocycler (Perkin Elmer Cetus, Norwalk, CT, U.S.A.), the PCR was carried out in 50 μl of reaction mixture containing 250 ng of genomic DNA, which was prepared from each patient's leukocytes as templates; 20 μM of primer; 200 μM each of dATP, dCTP, dGTP, and TTP, 1.25 units of Taq polymerase; and 0.5 unit of Perfect Match Polymerase Enhancer® (Stratagene, La Jolla, CA, U.S.A.). Buffer contained 50 mM KCl, 10 mM TrisCl (pH 8.3), and 1.5 mM $MgCl_2$. Reaction cycle was 30. The usual temperature settings for PCR were 94°C for 1 min of denaturation, 55 to 60°C for 2 min of anealing, and 72°C for 2 min of polymeration. Each pair of the amplification of the peripherin/RDS gene was determined in order that the size of each amplified fragment should not exceed 200 bp. For this size limitation, exon 1 was divided into 4 segments and exon 2 was divided into 2 segments in designing the oligonucleotide primer sets. The nucleotide sequence of each primer set is shown in Table 1.

Nonradioisotopic SSCP

A total of 3.0 μl of the PCR product was mixed with 3.0 μl of 95% deionized formamide containing 0.25% each of xylene cyanol and bromophenol blue. The mixture was heated at 95°C for 5 min, then cooled on ice before 4.0 μl of the mixture was loaded onto a 0.4 mm X 30 cm X 30 cm 8% acrylamide with 10% glycerol. Electrophoresis was done at room temperature at 20W for 7 hours. The gel was then silver stained with the use of Bio-rad silver stain kit.

Nucleotide Sequencing

The PCR products which had shown abnormal band shift on SSCP were subsequently subcloned using pBluescript or pGEM. Subcloned PCR products were then sequenced with an automated DNA sequencer (Pharmacia, Sweden) using a dideoxy chain termination

Table 1. Nucleotide sequence of primer sets for the amplification of the peripherin/*RDS* gene

Exon	Primer Sequence (5'->3')	Size of Amplified Region (bp)
1A	F: CCGGACTACACTTGGCAAGC R: CACATCGCTCCTCTTTCGGA	174
1B	F: GACTGTTCCTGAAGATTGAA R: ATAGCCAGGTACGGATTCAG	199
1C	F: TATGCCAGATGGAAGCCCTG R: TCTTCATGAAACACCTGCCA	192
1D	F: AGTACTACCGGGACACAGAC R: CTGACCCCAGGACTGGAAGC	181
2A	F: AAGCCCATCTCCAGCTGTCT R: GTCGTAACTGTAGTGTGCTG	163
2B	F: AGTATCAGATCACCAACAAC R: CCCAGCTGCCCAGGGCCTAC	178
3	F: TGCCTCTAAATCTCCTCTCC R: AACACTGAGAAATAGTGCAC	286

protocol with fluorated oligonucleotide primers. At least six subclones were sequenced in determining nucleotide sequence of the mutation to exclude artifacts in PCR.

Phenotypic Characterization of Patients with Mutations in the Peripherin/*RDS* Gene

We examined affected patients with detected mutations of the peripherin/*RDS* gene. The patients' medical histories included subjective recall of onset and history of night blindness, disturbance of central and/or side vision, and other visual symptoms. The subsequent ophthalmic examination included best corrected visual acuity, Goldmann's kinetic visual field examination, fundus examination, fluorescein angiography, electroretinography (ERG), dark adaptometry, and color vision testing.

Secondary Structure Analysis of Mutated Peripherin/*RDS*

To assess possible structural alterations of mutated peripherin/RDS proteins, we analyzed the change of secondary structure of these proteins using the protein structure algorithm described by Garnier and associates (18).

RESULTS

DNA Analysis

Of 58 unrelated Japanese patients (probands) with adRP and allied retinal degeneration with autosomal dominant inheritance, we detected two kinds of abnormal band shifts in exon 2 on SSCP (15,16,19). The same abnormal pattern was identified on SSCP in each probands' affected family members (16, 19). The alterations in the DNA sequence that were common to affected family members of each pedigree were identified as heterozygous transversional changes of C to A at the third nucleotide in codon 244 (AAC -> AAA, Family 1) and of A to C at the first nucleotide in the same codon (AAC -> CAC, Family 2). These base changes resulted in the replacement of an asparagine residue with a lysine residue in Family 1, designated as Asn244Lys (15,19), and with a histidine residue in Family 2, designated as Asn244His (16).

Clinical Features of Patients Associated with Asn244Lys And Asn244His

The phenotypic expressions associated with these mutations showed marked contrasts, although each mutation recorded intrafamilial similarity. We examined 8 affected patients in Family 1 (Asn244Lys), ranging in age from 9 years to 64 years, and 2 affected patients in Family 2 (Asn244His) whose ages were 46 and 53 years. The clinical features with Asn244Lys are summerized as night blindness, usually noticed by the patient in the early teens; decreased visual acuity, with an onset in the late thirties; diffuse pigmentary retinal degeneration in the mid-peripheral to peripheral retina; and bull's-eye maculopathy, which also appears in the late thirties. In addition, ERG assessments show nonrecordable amplitudes of rod-isolated responses and severely reduced amplitudes of cone-isolated responses that have already started at about age 9 years, even though the patient had had no complaint of difficulty with night vision. Visual field testing shows progressive visual field loss, from contraction of the II-4-e isopter in their twenties and thirties to general constriction in the late stage. The color vision is defective only in patients older than 50 years old. These clinical features and ERG findings indicate that the clinical course produced by Asn244Lys is categorized in rod-cone dystrophy.

On the other hand, the phenotypic expressions associated with Asn244His are characterized as cone-rod dystrophy. The 46-year-old proband has had poor visual acuity since childhood but has not experienced disturbance of night vision. The ERG assessment demonstrates nonrecordable cone-isolated responses and subnormal patterns of rod-isolated responses. Fundus examination and fluorescein angiogram disclose bull's-eye maculopathy. Visual field testing shows paracentral scotomas in both eyes with relatively well preserved peripheral visual fields. Color vision is defective by FM 100-hue test. Her 53-year-old brother has had a similar history in visual acuity. In addition, he has had night blindness and visual field disturbance for 3 years. The ERG assessment shows severely deteriorated responses of both rods and cones. Funduscopy shows that bull's-eye maculopathy was associated with diffuse bone-spicule pigmentation in the midperipheral area. These clinical features correspond to an advanced stage of cone-rod dystrophy. And these findings of cone-rod dystrophy in Family 2 can be categorized into cone-rod dystrophy type 2a by the subclassification of Szlyk and associates (20). Summary of clinical features in both families is shown in Table 2.

Secondary Structure of Mutated Peripherin/*RDS* Prtoteins with Asn244Lys and Asn244His

The results of secondary structure analysis of mutated peripherin/*RDS* proteins are shown in Figure 1. Secondary structure produced by these two mutations disclosed that Asn244His and Asn244Lys mutations give rise to slightly different alterations in the coil structure between codons 244 and 250 of the peripherin/*RDS*.

DISCUSSION

The present study regarding phenotypes with Asn244Lys and Asn244His mutations indicates two important findings concerning genotype-phenotype correlations associated

Table 2. Summary of clinical features in Family 1 (Asn244Lys) and Family 2 (Asn244His)

	Asn244Lys	Asn244His
Initial visual symptom	night blindness	decreased visual acuity
Initial fundus appearance	diffuse pigmentary degeneration in mid-peripheral and peripheral retina	bull's eye maculopathy
Additional fundus appearance in the late stage	bull's eye maculopathy	pigmentary degeneration
Rod-mediated ERG	non-recordable	reduced, the non-recordable
Cone-mediated ERG	reduced, then non-recordable	non-recordable
Clinical diagnosis	rod-cone dystrophy	cone-rod dystrophy

with mutations in the peripherin/*RDS* gene. The first finding is that mutations in codon 244 of the peripherin/*RDS* gene cause either cone-rod dystrophy or rod-cone dystrophy, depending on which amino acid is substituted for asparagine. Although the precise molecular mechanism producing such distinct differences remains uncertain, it is possible that secondary structures of peripherin/*RDS* differently altered by these two kinds of substitutions of amino acid residues in codon 244 are related to these different predominances in rod and cone photoreceptor degenerations. The second finding is that bull's-eye maculopathy is a common clinical feature associated with mutations in codon 244 of the peripherin/*RDS* gene. Proteins in the photoreceptor outer segment are normally subject to digestion by lysosomal enzymes in the retinal pigment epithelium (21). In addition, the macular region has the highest specific activity of lysosomal enzymes(22), indicating high rates of metabolic demands or turnovers in this area. Therefore, it is reasonable that peripherin/*RDS*, one of the structural proteins in the photoreceptor disk membranes of both rods and cones, is catabol-

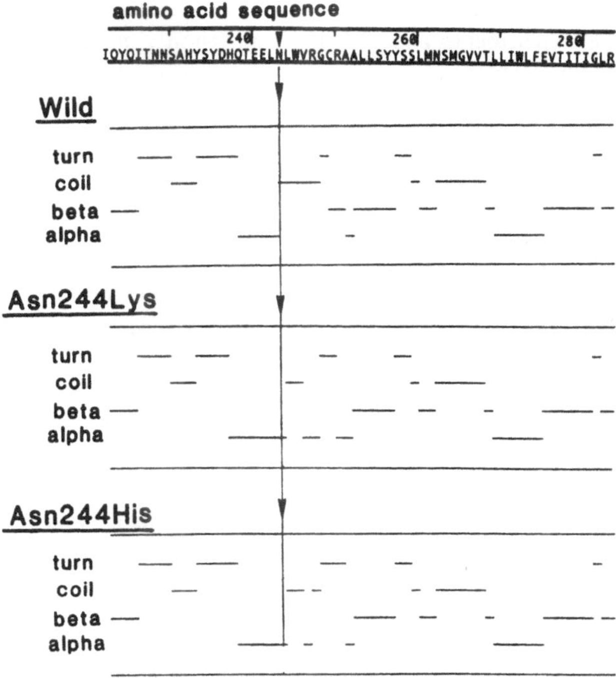

Figure 1. Secondary structure analysis of wild and mutated peripherin/*RDS* protein

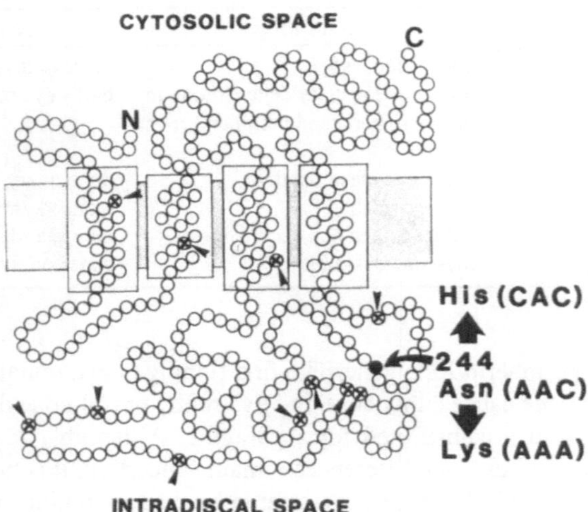

Figure 2. Molecular structure of peripherin/*RDS* and the location of codon 244. Arrow heads indicate the location of mutations previously reported.

ized by these enzymes. We have speculated that presence of mutations in codon 244 can change specificity of reactions to some extent between the substrate to be digested and these catabolic enzymes. Also, these metabolic changes can lead to accumulation of abnormal substances in the subretinal layer, particularly in the macular area over many years. The present finding that a bull's-eye macular lesion is seen in patients with Asn244Lys and with Asn244His mutations supports this speculation. Further molecular biological analysis, including transgenic experiments and in vitro expression analysis, may provide better understandings of the molecular mechanism producing thse clinical features.

Previous reports have described phenotypic characteristics associated with two different mutations in the same codon, i.e., codon 172 of the peripherin/RDS gene (8,11), indicating that both Arg172Trp and Arg172Gln mutations are associated with macular degeneration. These two mutations are similar in distribution and in the same nature of abnormality, although patients with the Arg172Gln mutation were more mildly affected than those with the Arg172Trp mutation. These findings imply that not only the location of mutations but also the kind of amino acid substitution can modify clinical features of hereditary retinal degeneration. Our findings that rod-cone dystrophy and cone-rod dystrophy are caused by different mutations in the same codon of the peripherin/RDS provide an important insight to the understanding of these phenotypes in terms of clinical classification. The traditional definition of retinitis pigmentosa has been restricted to rod-predominant photoreceptor degeneration (23,24), while some investigators have reported that retinitis pigmentosa consists of both the rod-cone type and the cone-rod type, the final status of which cannot be differentiated (25). Our findings provide direct evidence that an overlapped area exists between rod-cone dystrophy and cone-rod dystrophy and support the idea that these two conditions are on the same spectrum of one disease as allelic heterogeneity.

Although we have not yet obtained certain rules regarding genotype-phenotype correlations of the peripherin/*RDS* gene, further accumulations about phenotypic characteristics produced by specific mutations will augment our understandings of the mechanism of photoreceptor degeneration.

ACKNOWLEDGMENTS

This study was supported in part by grants from National Society for the Prevention of Blindness (Dr. Nakazawa), the Ministry of Education, Culture and Science of the Japanese Government: Grant-in-Aid for Scientific Research, B-05454468 (Dr. Nakazawa) and the Research Committee on Chorioretinal Degenerations, the Ministry for Health and Social Welfare of the Japanese Government (Dr. Tamai).

REFERENCES

1. Arikawa, K, Molday, R.S., and Wlliams, D.S., 1992, Localization of peripherin/*RDS* in the disk membranes of cone and rod photoreceptors: relationship to disk membrane morphogenesis and retinal degeneration, *J. Cell Biol.* 116: 659-667.
2. Connel, G., and Molday, R.S., 1990, Molecular cloning, primary structure and orientation of the vertebrate photoreceptor cell protein peripherin in the rod disc membrane. *Biochemistry* 29:4691-4698.
3. Connel, G., Bascom, R., Molday, L., Reid, D., McInnes, R.R., and Molday, R.S., 1991, Photoreceptor cell peripherin is the normal product of the gene responsible for the retinal degeneration in the RDS mouse. *Proc. Natl. Acad. Sci. U.S.A.* 88:723-726.
4. Travis, G.H., Brennan, M.B., Danielson, P.E., Kozak, C.A., and Sutcliffe, J.G., 1989, *Nature* 338:70-73.
5. Kajiwara, K., Hahn, L. B., Mukai, S., Travis, G. H., Berson, E. L., and Dryja, T. P., 1991, Mutations in the human retinal degeneration slow gene in autosomal dominant retinitis pigmentosa. *Nature* 354: 480-483.
6. Farrar, G. J., Kenna, P., Jordan, S. A., Kumar-Singh, R., Humphries, M. M., Sharp, E. M., Sheils, D., and Humphries, P., 1991, A three-base pair deletion in peripherin/*RDS* gene in one form of retinitis pigmentosa. *Nature* 354: 478-480.
7. Farrar, G. J., Kenna, P., Jordan, S. A., Kumar-Singh, R., Humpgries, M. M., Sharp, E. M., Sheils, D., and Humphries, P., 1992, Autosomal dominant retinitis pigmentosa: A novel mutation at the peripherin/*RDS* locus in the original 6p-linked pedigree. *Genomics* 14: 805-807.
8. Wells, J., Wroblewski, J., Keen, J., Inglehearn, C., Jubb, C., Eckestein, A., Jay, M., Arden, G., Bhattacharya, S., Fitzke, F., and Bird, A., 1993, Mutations in the human retinal degeneration slow (*RDS*) gene can cause either retinitis pigmentosa or macular dystrophy. *Nature Genet.* 3: 213-218.
9. Kajiwara, K., Berson, E. L., and Dryja, T. P., Screen for mutations in the entire coding sequence of the human *RDS*/peripherin gene in patients with hereditary retinal degeneration. *Invest. Ophthalmol. Vis. Sci.* 34: 1149.
10. Saga, M., Mashima, Y., Akeo, K., Oguchi, Y., Kudoh, J., and Shimizu, N., 1993, A novel Cys-214-Ser mutation in the peripherin/RDS gene in a Japanese family with autosomal dominant retinitis pigmentosa. *Human Genet.* 92: 519-521.
11. Wroblewski, J. J., Wells III, J. A., Eckstein, A., Fitzke, F., Jubb, C., Keen, T. J., Inglehearn, C., Bhattacharya, S., Arden, G. B., Jay, M., and Bird, A., 1994, Macular dystrophy associated with mutations at codon 172 in the human retinal degeneration slow gene. *Ophthalmology* 101: 12-22.
12. Kajiwara, K., Sandberg, M. A., Berson, E. L., and Dryja, T. P., 1993, A null mutation in the human peripherin/*RDS* gene in a family with autosomal dominant retinitis punctata albescens. *Nature Genet.* 3: 208-212.
13. Nichols, B. E., Shefield, V. C., Vandenburgh, K., Drack, A. V., Kimura, A., E., and Stone, E. M., 1993, Butterfly-shaped pigment dystrophy of the fovea caused by a point mutation in codon 167 of the *RDS* gene. *Nature Genet.* 3: 202-207.
14. Weleber, R. G., Carr, R. E., Murphy, W. H., Shefield, V. C., and Stone, E. M., 1993, Phenotypic variation including retinitis pigmentosa, pattern dystrophy, and fundus flavimaculatus in a single family with a deletion of codon 153 or 154 of the peripherin/*RDS* gene. *Arch. Ophthatmol.* 111: 1531-1542.
15. Kikawa, E., Nakazawa, M., Chida, Y., Shiono, T., and Tamai, M., 1994, A novel mutation (Asn244Lys) in the peripherin/*RDS* gene causing autosomal dominant retinitis pigmentosa associated with bull's-eye maculopathy detected by nonradioisotopic SSCP. *Genomics,* 20:137-139.
16. Nakazawa, M., Kikawa, E., Chida, Y., and Tamai, M., 1994, Asn244His mutation of the peripherin/*RDS* gene causing autosomal dominant cone-rod degeneration. *Human Mol. Genet.* 3: 1195-1196.
17. Nakazawa, M., Kikawa, E., Chida, Y., Shiono, T., and Tamai, M., 1993, Nonradioactive single strand conformation polymorphism (PCR-SSCP): A simplified method applied to a molecular genetic screening

of retinitis pigmentosa. In Hollyfield, J. G., LaVail, M. M., Anderson, R. E. (eds.): *Retinal Degeneration: Clinical and Laboratory Applications,* New York, Plenum Publishing Corp., pp. 181-188.

18. Garnier, J., Osguthorpe, D. J., and Robson, B., 1978, Analysis of the accuracy and implications of simple methods for predicting the second structure of globular proteins. *J. Mol. Biol.* 120:97-120.
19. Nakazawa, M., Kikawa, E., Kamio, K., Chida, Y., Shiono, T., and Tamai, M., 1994, Ocular findings in patients with autosomal dominant retinitis pigmentosa and transversion mutation in codon 244 (Asn244Lys) of the peripherin/*RDS* gene, *Arch. Ophthalmol.* 112:1567-1572.
20. Szlyk, J. P., Fishman, G. A., Alexander, K. R., Peachery, N. S., and Derlacki, D. J., 1993, Clinical subtypes of cone-rod dystrophy. *Arch. Ophthalmol.* 111: 781-788.
21. Hayasaka, S., 1983, Lysosomal enzymes in ocular tissues and diseases. *Surv. Ophthalmol.* 27: 245-258.
22. Hayasaka, S., Shiono, T., Hara, S., and Mizuno, K., 1981, Regional distribution of lysosomal enzymes in the retina and choroid of human eyes. *Albrecht von Graefes Arch. Klin. Ophthalmol.* 216: 269-273.
23. Deutman, A. F., 1977, Rod cone dystrophy: Primary hereditary pigmentary retinopathy, retinitis pigmentosa. In Krill, A. E. (ed.) *Krill's Hereditary Retinal and Choroidal Disease vol. II, Clinical Characteristics*, Hagerstown, Harper and Row, pp. 479-576.
24. Pagon, R. A. : Retinitis pigmentosa. *Surv. Ophthalmol.* 33: 137-177.
25. Heckenlively, J. R., 1988, Autosomal dominant retinitis pigmentosa. In Heckenlively, J. R. (ed.) *Retinitis Pigmentosa*, Philadelphia, Lippincott, pp. 125-149.

33

GENETIC STUDIES IN AUTOSOMAL RECESSIVE FORMS OF RETINITIS PIGMENTOSA

A. F. Wright,[1] D. C. Mansfield,[1] E. A. Bruford,[1] P. W. Teague,[1] K. L. Thomson,[1] R. Riise,[2] Jay. M,[3] M. A. Patton,[4] S. Jeffery,[4] A. Schinzel,[5] N. Tommerup,[6] and M. Fossarello[7]

[1] MRC Human Genetics Unit, Western General Hospital
Crewe Road, Edinburgh, United Kingdom

[2] Eye Department, Hamar Sjukehus
Hamar, Norway

[3] Department of Clinical Ophthalmology, University of London
Moorfields Eye Hospital, City Road, London, United Kingdom

[4] Department of Clinical Genetics, St George's Hospital Medical School
London, United Kingdom

[5] Institute for Medical Genetics, University of Zurich
Zurich, Switzerland

[6] Department of Clinical Genetics, Uinversity of Oslo
Oslo, Norway

[7] Department of Clinical Ophthalmology, University of Cagliari
Cagliari, Sardinia, Italy

INTRODUCTION

Retinitis pigmentosa is a heterogeneous group of progressive retinal degenerations sharing a common set of clinical characteristics comprising night blindness, constricted visual fields, pigment deposition in the outer retina and diminished or absent electroretinogram (ERG). It affects about 1 in 4000 of the general population and occurs as autosomal dominant, autosomal recessive, X-linked and sporadic forms (Heckenlively, 1988; Humphries et al., 1992). The proportion of the different forms of RP varies between populations. In addition to the above group of patients with isolated retinal degeneration there is a sizeable group in whom the retinal degeneration forms part of a genetic syndrome with non-ocular manifestations. In one study in the United Kingdom for example, the syndromal group formed 16% of index cases with RP (Bundey and Crews, 1984). The most common syndromal forms of RP are Usher's syndrome and Bardet-Biedl syndrome, both of which are inherited as autosomal recessive disorders.

Degenerative Diseases of the Retina, Edited by Robert E. Anderson et al.
Plenum Press, New York, 1995

Identifying the relative frequency of different types of RP is often a difficult task. The largest single group is the isolated or sporadic patients, usually comprising 40-50% of index cases (Jay, 1982; Bundey and Crews, 1984; Boughman and Fishman, 1983). This group includes a sizable number of autosomal recessive patients, three-quarters of whose siblings are expected to be normal, however when these are grouped together with definite or probable autosomal recessive patients, the segregation ratio is less than the 0.25 expected on a simple autosomal recessive hypothesis (Jay, 1982). The explanation for this is that either non-genetic forms of RP or patients with new dominant mutations are likely to be present in the isolated group. Recently, a digenic mode of inheritance for RP was described in which a combination of mutations at the peripherin/RDS and rom1 loci was necessary to produce the clinical picture of RP (Kajiwara et al., 1994). Oligogenic forms of RP may not be uncommon but are very hard to demonstrate. The results of prevalence studies in the United Kingdom illustrate the complexity of defining the proportions of genetic types of RP. In one study, Jay (1982) carried out a segregation analysis of a large series of out-patients and concluded that 41% of RP patients suffered from an autosomal recessive disorder whereas Bundey and Crews (1984) in a similar population only identified 8% of their RP patients as autosomal recessive in a defined population survey. However, the last authors did not carry out a segregation analysis and so did not allow for recessive families with only a single affected member. In some populations, the proportion of autosomal recessive RP is high, particularly in parts of Switzerland, Israel, Norway, Japan, Italy and India (Ammann et al., 1961; Faber, 1970; Grondahl, 1986; Tanabe, 1972; Fossarello et al., 1993).

The rate of location or identification of specific RP genes has accelerated in the last few years. To date, at least 6 genes causing autosomal dominant RP (on chromosomes 3, 6, 7p, 7q, 8, 19), 5 causing autosomal recessive RP (on chromosomes 1, 3, 4, 4 and 6) and two causing X-linked RP (RP2, RP3) have been located by linkage analysis or mutations identified in specific genes (Rosenfeld et al., 1994). Nine genes were first localised by linkage analysis, two by screening of candidate genes for mutation and one by both methods simultaneously. The extent of heterogeneity within RP has been a surprise. This raises a number of issues. Firstly, identification of genes becomes more complex since it becomes difficult to pool families for linkage studies. This is particularly so for autosomal recessive RP since the majority of families are too small to map. Some large inbred kindreds have been studied to overcome this problem which led to the successful localisation of recessive genes on chromosomes 1 and 6 (Knowles et al., 1994; van Soest et al., 1994). However, there are relatively few opportunities for studying such families. Secondly, since only a minority of patients are accounted for by the RP genes identified to date, how many genes must be identified to account for RP in most patients? The answer is likely to be over 50, based on a rough extrapolation of the present data regarding number of genes and percentage of patients accounted for in each type of RP. This is a formidable task and raises serious issues with regard to gene therapy and other less generic approaches to therapy.

The studies described in this paper represent an alternative attempt to overcome some of the problems of linkage in autosomal recessive forms of RP. The first study has the aim of identifying autosomal recessive RP genes in the Sardinian population, in which there is an unusually high proportion of recessive disease. The second study was designed to identify the genetic cause of an autosomal recessive syndromal form of RP, Bardet-Biedl syndrome. In both cases, it was assumed that genetic heterogeneity would be reduced if not eliminated, either by studying a relatively homogeneous ethnic population or by studying a relatively rare and clinically specific disorder such as Bardet-Biedl sydrome. Both of these assumptions are probably over-simplifications since the available evidence suggests that both groups are still genetically heterogeneous.

AUTOSOMAL RECESSIVE RP IN THE SARDINIAN POPULATION

Autosomal recessive RP is one of the most consistently severe forms of this disorder, with onset of night blindness usually in the first decade, progressing to severe visual handicap by the end of the fourth decade (Fossarello et al., 1993). It is also the most difficult form to identify with certainty. Families with one or more affected children and consanguineous parents are generally assumed to be of autosomal recessive type. However, in some populations the carrier frequency for a recessive disorder may be quite high (e.g. 1 in 20 Caucasians are carriers for cystic fibrosis), in which case the majority of families will not be identifiably related to each other. Consanguinity rates tend to rise as the carrier frequency falls. The "probable" autosomal recessive group includes those with two or more affected children of opposite sex and with normal parents and other relatives. This eliminates X-linked and autosomal dominant forms of RP except perhaps dominant families with very low penetrance. In this study, a previous epidemiological survey of 158 RP families from south-central Sardinia identified 54% as isolated cases and of the remainder, 75% of families were probable autosomal recessive with an unusually low percentage of autosomal dominant (5%) and X-linked families (Fossarello et al., 1993). The autosomal dominant RP showed a high proportion of rhodopsin mutations (50%) with high penetrance which are therefore unlikely to contaminate the recessive group (Fossarello et al., 1993). The recessive cases included 5 out of 29 non-syndromal RP families (17%) with consanguineous parents and the remainder had normal parents, showed no other family history within the past three or four generations and came from geographical areas known to show high rates of inbreeding in the recent or disant past. All families were native Sardinian at least for the past three or four generations.

Linkage Studies

A set of 11 clinically well characterised nuclear families with autosomal recessive RP was ascertained and sampled for a linkage study. The families contained 26 affected members (Figure 1) and came from 10 distinct geographical areas.

Lymphoblastoid cell lines were established from each individual and genomic DNA was extracted from blood samples. A total of 195 markers were typed and analysed for linkage in the families. Over 60% of the genome was excluded but no single marker provided unambiguous evidence of linkage. However a small number of chromosomal regions showed positive lod scores within subsets of families. The region which looked most interesting was first identified by a marker on the short arm of chromosome 14, D14S80, which showed an unexpectedly large number of families with few or no recombinants. Heterogeneity analysis was carried out which tended to confirm the presence of a subset of linked families although at a borderline significance level ($\chi^2 = 4.45$; $P = 0.054$). The lod score between arRP and D14S80 within the putative subgroup of 5 linked families was 3.33 at zero recombination.

In view of the borderline significance levels, other markers were studied from the region in order to carry out a multipoint analysis. The markers D14S264, D14S275, D14S64, D14S80, D14S262 and D14S252, all located in the distal part of the short arm of chromosome 14 were then analysed. The results again provided support for heterogeneity but at a borderline level of significance ($P = 0.065$). Linkage to this region was excluded in 6 families and the maximum likelihood was found with 24% of families linked to chromosome 14.

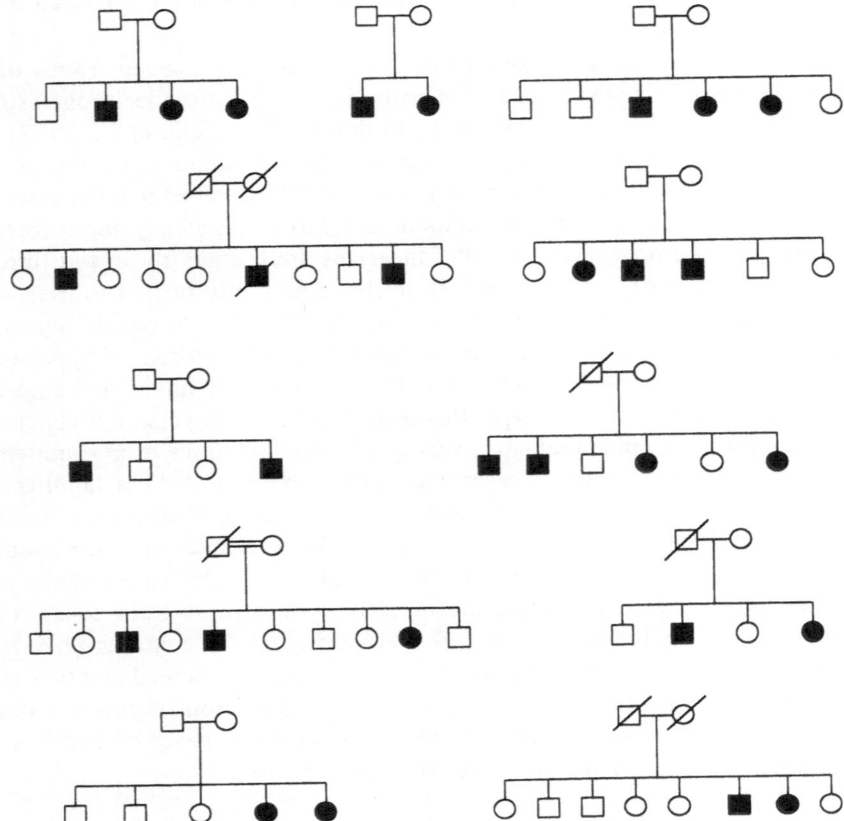

Figure 1. Pedigrees of Sardinian families with autosomal recessive RP used for linkage analysis.

Homozygosity for Chromosome 14 Markers in a Consanguineous Family

The most striking finding was a region of homozygosity in one of the consanguineous families (arRP1) extending over four adjacent markers in three affected individuals (Figure 2).

The results in Figure 2 show that there is homozygosity for markers D14S275, D14S64, D14S80 and D14S262, raising the possibility that this region of chromosome 14 is identical-by-descent (IBD) in the three affected sibs. The method of homozygosity mapping was proposed by Lander and Botstein (1987) to help map rare recessive disorders with small numbers of consanguineous families by taking advantage of the additional meioses available between the common ancestor and the affected members sharing alleles IBD for the disease allele and its flanking markers. For example, the parents of family arRP1 are second cousins, so that the affected children will on average share one-sixteenth (6.25%) of their genes, including both mutant copies of the disease gene from the common ancestor, on the assumption that the gene is rare in the general population and so is much more likely to have been inherited from the common ancestor. Identity-by-descent will be indicated by a region of homozygosity spanning the disease locus. The markers showing homozygosity in arRP1 span a genetic distance of 6 centiMorgans (cM) representing about 0.15% of the genome. The key question is whether this is due to the chance occurrence of homozygosity in adjacent markers or to common ancestry. If the homozy-

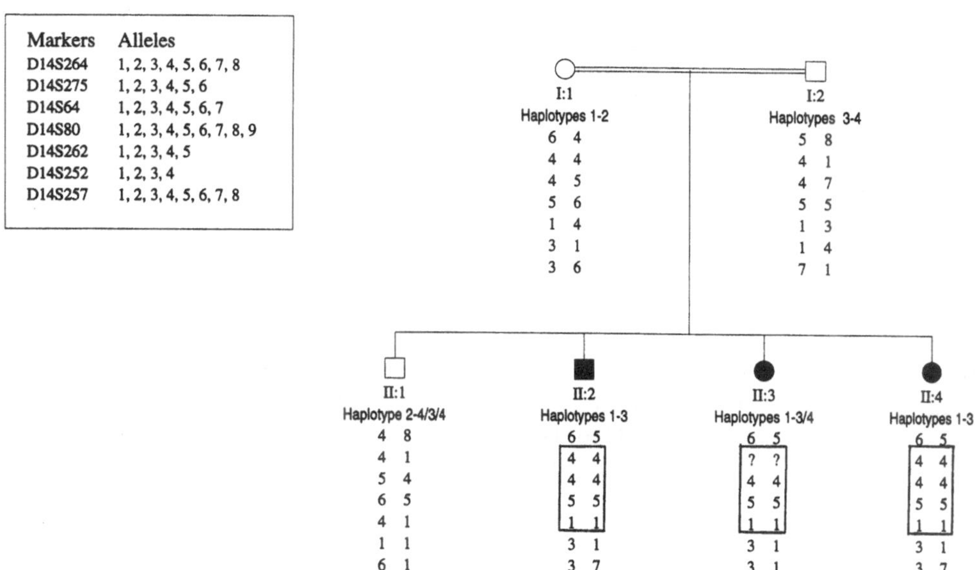

Figure 2. Marker homozygosity extending over D14S262, D14S80, D14S64 and D14S275 in a consanguineous family (arRP1) with autosomal recessive retinitis pigmentosa.

gosity includes rare alleles it is more likely to result from common ancestry and similarly if the disease gene is prevalent in the population, it is less likely to be IBD in a family of this sort. Assumptions therefore have to be made regarding the disease gene frequency in the population . Marker allele frequencies are more precisely known. In this case, all the alleles are common ones, although the expected proportion of individuals homozygous for all four markers is only about 0.02%. When the lod scores, or likelihood of linkage, are calculated in this family, the lods vary between 0.9 and 2.0, depending on the gene frequency assumptions and whether they are calculated by homozygosity mapping or linkage analysis programmes. The advantage of a second-cousin consanguineous marriage such as arRP1 is that it allows the analysis of linkage in eight meioses per individual compared with only two for a non-consanguineous marriage. There are therefore 24 meioses available for study in a family with as few as three affected members. This advantage is offset by the need for assumptions about the disease gene frequency and inferences on the basis of marker allele frequencies. The other four putative linked families were neither consanguineous nor homozygous for more than one marker spanning the region, consistent with the absence of a recent common ancestor. In summary, the overall evidence for linkage was increased by the observation of homozygosity spanning 4 adjacent markers in one family, but the linkage evidence was not definitive in this family.

Candidate Gene Studies

During the course of these studies, it came to our notice that a potentially interesting candidate gene is present on the same cosmid as marker D14S64, within the region of homozygosity (A. Swaroop, personal communication). This is the neural retina leucine zipper (NRL) gene, which is located in chromosomal region 14q11.1-q11.2 (Swaroop et al., 1992). The NRL gene is a highly conserved, retina - specific gene containing a basic motif, consistent with a DNA binding domain, a leucine zipper region, often utilised in protein

Table 1. Ethnic origin of families used in this study of Bardet-Biedl syndrome

UK (Edinburgh)	BB1, BB2
UK (London)	BB4, BB5, BB6, BB7, BB8, BB9, BB15, BB17, BB19, BB21
Norway	BB12, BB13, BB14, BB18
Denmark	BB23, BB24, BB25/26, BB27
Sweden	BB22
Switzerland	BB10, BB11
Egypt	BB20

Total = 25 families (24 kindreds).

dimerisation and consensus phosphorylation sites. Its function is not yet clear but it has been implicated in the control of rhodopsin expression by specifically binding to upstream regulatory sequences at the rhodopsin locus (Farjo et al., 1993). The gene is comparatively small, with 3 exons spanning 6.4 kb of DNA and is currently being screened for mutations in the five families showing linkage to this region.

BARDET-BIEDL SYNDROME STUDY

Bardet-Biedl syndrome is an autosomal recessive disorder characterised by retinitis pigmentosa, polydactyly, obesity, hypogonadism, renal anomalies and mental retardation (Green et al., 1989). The retinal degeneration comes under the general rubric of RP but often does not show the characteristic pigmentary deposits of RP. It is associated with a severe central and peripheral dystrophy frequently resulting in blindness by the end of the second or third decade. The prevalence of Bardet-Biedl syndrome is not known with accuracy but the comparative rarity of the disorder, together with the specificity of the clinical features, initially suggested that it might be genetically homogeneous. Although clinical heterogeneity is quite marked both within and between families, much of this could be attributed to allelic heterogeneity and genetic background effects. However, soon after commencing these studies, linkage was reported between Bardet-Biedl syndrome and the D16S408 and flanking markers in chromosomal region 16q21 in an extended Bedouin kindred (Kwitek-Black et al., 1993) and then in a second large Bedouin kindred to markers on the short arm of chromosome 3 (Sheffield et al., 1994). A third locus was reported in a subset of 17 out of 31 families using the PYGM/D11S913 markers in chromosomal region 11q13 (Leppert et al., 1994). In order to try and confirm these results we have analysed linkage to the 16q21 and 11q13 regions in a set of 26 Bardet-Biedl families, each containing at least two affected individuals, with a total of 57 affected members. The ethnic origin of the families is shown in Table 1.

(i) Two-Point lod Scores

The results of two-point linkage analysis between Bardet-Biedl syndrome and D11S913, D11S987, D11S916, D11S527, D11S906 and D11S901 are shown in Table 2a: the results with D16S408, D16S400 and D16S421 are shown in table 2b. The combined lod scores are strongly negative in both cases, excluding the possibility that all families are linked to one or other locus. The highest lod scores in the combined families are 2.61 ($\theta = 0.20$) with D11S527 and 2.49 ($\theta = 0.10$) with D11S987, suggesting the possibility of genetic heterogeneity.

Table 2. (a) Results of two-point linkage analysis with chromosome 11 markers in 26 Bardet-Biedl families. (b) Results of two-point linkage analysis with chromsome 16 markers in 26 Bardet-Biedl families. The lod scores and corresponding recombination fractions are shown

(a) Linkage analysis with chromosome 11 markers

Locus	0.00	0.01	0.05	0.1	0.2	0.3	0.4
D11S913	- ∞	-9.55	-2.50	-0.22	**0.85**	0.62	0.18
D11S987	- ∞	-3.75	1.24	**2.49**	2.34	1.30	0.35
D11S916	- ∞	-11.78	-3.17	-0.42	**0.92**	0.74	0.26
D11S527	- ∞	-6.77	0.32	2.29	**2.61**	1.60	0.49
D11S906	- ∞	-1.17	1.14	**1.65**	1.42	0.82	0.26
D11S901	- ∞	-13.67	-4.86	-1.87	-0.03	**0.26**	0.13

(b) Linkage analysis with chromosome 16 markers

Locus	0.00	0.01	0.05	0.10	0.20	0.30	0.40
D16S408	- ∞	-12.28	-4.20	-1.41	0.28	**0.41**	0.15
D16S400	- ∞	-16.39	-6.85	-3.35	-0.83	-0.11	**0.01**
D16S421	- ∞	-15.80	-6.73	-3.42	-0.99	-0.23	**-0.03**

(ii) Multipoint and Heterogeneity Analyses

Multipoint analysis was carried out on all the families using the data from the 6 chromosome 11 markers by serial 3-point analyses with the LINKMAP programme (Lathrop et al., 1984). Heterogeneity analysis of these data using the HOMOG programme (Terwilliger and Ott, 1994) showed evidence of genetic heterogeneity (P = 0.023) with 50% of families linked to chromosome 11q13 markers (P = 0.003). On the basis of these results, 17 families with positive lod scores for the chromosome 11 region were combined and re-analysed for genetic heterogeneity. The results showed that there was no longer evidence

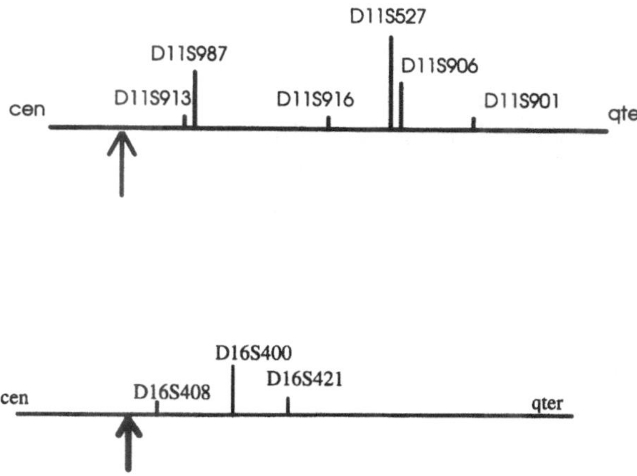

Figure 3. (a) Schematic diagram of genetic markers around chromosomal band 11q13 chromosome showing the position of the Bardet-Biedl syndrome gene located in this study. (b) Schematic diagram of genetic markers around chromosomal band 16q21 showing the position of the Bardet-Biedl syndrome gene identified in this study.

of heterogeneity (P = 0.5) and the evidence for linkage was increased ($\chi^2 = 26.16$, $p < 0.00001$). The maximum likelihood location of the Bardet-Biedl locus was 6 cM proximal to D11S913 (Figure 3a) with a lod score of 5.68. The approximate 95% confidence interval extends 15 cM proximally from D11S913. This localisation for the Bardet-Biedl syndrome locus is similar to that identified by Leppert et al. (1994) at the PYGM locus, 3 cM proximal to D11S913. Interestingly, a second possible location for the disease gene was found in the present study 18 cM away, at a site 1cM distal to marker D11S916 (P=0.11). This emphasizes the wide confidence limits when mapping a heterogeneous disease in a series of small families.

Similarly, heterogeneity analysis was carried out using the multipoint data obtained with chromosome 16q21 markers. The results showed evidence of genetic heterogeneity at a borderline significance level (P=0.053) with 27% of families linked to this region. A subset of 11 families showing positive lod scores with chromosome 16 markers was re-analysed for genetic heterogeneity which showed that there was no longer evidence of heterogeneity. The Bardet-Biedl syndrome locus was linked to a site 2 cM proximal to D16S408 (Figure 3 b) with a lod score of 5.45. The approximate 95% confidence interval extended 10 cM proximal to D16S408 and 2 cM distal to it. The Utah group previously found tight linkage to D16S408 but the closest flanking markers were D16S419 and D16S265 separated by 18 cM, defining the region of search (Leppert et al., 1994).

Unlinked Families and Clinical Differences

An estimated 23% of families are unlinked to both the 11q13 and 16q21 loci. There were no clinical features that distinguished families linked to chromosome 11, chromosome 16 or those unlinked to either. No evidence was found of linkage to the chromosome 3 markers reported by Sheffield et al. (1994) to be linked in a large Bedouin Bardet-Biedl syndrome kindred.

SUMMARY AND CONCLUSIONS

Autosomal Recessive RP in the Sardinian Population

The result of analysing over 200 microsatellite markers for evidence of linkage to 11 Sardinian autosomal recessive RP families was to highlight one chromosomal region potentially involved in this disease. Prevous studies excluded most of the known sites for recessive RP loci, including loci coding for the cyclic GMP phosphodiesterase beta subunit (McLaughlin et al., 1993), rhodopsin (Rosenfeld et al., 1992), cyclic nucleotide gated channel α subunit (McGee et al., 1994) and the chromosome 1q (van Soest et al., 1994) and 6p (Knowles et al., 1994) sites, in the majority of families. However, the linkage data supported but did not fully confirm a locus in chromosomal region 14q11.1-q11.2 in a subset of families. This result was based on heterogeneity analysis using marker D14S80 and the demonstration of a region of homozygosity spanning four adjacent markers (D14S275, D14S64, D14S80, D14S262) in a consanguineous pedigree. Current work is focusing on analysis of a candidate gene within the region. Future linkage studies will focus on a smaller number of extended families in order to circumvent the problem of locus heterogeneity.

Bardet-Biedl Syndrome

These results support the presence of genetic heterogeneity in Bardet-Biedl syndrome with loci on chromosomes 11, 16 and at least one other locus. The most common form of

Bardet-Biedl syndrome in this heterogeneous set of families is linked to the 11q13 locus. About 50% of families are 11q13-linked, 27% are 16q21-linked and 23% unlinked to either locus. The location of the 11q13 locus appears to be about 6 cM proximal to D11S913, which is close to that reported by Leppert et al. (1994). The location of the chromosome 16 locus was found to be 2cM proximal to D16S408, similar to that reported by Kwitek-Black et al. (1993) in a large consanguineous pedigree. There are no obvious clinical differences between the three genetic types of Bardet-Biedl syndrome identified.

ADVANTAGES AND DISADVANTAGES OF MAPPING AUTOSOMAL RECESSIVE RP GENES

These results illustrate both the advantages and disadvantages of mapping autosomal recessive genes by linkage analysis. The major disadvantage is the presence of locus heterogeneity which is extraordinarily high in retinitis pigmentosa. Even within the relatively rare syndromal form of recessive RP, Bardet-Biedl syndrome, at least 4 loci have been identified or inferred by linkage studies. Two of these studies were carried out on extended inbred kindreds which were large enough to confirm or exclude linkage by themselves. This study and that of Leppert et al. (1994) consisted of pooled families from more diverse ethnic backgrounds and both identified evidence for locus heterogeneity. Large samples are required for this approach to be successful and the fine mapping of these loci in heterogeneous sets is problematic. The main advantage of mapping recessive disorders lies in the use of homozygosity mapping methods when consanguineous pedigrees are available. This provides a powerful means of mapping loci even with comparatively small numbers of inbred families and probably provides the best means of advancing linkage studies in the presence of locus heterogeneity.

ACKNOWLEDGEMENTS

We would like to thank the British Retinitis Pigmentosa Society and Gift of Thomas Pocklington for their generous financial support.

REFERENCES

Ammann F, Klein D, Bohringer HR: J. Genet. Hum. 1961;10:99-127.
Boughman JA and Fishman GA.: Br J Ophthalmol 1983;67:449-454.
Bundey S and Crews SJ: J Med Genet 1984; 21:421-428.
Faber J: Retinitis pigmenosa in Israel. Statistical-clinical survey. MD Thesis, Jerusalem, Hebrew University, Vision Research Laboratory, 1970.
Farjo QA et al/: Invest Ophthlmol Vis Sci 1993;34:1457.
Fossarello M et al. : In JG Holleyfield, RE Anderson, N Orzalesi, (eds): Retinal Degeneration. New York, Plenum Press, 1993, pp.79-90.
Green JS et al. New Engl J Med 1989:321, 1002-1009.
Grondahl J:. Clin Genet 1986;29:17-41.
Heckenlively JR: "Retinitis Pigmentosa", Philadelphia, J.B. Lippincott Company, 1988, pp.1-269.
Humphries P et al.: Science 1992;256:804-808.
Jay M Br J Ophthalmol 1982;66:405-416.
Kajiwara K et al. Science 1994;264:1604-1608.
Knowles JA et al. Hum Molec Genet 1994;3:1401-1403.
Kwitek-Black et al.: Nature Genetics 1993;5:392.
Lander ES and Botstein D Science 1987; 236:1567-1570.

Lathrop GM et al.: Proc Natl Acad Sci USA 1984;81:3443-3446.
Leppert M et al.: Nature Genetics 1994;7:108.
Mansfield DC et al.: Genomics 1994; 24: 225-233.
McGee TL et al. : Invest. Ophthalmol. Vis. Sci. 1994;35:2154.
McLaughlin ME et al. : Nature Genet 1993;4:130-134.
Rosenfeld PJ et al.: J Med Genet 1994; 31:903-915.
Rosenfeld PJ et al.: Nature Genet 1992;1:209-213.
Sheffield VC et al. Hum Molec Genet 1994;3:1331-1335.
Swaroop A et al., 1992 Proc Natl Acad Sci USA 1992; 89:266-270.
Tanabe U: Jinrui Idengaku Zasshi 1972;16:119-154.
Terwilliger JD and Ott J.: In Handbook of Human Genetic Linkage. Johns Hopkins Univ Press, Baltimore, 1994.
van Soest S et al. Genomics 1994: 22, 499-504.

CLINICAL FEATURES OF AUTOSOMAL DOMINANT RETINITIS PIGMENTOSA ASSOCIATED WITH THE Ser 186TRP MUTATION OF RHODOPSIN

K. Rüther,[1] C. L. v. Ballestrem,[2] A. Müller,[2] S. Kremmer,[1] A. Eckstein,[1] E. Apfelstedt-Sylla,[1] A. Gal,[2,3] and E. Zrenner[1]

[1] Universitäts-Augenklinik
Schleichstrasse 12-16, D-72076 Tübingen, Germany
[2] Institut für Humangenetik, Medizinische Universität
Ratzeburger Allee 160, D-23538 Lübeck, Germany
[3] Institut für Humangenetik, Universitäts-Krankenhaus Eppendorf
Butenfeld 32, D-22529 Hamburg, Germany

SUMMARY

To date more than 60 different heterozygous mutations of the rhodopsin gene have been identified and suggested to be causative in autosomal dominant retinitis pigmentosa (adRP). Here we report two families segregating for an inherited retinal degeneration with most likely autosomal dominant inheritance of the trait. In the first family, a novel heterozygous missense mutation of the rhodopsin gene, predicting a Ser186Trp change in the protein, results in adRP with incomplete penetrance, as the 52-year-old grandmother, heterozygote for the mutation, does not show any clinical abnormalities whereas two other carriers suffer from retinal degeneration of moderate severity. In the second family, a novel point mutation in exon IV of the rhodopsin gene predicting the Gly284Ser alteration was identified. However, the mutation did not cosegregate with the disease phenotype in the family. Of the eight family members carrying this alteration, three have not shown any abnormality on clinical examination. More importantly, two clinically affected persons did not carry the rhodopsin alteration, which strongly suggests that the Gly284Ser change is non-pathogenic in the heterozygous state.

INTRODUCTION

Since the first report on the heterozygous Pro23His mutation of rhodopsin in patients/families with autosomal dominant retinitis pigmentosa (adRP; ref. 1) more than 60

Degenerative Diseases of the Retina, Edited by Robert E. Anderson et al.
Plenum Press, New York, 1995

further mutations of the rhodopsin gene have been described (for review see 2). These mutations may result in largely variable clinical phenotypes (3-10). However, a tight correlation between genotype and phenotype has not been established, yet. Therefore it is of interest to characterize the clinical phenotype associated with any novel rhodopsin mutation that in the long run may also be helpful in counselling of patients with respect to their visual prognosis.

Here we report the clinical data obtained on members of a family with adRP due to a novel heterozygous missense mutation of the rhodopsin gene. In addition, a heterozygous missense mutation in exon 4 is described that does not cosegregate with the disease phenotype in a large family with adRP.

Methods

For each patient a questionnaire was completed with respect to age of onset of night blindness, visual field restriction, glare sensitivity, and visual acuity. Clinical examination included visual acuity, slit-lamp examination, fundus photograph, colour vision testing, and electroretinogram (ERG). In some cases, dark adaptation testing was performed with a 106 minutes of arc stimulus presented in a modified kinetic perimeter. With regard to the ERG, rod responses (standard flash minus 2 log units neutral density filter), rod-cone responses (standard flash in the dark adapted eye), and the 30-Hz light adapted responses dominated by cones were recorded. In affected individuals, kinetic perimetry, in family members without subjective symptoms automatic perimetry was carried out (Tuebingen automatic perimeter). All methods have been described in detail previously (6,7).

Screening for mutations in the rhodopsin gene was carried out by single strand conformation polymorphism and heteroduplex analyses. Direct sequencing of DNA fragments amplified by polymerase chain reaction (PCR) was performed if abnormalities were detected by either of the two methods. Details of the molecular genetic methods used have been described elsewhere (11).

RESULTS

Family 1

The pedigree of the family is shown in Fig. 1. There are two affected persons in two consecutive generations.

The index patient (IV.1) was a 6-year-old girl. She reported having difficulties to adapt from bright to darkened environment. Visual fields were measured by kinetic perimetry and the Goldmann III-4e target and were found being essentially normal (Fig. 2), indicating a relatively good cone function. The rod-ERG was markedly reduced while the cone-ERG revealed clearly detectable potentials (table 1). On funduscopy, optic discs were pink and retinal vessels had a normal caliber. At the posterior pole, the foveolar reflex was present but the macula showed tiny defects of the pigment epithelium. In the periphery, there were only very few and tiny pigment deposits. It is worth mentioning that viusal acuity of the patient was only 0.5 OU, that is much less than that of her affected mother (0.8 OU at the age of 30), suggesting a more severe phenotype. A summary of the clinical findings is given in Table 1.

The retinal degeneration of the mother (III.2) of the index case was diagnosed at the age of six. At that time she complained about night vision problems and visual field losses. She noticed no problems with colour vision. At the age of 28, she realized a reduction of visual acuity. One year later, she had cataract surgery on her right eye. At the time of our

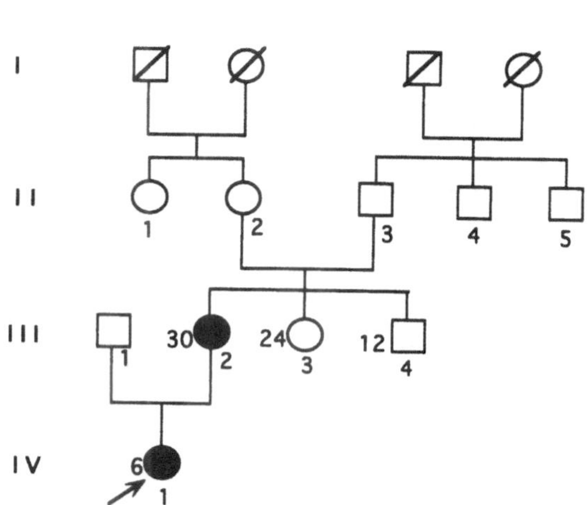

Figure 1. Pedigree of family 1. Left side numbers indicate age.

examination she was 30 years old. There was a considerable constriction of visual fields, more advanced in her right eye (Fig. 3). The ERG-responses of rods and cones were barely recordable (see table 1). Fundus examination revealed atrophic discs and severely attenuated vessels in both eyes. At the posterior pole and in the fundus periphery, a diffuse atrophy of the pigment epithelium was evident. There were only few bone spicule pigments, in part paravascular. Further clinical findings are again summarized in Table 1.

DNA analysis revealed both in the index patient and her mother a heterozygous missense mutation (TCG › TGG) at codon 186 of the rhodopsin gene predicting the amino acid exchange of serine-186 to tryptophan. The unaffected sister (III.3) of the mother did not carry the mutation. Similarly, the analysis of DNA samples of 60 unrelated and unaffected controls has failed to detect this particular gene alteration suggesting that it does not represent a frequent polymorphism. However, molecular genetic analysis on the clinically unaffected

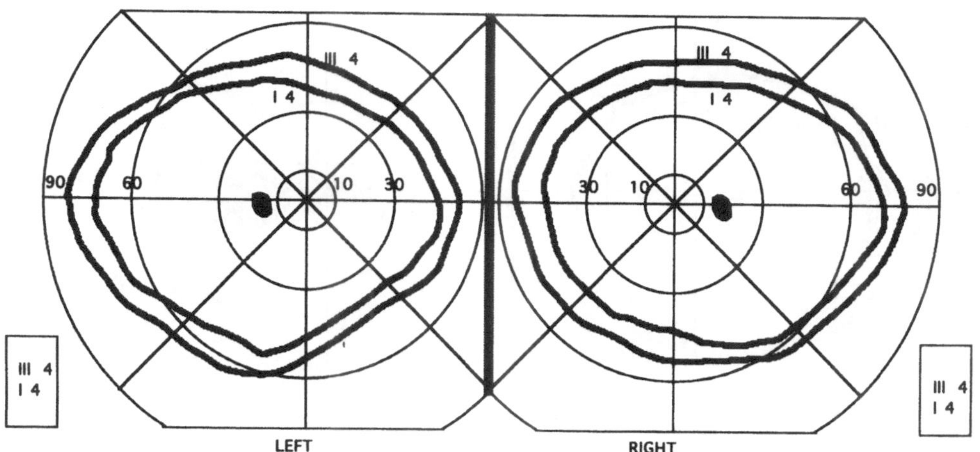

Figure 2. Kinetic perimetry of IV.1 in family 1.

Table 1. Summary of the clinical data of family 1

Patient	Visual acuity (distance)	Refraction (Dpt)	Colour vision	ERG rod response (μV)	ERG rod/cone response (μV)	ERG-30 Hz cone response (μV)
Mother	RE 0.8	RE +1.0 IOL	minimal changes	not detectable	R/L < 10	RE 7.0
(III.2)	LE 0.8	LE -0.75				LE 4.5
Daughter	RE 0.5	RE -0.75	moderate changes*	R/L < 5	R/L 50	RE 35
(IV.1)	LE 0.5	LE -0.75				LE 50
Grandmother	RE 0.8	RE -0.5	normal	RE 137	RE 380	RE 62
(II.2)	LE 0.8	LE 0		LE 146	LE 420	LE 82
				(normal)	(normal)	(normal)
Grandfather	RE 0.8	RE -0.25	moderate changes*	RE 88	RE 347	RE 121
(II.3)	LE 1.0	LE -0.75		LE 112	LE 356	LE 133
				(low normal)	(normal)	(normal)

*Mainly reduction of blue sensitivity.

maternal grandmother (II.2) disclosed the same mutation in heterozygous form. At the time of her clinical examination, the grandmother was 52 years old. She reported having difficulties to see in the dark during the last years. An increased glare sensitivity has also been noted for approximately one year. However, she did not note a reduction of visual fields. Indeed, visual field testing by automated perimetry showed essentially normal results with slightly reduced sensitivity in the right eye. There was a mild exotropia of the right eye. On slit lamp examination, lenses showed beginning age-related clouding. Fundus picture showed a mild granular structure of pigment epithelium but was essentially normal. The electroretinogram was normal both in amplitude and implicit time (Table 1 and not shown). In dark adaptation testing, after 30 minutes of darkness, rod absolute threshold of the left eye, measured 20° nasally on the horizontal meridian, was elevated by half a log unit.

The grandfather (II.3), not carrying the Ser186Trp mutation, was also examined clinically. He was 53 years old and reported also having dark vision problems and elevated glare sensitivity since the age of 52. Lenses were slightly cloudy and he made some errors

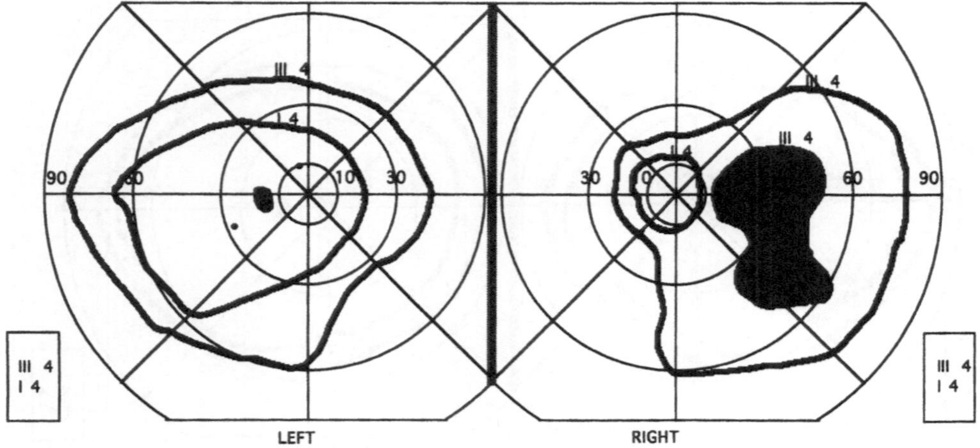

Figure 3. Kinetic perimetry of III.2 in family 1.

in the Panel-D-15 colour vision test pointing to a decrease of blue sensitivity. The macula showed first signs of age related degeneration, but otherwise the fundus was unremarkable. Rod absolute threshold as well as the ERG were normal (for all results see Table 1).

Family 2

In family 2 (Fig. 4), there are 7 affected individuals in three consecutive generations suggesting autosomal dominant inheritance of the trait. Nevertheless, X-chromosomal inheritance can not be ruled out completely, either, as there is no male to male transmission. Clinical examinations were performed only on persons from generations III and IV.

DNA samples of two affected members from the third generation were screened for rhodopsin gene mutation. The analysis revealed a nucleotide change in codon 284 (GGT › AGT) in exon IV predicting the replacement of glycin-284 by serine. The G to A transition destroys the recognition site for the restriction enzyme Ava II which is present in the wild-type rhodopsin sequence at this position. Therefore, the PCR-amplified exon IV fragment of 332 base pairs (bp) can not be cleaved into fragments of 171 bp and 160 bp if the mutation is present (see Fig. 5). DNA samples of a total of 12 members from generations II-IV were analyzed by Ava II digestion following PCR amplification of exon IV, to identify carriers of the alteration. Five individuals are heterozygous for the Gly284Ser change and affected by a retinal degeneration.

II.2 was examined only by DNA analysis and shown to carry the alteration in heterozygous form. Medical reports said that he was affected by RP. Apart from that he had also glaucoma. Two of the patients with the mutation (III.5, III.11) had a mild RP, as far as visual fields were concerned. Using a bright and large target (Goldmann III/4), visual fields were quite normal, whereas for smaller and less intense stimuli there was a marked reduction in sensitivity in both cases (see visual fields of III.11 in Fig. 6). In contrast, ERG-responses for rods and cones were barely detectable. Subjectively these individuals were unaware of any visual field constriction. Individual IV.15 was just five-years-old when he was examined. He showed subtle retinal pigment epithelium changes, a reduction of ERG rod potentials, but cone potentials in the normal range. His rod absolute threshold was elevated by about one log unit. Visual field testing was not reliable because of the reduced compliance. In the fifth patient with the mutation (III.4), the phenotype was rather different than in III.5 and III.11, in that the visual fields were much more constricted with temporal islands in both eyes (see visual fields in Fig. 7); however, ERG responses were comparable to III.5 and III.11. In contrast, patient III.1 had the same mild RP phenotype as III.5 and III.11, but does not carry the Gly284Ser alteration. The same was true for IV.14, a twelve-year-old girl, suffering from an early form of the disease. She had normal visual acuity, normal visual fields using the Goldmann III/4 target, but a severely reduced rod-ERG. The cone ERG was still within normal limits. On funduscopy she showed slightly attenuated vessels, a diffuse pigment epitheliopathy and some tiny pigment deposits. All five exons of the rhodopsin gene of III.1 adn IV.14 were sequenced but no alteration was found. Restriction analysis of DNA samples of 60 unrelated and unaffected controls failed to detect this mutation.

Three individuals (III.2, III.8, and III.9) were heterozygote for the rhodopsin alteration without presenting any signs of RP. Visual fields, dark adaptation, and the ERG were normal.

DISCUSSION

Polymorphisms in the coding region of rhodopsin gene are infrequent (see 12 and references therein). Most of them are “silent” (samesense) third-position nucleotide ex-

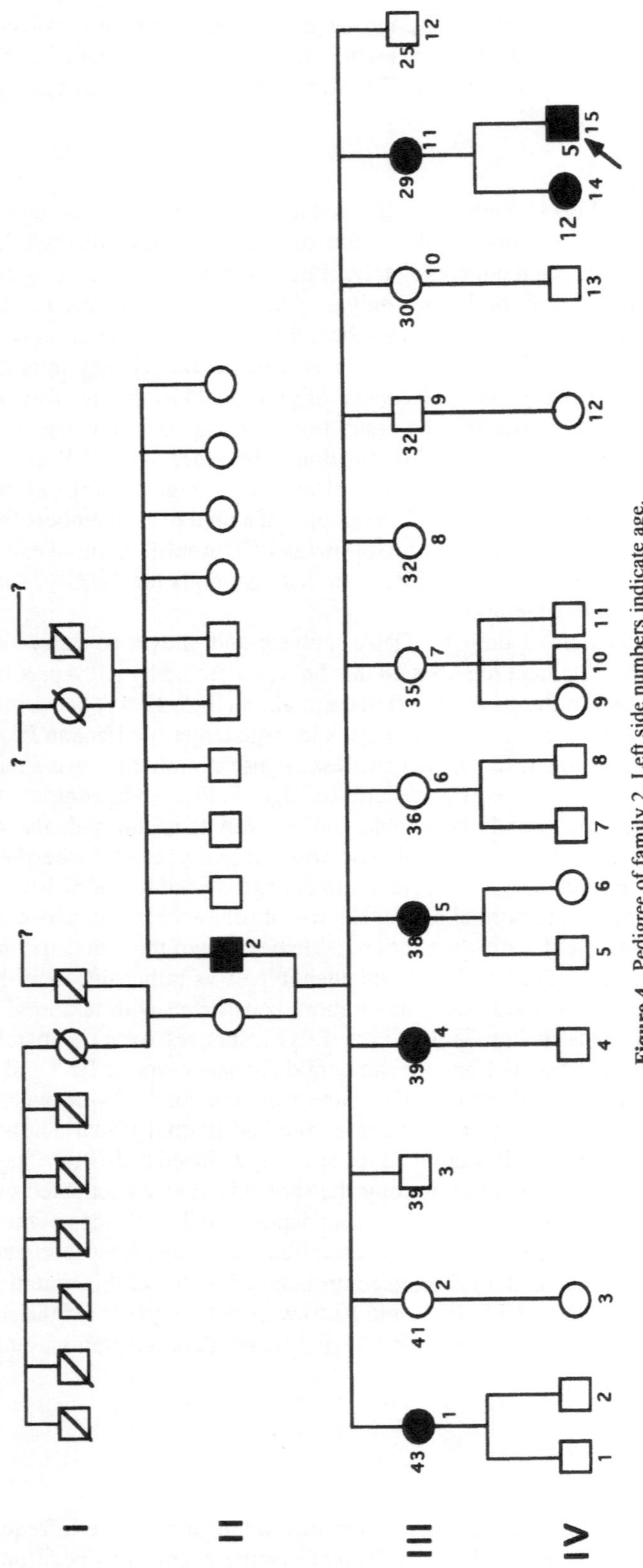

Figure 4. Pedigree of family 2. Left side numbers indicate age.

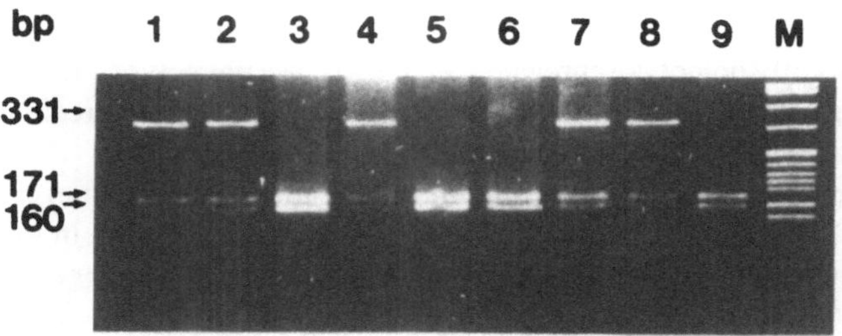

Figure 5. Detection of the heterozygous G to A transition of the first nucleotide of codon 284 by Ava II digestion of PCR-amplified exon IV of the rhodopsin gene. For explanation see text. Lanes 1,2,4,7 and 8: individuals II.2, III.2, III.4, III.5, and III.8; lanes 3,5,6 and 9: individuals III.1, III.3, IV.14, and an unaffected and unrelated control.

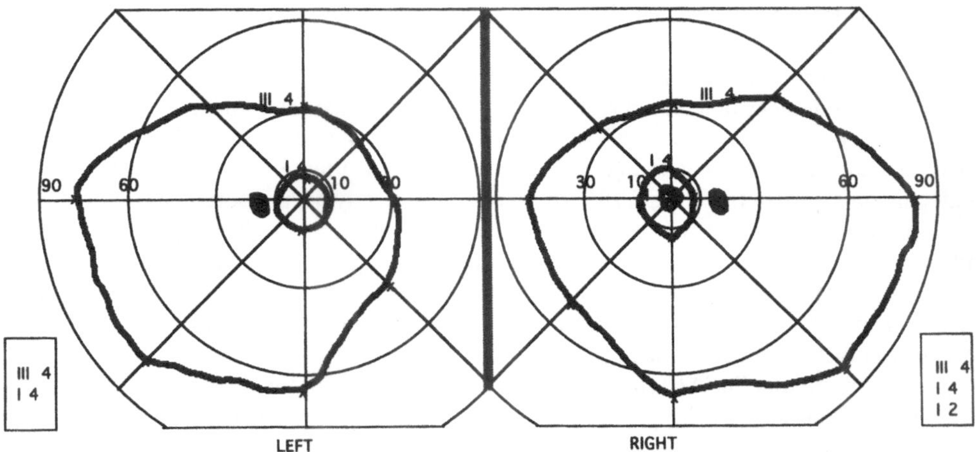

Figure 6. Kinetic perimetry in III.11.

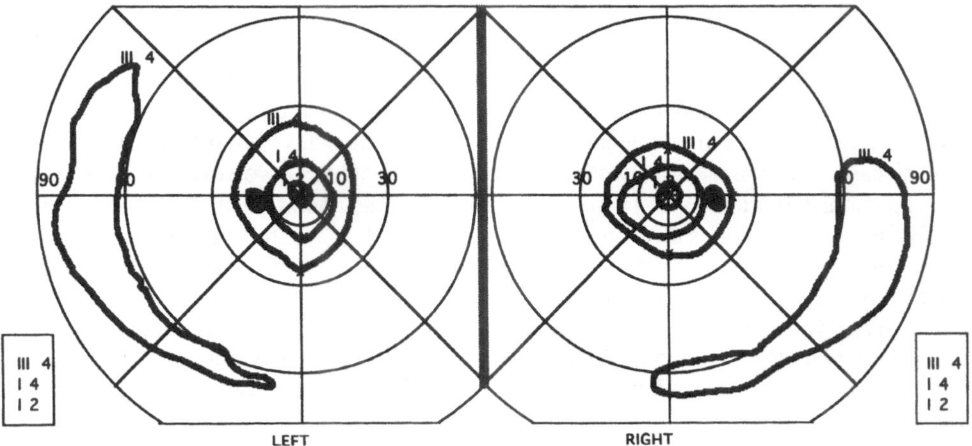

Figure 7. Kinetic perimetry in III.4.

changes. Heterozygous missense mutations of the rhodopsin gene not leading to retinal degeneration are extremely rare. There have been several reports on rhodopsin gene mutations which apparently do not cosegregate with the disease phenotype (13-16). However, the possibility of variable expression, incomplete penetrance, autosomal recessive inheritance, or digenic inheritance should always be considered in such cases.

In this report we described two families with adRP and mutations in the rhodopsin gene. In family 1, there are three heterozygous carriers of Ser186Trp in three generations. Two of them (mother and daughter) show signs of a moderate retinal degeneration. The third individual (grandmother) should be regarded clinically as unaffected. The disturbed night vision and elevated glare sensitivity might be due to the begining cataract. Although rod sensitivity is slightly elevated, by half a log unit, the ERG is completely normal without any signs of a rod defect. A granular appearance of the fundus, as observed in the grandmother, can also be seen in unaffected controls.

One explanation for the absence of retinal dystrophy in the grandmother can be that the disease (of all three family members) is due to a mutation of another gene and Ser186Trp is not pathogenic but represents a rare sequence variant (polymorphism). Residue 186 is in the second extracellular/intradiscal loop of rhodopsin next to cysteine-187, an amino acid which is part of the disulphide bond formed with cysteine-110. This disulphide bridge is thought to be very important for the proper folding, processing, intracellular transport, and function of rhodopsin. Indeed, there are numerous reports that the replacement of residues cysteine-110 or cysteine-187 as well as that of neighbouring amino acids is associated with adRP (see 17 and references therein). A different missense mutation of codon 186 of the rhodopsin gene, predicting the Ser186Pro change, has already been reported in a patient with RP (18). Sung and coworkers examined this latter alteration by in vitro expression studies and provided evidence that the protein is afunctional in that it does not reach the plasma membrane but stays in the endoplasmic reticulum, and fails to reconstitute with 11-cis-retinal (19,20). Nevertheless it remains to be proven whether or not the same is true for the Ser186Trp mutant described here.

Assuming that Ser186Trp is the primary genetic cause of the disease in family 1, the lack of RP-symptoms in the grandmother can be due to autosomal dominant RP with incomplete penetrance. Berson and coworkers examined extensively families with retinitis pigmentosa and apparently incomplete penetrance (13,14). As the authors found mild abnormalities in the ERG of subjectively unaffected individuals they suggested that in fact these cases represent variable expression of the mutation. Similarly, Moore and coworkers studied a large family with adRP in which several individuals had an affected parent and an affected offspring but were subjectively asymptomatic (21). Again the authors found evidence of mild abnormalities in rod and cone function and interpreted these cases as variation in expression rather than true non-penetrance of the gene mutation.

Of the 8 individuals heterozygote for the Gly284Ser change of rhodopsin in family 2, five are affected by a retinal degeneration. As three carriers have not presented any signs of the disease, this observation would suggest non-penetrance. However, there are two affected persons (III.1 and IV.15) who do not carry the gene alteration at all. Therefore, it is unlikely that this mutation is the primary genetic cause of the RP in this family. Furthermore, our findings suggest that Gly284Ser in heterozygous state is an apathogenic rare sequence variant (polymorphism). In fact, from the biochemical point of view, there is little difference between glycine and serine, as both are neutral aliphatic amino acids. In addition, codon 284 is in the third extracellular loop of rhodopsin, which is evolutionarily not conserved among receptors belonging to the rhodopsin family (see 22) suggesting that this residue/this region of rhodopsin is not absolutely essential for the function of the molecule.

Macke and coworkers (16) identified a heterozygous missense mutation prediciting the Val104Ile change of rhodopsin in a patient with Leber congenital amaurosis, an autosomal recessive trait. Also this mutation did not coinherit with retinal disease in the patient's family suggesting that Val104Ile is a polymorphism in heterozygous state. In this case, too, the wild-type and mutant amino acids are rather similar from biochemical point of view ("conservative" amino acid exchange). Again, valine-104 is in the first extracellular loop, a region of the rhodopsin molecule which is not strictly conserved during evolution in related proteins.

Nevertheless, it it possible that both Gly284Ser and Val104Ile are pathogenic in the homozygous state. Indeed, Kumaramanickavel and coworkers reported a family with autosomal recessive retinitis pigmentosa in which affected members were found to carry a homozygous Glu150Lys mutation in the rhodopsin gene (23). However, the patients' parents and siblings, heterozygote for the same alteration, did not show any abnormalities on ophthalmological and psychophysical examinations.

In conclusion, data presented in this report underlines that heterozygous missense mutations of rhodopsin should not necessarily result in retinal degeneration due to largely variable expression, incomplete penetrance, or the complete absence of pathogenicity of the mutation in question.

ACKNOWLEDGEMENTS

The study was supported by the German Research Council (DFG projects Ru 457/1-3, Ga 210/5-3 and Zr 1/7-3)

REFERENCES

1. Dryja, T.P., McGee, T.L., Reichel, E., Hahn, L.B., Cowley, G.S., Yandell, D.W., Sandberg, M.A., Berson, E.L., 1990, A point mutation of the rhodopsin gene in one form of retinitis pigmentosa, *Nature* 343:364-366.
2. Humphries, P., Kenna, P., Farrar, G.J., 1994, Autosomal dominant retinitis pigmentosa, in: Wright, A.F., Jay, B. (Eds), Molecular genetics of inherited eye disorders, *Harwood Academic Publishers*, Chur, pp. 153-172.
3. Fishman, G.A., Stone, E.M., Gilbert, L.D., Sheffield, V.C., 1992, Ocular findings associated with a rhodopsin gene codon 106 mutation, *Arch. Ophthalmol.* 110:646-653.
4. Fishman, G.A., Stone, E.M., Sheffield, V.C., Gilbert, L.D., Kimura, A.E., 1992, Ocular findings associated with rhodopsin gene codon 17 and codon 182 transition mutations in dominant retinitis pigmentosa, *Arch. Ophthalmol.* 110:54-62
5. Richards, J.E., Kuo, C.Y., Boehnke, M., Sieving, P.A., 1991, Rhodopsin Thr58Arg mutation in a family with autosomal dominant retinitis pigmentosa, *Ophthalmology* 98:1797-1805 (1991)
6. Apfelstedt-Sylla, E., Bunge, S., David, D., Ruether, K., Gal, A., Zrenner, E., 1993, Phenotypes of carboxyl-terminal rhodopsin mutations in autosomal dominant retinitis pigmentosa, in: Hollyfield, J.G. (Ed.) *Retinal Degeneration*, Plenum Press, New York, pp 115-125.
7. Apfelstedt-Sylla, E., Kunisch, M., Horn. M., Ruether, K., Gerding, H., Gal, A., Zrenner, E., 1993, Ocular findings in a family with autosomal dominant retinitis pigmentosa and a frameshift mutation altering the carboxyl terminal sequence of rhodopsin, *Br. J. Ophthalmol.* 77: 495-501.
8. Apfelstedt-Sylla, E., Theischen, M., Ruether, K., Meins, J., Grüning. G., Gal, A., Zrenner, E., 1995, Extensive intrafamilial and interfamilial phenotypic variation among patients with autosomal dominant retinal dystrophy and mutations in the human RDS/peripherin gene *Br. J. Ophthalmol.* 79:28-34.
9. Berson, E.L., Rosner, B., Sandberg, M.A., Weigel-DiFranco, C., Dryja, T.P., 1991, Ocular findings in patients with autosomal dominant retinitis pigmentosa and rhodopsin, proline-347-leucine *Am. J. Ophthalmol.* 111:614-623.
10. Berson, E.L., Rosner, B., Sandberg, M.A., Dryja, T.P., 1991, Ocular findings in patients with autosomal dominant retintis pigmentosa and a rhodopsin gene defect (Pro-23-His), *Arch. Ophthalmol.* 109:92-101.
11. Bunge, S., Wedemann, H., David, D., Terwillinger, D.J., Aulehla-Scholz, C., Sammans, C., Horn, M., Ott, J., Schwinger, E., Bleeker-Wagemakers, E.M., Schinzel, A., Gal, A., 1993, Molecular analysis and genetic

mapping of the rhodopsin gene in families with autosomal dominant retinitis pigmentosa, *Genomics* 17:230-233.

12. Al-Maghtheh, M., Gregory, C., Inglehearn, C., Hardcastle, A., and Bhattacharya, S.S., 1993, Rhodopsin mutations in autosomal dominant retinitis pigmentosa, *Human Mutation* 2:249-255.
13. Berson, E.L., Gouras, P., Gunkel, R.D., Myrianthopoulos, N.C., 1969, Dominant retinitis pigmentosa with reduced penetrance. *Arch. Ophthalmol.* 81:226-234.
14. Berson, E.L., Simonoff, E.A., 1979, Dominant retinitis pigmentosa with reduced penetrance: Further studies of the electroretinogram, *Arch. Ophthalmol.* 97: 1286-1291.
15. Kim, S.K., Haines, J.L., Berson, E.L., Dryja, T.P., 1994, Nonallelic heterogeneity in autosomal dominant retinitis pigmentosa with reduced penetrance, *Invest. Ophthalmol. Vis. Sci.* 35:1479.
16. Macke, J.P., Davenport, C.M., Jacobson, S.G., Hennessey, J.C., Gonzalez-Fernandez, F., Conway, B.P., Heckenlively, J., Palmer, R., Maumenee, I.H., Sieving, P. et al., 1993, Identification of novel rhodopsin mutations responsible for retinitis pigmentosa: implications for the structure and function of rhodopsin, *Am. J. Hum. Genet.* 53: 80-89.
17. Del Porto, G., Vingolo, E.M., David, D., Steindl, K., Wedemann, H., Forte, R., Iannaccone, A., Gal, A., Pannarale, M.R., 1993, Clinical features of autosomal dominant retinitis pigmentosa associated with the GLY-188-ARG mutation of the rhodopsin gene, in: Hollyfield, J.G. (Ed.) *Retinal Degeneration*, Plenum Press, New York, 91.
18. Dryja, T.P., Hahn, L.B., Cowley, G.S., McGee, T.L., Berson, E.L., 1991, Mutation spectrum of the rhodopsin gene among patients with autosomal dominant retinitis pigmentosa, *Proc. Natl. Acad. Sci. USA* 88:9370-9374
19. Sung, C.H., Schneider, B.G., Agarwal, N., Papermaster, D.S., Nathans, J., 1991, Functional heterogeneity of mutant rhodopsins responsible for autosomal dominant retinitis pigmentosa, *Proc. Natl. Acad. Sci. USA* 88:8840-8844.
20. Sung, C.H., Davenport, C.M., Nathans, J., 1993, Rhodopsin mutations responsible for autosomal dominant retinitis pigmentosa. Clustering of functional classes along the polypeptide chain, *J. Biol. Chem.* 268: 26645-26649.
21. Moore, A.T., Fitzke, F., Jay, M., Arden, G.B., Inglehearn, C.F., Keen, T.F., Bhattacharya, S.S., Bird, A.C., 1993, Autosomal dominant retinitis pigmentosa with apparent incomplete penetrance: a clinical, electrophysiological, psychophysical, and molecular genetic study, *Br. J. Ophthalmol.* 77: 469-70.
22. Hargrave, P.A., O'Brien, P.J., 1991, Speculations on the molecular basis of retinal degeneration in retinitis pigmentosa, in: Anderson, R.E., Hollyfield, J.H., LaVail, M.M. (Eds) Retinal Degenerations. *CRC Press Inc.* Boca Raton, pp. 517-528.
23. Kumaramanickavel, G., Maw, M., Denton, M., John, S., Srikumari, C.R.S., Orth, U., Oelmann, R., Gal, A., 1994, Missense rhodopsin mutation in a family with recessive RP, *Nature Genetics* 8: 10-11.

MUTATIONS IN THE GENE FOR THE B-SUBUNIT OF ROD PHOTORECEPTOR CGMP-SPECIFIC PHOSPHODIESTERASE (PDEB) IN PATIENTS WITH RETINAL DYSTROPHIES AND DYSFUNCTIONS

A. Veske,[1*] U. Orth,[1] K. Rüther,[2] E. Zrenner,[2] T. Rosenberg,[3] W. Baehr,[4] and A. Gal[1]

[1] Institut für Humangenetik, Universitäts-Krankenhaus Eppendorf
Butenfeld 32, D-22529 Hamburg, Germany
[2] Augenklinik der Universität, Abteilung II
Schleichstrasse 12-16, D-72076 Tübingen, Germany
[3] National Eye Clinic for the Visually Impaired
1 Rymarksvej, D-2900 Hellerup, Denmark
[4] Department of Ophthalmology, Cullen Eye Institute, Baylor College of Medicine
Houston, Texas 77030

SUMMARY

Rod photoreceptor cGMP-specific phosphodiesterase (PDE) is a key enzyme of the phototransduction cascade. Mutations in the gene encoding the β-subunit of PDE (βPDE) have been implicated in two distinct human retinal disorders; autosomal recessive retinitis pigmentosa (arRP) and autosomal dominant congenital stationary night blindness (adCSNB).

We examined the βPDE gene in 101 unrelated patients with arRP from Germany for sequence changes by single strand conformation polymorphism analysis. Band shifts were detected in 17 different gene fragments amplified by polymerase chain reaction. Direct sequencing revealed 11 single base substitutions that were considered being most likely polymorphisms. Of the remaining six, and most likely disease related exonic mutations, three (225C→T/Arg74Cys, 226insC, and 660T→C/Tyr219His) have been found in patients with only one (heterozygous) βPDE-mutation identified so far. In two patients, both βPDE-alleles

* On leave of absence from the Institute of Molecular and Cell Biology, University of Tartu, Estonia.

Degenerative Diseases of the Retina, Edited by Robert E. Anderson et al.
Plenum Press, New York, 1995

carried potentially pathogenic mutations; one patient was homozygote for the 1585T→C/Leu527Pro change while the other one was a compound heterozygote for 897C→T/Gln298stop and 2565C→G/Leu854Val. This latter patient, a 19-year-old female, presented with typical clinical, electrophysiological, and psychophysical symptoms of a rod-cone degeneration.

We have recently described a large multigeneration Danish pedigree with adCSNB, in which the photoreceptor dysfunction was cosegregating with a heterozygous His258Asn mutation of the βPDE gene. Here we report two further Danish pedigrees with adCSNB and the His258Asn mutation. Both families were shown to be unrelated to the original pedigree for (at least the last) 6 generations. Nevertheless, genotyping for 11 different intragenic DNA polymorphisms revealed that night blind subjects in the three families shared the same βPDE haplotype suggesting that the His258Asn allele originated from a common ancestor. Assuming that the three families were linked in the seventh generation, a maximum lod score of 28.615 at theta = 0.00 was obtained for the linkage relationship between the loci for adCSNB and βPDE. These data provide strong support for the hypothesis that the mutant βPDE is responsible for CSNB in this family, i.e. that the molecular defect is in the phototransduction cascade and at the photoreceptor level.

INTRODUCTION

Rod photoreceptor cGMP-specific phosphodiesterase (PDE) is a heterotetrameric enzyme ($\alpha\beta\gamma_2$) that has a central function in the phototransduction process (for review see ref. 1). cGMP hydrolysis by activated PDE results in closure of cGMP-gated cation channels leading to hyperpolarization of the cell membrane and signalling. Enzyme activity is associated with the two larger subunits of PDE, α and β, while γ-subunits have regulatory/inhibitory functions. Interestingly, mutations in the gene encoding the β-subunit of PDE (βPDE) have been implicated in two rather different human retinal disorders; autosomal recessive retinitis pigmentosa, a progressive degeneration of photoreceptors, and autosomal dominant congenital stationary night blindness, a nonprogressive photoreceptor dysfunction. Here we briefly summarize our data on the βPDE gene in these two disorders.

Methodology

Oligonucleotide primers and experimental conditions used for mutation screening and indentification following PCR amplification were essentially the same as given in the literature (2). For gel electrophoresis in single strand conformation polymorphism (SSCP) analysis, two different conditions were used routinely, and a third one occasionally to increase mutation detection efficiency (Fig. 1); (i) 8% acrylamide (cross linking 2.6), 10% glycerol, and 0.9x TBE buffer, (ii) 8% acrylamide (cross linking 1.3), 10% sucrose, and 0.9x TBE, or (iii) 6% acrylamide (cross linking 2.6), 5% glycerol, and 0.9x TBE buffer. Samples were electrophoresed in the same buffer at 15-25 W for 10-16 hours at room temperature or +4°C. Gels were silver stained.

The C→A transversion of the first nucleotide at codon 258 (His258Asn, H258N) was detected by the method of primer-specific restriction map modification ('designed mismatch' PCR). This missense mutation can be identified specifically in PCR fragments of exon 4 amplified by an oligonucleotide primer pair introducing a T to G mismatch in codon 257, which, together with the codon 258 mutation, creates a new Hinc II restriction site (for experimental details see ref. 3).

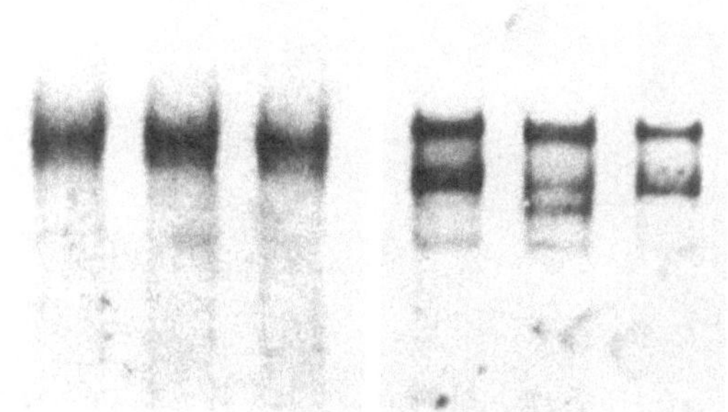

Figure 1. Detection of 'mobility shifts' by SSCP analysis of PCR-amplified DNA fragments of exon 22 using two different separation conditions. a) 8% acrylamide (cross linking 2.6) with 10% glycerol, b) 8% acrylamide (cross linking 1.3) with 10% sucrose. Lanes 1 and 3: wild type, lane 2: heterozygote for the point mutation 2565C→G (Leu854Val). Note that the 'band shift' is much more prominent by the second condition.

AUTOSOMAL RECESSIVE RETINITIS PIGMENTOSA

Retinitis pigmentosa (RP) is the most common form of inherited retinal dystrophies in men with an incidence of about 1 case in 3000-7000 individuals (4,5). In addition to X-linked, autosomal dominant (ad), and autosomal recessive (ar) RP, the majority of patients represents simplex cases, i.e. the mode of inheritance cannot be determined by family analysis. Within the three above mentioned monogenic types of RP, both allelic and non-allelic genetic heterogeneity has been demonstrated (6).

Three not yet identified arRP-genes have been recently assigned to chromosomes 1q (7,8), 6p (9), and 14q (10) by linkage studies. In addition, mutations of the gene encoding rhodopsin have been identified in two patients/families with arRP (11,12). Finally, several nonsense and missense mutations in the βPDE gene, mapped on chromosome 4p16.3, have been described recently in patients with arRP (13). Also, in animals (*rd/rd* mice and *rcd1* Irish setter), nonsense mutations in the βPDE gene have been found to be the genetic cause of retinal degeneration (16,17). It has been suggested that the loss of PDE activity results in excessive accumulation of cGMP which is deleterious for rod photoreceptor cells (18).

Mutations of the βPDE Gene in Patients with Autosomal Recessive Retinal Dystrophy

We examined the βPDE gene in 101 unrelated patients with arRP from Germany by amplifying each of the 22 exons together with the donor and acceptor splice sites and about 40 bp of intronic sequences at both sides (see Fig. 2).

SSCP band shifts were detected in 17 different PCR products from a total of 40 patients. All samples were analysed further by direct sequencing and, if feasible, by restriction digestion. Of the 17 band shifts, eleven were found in exons and six in introns. 'Silent' third-position nucleotide exchanges were detected in exon 2 [in 5 of 101 patients], exon 5 [6/101], exon 8 [1/101], exon 13 [6/101] and exon 19 [1/101; see Table 1]. A further variant was found in the donor splice site of exon 17 with GC instead of GT which has been reported originally (2) and represents the consensus donor splice site sequence. As only GC was found in the 6 independent samples, including two unaffected controls, examined in our laboratory, this might be the normal splice site for this exon. We have found a few additional minor differences when compared to the previously published human βPDE gene sequence (2,19) which are summarized in Tables 1 and 2. Several polymorphisms due to single base

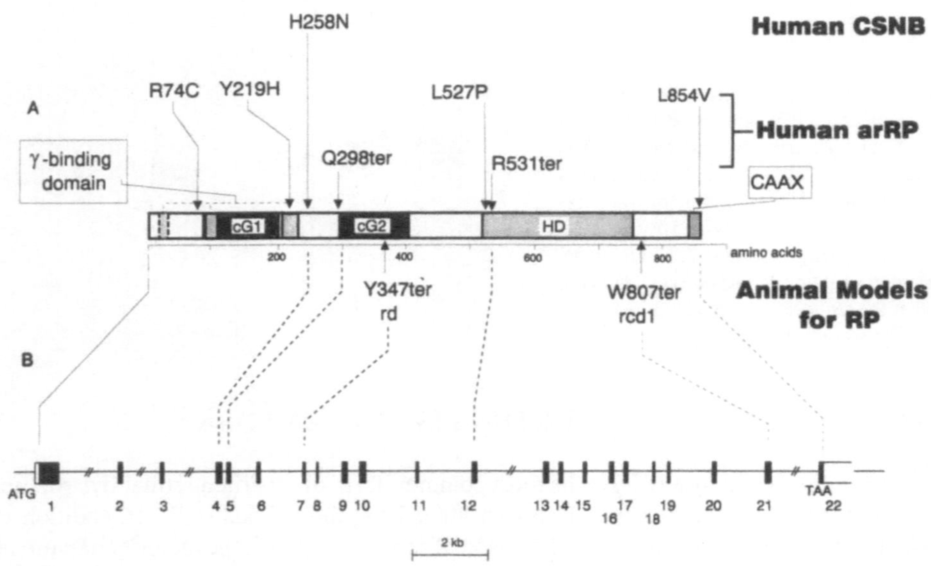

Figure 2. Mutations in the βPDE gene. A, domain structure of the PDE β-subunit. cG1 and cG2 denote the noncatalytic cGMP binding sites. HD, homologous domain associated with the active site of PDE. CAAX, motif signaling prenylation. The γ-binding domain is thought to be near the N-terminal portion of the subunit. The approximate location of the mutations is indicated by arrows. The mutations were taken from the references indicated in text. B, schematic representation of the human βPDE gene. Exons and their respective mutations are connected by broken lines.

Table 1. Sequence variants detected in the βPDE gene that are most likely polymorphisms

	Changes in the			
Exon - intron	DNA[a,b]	Protein	Restriction map[c]	Frequency
5'UTR	317A/G[a]			< 1.0%
exon 2	620C/T[b]	no (Asp205)		5.0%
intron 3	1128-11A/C[a]			< 1.0%
exon 5	920G/A[b]	no (Thr305)		7.0%
intron 7	1476-13C/A[a]			7.0%
exon 8	1088C/T[b]	no (Ser361)	+MvnI	< 1.0%
exon 13	1661T/C[b]	no (Arg552)	+MspI	6.0%
intron 13	2138+12C/T[a]			< 1.0%
exon 19	2201C/A[b]	no (Val732)		< 1.0%
3'UTR	2978+11G/A[a]			4.0%
	2978+21T/A[a]			4.0%

[a,b]Position of base pairs according to refs. 2 and 19, respectively.
[c]'+' and '-' refers, respectively, to gain or loss of the restriction site given.

Table 2. Summary of sequence variants of the human βPDE gene found in this study when compared to the sequence originally reported (2)

Exon/intron	Differences between base pairs/amino acids
5'UTR[a]	346insGA; 349insT; 352insG; 360insG; 367insG; 369insT
intron 4[a]	1268+3insC
exon 6[b,c]	947C→A (Gln315Lys)
	963C→A (Leu320Ile)
exon 17[b,c]	2097A→T + 2098T→A (Ile698Tyr)
intron 17[a]	2437+2T→C

[a,b]Position of base pairs according to refs. 2 and 19, respectively.
[c]The same sequence variants have also been found in an independent study (19).

substitutions were identified in the 5′- and 3′-untranslated regions (UTR) and in introns 4,7, and 13. Some of these variants have already been described (3,13-15). Clearly, none of the above mentioned changes in the βPDE gene can be considered causal for RP.

Of the six most likely disease related exonic mutations, three have been found in patients with only one (heterozygous) βPDE-mutation identified so far (Table 3 and Fig. 2).

The 1-bp insertion in exon 1 should lead to a shift in the reading frame and a premature termination of protein translation 315 bp downstream. A missense mutation, predicting Tyr219His (Y219H), was detected in exon 3. The corresponding region in the protein is thought to interact with γPDE and is part of the noncatalytic cGMP-binding domain. Therefore the substitution of a polar amino acid by a positively charged amino acid may negatively affect enzyme activity and specifity. Another missense mutation was detected in exon 1 predicting the Arg74Cys (R74C) change in the protein.

Although each of the above mentioned three mutations may result in drastic reduction or complete loss of enzyme activity, they are unlikely to be the cause of RP in the patients in heterozygous state. Clearly, we can not exclude compound heterozygosity, i.e. the presence of a second deleterious mutation on the other βPDE allele. On one hand, certain sequence changes may escape detection in SSCP analysis. On the other hand, we have analysed only the 22 exons of the βPDE gene together with about 40 bp in introns. The three patients may carry a second mutation in one of the introns outside the regions tested or in the βPDE gene promoter either of which may lead to an afunctional PDE allele and RP. Alternatively, the three patients may be heterozygote for the βPDE mutation just by chance while the retinal dystrophy is due to mutations of another arRP gene. Finally, a further explanation of the RP in the three patients could be digenic inheritance. A recent report has

Table 3. Sequence changes detected in the βPDE gene that are most likely pathogenic mutations

Exon	Changes in the DNA[a]	Changes in the protein	Restr. map[b]
exon 1	225C→T	Arg74Cys	-MvnI
	226insC	frameshift	+HaeII
exon 3	660T→C	Tyr219His	
exon 5	897C→T	Gln298ter	
exon 12	1585T→C	Leu527Pro	+MspI
exon 22	2565C→G	Leu854Val	

[a]Position of base pairs according to ref. 19.
[b]'+' and '-' refers, respectively, to gain or loss of the restriction site given.

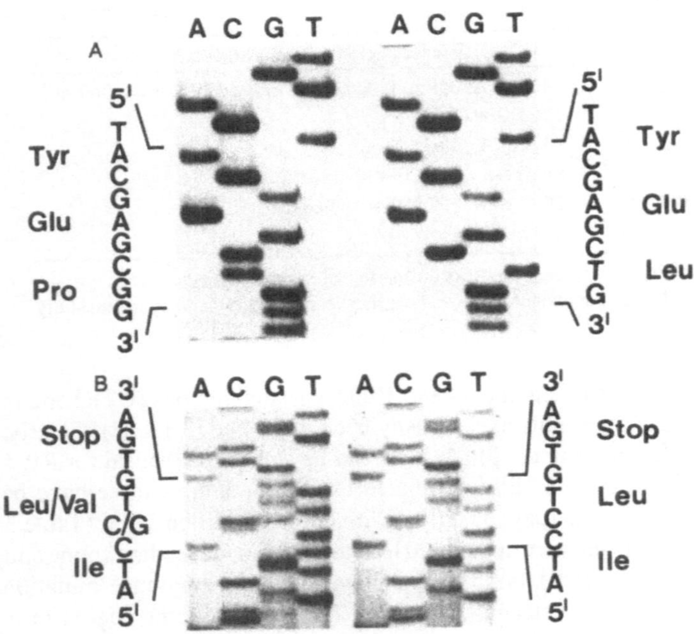

Figure 3. Partial nucleotide sequence of the βPDE gene in patients nos. 167 (A) and 147 (B).

suggested that heterozygous (recessive) mutations of two different genes (those encodin peripherin/RDS and ROM1) may cause RP due to noncomplementation (20).

In two patients, both βPDE-alleles carried potentially pathogenic mutations. A homozygous missense mutation was found in exon 12 of patient no. 167. The T→C substitution at position 1585 creates a new MspI restriction site and predicts a Leu527Pr (L527P) change in an evolutionarily conserved region of the protein (Fig. 2). The health sister of the patient is heterozygote for the mutated allele as evidenced by restriction analysi (not shown).

Patient no. 147 carries two different heterozygous mutations. A C→T transition a position 897 in exon 5 creates a nonsense codon (Gln298ter, Q298ter). This stop appear before the gene region encoding the second putative noncatalytic cGMP-binding domai and the domains essential for both the catalytic function and membrane binding. All togethe more than two-third of the polypeptide chain is eliminated which should result in a functiona null-allele. The same mutation has already been reported in another patient with arRP fron the USA (3). In addition to the amber mutation in exon 5, a C→G transversion was identifie at nucleotide 2565 in exon 22 of the same patient predicting the replacement of the conserve amino acid leucine-854 by valine (Leu854Val, L854V, Fig. 3).

Leu854 is the last residue of the βPDE polypeptide and highly conserved throughou evolution being found in mouse, dog and bovine βPDEs (see ref. 21). It has been suggeste that leucine-854 signals attachment of geranylgeranyl, a C_{20}-isoprenoid, which has bee shown necessary for rod outer segment membrane association of PDE (17). The biologica importance of geranylgeranylation has been demonstrated in choroideremia, an X-linke degenerative retinopathy due to deficiency in the enzyme responsible for geranylgeranyla tion of the small GTP-binding protein Rab (22). The L854V replacement may alter post translational prenylation to farnesylation. The consequences of this change in prenylation are not known, but are thought not to interfere with enzymatic activity.

Ophthalmological Findings in a Patient with Autosomal Recessive Retinitis Pigmentosa and Mutations in the βPDE Gene

Patient no. 147 is a 19-year-old female first diagnosed as having RP at the age of 12. Her night vision has been impaired since she could remember. She got her first glasses because of hyperopia when she was about 1 year of age. From the age of 12, she has increasingly suffered from pathological glare sensitivity. So far, she has not noticed loss of visual acuity or visual field. In her family, nobody else has been reported to be affected by any kind of retinal degeneration. It is worth mentioning that the patient has been suffering from multiple atopic manifestations including asthma since the age of 8, and from migraine since the age of 12 years.

On ophthalmological examination, visual acuity in both eyes was 20/40 with a refractive error of OD +4.25 comb. -2.25 cyl. axis 20° and OS +4.0 comb. -2.0 cyl. axis 165°. Reading ability was good. The patient had a microstrabismus of the left eye. Anterior segments and intraocular pressure were normal, no significant cataract could be observed.

On funduscopy, the optic discs were only mildly atrophic and retinal arterioles only moderately attenuated. The posterior pole showed absence of the foveolar reflex and wrinkling of the inner limiting membrane. There was a diffuse atrophy of the pigment epithelium at the unusually bright fundus. Only a few bone spicules were present and located at the superior and temporal part of the midperipheral retina. The fundus picture was quite symmetric.

The desaturated Farnsworth D-15 test revealed normal colour vision in both eyes. Kinetic perimetry (Tübinger perimeter, 30′ and 10′, 314 cd m^{-2}-target) in the right eye showed a mild peripheral constriction at the nasal-inferior part of the visual field with some additional losses within the remaining field. In the left eye, there was an additional loss in the superior visual field. The overall sensitivity was reduced, the 10′-target showed a nearly symmetric constriction. There has been a moderate progression of the visual field loss within the last 3 years.

Rod sensitivity was determined after 30 minutes of dark adaptation with a 106′ stimulus of the Tübinger perimeter. Stimulus location was 20° nasally on the horizontal meridian. Rod threshold was elevated by about 2.4 log units indicating a remaining adaptation capability of about 1.5 log units, contributed mainly by cones.

Rod-ERG-amplitudes were largely reduced but still detectable (about 10 μV on both sides). The photopic 30-Hz flicker response, which reflects cone activity, was also reduced with amplitudes of about 20% of the normal mean in both eyes.

In summary, the patient shows a typical form of a rod-cone degeneration. At the age of 19 years, cone function was preserved quite well.

A detailed ophthalmological description of the phenotype of patient no. 167 was unavailable.

CONGENITAL STATIONARY NIGHT BLINDNESS

Night blindness is a frequent and early symptom in retinal dystrophies, possibly reflecting the progressive disintegration of photoreceptors. However, congenital stationary night blindness(CSNB) is a separate, clinically and genetically heterogeneous condition, not accompanied by retinal dystrophy, and considered a pure functional defect. Recently, two different heterozygous missense mutations of the rhodopsin gene have been identified in two patients/families with congenital night blindness (23,24). Data of in vitro experiments suggest that the predicted amino acid substitutions (Gly90Asp and Ala292Glu) interfere with the formation of the salt bridge between residues Lys296 and Glu113 which stabilizes opsin transiently lacking its chromophore and keeps free opsin in an inactive conformation (24).

We have studied a large Danish family first described by Rambusch in 1909 (26), in which autosomal dominant CSNB is segregating. Both linkage to the rhodopsin locus and the presence of a disease related mutation in the rhodopsin gene have been excluded. In contrast, close linkage has been found between the locus for adCSNB and polymorphic DNA loci from distal 4p (27, and unpublished data). Subsequently, a heterozygous missense mutation (CAC→AAC, 777C→A) predicting the replacement of histidine-258 by asparagine (His258Asn, H258N) has been identified in the βPDE gene which showed a perfect cosegregation with the disease phenotype (with a maximum lod score of 22.6 at theta = 0.00; ref. 3). Although histidine-258 is a residue that is absolutely conserved in various photoreceptor PDEs during evolution, a direct proof of the pathogenic nature of the H258N exchange is still missing. Therefore, the identification of additional families with adCSNB and mutation in the βPDE gene is of great interest. Here we report the identification of two further Danish pedigrees with adCSNB and the His258Asn mutation.

His258Asn Mutation of βPDE in Two Danish Kindreds with Autosomal Dominant Congenital Stationary Night Blindness

We have identified two subjects with the clinical, electrophysiological, and psychophysical features typical for CSNB.

Proband 1 is a 41-year-old male with congenital stationary night blindness. He belongs to a large 8-generation adCSNB family with 41 recorded night blind members. The pedigree has been traced back to 1760 without finding a link to the 'Rambusch pedigree' for (at least the last) 6 generations.

Proband 2 is a 26-year old male. Congenital night blindness was a common feature in many of his relatives including his father, paternal grandfather and greatgrandmother suggesting an autosomal dominant inheritance of the trait (for further details see ref. 28). As the 'Rambusch family' originated from a region that is about 200 km north from the residence of the proband's family, it is likely that the two families are related to each other. Nevertheless, a direct link to the 'Rambusch pedigree' could not be established by genealogical studies on the last 6 generations of the proband's family for which reliable data were available.

As discussed before, the human βPDE gene contains a great number of nonpathogenic sequence variants (see Table 1). Due to this fact, each person carries an individual combination of these polymorphisms. This 'fingerprint' (haplotype) is a very stable and heritable feature which can be used for pedigree analysis. Assuming that the families of the two probands belong also to the 'Rambush pedigree', all night blind members in the three families should share the same haplotype of DNA sequence variants of the βPDE gene.

Table 4. Results of pairwise linkage analysis between the loci for autosomal dominant congenital stationary night blindness and the β-subunit of the rod cGMP-specific phosphodiesterase (777C→A) in the extended Danish pedigree

Θmax	z(Θmax)	z(0.00)	z(0.05)	z(0.10)	z(0.20)	z(0.30)	z(0.40)
0.00	28.615	28.615	26.388	24.045	18.957	13.245	6.829

We have genotyped probands 1 and 2 together with two night blind members (mother and son) of the 'Rambusch family' for the 11 DNA polymorphisms given in Table 1. Although the allele combinations carried by the four individuals were somewhat different for the 11 polymorphisms analysed, all four shared one haplotype (data not shown) suggesting that this allele originated from a common ancestor. Having this information, we designed a hypothetical extension of the 'Rambusch pedigree' by adding the families of probands 1 and 2. As genealogic studies excluded a relationship within the last six generations in both cases, for the sake of simplicity, we assumed that the three families were linked in the seventh generation. Using this extended pedigree, we reanalysed the linkage relationship between the C→A transversion at codon 258 and the CSNB phenotype. As shown in Table 4, a maximum lod score value of 28.615 was obtained at theta = 0.00 with a ± 1 lod unit confidence interval of 0.00-0.025.

In conclusion, the extended linkage data make the βPDE gene a strong candidate for adCSNB in the 'Rambush family'. Both the conservation of histidine-258 through evolution and the electrophysiological data obtained on patients with CSNB and the H258N mutation (see ref. 28) provide strong support for the hypothesis that the molecular defect responsible for CSNB is in the phototransduction cascade, i.e. at the photoreceptor level in this family. Nevertheless, as a direct proof of the pathogenic nature of H258N (by in vitro assays or in animal models) is still lacking, further genetic, biochemical, and electrophysiological studies are necessary to confirm the above hypothesis.

ACKNOWLEDGMENTS

This work was financially supported by the Deutsche Forschungsgemeinschaft (DFG, Germany), the Deutscher Akademischer Austauschdienst (DAAD), and the German Retinitis Pigmentosa Association (DRPV).

REFERENCES

1. Pfister, C.,Bennett, N., Bruckert, F., Catty, P., Clerc, A., Pages, F., and Deterre, P., 1993, Interactions of a G-protein with its effector: transducin and cGMP phosphodiesterase in retinal rods, *Cell. Signalling* 5:235-251.
2. Weber, B., Riess, O., Hutchinson, G., Collins, C., Lin, B., Kowbel, D., Andrew, S., Schappert, K., and Hayden, M.R., 1991, Genomic organization and complete sequence of the human gene encoding the β-subunit of the cGMP phosphodiesterase and its localisation to 4p16.3, *Nucleic Acids Res.* 19:6263-6268.
3. Gal, A., Orth, U., Baehr, W., Schwinger, E., and Rosenberg, T, 1994, Heterozygous missense mutation in the rod cGMP phosphodiesterase β-subunit gene in autosomal dominant stationary night blindness, *Nature Genet.* 7:64-68.
4. Haim, M., Holm, N.V. and Rosenberg, T., 1992, Prevalence of retinitis pigmentosa and allied disorders in Denmark. I. Main results, *Acta Ophthalmol. (Copenh)* 70:178-186.
5. Boughman, J.A., Conneally, P.M., and Nance, W.E., 1980, Population genetic studies of retinitis pigmentosa, *Am. J. Hum. Genet.* 32:223-235.
6. Humphries, P., Kenna, P., and Farrar J.G., 1992, On the molecular genetics of retinitis pigmentosa, *Science* 256:804-808.
7. Leutelt, J., Oehlman, R., Younus, F., van den Born, I., Weber, J.L., Denton, M.J., Mehdi, Q.S., and Gal, A., 1995, Autosomal recessive retinitis pigmentosa locus maps on chromosome 1q in a large consanguineous family from Pakistan, *Clin. Genet.* (in press)
8. van Soest, S., van den Born, I.L., Gal, A., Farrar, J.G., Bleeker-Wagemakers, L.M., Westerveld, A., Humphries, P., Sandkuijl, L.A., and Bergen, A.A., 1994, Assignment of a gene for autosomal recessive retinitis pigmentosa (RP12) to chromosome 1q31-q32.1 in an inbred and genetically heterogeneous disease population, *Genomics* 22:499-504.

9. Knowles, J.A., Shugart, L., Banerjee, P., Gilliam, T.C., Lewis, C.A., Jacobson, S.G., and Ott, J., 1994, Identification of a locus, distinct from RDS-peripherin, for autosomal recessive retinitis pigmentosa on chromosome 6p, *Hum. Mol. Gen.* 3:1401-1403.
10. Bruford, E.A., Mansfield, D.C., Teague, P.W., Barber, A., Fossarello, M., and Wright, A.F., 1994, Genetic linkage studies in autosomal recessive retinitis pigmentosa, *Am. J. Hum. Genet.* 55:(supplement) A181.
11. Rosenfeld, P.J., Cowley, G.S., McGee, T.L., Sandberg, M.A., Berson, E.L., and Dryja, T.P., 1992, A null mutation in the rhodopsin gene causes rod photoreceptor dysfunction and autosomal recessive retinitis pigmentosa, *Nature Genet.* 1:209-213.
12. Kumaramanickavel, G., Maw, M., Denton, M.J., John, S., Srisailapathy Srikumari, C.R., Orth, U., Oehlmann, R., and Gal, A., 1994, Missense rhodopsin mutation in a family with recessive RP, *Nature Genet.* 8:10-11.
13. McLaughlin, M.E., Sandberg, M.A., Berson, E.L., and Dryja, T.P., 1993, Recessive mutations in the gene encoding the β-subunit of rod phosphodiesterase in patients with retinitis pigmentosa, *Nature Genet.* 4:130-134.
14. Riess, O., Noerremoelle, A., Weber, B., Musarella, M.A., and Hayden, M.R., 1992, The search for mutations in the gene for the beta subunit of the cGMP phosphodiesterase (PDEB) in patients with autosomal recessive retinitis pigmentosa, *Am. J. Hum. Genet.* 51:755-762.
15. Riess, O., Noerremoelle, A., Collins, C., Mah, D., Weber, B., and Hayden, M.R., 1992, Exclusion of DNA Changes in the β-Subunit of the c-GMP phosphodiesterase gene as the cause for Huntington's disease, *Nature Genet.* 1:104-108.
16. Pittler, S.J., and Baehr W., 1991, Identification of a nonsense mutation in the rod photoreceptor cGMP phosphodiesterase β-subunit gene of the rd mouse, *Proc. Natl. Acad. Sci. USA* 88:8322-8326.
17. Suber, M.L. Pittler, S.J., Qin, N., Wright, G.C., Holcombe, V., Lee, R.H., Craft, C.M., Lolley, R.N., Baehr, W., and Hurwitz, R.L., 1993, Irish setter dogs affected with rod/cone dysplasia contain a nonsense mutation in the rod cGMP phosphodiesterase β-subunit gene, *Proc. Natl. Acad. Sci. USA* 90:3968-3972.
18. Portera-Cailliau, C., Sung, C.-H., Nathans, J., and Adler, R., 1994, Apoptotic photoreceptor cell death in mouse models of retinitis pigmentosa, *Proc. Natl. Acad. Sci.* USA 91:974-978.
19. Khramtsov, N.V., Feschenko, E.A., Suslova, V.A., Shmukler, B.E., Terpugov, B.E., Rakitina, T.V., Atabekova, N.V., and Lipkin, V.M., 1993, The human rod photoreceptor cGMP phosphodiesterase β-subunit, *FEBS Letters* 327:275-278.
20. Kajiwara, K., Berson, E.L., and Dryja, T.P., 1994, Digenic retinitis pigmentosa due to mutations at the unlinked peripherin/RDS and ROM1 loci, *Science* 264:1604-1608.
21. Collins, C., Hutchinson, G., Kowbel, D., Riess, O., Weber, B., and Hayden, M., 1992, The human β-subunit of rod photoreceptor cGMP phosphodiesterase: complete retinal cDNA sequence and evidence for expression in brain, *Genomics* 13:698-704.
22. Seabra, M.C., Brown, M.S., Slaughter, C.A., Südhof, T.C., and Goldstein, J.L., 1992, Purification of component A of Rab geranylgeranyl transferase: Possible identity with the choroideremia gene product, *Cell* 70:1049-1057.
23. Dryja, T.P., Berson, E.L., Rao, V.R. and Oprian, D.D., 1993, Heterozygous missense mutation in the rhodopsin gene as a cause of congenital stationary night blindness, *Nature Genet.* 4:280-283.
24. Sieving, P.A., Richards, J.E., Naarendorp, F., Bingham, E., Scott, K. and Alpern, M., 1995, Dark-light: Model for nightblindness from the human rhodopsin Gly-90 → Asp mutation, *Proc. Natl. Acad. Sci. USA* 92:880-884.
25. Rao, V.R., Cohen G.B. and Oprian, D.D., 1994, Rhodopsin mutation G90D and a molecular mechanism for congenital night blindness, *Nature* 367:639-642.
26. Rambusch, S.H.A., 1909, Den medfodte Natteblindheds Arvelighedsforhold, *Oversigt over det Kgl. Danske Videnskabernes Selskabs Forhandlinger* 3:337-347.
27. Gal, A., Xu, S., Piczenik, Y., Eiberg, H., Duvigneau, C., Schwinger, E. and Rosenberg, T., 1994, Gene for autosomal dominant congenital stationary night blindness maps to the same region as the gene for the β-subunit of the rod photoreceptor cGMP phosphodiesterase (PDEB) in chromosome 4p16.3, *Hum. Mol. Genet.* 3:323-325.
28. Rosenberg, T., Gal, A. and Simonsen, S.E., 1995, ERG findings in two patients with autosomal dominant congenital stationary night blindness and His258Asn mutation of the β-subunit of rod photoreceptor cGMP-specific phosphodiesterase. This volume.

MOLECULAR GENETIC STUDIES OF RETINAL DYSTROPHIES PRINCIPALLY AFFECTING THE MACULA

Kevin Evans,[1] Cheryl Y. Gregory,[2] Sujeewa Wijesuriya,[2] Marcelle Jay,[1] Amresh Chopdar,[3] and Shomi S. Bhattacharya[2]

[1] Department of Clinical Ophthalmology
Moorfields Eye Hospital
City Road, London, United Kingdom
[2] Department of Molecular Genetics
Institute of Ophthalmology
Bath Street, London, United Kingdom
[3] Department of Ophthalmology
Easy Surrey Hospital
Three Arch Road, Redhill, United Kingdom

INTRODUCTION

More than 3000 inherited disorders are known to afflict man. Amongst these, 372 distinct entities are associated with choroidoretinal dystrophies, 104 of which are solely ocular (1). For most, little information is as yet available on the underlying genetic or biological deficit. Developments in molecular genetics are improving this situation. Recently a number of retinal dystrophies have been assigned to refined chromosomal loci and in some cases specific gene mutations identified. Dystrophies which exclusively or principally affect the macular region of the human retina are an important subgroup. Characteristically there is earlier onset loss of central acuity with color vision deficits. These conditions contribute significantly to the incidence of blindness in developed countries such as the USA and UK, especially for onset of blindness in childhood (2). Since some macular diseases share similar histopathologic and clinical features, elucidation of the precise pathogenic mechanisms in selected examples may indirectly shed light on the pathogenesis of others. Therefore, with the aim of identifying genetic loci important in the pathogenesis of macular disease in general, a molecular genetic study was undertaken in seven pedigrees expressing different phenotypes that principally affect macular function.

Degenerative Diseases of the Retina, Edited by Robert E. Anderson et al.
Plenum Press, New York, 1995

PATIENTS AND METHODS

Clinical Studies

Seven families (I-VII), large enough for molecular genetic analysis and expressing autosomal dominant phenotypes with predominant macular dysfunction, were identified. Local Ethical Committee approval was obtained prior to enrolment into the study. All had undergone detailed clinical assessment prior to molecular genetic analysis. Examinations included standard visual acuity assessment with a Snellen chart, Farnsworth-Munsell 100-hue color vision testing and visual field assessment using a Humphrey automated perimeter. In addition, selected affected members of each family had undergone electrophysiological examination and fundus fluorescein angiography. Detailed phenotypic descriptions are presented elsewhere (3-6).

Genotyping and linkage analysis

Genomic DNA was extracted from peripheral blood lymphocytes. 100ng aliquots were radioactively genotyped using highly polymorphic microsatellite repeat markers as previously described (7). Radioactive PCR products were fractionated by 6% denaturing polyacrylamide gel electrophoresis, dried and autoradiographed.

The LINKSYS information management package version 3.1 (8) was used to prepare data for linkage analysis. Two-point lodscores were obtained using the MLINK subprogram of LINKAGE program version 5.1(9). Simulated multipoint analysis was undertaken using FASTMAP (10). Allele frequencies were calculated using information from PCR products of spouses within the families.

TIMP3 Exon 5 Sequencing

Two mutations, Tyr168Cys and Ser181Cys, in exon 5 of the TIMP3 gene on chromosome 22q have been shown to segregate with disease in two different Sorsby's Fundus Dystrophy pedigrees. To assess whether similar mutations occur in pedigrees I-IV,

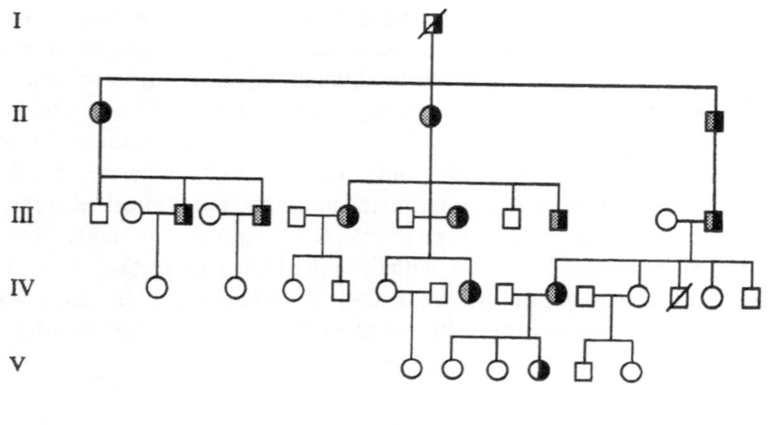

Figure 1. Family II, showing segregation of ocular phenotype with juvenile periodontitis.

genomic DNA was amplified using intronic primers known to flank exon 5 (supplied by B. Weber). PCR products were sequenced using the dideoxy nucleotide chain termination method and internal sequencing primers (11).

Myotonic Dystrophy Molecular Diagnosis

A CTG repeat sequence which is highly polymorphic in the normal population undergoes extensive expansion in myotonic dystrophy patients. To detect this repeat expansion in family III, 100ng samples of DNA were radioactively amplified by the polymerase chain reaction (PCR) using oligonucleotides 101 and 102, with standard PCR cycling conditions at an annealing temperature of 62°C (12). Radiolabelled products were separated on 6% denaturing polyacrylamide gels and detected by autoradiography. Unaffected individuals have between five and 27 copies. Myotonic dystrophy patients have expansion of the repeat containing region from 50 to several hundred repeats.

RESULTS

Families I-IV, Sorsby's Fundus Dystrophy

Four families expressing clinical and fundus-fluorescein angiographic features of Sorsby's fundus dystrophy were studied. Delayed choriocapillary filling on angiography was seen in younger affected members which had progressed to subretinal neovascularisation with a precipitous fall in visual acuity in older affected individuals. Previous histologic studies had identified deposition of confluent subretinal lipid-containing material at the level of Bruch's membrane as pathonomonic of the condition (13).

Family II and III were found in addition to express non-ocular phenotypes. A localised juvenile periodontitis (II) and myotonic dystrophy (III, confirmed on molecular genetic testing) were diagnosed and seemed to segregate with the eye condition (Fig 1). Significant linkage to the previously identified chromosome 22q locus was obtained in pedigrees I and II (Table 1a) (11). Insufficient DNA samples were obtained from family III and IV for linkage analysis. The identification of a number of critical recombinants in pedigree I has allowed for the refinement of this assignment to an 8cM region bounded by D22S275 and D22S278 (14). Direct sequencing of exon 5 of TIMP3 in families I-III identified Ser181Cys mutations segregating with disease. No mutation was identified in family IV, in particular mutations at codons 168 and 181 were specifically excluded.

Family V, Cone-Rod Retinal Dystrophy

A large, 8 generation family expressing an autosomal dominant retinal disease classified as a cone-rod retinal dystrophy from electroretinographic changes was studied. The phenotype presents with a severe visual acuity loss before 4 years of age progressing to no light perception in the fifth decade of life (4). Linkage analysis had refined the genetic locus to a 5 cM region on chromosome 19q13.3-13.4 flanked by D19S219 and D19S246 (15) Three previously unavailable microsatellite markers have been localised to this region. Each was used to genotype members of pedigree V and lod scores calculated (Table 1a). The critical recombinants identified by D19S219 and D19S246 were not recombinant with these markers although all were fully informative for these individuals. Simulated multipoint analysis with a 1-lod confidence interval however suggested that the disease locus is most likely to be in the distal 1.5cM of the region (Figure 2).

Table 1a. Two-point lod scores for markers mapping to known candidate loci in families I,II, V-VII

			Recombination fraction					
Family	Marker	Disease locus	0.0	0.1	0.2	0.3	Zmax	θmax
I	D22S275	SFD	-inf	3.8	3.21	2.23	3.83	0.08
	D22S280	"	7.09	5.84	4.46	2.94	7.09	0.00
	D22S281	"	6.19	5.03	3.77	2.43	6.19	0.00
	D22S278	"	-inf	1.52	1.33	0.75	1.52	0.01
II	D22S280	SFD	-inf	0.08	0.39	0.35	0.40	0.23
	D22S281	"	-inf	1.15	0.96	0.62	1.15	0.01
	D22S278	"	1.20	0.90	0.61	0.34	1.20	0.00
V	D19S219	CRD	-inf	7.24	5.69	3.87	7.92	0.03
	D19S606	"	8.19	6.62	4.99	3.30	8.19	0.00
	AFMa283yh1	"	15.91	13.17	10.15	6.81	15.91	0.00
	D19S604	"	15.35	12.78	9.94	6.80	15.35	0.00
	D19S246	"	-inf	11.72	9.17	6.24	13.11	0.02

SFD=Sorsby's fundus dystrophy, CRD=cone-rod dystrophy, NCMD=North Carolina macular dystrophy, CMO=dominant cystoid macular dystrophy, MD=macular dystrophy, AVMD=atypical vitelliform macular dystrophy

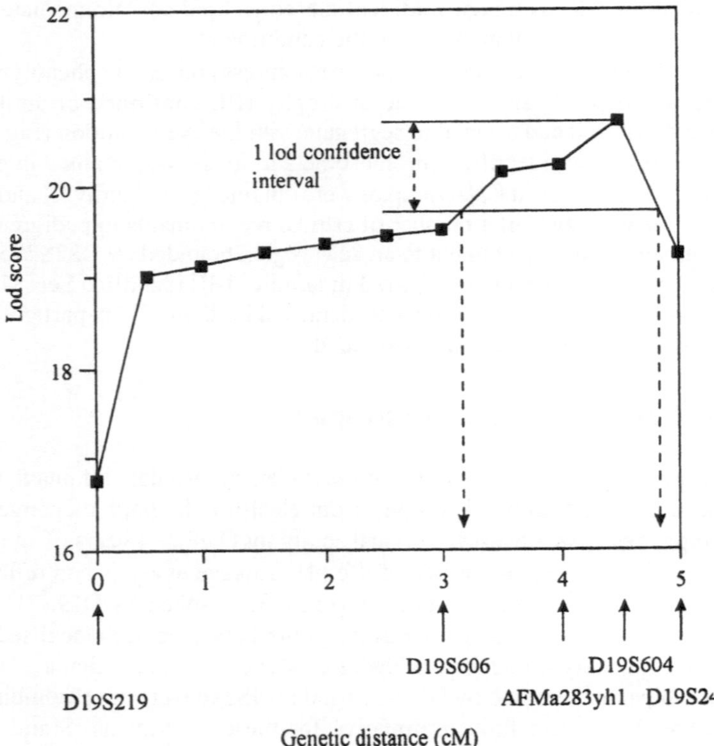

Figure 2. Simulated multipoint linkage analysis in family V

Table 1b. Two-point lod scores for markers mapping to known candidate loci in families I,II, V-VII

Family	Marker	Disease locus	Recombination fraction				Zmax	Exclusion
			0.05	0.10	0.20	0.30		
VI	D1S207	Stargardt's	-1.9	-1.0	-0.4	0.1	0.06	4
	RDS	MD	-2.9	-1.6	-0.6	-0.1	-	9
	D6S294	"	0.3	0.5	0.5	0.3	0.51	<1
	D6S252	NCMD	0.8	0.9	0.7	0.4	0.92	<1
	D6S251	"	-1.9	-0.9	-0.2	0.1	0.06	4
	D7S493	CMO	-2.7	-1.6	-0.7	-0.2	-	9
	D7S516	"	0.1	0.6	0.9	0.6	0.90	1
	D8S272	AVMD	-9.5	-5.5	-2.2	-0.8	-	21
	D11S527	Best's disease	-2.5	-1.3	-0.3	0.0	0.10	8
	D11S956	"	0.3	0.5	0.5	0.3	0.60	2
	D13S173	Stargardt's	-2.6	-1.5	-0.5	-0.1	-	8
	D19S219	CRD	-2.5	-1.4	-0.5	-0.1	-	8
	D19S246	"	-1.1	-0.9	-0.6	-0.3	-	1
VII	D1S207	Stargardt's	-1.8	-1.1	-0.4	-0.2	-	4
	RDS	MD	-1.8	-0.5	-0.1	0.1	0.09	3
	D6S252	NCMD	-1.2	-0.5	-0.3	-0.1	-	2
	D6S251	"	-2.5	-1.4	-0.5	-0.1	-	8
	D6S268	Stargardt's	-2.8	-1.5	-0.4	0.0	0.07	7
	D7S493	CMO	-2.8	-1.6	-0.6	-0.2	-	8
	D8S272	AVMD	-8.2	-4.7	-1.9	-0.7	-	15
	D11S527	Best's disease	-4.6	-2.7	-1.2	-0.5	-	14
	D13S173	Stargardt's	-2.4	-1.2	-0.3	0.0	0.07	6
	D19S219	CRD	-2.7	-1.4	-0.4	0.0	0.05	7
	D22S281	SFD	-1.5	-0.8	-0.2	0.0	0.03	3

SFD=Sorsby's fundus dystrophy, CRD=cone-rod dystrophy, NCMD=North Carolina macular dystrophy, CMO=dominant cystoid macular dystrophy, MD=macular dystrophy, AVMD=atypical vitelliform macular dystrophy

Two potential candidate genes for cone-rod retinal dystrophy have been mapped to chromosome 19q. A voltage-gated potassium channel gene, Kv3.3 has mapped to chromosome 19q13.3. Such cation channel genes have been shown to be present in adult mouse retina (16,17). In addition a novel retina-specific homeobox gene has been mapped to the 19q (R. McInnes, personnal communication). Future work will focus on mutation detection of these candidates.

Families VI and VII, Macular Dystrophy Simulating North Carolina Macular Dystrophy

A variable phenotype was seen in family VI who were of Indian origin. Clinical features similar to those seen in fundus flavimaculatus, pattern dystrophy and dominant drusen were seen in different affected individuals. One individual had a fundus appearance and symptoms similar to those seen in stage 3 North Carolina macular dystrophy (NCMD) (5,18). Family VII expresses a variant of central areolar choroidal dystrophy. A number of affected individuals had features similar to stage 3 NCMD and it has been suggested that central areolar choroidal dystrophy is a type of NCMD (6). Although electrophysiologic abnormalities (VI and VII) and progression of retinal pathology (family VII) differentiate

Table 2. Known chromosomal localisations of retinal dystrophies that principally affect the macula. For abbreviations see legend of Table 1

Dystrophy	Genomic locus	Dystrophy	Genomic locus
Stargardt's macular dystrophy (20)	1p21-13	Best's disease (26)	11q13
Macular dystrophy (21)	6p21	Stargardt's macular dystrophy (27)	13q34
Stargardt's macular dystrophy (22)	6q11-15	Cone-rod dystrophy (28)	17q11
NCMD (18)	6q14-16	Cone-rod dystrophy (29)	18q21
Cone dystrophy (23)	6q25-26	Cone-rod dystrophy (15)	19q13
Cystoid macular dystrophy (24)	7p15-21	Sorsby's fundus dystrophy (11)	22q13-qter
AVMD (25)	8q24	Cone dystrophy (30)	Xp21-11

these phenotypes from that attributed to North Carolina macular dystrophy, the possibility that these conditions are allelic remained.

In both pedigrees, significant linkage to the NCMD locus on chromosome 6q (19) was excluded. In addition, linkage to loci associated with other retinal dystrophies principally affecting the posterior pole on 1p (20), 6p (21), 6q (22-23), 7p (24), 8q (25), 11q (26), 13q (27), 19q (15) and 22q (11) were excluded (Table 1b). These two families therefore represent new dominantly inherited macular dystrophies with features of a number of diseases which have already been genetically assigned, illustrating genetic and phenotypic heterogeneity in macular disease.

DISCUSSION

A number of chromosomal loci identified by genetic linkage analysis in pedigrees expressing retinal dystrophies that principally affect the macular are now known (Table 2). Linkage analysis in other pedigrees (e.g. families VI and VII) has suggested that even more, important loci await identification. In addition, mutations of specific genes are being identified, in particular mutations of peripherin/rds and TIMP3. As work on other linked pedigrees progresses, (e.g. family V), new candidate genes will be highlighted. Our studies in families I-III imply a founder effect in most families with Sorsby's fundus dystrophy in the UK. Exclusion of TIMP3, codon 168 and 181 mutations in pedigree IV however suggests that this is not true for all families and other TIMP3 mutations or mutations in other genes may also be responsible. The association with non-ocular disease in family III was most likely coincidental. However TIMP1 mutations have previously been identified in individuals with periodontal disease (31). Further clinical studies in family II may therefore also prove a role for TIMP3 mutations in periodontitis.

A number of epidemiological investigations, retrospective surveys (32), twin studies (33) and case control studies (34,35) have suggested a genetic influence in age-related macular degeneration. One approach to identifying important predisposing genetic mutations for this condition may be the screening of genes identified by linkage analysis in families expressing macular dystrophies. Therefore future studies on retinal dystrophies that principally affect the macula will not only shed light on the aetiology of retinal diseases inherited as classical mendelian traits but may also identify the genomic loci of the most important candidate genes predisposing to this commoner cause of blindness in adults in developed countries.

ACKNOWLEDGMENTS

The authors would like to thank the Medical Research Council (KE) and The Wellcome Trust (CYG & SW) for supporting this work.

REFERENCES

1. McKusick, V.A., *"Mendelian inheritance in man, 9th ed,"* Johns Hopkins University Press, Baltimore (1992).
2. Elston, J., Epidemiology of visual handicap in childhood, *in*: *"Pediatric Ophthalmology,"* D Taylor, ed., Blackwell Scientific Publications, London, (1992).
3. Polkinghorne, P.J., Capon, M.R., Berninger, T., Lyness, A.L., Sehmi, K., and Bird, A.C., 1989, Sorsby's fundus dystrophy. A clinical study, *Ophthalmology* 96:1763-1768.
4. Evans, K., Duvall-Young, J., Arden, G.B., and Bird, A.C., 1995, Chromosome 19q cone-rod retinal dystrophy, ocular phenotype, *Arch. Ophthalmol.* 113:195-201.
5. Holz, F.G., Evans, K., Gregory, C.Y., Bhattacharya, S.S., and Bird, A.C., 1995, Autosomal dominant macular dystrophy simulating North Carolina macular dystrophy, *Arch. Ophthalmol.* 113:178-184.
6. Chopdar, A., 1993, A variant of central areolar choroidal dystrophy, *Ophthal. Paed. Genet.* 14:151-164.
7. Evans, K., Fryer, A., Inglehearn, C., Duvall-Young, J., Whittaker, J.L., Gregory, C.Y., Butler, R., Ebenezer, N., Hunt, D.M., and Bhattacharya, S.S., 1994, Genetic linkage of cone-rod retinal dystrophy to chromosome 19q and evidence for segregation distortion, *Nature Genet.* 6:210-213.
8. Attwood, J., and Bryant, S., 1988, A computer programme to make analysis with LIPED and LINKAGE easier to perform and less prone to input error, *Ann. Hum. Genet.* 52:259.
9. Lathrop, G.M., Lalouel, J.M., Julier, C., Ott, J., 1984, Strategies for multipoint linkage analysis in humans, *Proc. Natl. Acad. Sci. USA* 81:3443-6.
10. Curtis, D., and Gurling,H., 1993, A procedure for combining two-point lod scores into a summary multipoint map, *Hum. Hered.* 43:173-185.
11. Weber, B.H.F., Vogt, G., Pruett, R.C., Stohr, H., and Felbor, U., 1994, Mutations in the tissue inhibitor of metalloproteinases-3 (TIMP3) in patients with Sorsby's fundus dystrophy, *Nature Genet.* 8:352-355.
12. Brook, J.D., McCurrach, M.E., Harley, H.G., Buckler, A.J., Church,D., Aburtani, H., Hunter, K., Stanton, V.P., Thirion, J.P., Hudson, T., Sohn, R., Zemelman, B., Snell, R.G., Rundle, S.A., Crow, S., Davies, J., Shelbourne, P., Buxton, J., Jones, C., Juxonen, V., Johnson, K., Harper, P.S., Shaw, D.J., and Houseman, D.E., 1992, Molecular basis of myotonic dystrophy: expansion of a trinucleotide (CTG) repeat at the 3'-end of a transcript encoding a protein kinase family member, *Cell* 68:799-808.
13. Capon, M.R., Marshall, J., Krafft, J.I., Alexander, R.A., Hiscott, P.S., and Bird, A.C., 1989, Sorsby's fundus dystrophy. A light and electron microscopic study, *Ophthalmology* 96:1769-77.
14. Gregory, C.Y., Wijesuriya, S., Evans, K., Jay, M.R., Bird, A.C., and Bhattacharya, S.S., 1994, Refinement of Sorsby's fundus dystrophy between D22S273 and D22S280. *J. Med. Genet.* In Press.
15. Gregory, C.Y., Evans, K., Whittaker, J.L., Fryer, A., Weissenbach, J., and Bhattacharya, S.S., 1994, Refinement of the cone-rod retinal dystrophy locus on chromosome 19q, *Am. J. Hum. Genet.* 55:1061-1063.
16. Ghanshani, S., Pak, M., McPherson, J.D., Strong, M., Dethlefs, B., Wasmuth, J.J., Salkoff, L., Gutman, G.A., and Chandy, K.G., 1992, Genomic organisation, nucleotide sequence, and cellular distribution of a Shaw-related potassium channel gene, Kv3.3, and mapping of Kv3.3 and Kv3.4 to human chromosomes 19 and 1, *Genomics* 12:190-196.
17. Klumpp, D.J., Song, E-J., Sheng, M., Jan, L.Y., and Pinto, L.H., 1994, Potassium channel show differential expression and subcellular localisation within the mouse retina, *Invest. Ophthalmol. Vis. Sci.* 35(Suppl) p1491.
18. Small, K.W., Killian, J., and McLean, W.C., (1991). North Carolina dominant progressive foveal dystrophy: how progressive is it? *Br. J. Ophthalmol.* 75:401-406.
19. Small, K.W., Weber, J.L., Roser, A., Lennon, F., Vance, J.M., and Pericak-Vance, M.A., 1992, North Carolina macular dystrophy is assigned to chromosome 6, *Genomics* 12:681-685.
20. Kaplan, J., Gerber, S., Larget-Piet, D., Rozet, J-M., Dollfus, H., Dufier, J-L., Oent, S., Postel-Vinay, A., Janin, N., Briard, M-L., Frezal, J., and Munnich, A., 1993, A gene for Stargardt's disease (*fundus flavimaculatus*) maps to the short arm of chromosome 1, *Nature Genet.* 5:308-311.

21. Wells, J., Wroblewski, J., Keen, J., Inglehearn, C., Jubb, C., Eckstein, A., Jay, M., Arden, G., Bhattacharya, S.S., Fitzke, F., and Bird, A.C., 1993, Mutations in the human retinal degeneration slow (RDS) gene can cause either retinitis pigmentosa or macular dystrophy, *Nature Genet.* 3:202-207.
22. Stone, E.M., Nichols, B.E., Kimura, A.E., Weingeist, T.A., Drack, A., and Sheffield, V.C., 1994, Clinical features of a Stargardt-like dominant progressive macular dystrophy with genetic linkage to chromosome 6q, *Arch. Ophthalmol.* 112:765-772.
23. Tranebjaerg, L., Sjo, O., and Warburg, M., 1986, Retinal cone dysfunction and mental retardation associated with a de novo balanced translocation 1:6(q44;q27), *Ophthalmic Paediatr. Genet.* 7:167-173.
24. Kremer, H., Pinckers, A., van den Helm, B., Deutman, A.F., Ropers, H-H., and Mariman, E.C.M., 1994, Localisation of the gene for dominant cystoid macular dystrophy on chromosome 7p, *Hum. Mol. Genet.* 3:299-302.
25. Ferell, R.E., Hittner, H.M., and Antaszyk, J.H., 1983, Linkage of atypical vitelliform macular dystrophy (VMD-1) to the soluble glutamate pyruvate transaminase (GPT1) locus, *Am. J. Hum. Genet.* 35:78-84.
26. Stone, E.M., Nichols, B.E., Streb, L.M., Kimura, A.E., and Sheffield, V.C., 1992, Genetic linkage of vitelliform macular degeneration (Best's disease) to chromosome 11q13, *Nature Genet.* 3:213-218.
27. Zhang, K., Bither, P.P., Park, R., Donoso, L.A., Seidman, J.G., and Seidman, C.E., 1994, A dominant Stargardt's macular dystrophy locus maps to chromosome 13q34, *Arch. Ophthalmol.* 112:759-764.
28. Kylstra, J.A., and Aylsworth, A.S., 1993, Cone-rod retinal dystrophy in a patient with neurofibromatosis type 1, *Can. J. Ophthalmol.* 28:79-80.
29. Warburg, M., Sjo, O., and Fledelius, H.C., 1991, Deletion mapping of a retinal cone-rod dystrophy: assignment to 18q211, *Am. J. Med. Genet.* 39:288-293.
30. Meire, F.M., Bergen, A.A.B., De Rouck, A., Leys, M., and Delleman, J.W., 1994, X linked progressive cone dystrophy, *Br. J. Ophthalmol.* 78:103-108.
31. Meikle, M.C., Hembry, R.M., Holley, J., Horton, C., McFarlane, C.G., and Reynolds, J.J., 1994, *J. Periodontal Res.* 29:118-126.
32. Silvestri, G., Johnson, P.B., and Hughes, A.E., 1994, Is genetic predisposition an important risk factor in age-related macular degeneration, *Eye* 8:564-568.
33. Dosso, A.A., and Bovet, J., 1992, Monozygotic twin brothers with age-related macular degeneration, *Ophthalmologica* 205:24-28.
34. Hyman, L.G., Lilienfield, A.M., Ferris, F.L., and Fine, S.L., 1983, Senile macular degeneration: a case-control study, *Am. J. Epidemiol.* 118:213-227.
35. Piguet, B., Wells, J.A., Palmvang, I.B., Wormald, R., Chisholm, I.H., and Bird, A.C., 1993, Age-related Bruch's membrane change: a clinical study of the relative role of heredity and environment, *Br. J. Ophthalmol.* 77:400-403.

MOLECULAR ANALYSIS OF THE HUMAN GAR1 GENE

A Candidate Gene for Retinal Degeneration

Michelle D. Ardell,[1] Ajay Makhija,[1] and Steven J. Pittler[1,2]

[1] Department of Biochemistry and Molecular Biology
[2] Department of Ophthalmology
University of South Alabama College of Medicine
Mobile, Alabama 36688-0002

I. INTRODUCTION

A bovine glutamic acid rich protein (garp) was first reported as a soluble protein of unknown function which copurified with cGMP phosphodiesterase from rod outer segments (1). Characterization of bovine retina cDNA clones encoding garp identified a 65 kD protein with a high overall glutamate composition (24%), including one 109 amino acid domain near the C-terminus especially enriched (68/109 residues) in glutamate. This domain contained four repeats of a glutamate-rich 11 amino acid residue sequence (undecapeptide) and two repeats of a 26 amino acid peptide also rich in glutamic acid. Antibodies against a decamer corresponding to the undecapeptide from the glutamate rich region recognized a 65 kD protein in crude rod outer segment extracts. Northern analysis of poly A^+ RNA from a variety of bovine tissues identified a 2.4 kb transcript only in the retina sample, suggesting that the expression may be retina-specific. More recently it was shown that garp is part of the 240 kD integral membrane protein complex that forms part of the cGMP-gated cation channel of bovine rod outer segments, thus identifing it as a third subunit (γ) of the channel (2).

The photoreceptor cGMP-gated cation channel, an integral component of the visual transduction cascade, was first isolated as a 63 kD polypeptide (now termed α) and functionally reconstituted into proteoliposomes (3). The cDNAs encoding the bovine (4), mouse and human (5) α subunits have been cloned and sequenced, and the bovine protein functionally expressed in *Xenopus* oocytes (4). The human α gene has also been characterized and localized to chromosome 4 (5,6), and defects in this gene responsible for certain forms of autosomal recessive retinitis pigmentosa have recently been reported (7). A second subunit (β) has been identified that does not form functional channels alone (8), but when coexpressed with the α subunit introduces the rapid channel flickerings and confers sensitivity to the calcium channel blocker cis-l-diltiazem, which are characteristics of the native channel (8). Bovine garp (γ) is covalently linked to the β subunit, comprising a 240 kd

Degenerative Diseases of the Retina, Edited by Robert E. Anderson et al.
Plenum Press, New York, 1995

complex that associates with the α subunit to form the native channel (2). Immunological analysis of purified human cation channel also identified two polypeptides of 63 and 240 kD, however, while it was demonstrated that the 63 kd protein is the bovine α homolog, and that the 240 kd polypeptide consists in part, of the β subunit, the complete composition of the complex was not established (9). In this chapter, we report on the characterization of the GAR1 gene encoding the human homolog of bovine γ, and discuss its potential as a candidate gene for retinal degenerations.

II. PRIMARY STRUCTURE AND GENE ORGANIZATION OF HUMAN GAR1

The conceptual amino acid sequence for human garp has been reported from cDNA cloning and DNA sequence analysis (10). Figure 1 shows the nucleotide and conceptual amino acid sequence deduced from cDNA clones and primer extension. The cDNA contains a coding region of 897 nucleotides, corresponding to 299 amino acid residues. Additionally, the sequence contains 68 and 697 bp of untranslated region at the 5′ and 3′ ends respectively. The 3′ untranslated region contains a putative polyadenylation signal beginning at nt 1541 and an Alu-like repetitive element at nt 1197-1318. PCR analysis of human first strand cDNA with an anchoring primer indicates that the 3′ untranslated region is complete.

The primary structures of human and bovine garp show only a moderate degree of overall homology (60%, excluding gaps), however areas of much higher homology exist (Figure 2). Comparison of the predicted protein sequences displayed 90% identity within the first 31 amino terminal residues. Since this sequence does not demonstrate any of the characteristics of known leader sequences, it is likely to be present in the mature proteins. No significant homologies to any other proteins were identified in the current SwissProt protein database. Kyle-Doolittle hydropathy index analysis indicated an absence of trans-membrane domains in human garp, consistent with the observation that bovine garp was initially isolated as a cytosolic protein (1). The human and bovine proteins are both rich in glutamate residues, leading to similar pIs for the two proteins (4.3 and 3.8, respectively). Thus, while the human protein is much smaller than the bovine protein, and lacks the C-terminal glutamate rich region, they are clearly homologous. Northern analysis of human retinal RNA identified a transcript of 2.0 kb, slightly shorter than that reported for bovine garp (10). Exhaustive screening of a human retinal cDNA library (gift of J. Nathans) with probes that amplified the bovine glutamate rich segment failed to yield any clones. These results suggest that a longer homolog may not exist in human retina, and that the protein encoded by the GAR1 gene is the γ subunit of the rod cation channel.

In order to further confirm the cDNA sequence of human garp, and to determine the intron/exon organization of the gene for candidate gene screening, clones were isolated from a human genomic library constructed in λEMBL-3. Analysis of overlapping clones revealed the presence of 11 introns and 12 exons that span the protein coding region (Figure 1). Characterization of the upstream region is currently in progress.

III. POSSIBLE FUNCTIONS OF THE γ SUBUNIT

The role of the γ subunit in channel function is unknown. However, amino acid sequence analysis revealed that the protein contains nine potential phosphorylation sites, making it possible that phosphorylation of the γ subunit could be involved in regulation of the cation channel's activity. This would be consistent with the studies

```
                                                    CCAGCTAC  -61
GAGTGGCAGCAAGAAGAAGGCAATTCCTGGCTGGCGGTTGGCATCTAAGCAGGCATCAGG  -1
                                              *

ATGTTGGGCTGGGTCCAGAGGGTGCTGCCTCAGCCCCCAGGGACCCCTCGGAAGACCAAG  60
 M  L  G  W  V  Q  R  V  L  P  Q  P  P  G  T  P  R  K  T  K-20

ATGCAGGAGGAAGAGGAAGTGGAACCAGAGCCAGAGATGGAGGCAGAGGTGGAACCAGAA  120
 M  Q  E  E  E  E  V  E  P  E  P  E  M  E  A  E  V  E  P  E-40
                                    |2.6|
CCGAATCCTGAGGAGGCCGAGACAGAGTCCGAGTCCATGCCCCCGAAGAGTCATTCAAG  180
 P  N  P  E  E  A  E  T  E  S  E  S  M  P  P  E  E  S  F  K-60
                                  |.28|
GAGGAGGAAGTGGCTGTGGCAGACCCAAGCCCTCAGGAGACCAAGGAGGCTGCCCTTACT  240
 E  E  E  V  A  V  A  D  P  S  P  Q  E  T  K  E  A  A  L  T-80
                                                   |1.0|
TCCACCATATCCCTCCGGGCCCAGGGCGCTGAGATTTCTGAAATGAATAGTCCCAGCCAC  300
 S  T  I  S  L  R  A  Q  G  A  E  I  S  E  M  N  S  P  S  H-100

AGGGTACTGACCTGGCTCATGAAGGGCGTAGAGAAGGTGATCCCGCAGCCTGTTCACAGC  360
 R  V  L  T  W  L  M  K  G  V  E  K  V  I  P  Q  P  V  H  S-120
                  |0.2|                            |0.2|
ATCACGGAGGACCCGGCTCAGATCCTGGGGCATGGCAGCACTGGGGACACAGGGTGCACA  420
 I  T  E  D  P  A  Q  I  L  G  H  G  S  T  G  D  T  G  C  T-140
                                    |1.6|
GATGAACCCAATGAGGCCCTTGAGGCCCAAGACACTAGGCCTGGGCTGCGGCTGCTTCTG  480
 D  E  P  N  E  A  L  E  A  Q  D  T  R  P  G  L  R  L  L  L-160
                                                      |0.3|
TGGCTGGAGCAGAATCTGGAAAGAGTGCTTCCTCAGCCCCCCAAATCCTCTGAGGTCTGG  540
 W  L  E  Q  N  L  E  R  V  L  P  Q  P  P  K  S  S  E  V  W-180
                                          |.36|
AGAGATGAGCCTGCAGTTGCTACAGGTGCTGCCTCAGACCCAGCGCCTCCAGGACGCCCC  600
 R  D  E  P  A  V  A  T  G  A  A  S  D  P  A  P  P  G  R  P-200

CAGGAAATGGGGCCCAAGCTGCAGGCCCGGGAGACCCCCTCCCTGCCCACACCCATCCCC  660
 Q  E  M  G  P  K  L  Q  A  R  E  T  P  S  L  P  T  P  I  P-220

CTGCAGCCCAAGGAGGAACCCAAGGAGGCACCAGCTCCAGAGCCCCAGCCCGGCTCCCAG  720
 L  Q  P  K  E  E  P  K  E  A  P  A  P  E  P  Q  P  G  S  Q-240
                                       |1.1|
GCCCAGACCTCCTCCCTGCCACCAACCAGGGACCCTGCCAGGCTGGTGGCATGGGTCCTG  780
 A  Q  T  S  S  L  P  P  T  R  D  P  A  R  L  V  A  W  V  L-260
                                                      |1.1|
CACAGGCTGGAGATGGCCTTGCCGCAGCCAGTGCTACATGGGAAAATAGGGGAACAGGAG  840
 H  R  L  E  M  A  L  P  Q  P  V  L  H  G  K  I  G  E  Q  E-280
                                  |1.9|
CCTGACTCCCCTGGGATATGTGATGTGCAGACCAGGGTGATGGGAGCTGGAGGTCTCTGA  900
 P  D  S  P  G  I  C  D  V  Q  T  R  V  M  G  A  G  G  L  *-299

AATAAGGAAGAAAGGGAATCTGGGAGAGCTCAGATGGTCACATGGATGGAAGGAAAGAAG  960
GATGCCCTGAAGAGAAGACCTTCCCGGGGGAGGTGGCCACTGACACCCCACCCTATTTGA  1020
ACAGCGAACCCCTCCCTCTCCACACACTCCCAGGCAGGACAAGGGGAGCCAGAGTCACCT  1080
GCACCAGCCCCAGAGCCTCCCAGACCAGAACAAGGGGAAGGCCAGCCATGGGCAGCCTGC  1140
AAGATTTAGGCTAGACAGCAGGACTTACCCCCAGAGGGCTGTGGAAAAGGCCCATAGCCG  1200
GGCGTGGTGGCTCACGCCTGTAACCCCTGACCTTTGGGAAGCCAGGGTCGGAAGACTCTC  1260
TTGAGCCCCGAAGTTCGAGACCACCGTGGGCAACACAGTGAGCCCTGTCTCTAAAAAACC  1320
TTTTTTAAATTAATAAATTAAAAAGCCCCATGGATGGAGGACTCAGTATTGAGCATCTCT  1380
TTGAGAGGACGCGTGCACAGCCCCCAGTGTGGTACCTGGCACCGTCAGCACCTCGACAGG  1440
ATACAGTTTTTCCCAGAAGAGGTTCTCCCTAGGCCTGGCCACACTCTCTCCTTCCAAGGC  1500
TTGGGGACACCAATGAGCAGCAACAAATGTTTTCATTAACATTAAAAGAGTGTAAATGAG  1560
                                        ------
CACATCAAAAAAAAAAA-1576
```

Figure 1. Structure and sequence of human cGMP-gated cation channel γ subunit. In frame stop codons (*) that delimit the open reading frame are marked below their site of occurrence. Nucleotides are numbered starting with the predicted initiation codon ATG as 1. A likely poly A site is underlined. Boxes indicate the size and location of introns in the genomic clones.

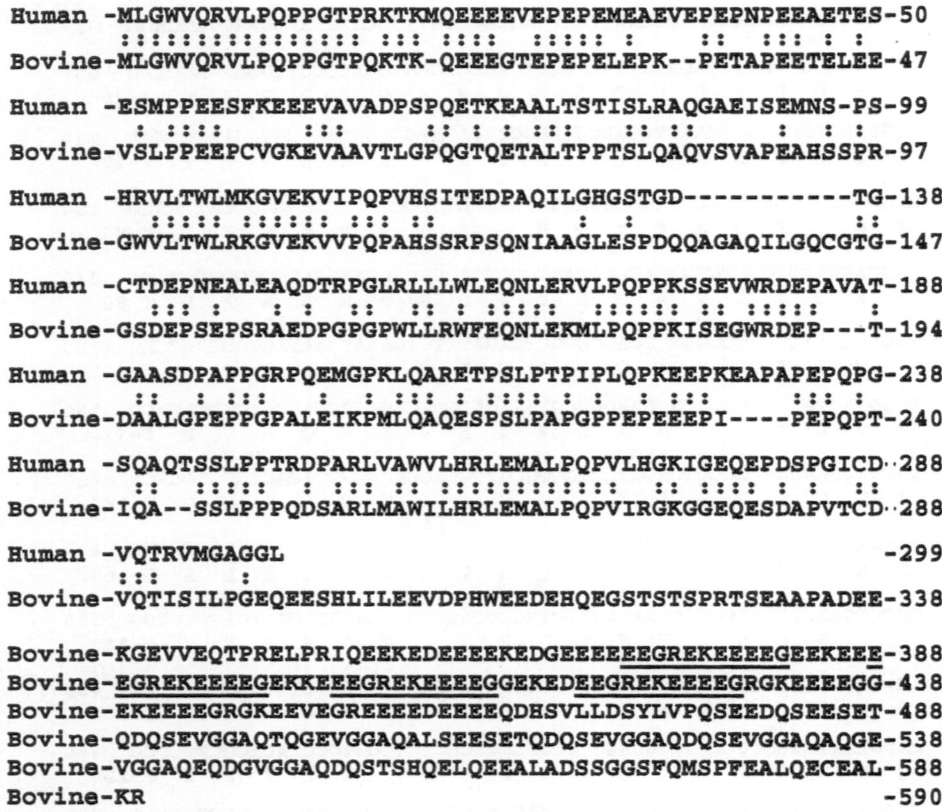

Figure 2. Amino acid sequence comparison of human and bovine γ subunits. Identical residues are indicated by :. Numering is according to Figure 1 for the human sequence, and according to Sugimoto, et al. for bovine (1). Gaps were introduced to yield optimal alignment. The undecapeptide glutamate rich repeat used for generation of polyclonal antibodies is underlined.

on excised frog photoreceptor membrane patches which indicated that phosphorylation of the channel may occur at two distinct sites (11). Since studies with the α and β subunits of the channel revealed that only the α subunit could be phosphorylated (12), it is reasonable to propose that the γ subunit could provide a second phosphorylation site.

Another possible role for the γ subunit arises from the observation that garp cDNA clones were orginally isolated from an expression library screened with an antibody generated from highly purified preparations of cGMP phosphodiesterase. This finding suggests that garp may tightly associate with the PDE, and is intriguing in view of the report that PDE can directly activate the cation channel in excised patches of bovine rod outer segment membranes (13). Hence, it is possible that PDE directly activates the channel via the γ subunit.

Alternatively, because garp is a highly acidic polypeptide (pI=4.3 in human) and the α subunit of the channel contains a domain highly enriched in basic residues, it is possible that interaction with the acidic garp acts to form a gate for ion flow. Another possibility is that due to garp's homology to neurofilaments (see below), it may act as part of a cytoskeletal system that helps to anchor the channel in the rod outer segment plasma membrane.

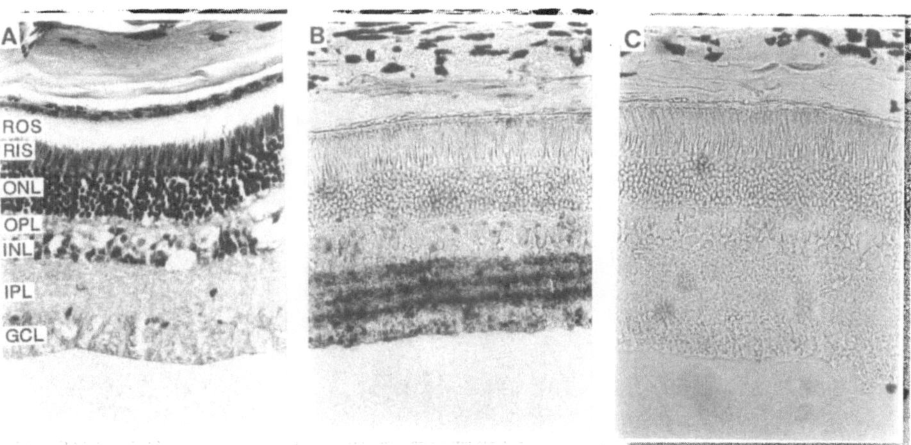

Figure 3. Immunocytochemistry of bovine retina sections with anti-garp polyclonal antibody. *a*, light micrograph of paraffin embedded retina. H&E stain, x80. ROS = rod outer segments, RIS = rod inner segments, ONL=outer nuclear layer, OPL = outer plexiform layer, INL = inner nuclear layer, IPL=inner plexiform layer, GCL=ganglion cell layer. *b*, Retinal section labeled with a 1:100 dilution of anti-garp polyclonal antibody, or *c*, the preimmune sera. The sections were stained using the Vector ABC biotinylation kit as per manufacturer's instructions (Vector Laboratories). x80. Note marked labelling of the inner plexiform layer.

IV. POLYCLONAL ANTIBODIES TO THE BOVINE GAR PROTEIN

Antipeptide antibodies serve as valuable probes of protein structure and cellular location. Previously, Sugimoto et. al. generated polyclonal antibodies against a decapeptide corresponding to a repeated undecapeptide in the glutamate-rich C-terminal region of bovine garp, and demonstrated that this antibody positively recognized a 65 kD protein in crude bovine ROS extracts (1). As the first step in determining if a closer homolog to bovine garp exists in human retina (one which contains the glutamate rich repeat), we performed control immunocytochemistry experiments with bovine retinal sections (Figure 3). Surprisingly, intense staining of only the inner plexiform layer was observed, with no staining of the photoreceptors. No labeling was seen when pre-immune sera was used as a control. Since the cGMP-gated cation channel has been shown to localize exclusively to the plasma membrane of photoreceptors (14), it was necessary to determine what protein was being recognized in the inner plexiform layer. A search of the SwissProt data base for proteins with homology to the peptide used to generate the antibody revealed that mammalian neurofilament L (NF-L) shared significant homology with the garp epitope. To test if the antibody cross-reacts with NF-L present in the nerve axons of the inner plexiform layer, Western analysis was performed with purified total neurofilaments (NF-L, NF-M, and NF-H, Figure 4). The antibody selectively recognized NF-L under the conditions employed. No cross reactivity was observed when the primary antibody was omitted. These results suggest that the garp epitope may be masked in photoreceptors, and they further indicate that this antibody cross reacts with NF-L. Therefore, any results obtained with the garp antipeptide antibody should be interpreted in light of these new findings. We are currently trying to develop antibodies against other epitopes that should recognize both human and bovine garp.

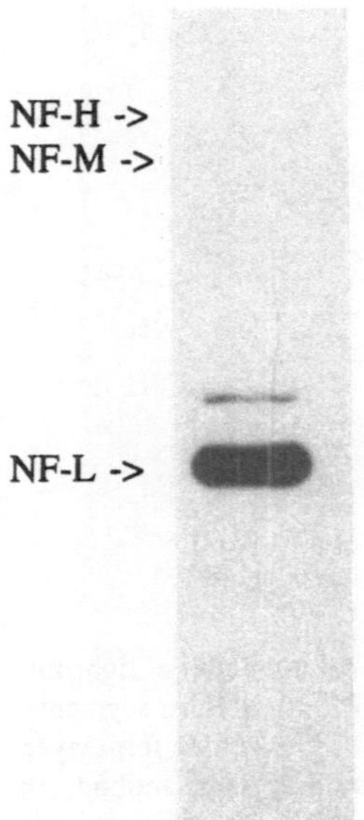

Figure 4. Western blots of total purified neurofilaments. Five hundred nanograms of total neurofilaments (NF-L, -M, and -H) were separated by SDS-PAGE. Immunoblotting was performed with a 1:50,000 dilution of garp antisera. Antibody binding was then detected using horseradish peroxidase-conjugated goat anti-rabbit IgG (Sigma) and the ECL chemiluminescence detection kit (Amersham).

V. CHROMOSOME MAPPING OF HUMAN GAR1

As a component of the phototransduction cascade, the gene encoding the γ subunit of the rod cGMP-gated cation channel is a candidate for retinal degenerations. To determine if the gene is located in the region of any known retinal disease loci, we first performed chromosome localization. PCR of a panel of human/rodent somatic cell hybrids (BIOS Corp.) was performed using a primer pair that amplifies a 2.8 kb segment, including the first (2.6 kb) intron. While at least three discordancies were observed with every other chromosome, perfect concordance was only observed with chromosome 16 (Figure 5), thus indicating that the GAR1 locus resides on chromosome 16, making it the first photoreceptor gene mapped to this chromosome.

Since highly homologous mRNAs for cyclic nucleotide gated channels have been reported in several tissues, including retina and kidney (15), it is reasonable to suggest that defects in the protein product of the GAR1 gene could produce pleiotropic effects, including retinal degeneration and neuropathy. One such disorder is Bardet-Biedl syndrome, a multi-

hybridization/ chromosome	Chromosome number																							
	1	2	3	4	5	6	7	8	9	10	11	12	13	14	15	16	17	18	19	20	21	22	X	Y
+/+	0	0	1	0	2	0	0	1	0	1	0	0	0	0	0	2	0	0	0	0	0	0	0	0
-/-	19	21	20	21	3	19	21	19	19	21	20	19	17	16	20	22	21	19	16	20	16	19	20	19
+/-	2	2	1	2	0	2	2	1	2	1	2	2	2	2	1	0	2	2	2	2	2	2	2	2
-/+	3	1	3	2	20	4	2	4	3	2	3	4	6	7	3	0	2	4	7	3	7	4	3	4
discordant hybrids	5	3	4	4	20	6	4	5	5	3	5	6	8	9	4	0	4	6	9	5	9	6	5	6
Informative hybrids	24	24	25	25	25	25	25	25	24	25	25	25	25	25	24	24	25	25	25	25	25	25	25	25

Figure 5. Chromosome localization of human GAR1 gene in human/hamster somatic cell hybrids. The number of hybrids which are concordant (+/+ or -/-) and discordant (+/- or -/+) with the human γ sequence are given for each chromosome. Hybrids for which a particular chromosome was present in less than or equal to 10% of cells were excluded from consideration.

systemic disorder primarily characterized by severe retinitis pigmentosa, renal abnormalities, obesity, polydactyly and/or syndactyly and sometimes mental retardation and other systemic abnormalities (16). Due in part to the heterogeneous nature of this disorder, the precise molecular defect(s) are not known. It has been shown, however, that Bardet-Biedl syndrome transmits as an autosomal recessive trait, of which one form is linked to chromosome 16. Our finding that the γ subunit of the rod cation channel is localized to the same chromosome makes GAR1 an excellent candidate for this disease. Studies are currently underway to determine if a defect in this gene is involved in the etiology of Bardet-Biedl syndrome.

VI. ACKNOWLEDGMENTS

We thank Luanne Oliveira and Ileana Aragon for expert technical assistance. We also thank Dr. Y. Sugimoto for the generous gift of bovine garp antibody and preimmune sera, and Dr. D. Dong for the gift of purified neurofilaments. We also thank Dr. Steve Fliesler for the gift of frozen acrylamide embedded fixed bovine tissue. This work was supported in part by grants from the NIH/NEI EY09924, NSF IBN9321132, and the Lions/USA Eye Research Institute (to SJP). MDA was supported in part by the Lions/University of South Alabama Eye Research Institute and is currently supported by an NEI/NRSA fellowship F32 EY06579. AKM was supported by a College of Medicine/National Institutes of Health medical student summer research fellowship.

VII. REFERENCES

1. Sugimoto, Y., Yatsunami, K., Tsujimoto, M., Khorana, H.G., and Ichikawa, A. (1991). The amino acid sequence of a glutamic-acid rich protein from bovine retina as deduced from the cDNA sequence. Proc. Natl. Acad. Sci. USA 88:3116-3119.
2. Illing, M., Colville, C.A., Williams, A.J., and Molday, R.S. (1994). Sequencing, cloning, and characterization of a third subunit of the cyclic nucleotide-gated channel complex of rod outer segments. ARVO ABSTRACTS. Invest. Ophthalmol. Vis. Sci. Suppl., vol 35, 1022.
3. Cook, N.J., Hanke, W., and Kaupp, U.B. (1987). Identification, purification, and functional reconstitution of the cyclic GMP-dependent channel from rod photoreceptors. Proc. Natl. Acad. Sci. USA. 84:585-589.
4. Kaupp, U.B., Niidome, T., Tanabe, T., Terada, S., Bönigk, W., Stühmer, Cook, N.J., Kangawa, K., Matsuo, H., Hirose, T., Miyata, T., and Numa, S. (1989). Primary structure and functional expression from complementary DNA of the rod photoreceptor cyclic GMP-gated channel. Nature 342:762-766.
5. Pittler, S.J., Lee, A.K., Altherr, M.R., Howard, T.A., Seldin, M.F., Hurwitz, R.L., Wasmuth, J.J., and Baehr, W. (1992). Primary structure and chromosomal localization of human and mouse rod photoreceptor cGMP-gated cation channel. J. Biol. Chem. 267:6257-6262.
6. Dhallan, R.S., Macke, J.P., Eddy, R.L., Shows, T.B., Reed, R.R., Yau, K.-W., and Nathans, J. (1992). Human rod photoreceptor cGMP-gated channel: amino acid sequence, gene structure, and functional expression. J. Neurosci. 12:3248-3256.
7. McGee, T.L., Lin, D., Berson, E.L., and Dryja, T.P. (1994). Defects in the rod cGMP-gated channel gene in patients with retinitis pigmentosa. ARVO ABSTRACTS. Invest. Ophthalmol. Vis. Sci. Suppl., vol 35, 2154.
8. Chen, T.-Y., Peng, Y.-W., Dhallan R.S., Ahamed, B., Reed, R.R., and Yau, K.-W. (1993). A new subunit of the cyclic nucleotide-gated cation channel in retinal rods. Nature 362:764-767.
9. Chen, T.-Y., Illing, M., Molday, L.L., Hsu, Y.-T, Yau, K.-W., and Molday, R.S. (1994). Subunit 2 (or β) of retinal rod cGMP-gated cation channel is a component of the 240-kDa channel-associated protein and mediates Ca^{+2}-calmodulin modulation. Proc. Natl. Acad. Sci. USA 91:11757-11761.
10. Ardell, M.D., Makhija, A.K., Oliveira, L., Miniou, P., Viegas-Pequignot, E., and Pittler, S.J. cDNA, gene structure, and chromosomal localization of human GAR1 (CNCG3L), a homolog of the third subunit of bovine photoreceptor cGMP-gated channel. (1995) In press, Genomics.

11. Gordon, S.E., Brautigan, D.L., and Zimmerman, A.L. (1992). Protein phosphatases modulate the apparent agonist affinity of the light-regulated ion channel in retinal rods. Neuron 9:739-748.
12. Liu, M.-Y., Li, J., and Yau, K.W. (1994). Phosphorylation of the N-terminal domain of human rod cyclic nucleotide-gated channel by protein kinase C and the cAMP-dependent protein kinase. ARVO ABSTRACTS. Invest. Ophthalmol. Vis. Sci. Suppl. 35:1474.
13. Bennett, N., Ildefonse, M., Crouzy, S., Chapron, Y., and Clerc, A. (1989). Direct activation of cGMP-dependent channels of retinal rods by the cGMP phosphodiesterase. Proc. Natl. Acad. Sci. USA 86:3634-3638.
14. Cook, N.J., Molday, L.L., Reid, D., Kaupp, U.B., and Molday, R.S. (1989). The cGMP-gated channel of bovine rod photoreceptors is localized exclusively in the plasma membrane. J. Biol. Chem. 264:6996-6999.
15. Ahmad, I., Redmond, L.J., and Barnstable, C.J. (1990). Developmental and tissue-specific expression of the rod photoreceptor cGMP-gated ion channel gene. Biochem. Biophys. Res. Commun. 173:463-470.
16. Schachat, A.P. and Maumenee, I.H. (1982). The Bardet-Biedl syndrome and related disorders. Arch. Ophthalmol. 100:285-288.

38

GUANYLATE CYCLASE-ACTIVATING PROTEIN (GCAP)

A Novel Ca^{2+}-Binding Protein in Vertebrate Photoreceptors

Wolfgang Baehr,[1] Iswari Subbaraya,[1] Wojciech A. Gorczyca,[2] and Krzysztof Palczewski[2]

[1] Department of Ophthalmology
Baylor College of Medicine
Houston, Texas 77030

[2] Departments of Ophthalmology and Pharmacology
University of Washington
Seattle, Washington 98195

INTRODUCTION

Vertebrate photoreceptors respond to light by a transient hyperpolarization of the plasma membrane. Hyperpolarization is achieved when cGMP-gated cation channels, located in the plasma membrane, are closed after hydrolysis of the gating ligand, cGMP (1,28). To re-open the channels, a prerequisite to return photoreceptors to the dark-adapted state, cGMP has to be produced at an accelerated rate by a guanylate cyclase (GC). As shown by numerous electrophysiological and biochemical experiments, activation of GC is regulated by Ca^{2+} ions.

In dark-adapted rod photoreceptor outer segments, cytoplasmic cGMP levels are high (~5μM (2)), cGMP phosphodiesterase (PDE) activity is low, and about 1-5% of cGMP-gated channels located in the plasma membrane are kept open. Influx of cations through the open channels maintains cytoplasmic free Ca^{2+} levels at ~300nM (12,17) and photoreceptors depolarized. Illumination of rhodopsin triggers an enzymatic cascade which activates PDE resulting in rapid hydrolysis of cytoplasmic cGMP (24) and closure of cGMP-gated cation channels (Fig. 1). The change in conductance hyperpolarizes the photoreceptor cell, which is the beginning of an electrical impulse to be sent to the brain. While cation channels are closed, the light-insensitive Na^+/K^+, Ca^{2+} exchangers, also located in the plasma membrane, continue to extrude Ca^{2+}, and cytoplasmic free Ca^{2+} concentration drops.

The drop in free Ca^{2+} after photobleaching is the signal for the photoreceptor GC (9,23) to accelerate the rate of cGMP synthesis. GC activation is mediated by a membrane-associated Ca^{2+}-binding protein termed guanylate cyclase activating protein (GCAP) (10). A second, less well characterized Ca^{2+}-binding protein, termed p24, closely related to GCAP and to other Ca^{2+}-binding proteins of the calmodulin superfamily, has been shown to also

Degenerative Diseases of the Retina, Edited by Robert E. Anderson et al.
Plenum Press, New York, 1995

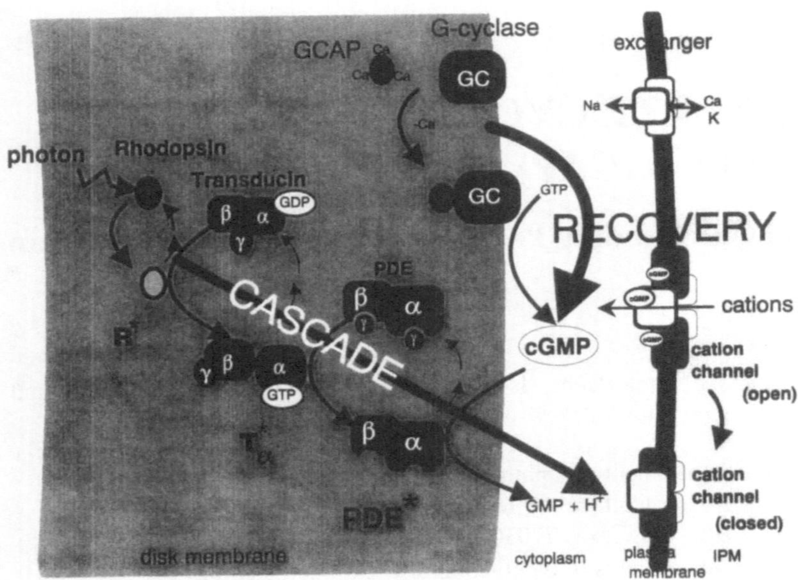

Figure 1. Signal transduction and recovery in rod photoreceptors. Only the key components of the phototransduction cascade (rhodopsin, transducin, cGMP phosphodiesterase, cGMP-gated channel) and their interactions are shown. The large black arrow symbolizes the recovery pathway consisting of activation of guanylate cyclase by GCAP in low Ca^{2+}. The ROS disc membrane is viewed from the top, with part of the disc membrane surface shown in gray, the cytoplasm in white, the plasma membrane in dark gray, and the interphotoreceptor matrix in light gray. For simplicity, the activated PDE (PDE*) is shown as a two subunit core, although PDE* may consist of a complex of PDE and transducin subunits, the GC is shown as a monomeric unit (the functional GC may be a homodimer), and the cGMP-gated channel as a heterodimer (the subunit distribution is not known).

activate GC in low Ca^{2+} (3). While GCAP is present exclusively in photoreceptors, the cellular or subcellular distribution of p24 is currently not known. In this chapter we review the properties and function of bovine GCAP, the primary sequences of vertebrate GCAPs, the human and mouse GCAP gene structures, and discuss implications for involvement of the GCAP gene in retinal disease.

ISOLATION OF BOVINE GCAP

Two methods were developed to isolate GCAP either from ROS or from the retina. The first procedure is based on conventional column chromatography, the second on binding to immobilized monoclonal antibody. Briefly, in the first procedure (10), a ROS supernatant containing GCAP is prepared from ROS membranes by high speed centrifugation, dialyzed against water, and loaded on a DEAE Sepharose in the presence of 5 mM BTP (1.3 bis[tris(hydroxymethyl)methylamino]propane), pH 7.5, and 100 mM NaCl. Bound proteins are eluted with a linear NaCl gradient (100 to 350 mM) in 5 mM BTP, pH 7.5. Fractions that contain GCAP, eluting at 220 mM NaCl, are pooled and loaded on a hydroxylapatite column. GCAP is eluted with a linear gradient of KH_2PO_4 (0 to 60 mM) and directly decreasing concentration of NaCl (from 100 to 0) in 10 mM BTP, pH 7.5 at 30 mM KH_2PO_4 and 50 mM NaCl. Fractions containing GCAP are mixed with acetonitrile to a final concentration of 15%, and the sample is loaded on a C-4 HPLC column equilibrated with 30% acetonitrile in 5 mm BTP, pH 7.5. Finally, GCAP is eluted with a linear gradient of acetonitrile (30% to 60%) in 5 mM BTP, pH 7.5.

In the second, immunoaffinity procedure (11), a retinal extract containing GCAP is prepared (equivalent to 50 bovine retinas/ 25 ml of water, containing 1 mM benzamidine) as described above. The extract is loaded onto an antibody-Sepharose column (Mab G-2; lx2 cm) equilibrated with 10 mM BTP buffer, pH 7.5. The monoclonal antibody was raised against GST-GCAP fusion proteins expressed in *E. coli* and purified from inclusion body. The column is washed with 10 mM BTP buffer, pH 7.5, containing 200 mM NaCl, and then with 10 mM BTP, pH 7.5. GCAP is eluted with 0.1 M of glycine, pH 2.5, and immediately neutralized with 1 M Tris/HCl, pH 8.4.

Biochemical Properties of native and expressed gcap

Purified native GCAP is an acidic, soluble, but distinctly hydrophobic polypeptide with a mobility 20 kDa in SDS-PAGE in the presence of Ca^{2+} and 23-24 kDa in the presence of EGTA. The amino acid analysis (from the cloned cDNA, see below) indicates a larger than average distribution of hydrophobic amino acids. The hydrophobicity is further increased by heterogenous acylation of the N-terminal Gly. GCAP is a Ca^{2+}-binding protein, as demonstrated by Ca^{2+}-sensitive changes in tryptophan fluorescence emission (22). GCAP is fairly stable when frozen in low ionic strength buffer, however, is very susceptible to denaturation when highly purified. GCAP can be expressed in bacteria either as a fusion protein with the C-terminal portion of glutathione S-transferase (GST), or as a nonfusion protein using pET vectors, but in either case is found to be stored in insoluble inclusion bodies and cannot be regenerated to biological activity. When expressed in insect cells using baculovirus transfer vectors, GCAP is soluble, identical to the native protein, and fully functional in terms of low Ca GC activation.

PRIMARY SEQUENCES OF VERTEBRATE GCAPs

An endlabelled degenerate oligonucleotide, reversely translated from a tryptic peptide sequence, was used as a probe to isolate several cDNA clones from a bovine retina library (22). The clones predict a GCAP polypeptide of 205 amino acid residues with a molecular mass of 23,524. The sequences of 17 tryptic peptides were identified in the deduced sequence, verifying the identity of the clones. The predicted amino acid sequence contains four EF hand motifs, three of which are most likely functional and capable of chelating Ca^{2+}. Genomic Southern blotting of eight vertebrate species (human, monkey, rat, mouse, dog, cow, rabbit, chicken), probed with bovine GCAP under stringent hybridization conditions, indicated that the GCAP sequences are well conserved among various species. Cloning of human, mouse, frog and chicken GCAP (Fig. 2) showed that the sequence similarity at the amino acid level among mammalian GCAPs is higher than 90% (among vertebrate and mammalian species higher than 70%). The lengths of the five cloned GCAP sequences vary from 199 (chicken) to 205 (bovine, frog) amino acids. A C-terminal divergent domain containing insertions/deletions is responsible for the variation in length. The N-terminal halves of vertebrate GCAP sequences are unusually well conserved showing only very few non-conservative substitutions. As judged by peptide competition experiments, the N-terminal residues 2 - 57 are most likely involved in GC activation (22).

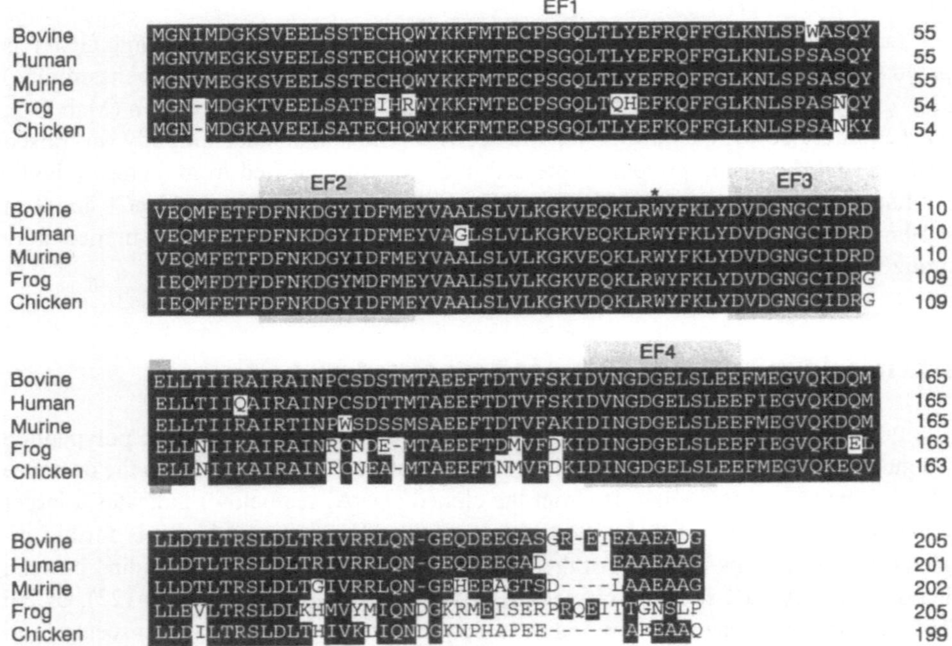

Figure 2. Amino acid sequence alignment of bovine, human, mouse, frog and chicken GCAP. Amino acids are depicted in single letter symbols. For best fit, several gaps were introduced (shown by hyphens). Residues printed in white on black are identical or represent conservative substitutions (L=I=V=M; Y=F; E=D; R=K). Deviating residues are shown in black on white. The four EF hand helix-loop-helix structures (EF1-4) are shaded. EF-1 is thought to be unable to chelate Ca^{2+}.

GCAP IS A MEMBER OF A LARGE SUPERFAMILY OF Ca^{2+}-BINDING PROTEINS

GCAP is a novel member of the large and diverse superfamily of Ca^{2+}-binding proteins which includes calmodulin, parvalbumin, troponin C, calbindins, recoverin-like proteins, and many others (20). Sequence alignment with a limited number of Ca^{2+}-binding proteins (recoverin, frequenin, and calmodulin) shows up to 32 identically conserved amino acid residues, most of them grouped around the four EF hand structures (26). The sequence similarity between the two photoreceptor Ca^{2+}-binding proteins recoverin and GCAP is only approximately 36% even when conservative substitutions are taken into account. As an illustration, we computed a dendrogram in which vertebrate GCAPs and recoverins are compared. As shown in Fig. 3, GCAPs and recoverins form two related subfamilies contained within the calmodulin superfamily.

GCAP FUNCTION

The functional activity of GCAP is assayed utilizing the low Ca^{2+} v-stimulation of ROS GC (mol. mass 110,000). The GC stimulation in low Ca^{2+} by native and insect cell-expressed GCAP is approximately 5-10 fold, while high Ca^{2+} abolishes this effect.

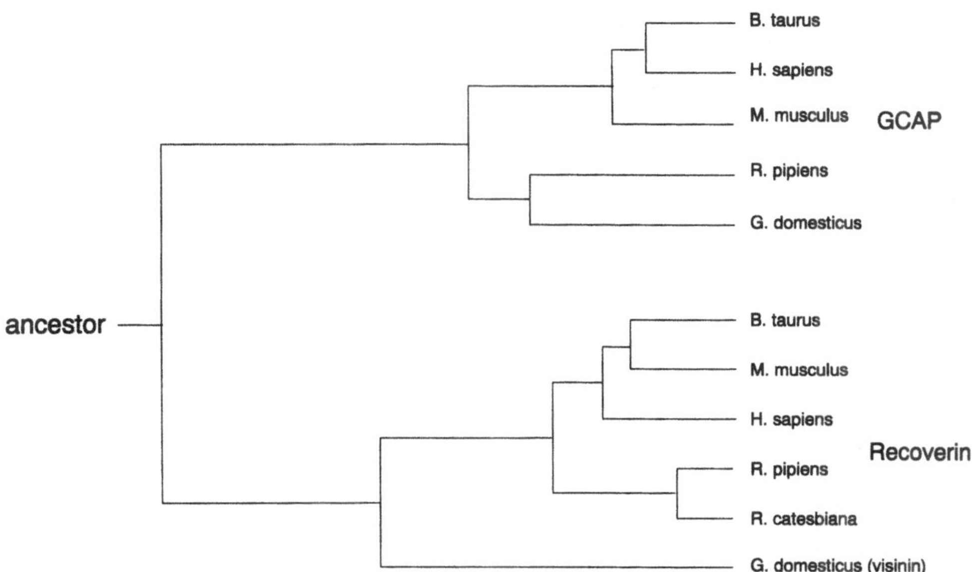

Figure 3. Dendrogram of vertebrate GCAPs and recoverins. The dendrogram was computed for protein sequence similarities using preset values for k-tuple and window size, gap penalty, and filtering level (PC-Gene, Intelligenetics, Inc.). The bovine, human, mouse and frog GCAP sequences are from (22). The recoverin sequences are from (4) (bovine), (19) (human), (25) (mouse), and (15) (*R. catesbiana*). The chicken GCAP sequence (Semper-Rowland et al., in preparation) and the *R. pipiens* recoverin sequence (Subbaraya et al., submitted) are unpublished.

Crosslinking with dithio-bis(sulfosuccinimidyl)proprionate (DTSSP), both at low and high Ca^{2+}, yielded a high molecular weight product of ~400 kDa which is antigenic with GCAP and GC antibodies, suggesting that the complex is formed by several GC molecules (11). It appears that the GCAP molecule exerts its action as a monomeric protein, although it may interact with several GC molecules (Fig. 4). The inhibitors of the GC stimulation by GCAP include hydrophobic molecules (such as benzamidine), N-terminal peptides derived from the cloned sequence of GCAP, and a monoclonal antibody that was raised against bacterially expressed GCAP (22).

PHOTORECEPTOR SPECIFIC EXPRESSION

We previously showed by Northern blotting of RNA from various tissues, that the GCAP gene is expressed in the retina, but not in heart, brain, placenta, lung, liver, muscle, kidney, or pancreas. We further showed by in-situ hybridization with digoxygenin-labelled antisense RNA in bovine and primate retina that GCAP is expressed specifically and discretely in photoreceptors cells, and that GCAP mRNA can be detected primarily in the myoid region of both rods and cones where protein synthesis occurs. Since in cone photoreceptors cytoplasmic Ca^{2+} regulates GC with features indistinguishable from those in rods (18), the presence of an identical or closely related GC/GCAP system in this cell type is not surprising. Immunocytochemical localization with a polyclonal antibody (11) revealed presence of GCAP in both rod and cone outer segments. In some tissue preparations, cone synapses also stained intensely.

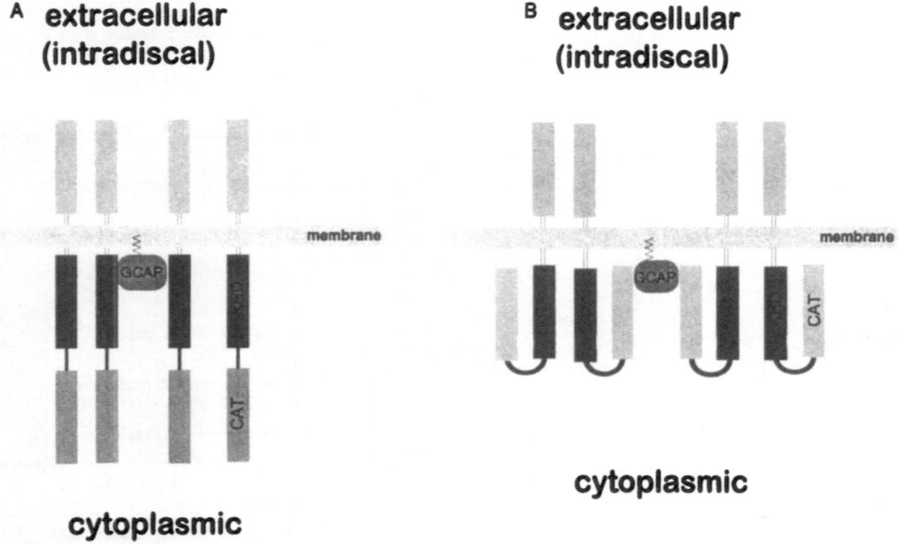

Figure 4. Hypothetical models of interaction between GC and GCAP. The four domains (extracellular, transmembrane, protein kinase-like domain KHD, catalytic domain CAT) of particulate GC (8) are shown in various shadings of grey. GCAP is pictured as membrane associated, activating GC at the cytoplasmic face. As two of several possibilities, GCAP is shown to interact with a dimer of GC through the KHD (A) or through the catalytic domain (B).

THE HUMAN AND MOUSE GCAP GENES

The human and mouse GCAP gene structures were elucidated by cloning and complete sequencing of subcloned genomic fragments (Fig. 5). The gene is split into four exons distributed over 6 kb (human) and 4 kb (mouse) of genomic DNA. In human, intron a is 4.8 kb in length, and contains two antisense singleton 300 bp Alu repeats with 90% similarity to AluSX (14) and a 50 bp T_nG polymorphism. Introns b and c are 373 and 341 bp, and the two central exons are 146 and 94 bp each. The 6 splice junctions conform to the donor/acceptor consensus sequences (21). In mouse, the intron postions are exactly as in the

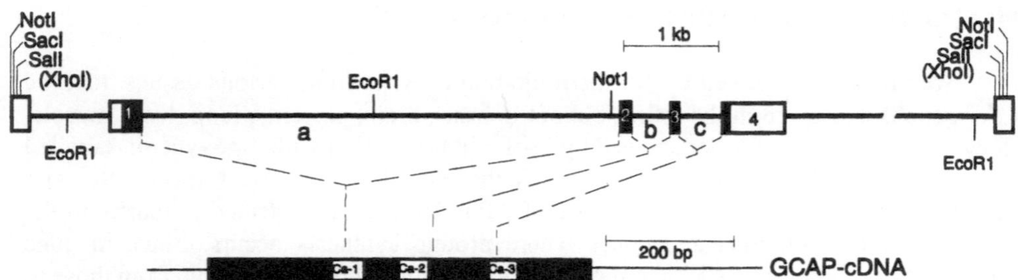

Figure 5. Human Gene Structure. Graphical representation of the gene. Exons of the GCAP gene are depicted as boxes, the coding portions are filled. Introns (a-c) and flanking sequences are shown as lines. The genomic vector is λFixII, only its multiple cloning site is shown. The lower part depicts the position of introns within the cDNA domain structure.

human gene, and introns are approximately the same length. While the sequences and EF hand structures are related, there is no conservation of intron/exon boundaries between the GCAP gene and the human recoverin gene containing three exons (19), or the human calmodulin gene CaMI containing six exons (6).

CHROMOSOMAL LOCALIZATION

The chromosomal localization of the GCAP gene was determined by amplification of an exon 1 fragment in human hamster hybrid panels. Only panels containing human exon 6 yielded the expected product. Sublocalization to the short arm of chromosome 6 was unequivocally determined on banded metaphase spreads. Assignment of the GCAP1 locus to region 6p21.1 was confirmed by co-hybridization of a chromosome 6 centromere-specific probe. Consistent signals were not observed on any other chromosomes. The GCAP1 locus is near RDS on 6p21.1-cen. By synteny with RDS (27) and other loci near p21.1, the mouse gcap1 locus is predicted to reside on chromosome 17.

GCAP AS A CANDIDATE GENE FOR HUMAN DISEASE

By virtue of its photoreceptor-specific expression, GCAP is a candidate gene for mammalian retinal dystrophies. In a variety of animal and human models, mutant genes expressing defective components involved in phototransduction or ROS disc structure, have been shown to cause cell death and retinal degenerations (13). Since GCAP plays a key role in return of the photoreceptor to the dark state after illumination, a defect in the GCAP gene may disturb activation of cyclase and possibly disable accelerated cGMP synthesis essential for depolarization of photoreceptors. A predicted phenotype for a GCAP missense mutation may be non-progressive night blindness in which cGMP levels in dark-adapted photoreceptors are abnormally low, as has been observed in families with defects in the rhodopsin gene causing constitutive activation of transducin (5) or in a large Danish pedigree in which a defect in the PDEβ gene causes persistent hydrolysis of cGMP (7). More severe phenotypes like RP and macular degeneration in which photoreceptors progressively degenerate would be consistent with a null allele and loss of GCAP function. Recently, a new locus for autosomal recessive retinitis pigmentosa, distinct from RDS, has been identified on 6p (16) in a dominican pedigree, and initial mapping studies indicate that this locus is within 2 cM of GCAP and may be identical with the GCAP locus (Poulabi Banerjee, James Knowles, and Conrad Gilliam, personal communication).

ACKNOWLEDGMENTS

We thank Claudia Ruiz, Bharati Helekar, Zinyu Zhao for expert technical assistance. This research was supported by USPHS grants EY08061, EY08123, awards from the Research to Prevent Blindness, Inc., to the Departments of Ophthalmology at the University of Washington and at Baylor College of Medicine, and an award from the RP Foundation to W.B. K.P. is a Jules and Doris Stein Research to Prevent Blindness professor.

REFERENCES

1. Baylor, D. A. (1987) Photoreceptor signals and vision. *Invest. Ophthalmol. & Vis. Sci.* **28**, 34-49

2. Cote, R. H. and Brunnock, M. A. (1993) Intracellular cGMP concentration in rod photoreceptors is regulated by binding to high and moderate affinity cGMP binding sites. *J. Biol. Chem.* **268**, 17190-17198
3. Dizhoor, A. M., Lowe, D. G., Olshevskaya, E. V., Laura, R. P., and Hurley, J. B. (1994) The human photoreceptor membrane guanylyl cyclase, RetGC, is present in outer segments and is regulated by Ca^{2+} and a soluble activator. *Neuron* **12**, 1-20
4. Dizhoor, A. M., Ray, S. , Kumar, S. , Niemi, G. , Spencer, M. , Brolley, D. , Walsh, K. A., Philipov, P. P., Hurley, J. B., and Stryer, L. (1991) Recoverin: a Ca^{2+} sensitive activator of retinal rod guanylate cyclase. *Science* **251**, 915-918
5. Dryja, T. P., Berson, E. L., Rao, V. R., and Oprian, D. D. (1993) Heterozygous missense mutation in the rhodopsin gene as a cause of congenital stationary night blindness. *Nature Genet.* **4**, 280-283
6. Fischer, R. , Koller, M. , Flura, M. , Mathews, S. , Strehler-Page, M. -A., Krebs, J. , Penniston, J. T., Carafoli, E. , and Strehler, E. E. (1988) Multiple divergent mRNAs code for a single human calmodulin. *J. Biol. Chem.* **263**, 17055-17062
7. Gal, A. , Orth, U. , Baehr, W. , Schwinger, E. , and Rosenberg, T. (1994) Heterozygous missense mutation in the rod cGMP phosphodiesterase β-subunit gene in autosomal dominant stationary night blindness. *Nature Genet.* **7**, 64-68
8. Garbers, D. L. and Lowe, D. G. (1994) Guanylyl cyclase receptors. *J. Biol. Chem.* **269**, 30741-30744
9. Goraczniak, R. M., Duda, T. , Sitaramayya, A. , and Sharma, R. K. (1994) Structural and functional characterization of the rod outer segment membrane guanylate cyclase. *Biochem. J.* **302**, 455-461
10. Gorczyca, W. A., Gray-Keller, M. P., Detwiler, P. B., and Palczewski, K. (1994) Purification and physiological evaluation of a guanylate cyclase activating protein from retinal rods. *Proc. Natl. Acad. Sci. U.S.A.* **91**, 4014-4018
11. Gorczyca, W. A., Polans, A. S., Surgucheva, I. , Subbaraya, I. , Baehr, W. , and Palczewski, K. (1995) Guanylyl cyclase activating protein: A Ca^{2+} -sensitive regulator of phototransduction. Submitted
12. Gray-Keller, M. P., Polans, A. S., Palczewski, K. , and Detwiler, P. B. (1993) The effect of recoverin-like Ca^{2+}-binding proteins on the photoresponse of retinal rods. *Neuron* **10**, 523-531
13. Humphries, P. , Kenna, P. , and Farrar, G. J. (1992) On the molecular genetics of retinitis pigmentosa. *Science* **256**, 804-808
14. Jurka, J. and Milosavljevic, A. (1991) Reconstruction and analysis of human Alu genes. *J. Mol. Evol.* **32**, 105-121
15. Kawamura, S. , Hisatomi, O. , Kayada, S. , Tokunaga, F. , and Kuo, C. (1993) Recoverin has S-modulin activity in frog rods. *J. Biol. Chem.* **268**, 14579-14582
16. Knowles, J. A., Shugart, Y. , Banerjee, P. , Gilliam, T. C., Lewis, C. A., Jacobson, S. G., and Ott, J. (1994) Identification of a locus, distinct from RDS-peripherin, for autosomal recessive retinitis pigmentosa on chromosome 6p. *Hum. Mol. Genet.* **3**, 1401-1403
17. Lagnado, L. and Baylor, D. (1992) Signal flow in visual transduction. *Neuron* **8**, 995-1002
18. Miller, J. L. and Korenbrot, J. I. (1994) In retinal cones, membrane depolarization in darkness activates the cGMP-dependent conductance. *J. Gen. Physiol.* **101**, 933-961
19. Murakami, A. , Yajima, T. , and Inana, G. (1992) Isolation of human retinal genes: Recoverin cDNA and gene. *Biochem. Biophys. Res. Commun.* **187**, 234-244
20. Nakayama, S. , Moncrief, N. D., and Kretsinger, R. H. (1992) Evolution of EF-hand Ca^{2+} -modulated proteins. II. domains of several subfamilies have diverse evolutionary histories. *J. Mol. Evol.* **34**, 416-448
21. Padgett, R. A., Grabowski, P. J., Konarska, M. M., Seiler, S. , and Sharp, P. A. (1986) Splicing of messenger RNA precursors. *Ann. Rev. Biochem.* **55**, 1119-1150
22. Palczewski, K. , Subbaraya, I. , Gorczyca, W. A., Helekar, B. S., Ruiz, C. C., Ohguro, H. , Huang, J. , Zhao, X. , Crabb, J. W., Johnson, R. S., Walsh, K. A., Gray-Keller, M. P., Detwiler, P. B., and Baehr, W. (1994) Molecular cloning and characterization of retinal photoreceptor guanylyl cyclase activating protein (GCAP). *Neuron* **13**, 395-404
23. Shyjan, A. W., de Sauvage, F. J., Gillett, N. A., Goeddel, D. V., and Lowe, D. G. (1992) Molecular cloning of a retina-specific membrane guanylyl cyclase. *Neuron* **9**, 727-737
24. Stryer, L. (1986) Cyclic GMP cascade of vision. *Ann. Rev. Neurosci.* **9**, 87-119
25. McGinnis, J.F., Stepanik, P.L.,Baehr, W., Subbaraya, I., Lerious, V. (1992) Cloning and sequencing of the 23kDa mouse photoreceptor cell-specific protein. *FEBS Letters* **302**, 172-176
26. Subbaraya, I. , Ruiz, C. C., Helekar, B. S., Zhao, X. , Gorczyca, W. A., Pettenati, M. J., Rao, P. N., Palczewski, K. , and Baehr, W. (1994) Molecular characterization of human and mouse photoreceptor guanylate cyclase activating protein (GCAP) and chromosomal localization of the human gene. *J. Biol. Chem.* **269**, 31080-31089

27. Travis, G. H., Christerson, L. , Danielson, P. E., Klisak, E. , Sparkes, R. S., Hahn, L. B., Dryja, T. P., and Sutcliffe, J. G. (1991) The human retinal degeneration slow (rds) gene: Chromosome assignment and structure of the mRNA. *Genomics* **10**, 733-739
28. Yau, K. -W. and Baylor, D. A. (1989) Cyclic GMP-activated conductance of retinal photoreceptor cells. *Ann. Rev. Neurosci.* **12**, 289-327

ABNORMAL CONE RECEPTOR ACTIVITY IN PATIENTS WITH HEREDITARY DEGENERATION

Donald C. Hood[1] and David G. Birch[2]

[1] Department of Psychology
Columbia University
New York, New York
[2] Retina Foundation of the Southwest
Dallas, Texas

INTRODUCTION

Although often thought of as a disease that primarily affects the rods, retinitis pigmentosa (RP) delays the implicit times of the cone ERGs very early in progression (see ref. 1 for review). The first two panels of Figure 1 show the cone responses recorded as part of a standard clinical protocol from a normal subject and four patients with RP. Cone ERGs are recorded by presenting a single white flash on a background that suppresses the rods (left panel) or by presenting a 30 Hz pulse train, too fast for the rods to follow, to the dark adapted eye (right panel). The vertical dashed lines in Fig. 1 mark the peak times for the normal observer and the patients' records were selected to illustrate a range of implicit times. In general, patients with RP show delays in the peak response to both the single flashes and the 30 Hz flicker. These delays have been attributed, at least in part, to cone photoreceptor abnormalities, specifically to a decrease in quantal absorption by the cone outer segments (1-3). Other ERG evidence is more difficult to reconcile with only a change in quantal absorption (4-10).

Here we summarize evidence that RP affects both cone phototransduction and processing beyond the outer segment. A model of cone phototransduction was fitted to the a-waves of thirteen patients with RP and five with cone-rod dystrophy (CRD). There are changes in the sensitivity of cone phototransduction, but these changes alone cannot account for the delays in the cone ERGs.

METHODS

Subjects

Thirteen patients with RP ranging in age from 11 to 67 participated in this study. They were classified as: 8 adRP, 3 simplex, 1 X-linked, and 1 recessive. Six of the patients with

Degenerative Diseases of the Retina, Edited by Robert E. Anderson et al.
Plenum Press, New York, 1995

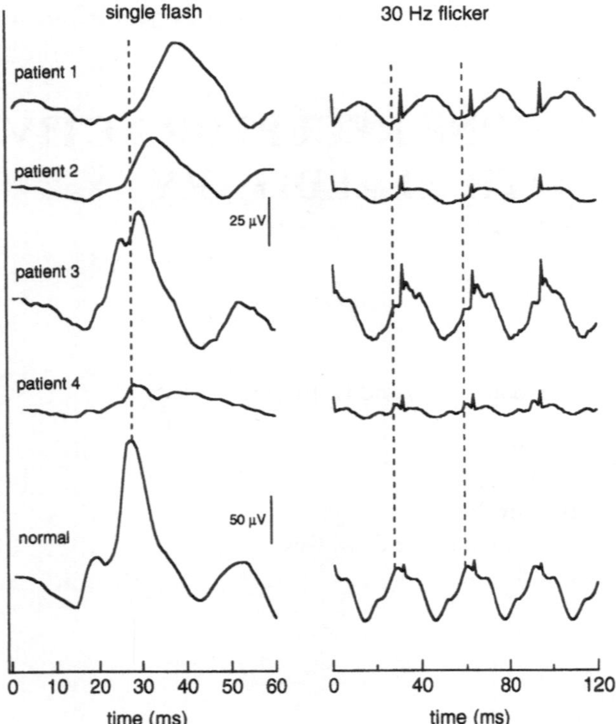

Figure 1. Responses to standard protocol stimuli.

adRP had a rhodopsin mutation [pro23his (3); leu46arg (2); and splice-17y (1)]. In addition, five patients with cone-rod dystrophy (CRD), ranging in age from 9 to 67 years, also participated.

The control group for the high intensity analysis consisted of eight subjects with normal color vision, normal full-field ERGs and normal ophthalmological examinations. They ranged in age from 33 to 51 yr (mean age = 42.5). The control group for the standard protocol included 62 subjects of ages 5 to 60 years. Norms by age have previously been presented (11).

All patients had been diagnosed by ophthalmologists specializing in retinal disease. The tenets of the Declaration of Helsinki were followed and all subjects gave written informed consent after a full explanation of the procedures was given.

Recording Techniques

The methods used for obtaining full-field ERGs are relatively standard (12). Responses were obtained from the anesthetized cornea with a Doran bipolar contact lens electrode. Signals were amplified (factor of 10,000; 3 dB down at 2 and 10,000 Hz) and averaged.

Standard Protocol Stimuli

In all patients, responses were obtained to a 30 Hz, 1.3 log td-s light and to a single flash (1.9 log td-s) upon a background of 3.2 log td. The implicit time and amplitude of the primary positive component were measured.

High Energy Single Flash Protocol

Cone ERGs were recorded to brief, red (W26) flashes presented upon a steady adapting field (2.6 to 2.9 log td). Although these fields have been shown to suppress rod activity to the stimuli used here (13), blue flashes were also presented to assure that the cone a-wave was isolated. There was no sign of rod activity in three patients (the two with the leu46arg mutation and the xlRP), thus cone ERGs were obtained in the dark adapted eye with white flashes to extend the range of flash intensities available. Six to eighteen responses were averaged at each flash energy.

Dark-adapted rod ERGs were recorded from 11 of the patients (7 RP and 4 CRD) and 5 of the normal controls. The general procedures have been previously described (14).

Theoretical Analysis of Phototransduction

A model of the activation phase of cone phototransduction was fitted to the leading edge of the cone a-wave (14). This model is based upon the Lamb and Pugh model (15) of rod photocurrent and includes the addition of a lowpass filter to account for the capacitance effects of the extensive cone outer segment membrane. In particular, the model includes eq 1 below followed by an exponential filter with a time constant τ. The part of the model common to both the rod and cone a-wave is

$$P3(i,t) = \{1 - \exp[-i \cdot S \cdot (t - t_d)^2]\} \cdot Rm_{p3} \text{ for } t > t_d \quad (1)$$

where the amplitude P3, named after Granit's receptoral component, is a function of flash energy i and time t after flash onset. S is a sensitivity parameter that scales flash energy i; Rm_{P3} is the maximum amplitude; and t_d is a brief delay. The leading edge of the cone a-waves of all subjects was fitted by setting t_d equal to 1.7 ms and τ equal to 1.8 ms, the parameters for a group of normals, and estimating the values of S and Rm_{P3} for best fit (see ref. 14 for details).

Equation 1 was fitted to the rod a-waves as previously described (16).

RESULTS AND DISCUSSION

Responses to Standard Protocol Stimuli

All the patients showed smaller and slower responses to the standard clinical stimuli. The implicit times for the 30 Hz flicker ranged from 30.5 to 45.6 ms (mean = 38.0) and were longer than the mean normal value of 28.0 ms +/- 1.8 for all the patients. Likewise the single flash implicit times ranged from 28.8 to 44.5 ms (mean = 35.0) and were longer than the mean normal value of 27.8 ms +/- 1.5.

Responses to Higher Energy Flashes

Figure 2 shows the cone ERGs to the red flashes upon the 2.9 log td field for a normal and a patient with RP. The background field was similar to the field of 3.2 log td used for the light-adapted single flash in the clinical protocol (left panel of fig. 1). The flashes, however, were more intense starting 0.3 log units higher and going up to 3.7 log td-s. The a-wave grows in amplitude reaching a maximum by about 3.0 log td-s, while the peak-to-peak amplitude of the ERG decreases over this same range. At the highest intensities, the cone system appears to be saturated. The amplitudes of the ERGs to the top two or three

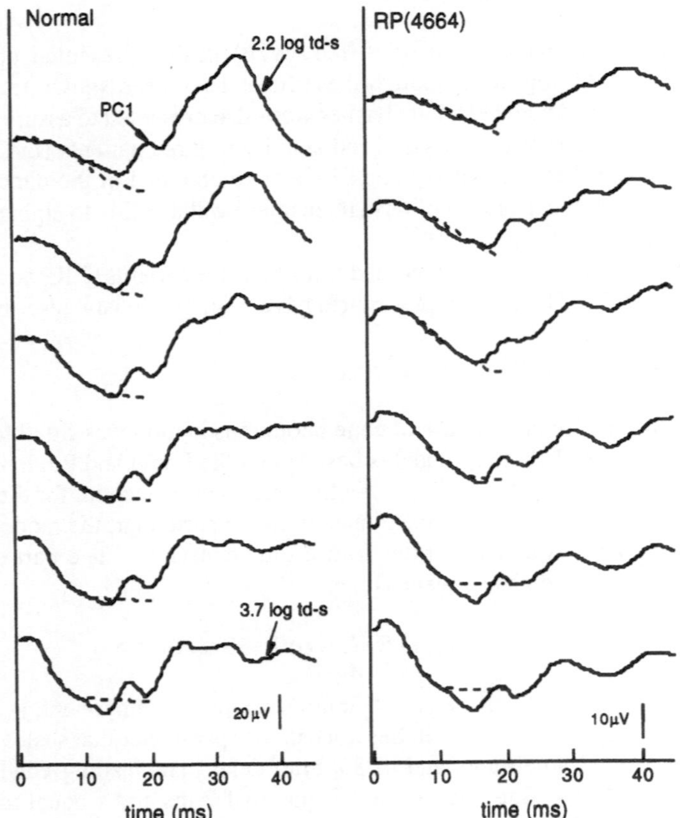

Figure 2. Responses to red flashes on a white background.

flash energies are very similar and only minor changes in timing appear to distinguish them The patient's responses are smaller and, as will be clearer below, slower.

Parameters of Cone Phototransduction

The dashed curves in Fig. 2 show the fit of the model (14) of cone receptor activity to the a-waves. All a-waves were fitted up to 11 ms, but the theoretical curves are shown for the first 20 ms. The model fits the patient data well. Two parameters, S (sensitivity) and Rm_{P3} (maximum a-wave amplitude), were obtained from these fits. The patient had values of S and Rm_{P3} that were 0.35 and 0.32 log units below the mean normal values.

Figure 3 (left panels) shows histograms of the values of log S and log Rm_{P3} for all the subjects in this study. The vertical dashed lines indicate the lower limit of the range of normal values. The patients have smaller values of cone Rm_{P3} (lower left panel), hardly surprising given that these diseases eventually destroy cone outer segments. However, the sensitivity parameter, S, is also affected for the cones. It is lower in this group of patients, on average by 0.43 log unit. And, all but two patients, one RP and one CRD, had S(cone) values that were below the range of the normal values.

Rod ERGs were available for 11 of the patients (7 RP and 4 CRD) and 5 of the normal subjects. The parameters of best fit are shown in the middle panel of Fig. 3. As previously reported (16), some of these patients show abnormal S values and some do not. All, however,

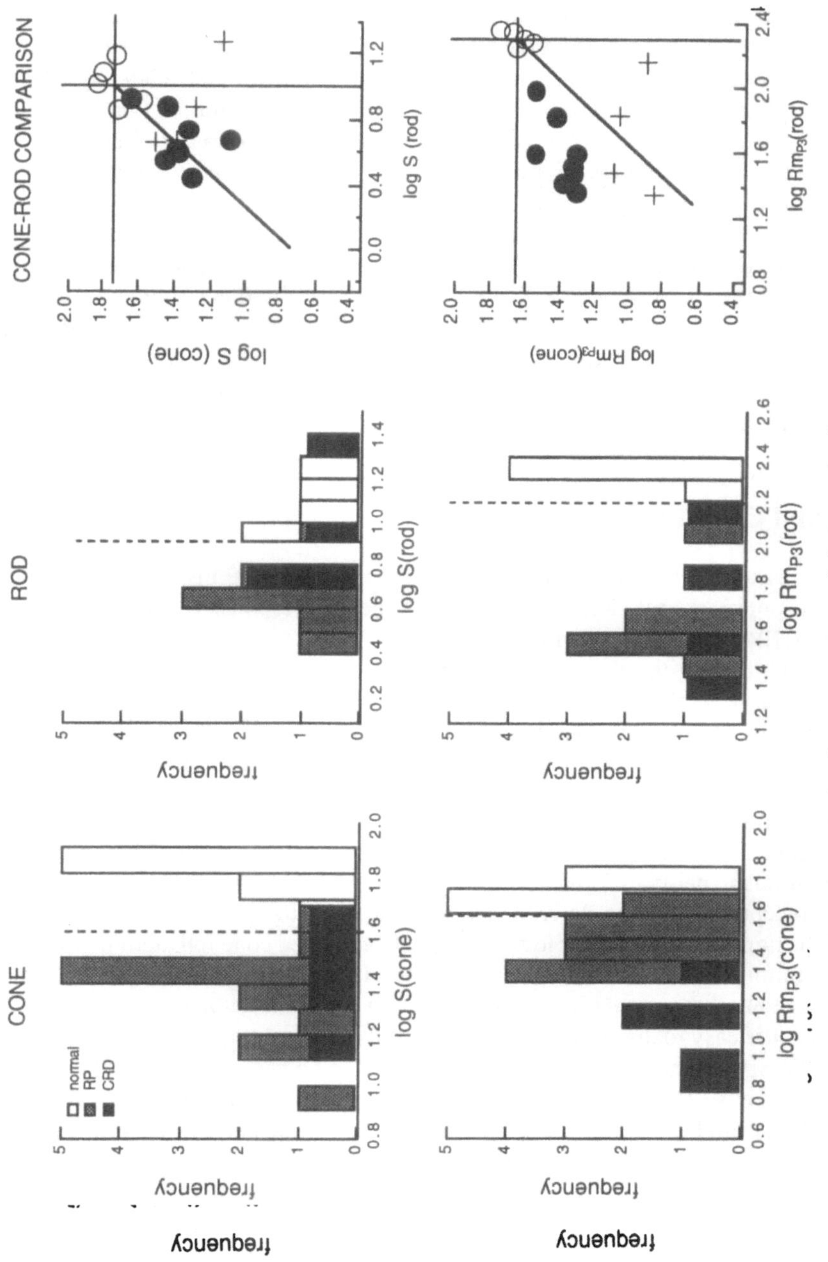

Figure 3. Rod and cone parameters of phototransduction.

show significantly decreased Rm_{P3} values consistent with a disease that is destroying rod outer segments.

In the right panels in Fig. 3, the individuals' rod and cone parameters are compared for the patients with RP(filled circles) and CRD (plusses). The diagonal lines mark the loci of equal changes in rod and cone parameter. If we assume that to a first approximation the Rm_{P3} values provide a measure of the total amount of functional, outer segment membrane (14, 17), then the values of Rm_{P3} (lower right panel) confirm that relatively more rod, as opposed to cone, outer segment membrane is missing in the RP retina. For the patients with CRD, the cone outer segment loss as measured by Rm_{P3} is about as extreme, or in one case, more extreme than the rod loss. Further, there appears to be a moderate correlation between cone and rod outer segment loss.

The sensitivity, S, of cone transduction is at least, if not more, affected than the sensitivity of rod function in both groups of patients (upper right panel). This is surprising, especially for the patients with RP where, as expected, the outer segment damage of the population of rods is greater than the damage of the cones. The cause(s) of the depressed value of cone sensitivity is not clear, but it is likely that it is, in part, due to an abnormal activation phase of most of the functioning cones in the retinas of these patients (see below). This abnormality is likely a change that is secondary to the rod degeneration in the patients with RP.

Abnormal Cone INL Activity

By assuming that the cone model describes the first 30 ms of the underlying cone receptor response, we can derive an estimate of the INL response (i.e. the non receptor part of the cone ERG). The curves in Fig. 4 (left panel) are derived INL responses obtained by computer subtracting the response of the cone model (dashed curves in Fig. 1) from the ERGs. The curves are for three flash energies and for the normal (solid) and patient (dashed) subjects in Fig. 2. A second patient with RP is also shown to illustrate the range of INL changes seen in these patients. The patients' derived cone INL responses are smaller and of a fundamentally different shape. The various sub-components of the ERG appear delayed in time as previously reported (2).

Notice in eq 1 that a decrease in the sensitivity of transduction, S, is equivalent to a decrease in flash energy, i. Thus, if the only effect of RP was to decrease the sensitivity of the cone receptors, then the derived INL responses for the patients should correspond to those of the normals to flashes of lower energy. The derived cone INL activity of the patients cannot be mimicked in normals by decreasing the effectiveness of the flash. Figure 4 (middle panel) shows another way to illustrate this point. In all normal subjects, the potential labeled PC1 in Fig. 1 is easy to discern. This is undoubtedly the potential that others have identified as an oscillatory potential (e.g. 18-20). It is thought to be generated in the INL probably by the amacrine cells (21). The implicit time of this potential decreases with increases in flash energy (18) as can be seen in Fig. 4 (middle) where the open symbols show the mean and standard deviation for the group of normal subjects. The other symbols show the implicit time of PC1 in three patients in whom this component was easily discernible. (These three patients had log S(cone) values 0.2 to 0.4 lower than normal.) All three patients show marked delays in PC1. For 17 out of 18 patients, the PC1 component of the cone ERG was delayed.

The delays in PC1 cannot be attributed to a decrease in the effective intensity secondary to a decrease in S(cone). Adjusting the effective flash energy by the change in S(cone) produces the dashed curves. The change in phototransduction sensitivity is clearly insufficient to explain the delayed PC1 in these patients. This argument does not take into consideration the fact that the change in S(cone) also affects the adapting field intensity. However, the changes in adapting field intensity have little effect on the implicit time of

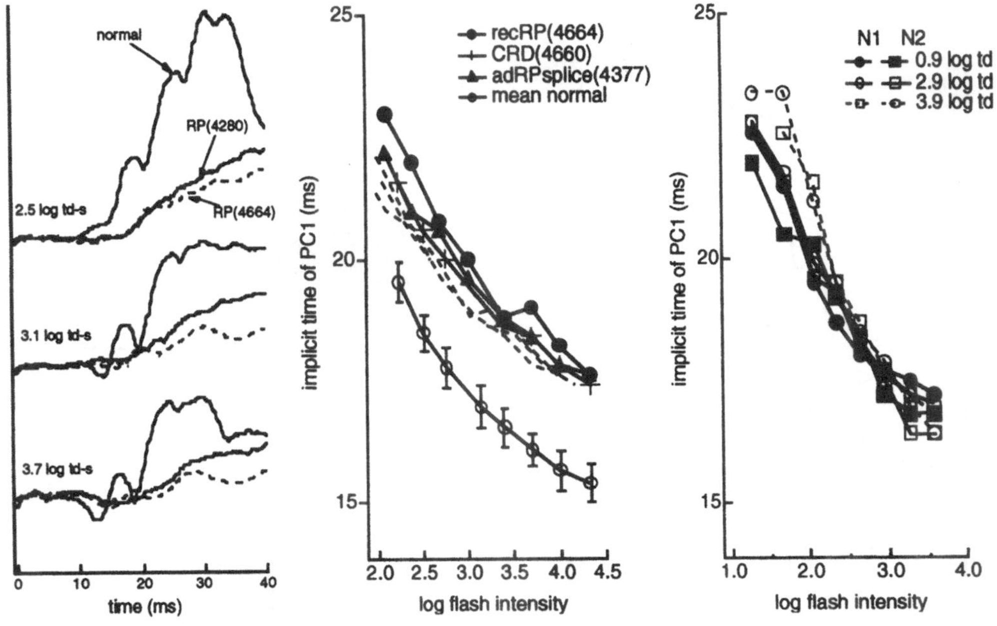

Figure 4. Measures of INL activity.

PC1. This is shown in the right panel of Fig. 4 where the implicit time of PC1 is shown for a three log unit change in the adapting field. For two normal subjects, there are small decreases in implicit time with lower adapting field intensity, a direction opposite to the changes seen in the patients. Thus a decrease in the effectiveness of the flash and the adapting field cannot account for the slower response in the patients.

A Comparison of Single Flash and Flicker ERGS

Figure 5 shows the cone ERGs from Fig. 1 on the same time axis. The timing of the peak of the single flash ERG (dashed curves) appears correlated with the amplitude and timing of the flicker ERG. That is, the implicit time of the flicker response appears to increase together with the implicit time of the peak of the response to the single flash. The filled symbols in the right panels show both the amplitude (upper) and the implicit times (lower) of these two cone ERG responses for the patients in this study. These measures appear reasonably well correlated.

However, these are not the best comparisons to make between the responses to single flashes and 30 Hz flicker. The conditions for the single flash differ in two ways; the flashes are about 0.6 log unit brighter and they are presented upon a 3.2 log td field. A flicker condition, a train of more intense (1.9 log td-s) flashes on a 3.2 log td background, equivalent to the single flash protocol has been added to the standard protocol to allow for more direct comparisons. The open symbols show a comparison of these light adapted 30 Hz responses and single flash responses for the patients in the present study (circles) and 26 other patients (squares) from the files of the Retina Foundation of the Southwest. The dashed lines show the means of the group of 62 normals and the associated bars are one standard deviation. The solid diagonal lines have a slope of one and denote equal implicit times and amplitudes. For the normals, the peak amplitude to the single flash is somewhat larger (110.4 vs. 78.0 μV) and slower (27.8 vs. 24.2 ms) than the peak of the 30 Hz response, presumably because

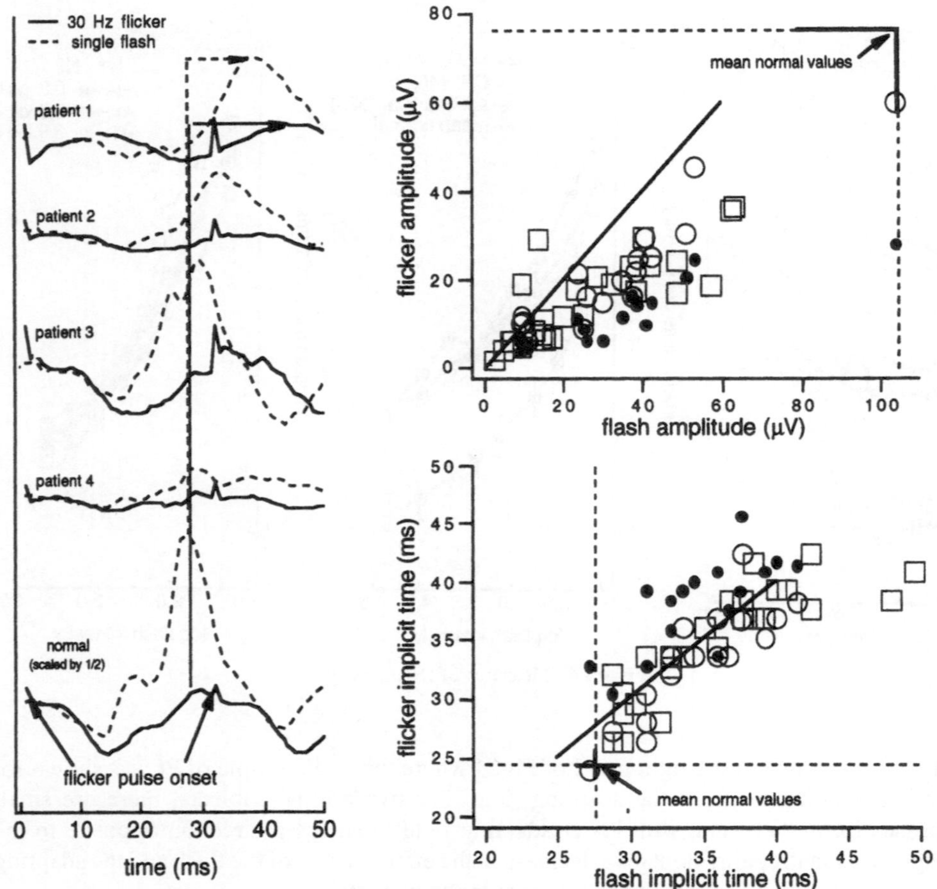

Figure 5. A comparison of the amplitude and timing of the peak response of the single flash ERG to the amplitude and timing of the flicker ERG.

of adaptation due to the flash train. For the patients, both measures are highly correlated. Notice, in particular, that the patients implicit times fall along the line of equal times from about 31 to 40 ms. At longer times, the flicker implicit times asymptote around 40 ms presumably because the a-wave to the next flash in the train cuts the response to the previous flash short. In any case, the responses to the single flash and flicker provide nearly equivalent information. And, whatever is delaying the peak response to the single flash is also delaying the peak of the flicker response.

Why Are the Cone ERGs Delayed?

The safest conclusion is that the delays seen in both the single flash and flicker cone responses (Fig. 1 left panels) are due to both outer segment sensitivity (S) changes and to the slowing of the INL responses.

Algebraically, the parameter S can be related to the amplification constant in the Lamb and Pugh model (15) of the activation phase of phototransduction by

$$\log A = \log S + \log 2 - \log k$$

where k is the factor for converting from td-sec to isomerizations/cone. In the normal human retina, the amplification, A, of the rods and cones is about the same (14). But to simply calculate A from S, or to assume a smaller S implies a smaller A, is to make a number of assumptions (16) including that the disease process treats the cones all evenly and does not affect k. We can rule out the most obvious violations of these assumptions, for example abnormal pre-retinal screening. And, even though the effects of the disease are undoubtedly uneven across the retina (e.g. 22), most of the cones must have depressed values of S to produce the changes we measure (23-25). Thus, we tentatively conclude that the diseases known collectively as RP, including those due to mutations that alter the structure of the rhodopsin molecule, affect the activation phase of cone transduction.

The most common explanation for the change in the timing of the cone ERG with RP is a decrease in quantal absorption secondary to a shortening of cone outer segments seen anatomically (see ref. 1 for references). A decreased quantal catch due to a shortened outer segment will not affect S (16, 17). The changes in S(cone) are probably secondary to rod receptor damage, at least in the patients with RP. It is not known how the degeneration of the rods affects transduction in the cones but it may be related to the other morphological changes that have been reported in the cones and/or to some factor or agent produced by rod degeneration.

The analysis in Fig. 4 (middle) suggests that the slowing of the INL component may be a far more prominent factor than the change in S in producing delayed cone ERGs. These delays are likely caused by changes beyond the cone receptor, although we cannot exclude changes in the steps between transduction in the cone outer segment and the release of transmitter at the cone synapse.

ACKNOWLEDGMENTS

Supported in part by National Eye Institute grants R01-EY-02115 (DCH), R01-EY-05235 (DGB), and R01-EY-09076 (DCH) and by a grant from the Foundation Fighting Blindness, Baltimore, MD (DGB). The screening for the rhodopsin mutations was conducted by Dr. Stephen Daiger at the Molecular Biological Laboratory, UT Health Science Center, Houston, TX.

REFERENCES

1. Berson, E., 1993, Retinitis pigmentosa: the Friedenwald lecture, Invest. Ophthal. and Vis. Sci, 34:1659-1676.
2. Sandberg, M., Sullivan, P., & Berson, E., 1981, Temporal aspects of the dark-adapted cone a-wave in retinitis pigmentosa, Invest. Ophthal. and Vis. Sci, 21:765-769.
3. Gouras, P., & MacKay, C. J., 1989, Light adaptation of the electroretinogram, Invest. Ophthal. and Vis. Sci, 30:619-624.
4. Berson, E., Gouras, P., & Hoff, M., 1969,. Temporal aspects of the electroretinogram. Arch. of Ophthal., 81:207-214.
5. Massof, R., Johnson, M., Sunness, J., Perry, C., & Finkelstein, D., 1986, Flicker electroretinogram in retinitis pigmentosa. Doc. Ophthal, 62:231-245.
6. Seiple, W., Siegel, I., Carr, R., & Mayron, C., 1986, Evaluating macular function using the focal ERG. Invest. Ophthal. and Vis. Sci, 27:1123-1130.
7. Birch, D. G., & Sandberg, M. A., 1987, Dependence of cone b-wave implicit on rod amplitude in retinitis pigmentosa. Vis. Res. 27:1105-1112
8. Seiple, W., Greenstein, V., & Carr, R., 1989, Losses of temporal modulation sensitivity in retinal degenerations. Brit. J. of Ophthal, 73:440-447.

9. Miller, S., & Sandberg, M., 1991, Cone electroretinographic change during light adaptation in retinitis pigmentosa. Invest. Ophthal. and Vis. Sci, 32:2536-2541.
10. Falsini, B., Iarossi, G., Porciatti, V., Merendino, E., Fadda, A., Cermola, S., and Buzzonetti, L., 1994, Postreceptoral contribution to Macular dysfunction in retinitis pigmentosa. Invest. Ophthal. and Vis. Sci, 35:4282-4290.
11. Birch, D. G., & Anderson, J., 1992, Standardized full-field electroretinography: Normal values and their variation with age. Arch. of Ophthal., 110:1571-1576.
12. Birch, D. G., & Fish, G. E., 1987, Rod ERGs in retinitis pigmentosa and cone-rod degeneration. Invest. Ophthal. and Vis. Sci, 28:140-150.
13. Hood, D. C., & Birch, D. G., 1993, Human cone receptor activity: The leading edge of the a-wave and models of receptor activity. Vis. Neurosci, 10:857-871.
14. Hood, D., & Birch, D., 1995, Phototransduction in human cones measured using the a-wave of the ERG. Vis. Res., In press.
15. Lamb, T. D., & Pugh, E. N., 1992, A quantitative account of the activation steps involved in phototransduction in amphibian photoreceptors. J. of Physiol. 499:719-758.
16. Hood, D. C., & Birch, D. G., 1994, Rod phototransduction in retinitis pigmentosa: Estimation and interpretation of parameters derived from the rod a-wave. Invest. Ophthal. and Vis. Sci, 35:2948-2961.
17. Breton, M., Schueller, A., Lamb, T., & Pugh, J., EN, 1994, Analysis of ERG a-wave amplification and kinetics in terms of the G-protein cascade of phototransduction. Invest. Ophthal. and Vis. Sci, 35:295-309.
18. Lachapelle, P., Little, J. M., & Polomeno, R. C., 1983, The photopic electroretinogram in congenital stationary night blindness with myopia. Invest. Ophthal. and Vis. Sci, 24:442-450.
19. Peachey, N. S., Alexander, K. R., Derlacki, D. J., Bobak, P., & Fishman, G. A., 1990, Light adaptation, rods, and the human cone flicker ERG. Vis. Neurosci, 8:142-150.
20. Kergoat, H., & Lovasik, J. V., 1990, The effects of altered retinal vascular perfusion pressure on the white flash scotopic ERG and oscillatory potentials in man. Electroencephal and Clin. Neurophysiol., 75:306-322.
21. Heynen, H., Wachtmeister, L., and van Norren, D., 1985, Origin of the oscillatory potentials in the primate retina. Vis. Res., 25:1365-1373.
22 Nusinowitz, S., & Birch, D. G., 1995, The topography of rod and cone sensitivity loss in retinal perimetry.
23. Hood, D.C., Shady, S. Birch, D. G., 1993, Heterogeneity in retinal disease and the computational model of the human rod response. J. Opt. Soc. Amer., 10:1624-1630.
24. Hood, D. C., Shady, S, & Birch, D. G., 1994, Understanding changes in the b-wave of the ERG caused by heterogeneous receptor damage. Invest. Ophthal. and Vis. Sci, 35:2477-2488.
25. Shady, S., Hood, D. C. & Birch, D. G., 1995, Rod phototransduction in retinitis pigmentosa: Distinguishing alternative mechanisms of degeneration. Invest. Ophthal. and Vis. Sci, In press.

ABNORMAL ROD PHOTORECEPTOR FUNCTION IN RETINITIS PIGMENTOSA

David G. Birch[1,2] and Donald C. Hood[3]

[1] Retina Foundation of the Southwest and
Dallas, Texas
[2] Department of Ophthalmology
University of Texas Southwestern Medical School
Dallas, Texas
[3] Department of Psychology
Columbia University
New York, New York

INTRODUCTION

Over the years, a clinical diagnostic protocol has evolved for the full-field electroretinogram (ERG) and recent attempts to standardize the protocol (1) have largely been successful. As part of the standard ERG protocol, a single rod flash is obtained from a patient in a full-field dome. The protocol response is useful for establishing the retinal basis of visual loss, and can provide information on the rate of progression in hereditary retinal degeneration. However, the peak-to-peak amplitude of a single response is a snapshot of rod behavior that reflects amplitude at a single point in time following a single stimulus intensity. The purpose of this paper is to show that considerably more information about rod function can be obtained by analyzing the dynamic behavior of rod responses over both time and intensity. In particular, we have developed two approaches for analyzing the human rod photoresponse. Both approaches require high intensity stimuli to reveal the underlying photoreceptor component of the ERG and both can be used to test specific hypotheses regarding mechanisms of degeneration in hereditary retinal diseases.

The standard rod response is often non-recordable in patients with retinitis pigmentosa, but sometimes a small positive deflection is evident. A single response is not very useful for discriminating among alternate possible mechanisms of degeneration. Progressive decline in peak-to-peak ERG amplitude over several years of follow up, as shown in Figure 1, could be due to any number of factors. Some can be controlled or compensated for, such as changes in pupil size and lens yellowing. Others, such as inner retinal damage, are less likely to play a large role in hereditary retinal degenerations. However, ERG declines in retinitis pigmentosa could be due to shortened rod outer segments resulting in reduced quantal catch, to random loss of rods from throughout the field, to regional death of rods leading to large

Degenerative Diseases of the Retina, Edited by Robert E. Anderson et al.
Plenum Press, New York, 1995

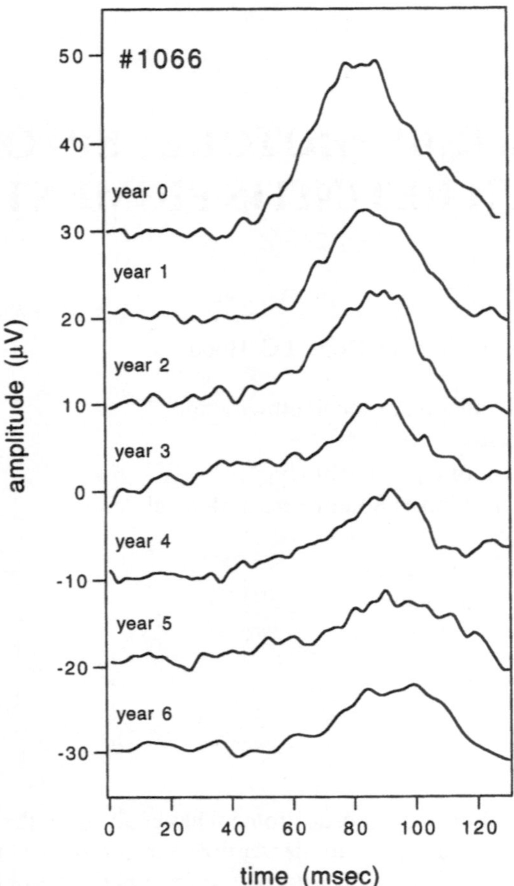

Figure 1. Rod ERG response to standard protocol flash of -0.1 log scot td-sec. A decline in peak-to-peak amplitude is clearly evident over six years of follow-up. The patient is a 43 year old female with an autosomal recessive form of retinitis pigmentosa.

areas of field loss, to abnormalities within one of the many stages of the phototransduction cascade, or to some combination of all. Evaluating the rod ERG as a function of intensity is a first step in discriminating among these possibilities.

As shown in Figure 2, the standard protocol response (dark curve) is but one of a series of responses that can be obtained by varying intensity. Rod b-wave amplitude grows approximately linearly with retinal illuminance over a range of greater than 4 log units. At around 1.0 log scot td-sec, growth in b-wave amplitude slows as the underlying b-wave generators reach their maximal response. The a-wave of the rod ERG first becomes prominent at intensities causing maximal b-wave response and further growth in peak-to-peak amplitude is due primarily to growth of the a-wave. Log peak-to-peak rod amplitude versus log retinal illuminance is shown in Figure 3 for the seven visits in the patient shown in Figure 1. Solid lines fit to each visit are saturating exponentials of the Michaelis-Menton variety, often referred to as Naka-Rushton functions in their application to visual responses. According to this relationship:

$$V/V_{max} = I^n / I^n + k^n ,$$

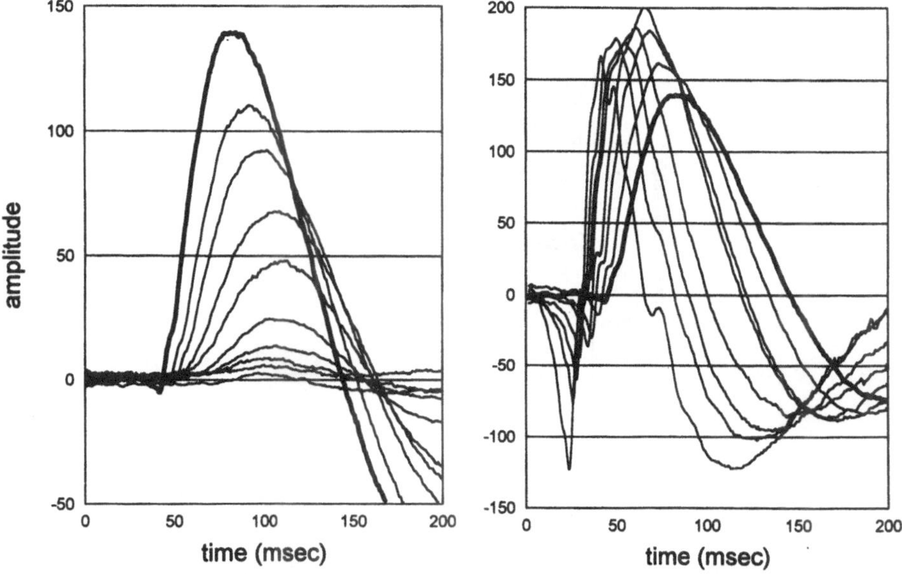

Figure 2. Standard protocol response (darkened curve in each panel) is on the linear portion of the intensity-response relationship. Below the protocol intensity (left panel), responses decline with intensity to near psychophysical threshold. Above the protocol intensity (right panel), b-wave amplitude saturates and the a-wave predominates.

where V is peak-to-peak amplitude at intensity I, V_{max} is the maximum amplitude, k is intensity at half-saturation, and n is a slope constant (typically 1.0). For the patient shown in Figure 3, there is a relatively systematic decrease in V_{max} over the 6 years of follow up and a clear trend for the curves to be displaced horizontally (increased log k) over time. The change in log V_{max} sums with the change in log k to give the change in log rod ERG threshold, defined as the retinal intensity necessary for a 2.0 μV b-wave response.

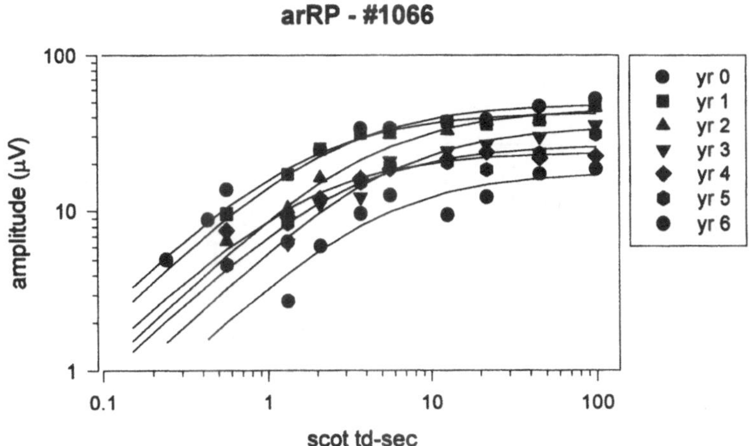

Figure 3. Log peak-to-peak amplitude versus log retinal illuminance over six years of testing in patient #1066. Solid lines fit to each visit are the relationship $V/V_{max} = I^n/I^n = k^n$ with n set to 1.0. Both V_{max} and log k change over time.

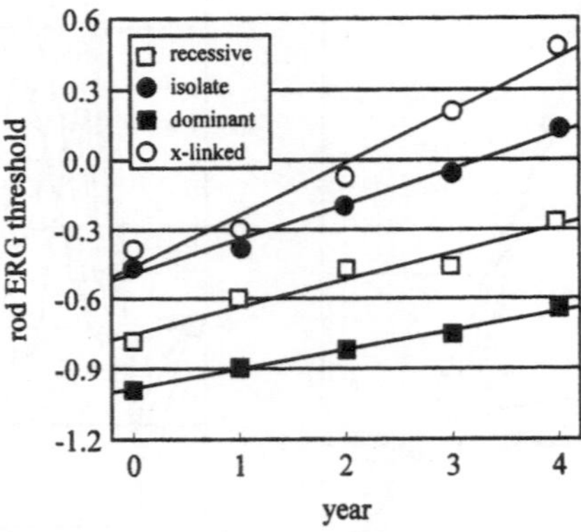

Figure 4. Yearly change in rod ERG threshold for 65 patients with retinitis pigmentosa tested over a 5 year period. Significant differences exist in the rates of change among the various inheritance patterns.

To document the natural history of rod loss in retinitis pigmentosa, yearly progression in rod ERG function was studied over a five year interval in 65 patients (2). Average yearly change in rod b-wave threshold is shown as a function of inheritance pattern in Figure 4. A repeated measures analysis of variance showed that there was a significant main effect of years (F = 54.6; p<0.0001) and a borderline main effect of inheritance pattern (F = 2.55, p = 0.06). Furthermore, the analysis revealed a significant interaction between inheritance pattern and years (F = 2.02; p = 0.02). Post-hoc analysis showed that the yearly rate of rod b-wave threshold elevation was significantly higher in xlRP (0.22 log/year; 60%) than in adRP (0.1 log/year; 20%) patients. Patients with recessive and isolated patterns of inheritance were intermediate, with rod threshold rising 0.14 (30%) each year. Overall, the annual increase in rod b-wave threshold across all patients with retinitis pigmentosa averaged 0.16 (32%). As shown in Figure 5, the threshold change was due to both an annual increase in log k of 0.09 log unit (19%) and an annual decrease in log V_{max} of 0.07 log unit (15%). The

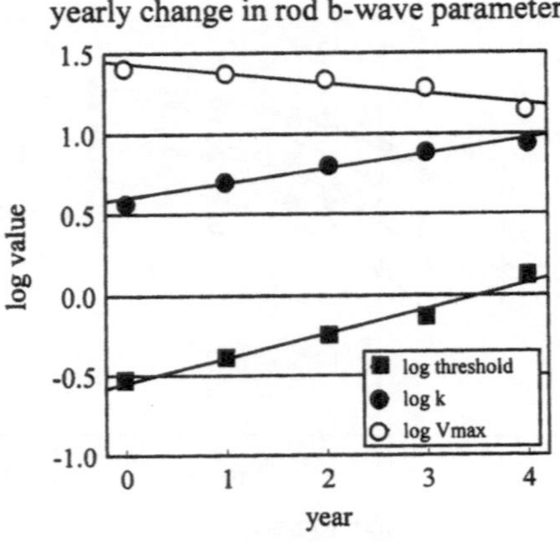

Figure 5. Yearly change in rod b-wave parameters determined from the Naka-Rushton relationship in 65 patients with retinitis pigmentosa. The annual increase in log threshold was due to both a yearly increase in log k and a yearly decrease in log V_{max}.

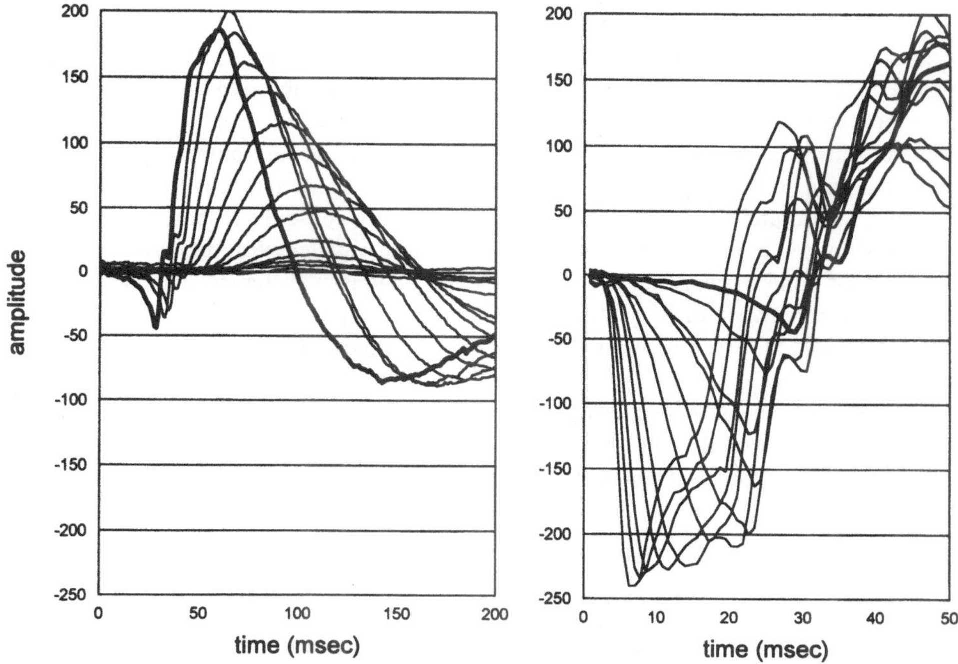

Figure 6. Rod ERGs from a normal subject elicited by stimuli ranging from -3.6 to 4.4 log scot td-sec. The dark curve is the rod response elicited by the maximal stimulus available on most commercial systems (2.0 log scot td-sec). Above this intensity, responses are dominated by the a-wave (right panel).

annual decrease in V_{max} can only be due to progressive, regional loss of b-wave generators. Because of the high convergence ratio of rod photoreceptors onto bipolar cells, it should be possible to elicit a maximum response from the bipolar with sufficient light despite random loss of rods within the receptive field and/or decreased quantal absorption of individual rods. The maximum bipolar cell response (and V_{max}, the measured ERG parameter) will decrease only when virtually all the rods in a bipolar cell receptive field cease functioning. There should, therefore, be a strong correlation between V_{max} of the rod b-wave and the area of functioning rod visual field. Indeed, strong correlations are present when precautions are taken to isolate rod-mediated parameters in both the ERG and the visual field (3-5). Similarly, the annual rate of decline in log V_{max} of the ERG is comparable to the yearly rate of decline in log rod visual field area (2).

The annual increase in log k is more difficult to attribute to underlying mechanisms. Changes in sensitivity at the level of the b-wave can result from a large number of abnormalities that are best studied by recording from the human photoreceptors themselves. At intensities higher than about 2.0 log scotopic troland-secs, the timing of the a- and b-waves is such that a substantial portion of the leading edge of the a-wave is revealed. At even higher intensities, a-wave amplitude reaches a maximum prior to the onset of the b-wave, revealing the actual saturated maximum of the underlying photoreceptor response. One approach to analyzing the human rod photoresponse is to fit current physiological models to the leading edge of the a-wave (6-11). A second approach is to "probe" the rod photoreceptor response with a double flash procedure (12).

To obtain high-intensity responses for both approaches, we record full-field ERGs in a custom Ganzfeld dome incorporating a xenon gas tube within a flash head (Model 2100,

Novatron, Dallas, Tx). The flash is driven by a 1600 W/sec power pack (Novatron, Dallas, Tx) and is capable of delivering relatively narrow-band short wavelength flashes of up to 4.5 log scot td-secs and achromatic flashes of up to 6.0 log scot td-sec. A second, identical high-intensity unit is mounted in the dome for some of the double-flash experiments. The precise timing of each flash is controlled by computer.

As shown in Figure 6, the rod ERG to intensities above approximately 3.0 log scot td-sec are dominated by the a-wave. Because cones also generate sizeable potentials at these intensities, techniques have been developed for isolating rod-only responses (9,13). The responses shown in Figure 5, for example, are from short-wavelength stimuli after computer-subtracting the cone responses to photopically-matched long-wavelength stimuli. The qualitative similarity between the a-wave and photoreceptor responses has been appreciated for many years (14-15). More recently, we (6-9,11) and others (10,16) have shown that the leading edge of the rod a-wave can be fit with the same class of model that has been used to characterize responses from isolated rod photoreceptors. The recent model of Lamb and Pugh (17) is particularly useful in this regard, since it contains parameters that can be directly related to activation mechanisms of transduction. To adapt the model to the ERG, we have

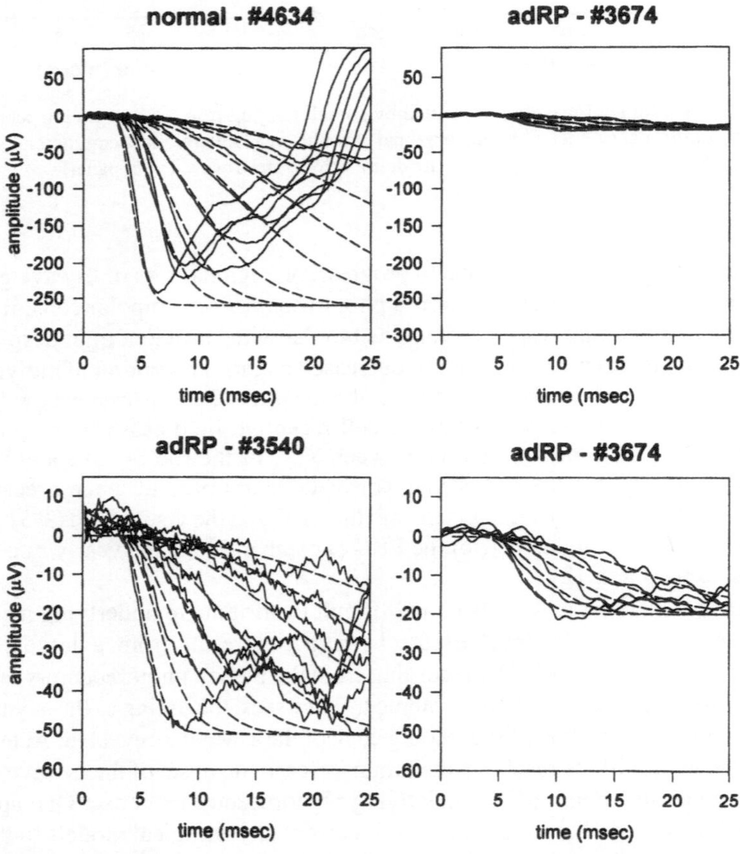

Figure 7. Leading edge of the a-wave in normal subject and two patients with autosomal dominant retinitis pigmentosa due to the pro-23-his mutation. Patient #3674 is shown twice; on the same scale as the normal subject (above) and magnified X5 (below). The dashed curves in each panel show the fits of the model to the leading edge. Both S and Rm_{P3} are reduced in these patients.

used the following equation to describe the response as a function of time (t) and intensity (I):

$$R(i,t) = [1 - \exp(-i \cdot S \cdot (t - td)] (R_{max})$$

where R_{max} is the maximum amplitude, S is a sensitivity parameter and td is a brief delay prior to response onset (reflecting time constants of both the recording equipment and the various stages of transduction). R_{max} primarily reflects the total number of gated channels in the retina and will be decreased by either a reduced number of rods or a reduced number of outer segment disks. S, on the other hand, reflects the gain of transduction. In normal subjects, where all rods have approximately equivalent parameters, S directly reflects the amplification (A) of individual rods. Using currently accepted values to convert to quantal absorption, we have shown that the amplification constant derived from the ERG is within the range of previously reported values for A (18). In patients with retinitis pigmentosa, where there is typically considerable heterogeneity among rods across the retina, it is more difficult to directly relate S to A. We have shown, however, that patients must have a sizable number of rods with A values depressed by at least the value of log S (11). Conversely, a normal value of S can only occur when the vast majority of rods have normal values of A.

The first 25 msec of the a-wave is shown for a normal subject and two patients with retinitis pigmentosa in Figure 7. The patients are from two different families with the pro-23-his rhodopsin mutation. Data from patient 3674 is shown twice: in the upper right panel at the same gain as normal to underscore the small amplitude of these signals in some patients and in expanded form below. The dashed curves in each panel show the fits to the leading edge with a single set of parameters for the intensity series. Despite the small amplitudes, the fit to the model is quite good. For both patients with the pro-23-his mutation, the values of S and Rmax are significantly lower than normal. Summary data for 50 patients

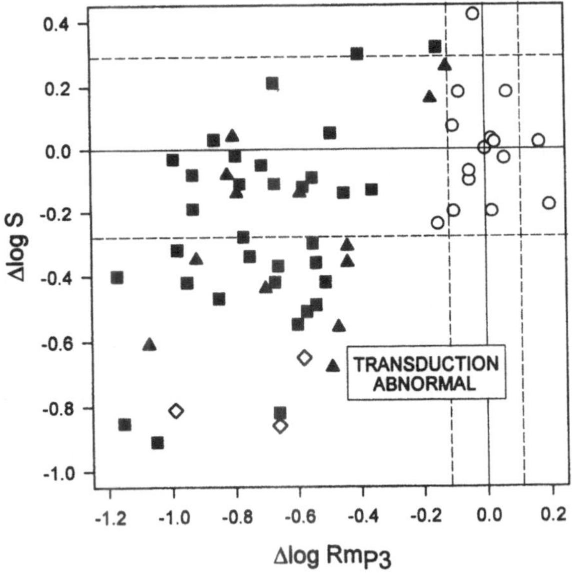

Figure 8. A comparison of the values of Δlog S and Δlog Rm_{P3}. The symbols show the values for the normal subjects (open circles) and patients with RP (filled squares) and cone-rod degeneration (filled triangles). Patients with autosomal dominant retinitis pigmentosa due to the pro-23-his mutation are shown as open diamonds. The value of zero Δlog S or Δlog Rm_{P3} refers to a value equal to the mean of the normal subjects.

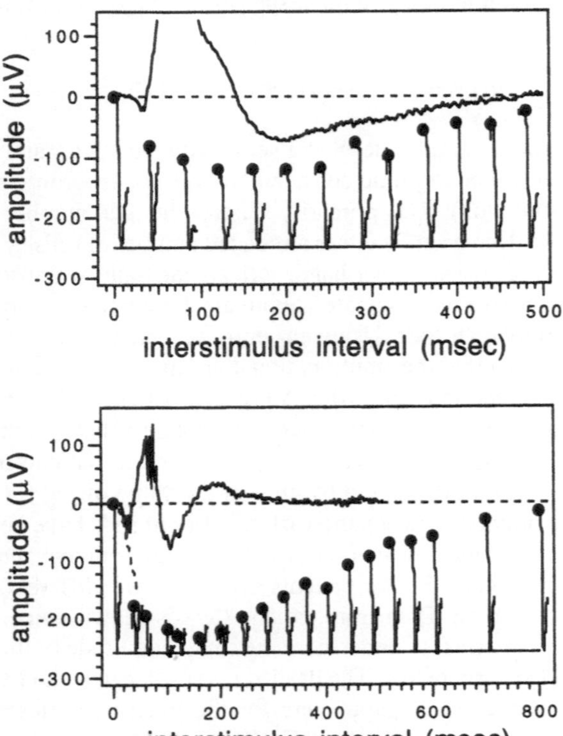

Figure 9. Recovery functions derived from double flash technique with 1.0 log scot td-sec test flash (above) and 1.7 log scot td-sec test flash (below). "Hook" responses pegged to the data points are rod responses to 4.2 log scot td-sec probe flashes. The level of hyperpolarization at each ISI following the test flash is the difference between the probe amplitude at that time and the saturated amplitude. Full ERGs in each panel are the responses to each test flash alone.

with hereditary retinal degeneration are shown in Figure 8. Plotted are the difference values (relative to mean normal) for log S on the vertical axis against log R_{max} on the horizontal axis. All values are plotted relative to mean normal; the dashed line indicates the values above which 95% of normal subjects fall. All patients show significant reductions in R_{max}. It is highly likely that the decrease in R_{max} is primarily due to a loss of outer segment membrane through loss of receptors or a decrease in rod outer segment length. Many patients have values of log S within the normal range, suggesting that this subset of patients has no significant alteration in activation mechanisms of transduction. Approximately half of the patients show abnormally low values of log S. This group of patients must have a sizeable number of rods with an abnormal transduction process in the form of decreased amplification.

A second way to probe the underlying photoreceptor response is with the double-flash procedure. This technique is based on the fact that it is possible to directly measure the maximal, saturated amplitude to a probe flash of 4.0 or brighter log scot td-sec. When the probe flash is presented after a test flash, the amplitude of the response to the probe will be the difference between the level of hyperpolarization due to the test flash and the maximal, saturated amplitude. An example of the application of this procedure in a normal subject is shown in Figure 9. The top trace shows the ERG in response to a test flash of 1.0 log scot td-sec. This intensity is at about the half-saturation value for the rod

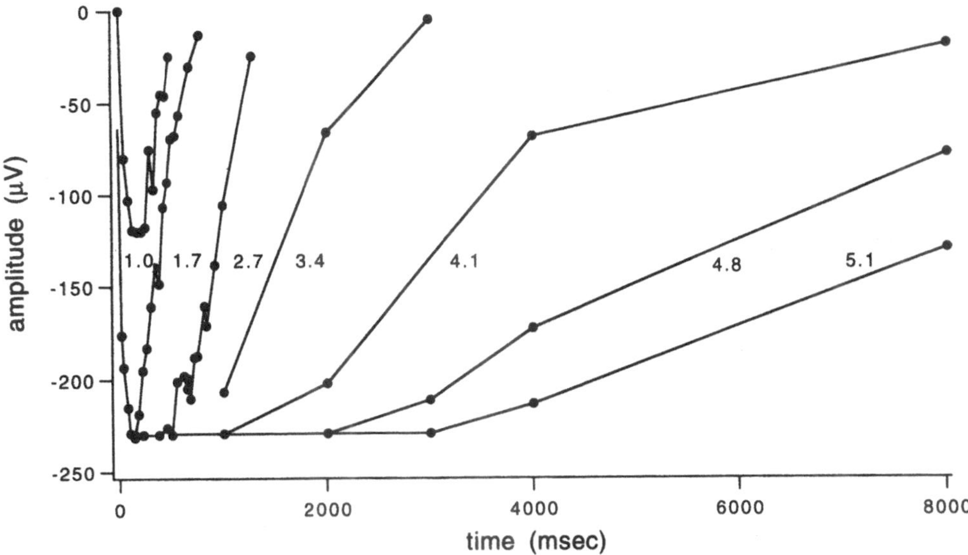

Figure 10. Recovery functions derived from the double flash techniques for test flashes ranging from 1.0 to 5.1 log scot td-sec. With increasing intensity of test flash (I_8), there is an increase in critical time (T_C) prior to initiation of recovery.

photoreceptors. The a-wave to this intensity is extremely small; only a few msecs of the leading edge is present before the intrusion of the b-wave. The "hook" responses shown below are the first 20 msec of the response to a probe flash of 4.2 log scot td-sec at interstimulus intervals (ISI) indicated on the abscissa. When presented alone (ISI = 0), the probe flash produces the maximal response of -250 μV. The amplitude at any given ISI depends upon the level of hyperpolarization of the rods at that instant following the test flash. The solid circles at each ISI are the differences between the probe flash amplitude and the saturated amplitude. As predicted, the double flash technique reveals an underlying photoreceptor response to a test flash of 1.0 log scot td-sec that has a time to peak of approximately 180 msec and reaches about one-half the saturated amplitude (18). The bottom half of Figure 9 shows a similar analysis for a more intense test flash of 1.7 log scot td-sec. Here the underlying response reaches saturation, then shows a prolonged course of recovery over the next several hundred milliseconds. Although these measures are time-consuming, in principle it is possible to utilize this technique to study underlying mechanisms of activation and inactivation over a wide range of test intensities in normal subjects and patients with hereditary retinal degeneration.

It is known from recordings in isolated rod photoreceptors from a number of species (19-20), including humans (18), that increases in test flash intensity above saturation produce progressively longer delays in the initiation of recovery from saturation. Whereas amplitude saturation reflects the complete closure of all gated sodium channels in the rod outer segment, the progressive delays in recovery from saturation with increasing intensities reflect the saturation of "upstream" activation stages. Pepperberg and co-workers (21) have shown that the critical time prior to the initiation of recovery, T_C, increases in proportion to the natural logarithm of test flash intensity, I_f, above intensities known to produce complete, transient activation of transducin. This linear T_C vs. ln I_f function above the transducin saturation level implies that the slope of this function is a measure of the lifetime of an intermediate "upstream" from transducin, namely activated rhodopsin, R*.

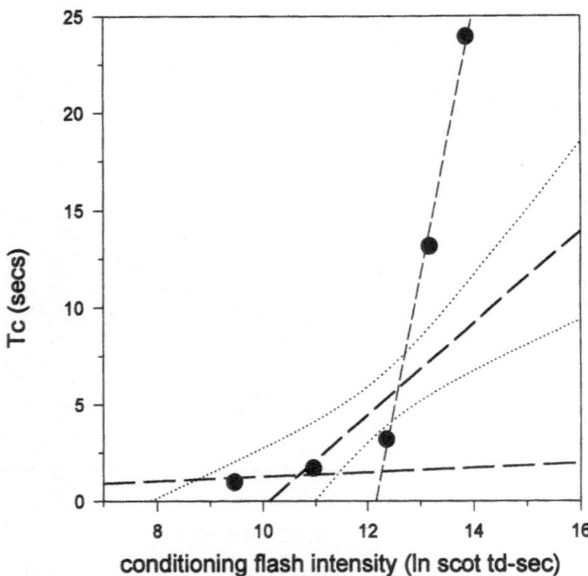

Figure 11. Critical time (T_C) as a function of test flash intensity. Solid symbols are means from three patients with the pro-23-his mutation. The dashed lines are the means of five normal subjects and the dotted lines show the 95% confidence interval for the slope. Over the range from 10 to 13.9 ln scot td-sec, T_C varied with the ln test flash intensity, but the slope in patients (12.1 sec) was five times greater than normal (2.3 sec).

The double-flash ERG technique described above can be used to measure the intensity-dependence of inactivation in human subjects. Figure 10 shows recovery functions for a series of test flashes (I_f) in a normal subject. Like responses from single human rods (18), these derived responses grow in amplitude and saturate. With increasing I_f above amplitude saturation, there is an increase in the critical time prior to the initiation of recovery (T_C). The recovery curves leading to the determination of T_C also appear to have an intensity-dependent time constant. Critical times based on results for five normal subjects are plotted as a function of I_f in Figure 11. For conditioning flash intensities up to 10.5 ln scot td-sec, T_C exhibits a relatively weak, approximately linear dependence on ln I_f. Beginning at about 11 ln scot td-sec and continuing up to the highest I_f examined (13.9 ln scot td-sec), T_C grows much more rapidly with ln I_f. This higher branch of the T_C vs, ln I_f function is well described by a straight line of slope 2.3 sec; 95% normal limits are shown as dotted curves. The solid symbols show results from similar measures in three patients with retinitis pigmentosa due to the pro-23-his rhodopsin mutation. These patients are identified with open diamonds in Figure 8. The patients have normal value of T_C for conditioning flashes up to 10.9 ln scot td-sec. However, all three patients (including one patient tested on two occasions) have values of T_C that exceed the normal confidence limit at high conditioning flash intensities. The dashed line indicates the average slope of the relationship between T_C and I_f for the three patients. Although the upper limb is based on a limited number of measures per patient, the slope of the line (average 12.1 sec) is clearly steeper than normal (2.3 sec).

These results in patients with autosomal dominant retinitis pigmentosa due to the pro-23-his mutation are presented as an example of how the techniques described here for analyzing the a-wave can contribute to our understanding of disease mechanisms. Clearly, these patients have decreased numbers of functioning outer segment disks due to cell loss and/or shortened outer segments of remaining cells. However, decreased numbers of disks alone cannot explain either the abnormally low gain of activation (S) or the slowed inactivation, as indexed by an upper branch of the recovery function that is about five times the normal value. If the density of cGMP-gated channels per unit rod outer segment length is normal in the patients tested here, the decrease in S implies abnormalities within the activation stages of transduction. Moreover, to the extent that the deactivation kinetics of R* determine the upper-branch $\Delta T_C/\Delta \ln I_f$ slope, our findings suggest a prolonged lifetime of R*. These abnormalities in activation and inactivation stages of transduction would not be expected if the mutant rhodopsin was confined to the rod inner segment. Hence, they provide strong evidence that abnormal rhodopsin resulting from the pro-23-his mutation is present and functioning in the rod outer segments of these patients.

Our long-term goal is to utilize the human ERG for studying rod photoreceptor activity in retinitis pigmentosa. We hope to utilize the paradigms described here to pursue genotype-phenotype correlations in the wide variety of patients with mutations in rhodopsin and other visual proteins.

REFERENCES

1. Marmor, M. F., Arden, G. B., Nilsson, S. E. G., et al., 1989, Standard for clinical electroretinography, *Arch. Ophthalmol.* 107:816-819.
2. Birch, D. G. and Anderson, J. L., 1993, Yearly rates of rod and cone functional loss in retinitis pigmentosa and cone-rod degeneration, *In* Digest of Topical Meeting on Noninvasive Assessment of Visual System, Washington, DC, Optical Society of America, vol. 3, pp. 334-337.
3. Birch, D. G., Herman, W. K., deFaller, J. M., et al., 1987, The relationship between rod perimetric thresholds and full-field rod ERGs in retinitis pigmentosa, *Invest Ophthalmol. Vis. Sci.* 28:954-965.
4. Hood, D.C., Shady, S., Birch, D.G., 1994, Understanding changes in the b-wave of the ERG caused by heterogeneous receptor damage, *Invest. Ophthalmol. Vis. Sci.* 35: 2477-2488.
5. Shady, S., Hood, D. C., Birch, D. G., In Press, Rod phototransduction in retinitis pigmentosa: distinguishing alternative mechanisms of degeneration, *Invest. Ophthalmol. Vis. Sci.*
6. Hood, D. C. and Birch, D. G., 1990, The relationship between models of receptor activity and the a-wave of the human ERG, *Clin. Vis. Res.* 5: 293-297.
7. Hood, D. C. and Birch, D. G., 1990, The a-wave of the human electroretinogram and rod receptor function, *Invest. Ophthal. Vis. Sci.* 31: 2070-2081.
8. Hood, D. C. and Birch, D. G., 1990, A quantitative measure of the electrical activity of human rod photoreceptors using electroretinography, *Vis. Neurosci.* 5: 379-387.
9. Hood, D. C. and Birch, D. G., 1993, Light adaptation of the human rod receptors: the leading edge of the human a-wave and models of rod receptor activity, *Vision. Res.* 33:1605-1618.
10. Breton, M. E., Schueller, A. W., Lamb, T. D., et al., 1994, Analysis of ERG a-wave amplification and kinetics in terms of the G-protein cascade of phototransduction, *Invest. Ophthalmol. Visual. Sci.* 35:295-309.
11. Hood, D. C. and Birch, D. G., 1994, Rod phototransduction in retinitis pigmentosa: estimation and interpretation of parameters derived from the rod a-wave, *Invest. Ophthalmol. Visual. Sci.* 35:2948-2961.
12. Birch, D. G, Hood. D. C., Nusinowitz, S., et al., In Press, Abnormal activation and inactivation mechanisms of rod transduction in patients with autosomal dominant retinitis pigmentosa and the Pro-23-His mutation, *Invest. Ophthalmol. Vis. Sci.*
13. Birch, D. G. and Fish, G. E., 1987, Rod ERGs in retinitis pigmentosa and cone-rod degeneration, Invest. *Ophthalmol. Vis. Sci.* 28:140.
14. Granit, R., 1933, The components of the retinal action potential in mammals and their relation to the discharge in the optic nerve, *J. Physiol.* 77:207-239.
15. Brown, K. T., 1968, The electroretinogram: Its components and their origin. *Vision Res.* 8:633-678.
16. Cideciyan, A. V. and Jacobson, S. G., 1993, Negative electroretinograms in retinitis pigmentosa, *Invest. Ophthalmol. Vis. Sci.*, 34:3253-3263.
17. Lamb, T. D. and Pugh, E. N. Jr., 1992, A quantitative account of the activation steps involved in phototransduction in amphibian photoreceptors, *J. Physiol.* 499:719-758.

18. Kraft, T. W., Schneeweis, D. M., and Schnapf, J. L., 1993, Visual transduction in human rod photoreceptors, *J. Physiol.* 464:747-765.
19. Lagnado, L. and Baylor, D. A., 1992, Signal flow in visual transduction. *Neuron* 8:995-1002.
20. Yau, K-W., 1994, Phototransduction mechanism in retinal rods and cones. The Friedenwald Lecture, *Invest. Ophthalmol. Vis. Sci.* 35:9-32.
21. Pepperberg, D. R., Cornwall, M. C., Kahlert, M., et al., 1992, Light-dependent delay in the falling phase of the retinal rod photoresponse, *Visual. Neurosci.* 8:9-18.

SCOTOPIC THRESHOLD RESPONSES AND ROD INTENSITY-RESPONSE FUNCTIONS AS SENSITIVE INDICATORS OF THE CARRIER STATUS IN X-LINKED RECESSIVE RETINITIS PIGMENTOSA

A. Iannaccone,[1] E. M. Vingolo,[1] R. Forte,[1] P. Tanzilli,[1] B. Grammatico,[2] C. De Bernardo,[2] E. Rispoli,[1] G. Del Porto,[2] and M. R. Pannarale [1]

[1] University "La Sapienza", Institute of Ophthalmology
Department of Ocular Electrophysiology, Center for Inherited Degenerative Retinal Disorders
Viale del Policlinico 1, 00161 Rome, Italy
[2] University "La Sapienza", Medical Genetic Section of the Experimental Medicine Department
c/o Ospedale L. Spallanzani
Via Portuense 292, 00149 Rome, Italy

INTRODUCTION

In previous studies several authors described modifications of ERG responses in X-linked recessive retinitis pigmentosa (xLRP) carriers (1-8). Investigations on the intensity-response functions of the rod ERG also demonstrated reduction of V_{max} in subjects with ophthalmoscopic evidence of the carrier status (pigmentary changes and/or tapetal-like reflex) (4). In patients affected with RP abnormalities of the Oscillatory Potentials (OPs) have also been demonstrated (9,10), suggesting a coexisting impairment of the inner retinal layers. Abnormalities of rod sensitivity were also found at the psychophysical level to flickering stimuli (11).

The Scotopic Threshold Response (STR) is a negative corneal potential elicited by a very dim light stimulus (about one log unit above the absolute psychophysical threshold) in a fully dark adapted eye. The resulting response is a slow wave easily differentiated from the Granit's fast PIII or a-wave component (12). The generators of this potential are in the IPL (13,14). Further studies (15-19) suggested that this potential could result from a Müller cell response to light-evoked K^+ fluxes (20) originating in a class of rod-driven depolarizing amacrine cells operating with a narrow intensity range near absolute threshold.

Degenerative Diseases of the Retina, Edited by Robert E. Anderson et al.
Plenum Press, New York, 1995

Recent studies on other retinal pathologies, such as diabetic retinopathy, x-linked juvenile retinoschisis, Goldmann-Favre syndrome and congenital stationary night blindness (CSNB) (21-26), have demonstrated that the STR is a sensitive and reliable measure of the inner plexiform layer (IPL) function. Recent work from our department (27-29) suggested that STR is even more sensitive than other electrofunctional parameters, such as OPs, in assessing IPL conditions, in that it showed alterations in mild myopia or in diabetic patients with no ophthalmoscopical evidence of diabetic retinopathy despite normality of the other ERG tests.

The aim of this investigation was to determine the most sensitive ERG parameter in identifying the carrier condition for xLRP, a disease that typically affects rod function, in the presence of normal maximal ERG responses. Therefore we tried to test completely the rod retinal function in xLRP carriers performing an intensity-response analysis, to evaluate rod amacrine-mediated response through the STR, rod bipolar-mediated response through the maximum b-wave amplitude ($\log V_{max}$), and rod pathway function and sensitivity through the other parameters of the Naka-Rushton equation (semisaturation constant, logK, and slope of the function, n).

SUBJECTS & METHODS

Two groups of women were investigated. The first one included 10 patients from 3 pedigrees, 6 of whom obligate xLRP carriers, and 4 from xLRP pedigrees with 50% chance of being carrier, in whom ophthalmoscopy disclosed tapetal-like reflex or, in some cases, also mild retinal pigmentary changes, and their carrier status was verified by molecular biology tests. Age was 32.29 ±14.78 years (mean ± SD). Two additional female patients at 50% risk of carrier status with mild doubtful fundus changes were also investigated. Molecular biology, despite suspicious fundus findings, verified their healthy homozygous state. No significant abnormalities of the maximal ERG response parameters (i.e., neither amplitude nor latencies) were evident in either one of these subjects, according to our standard age- and sex-matched normal reference values. The second group included 10 healthy, age- and sex-matched subjects of 28.29±9.66 years of age, on average.

Clinical investigations on each patient from the xLRP pedigrees were conducted according to a standardized national protocol for the study of retinitis pigmentosa and allied diseases, previously described in greater detail (30), that included a complete ophthalmological examination, with determination of morphological (slit lamp anterior segment examination, ophthalmoscopy, biomicroscopy of vitreous and of the posterior pole with Volk +78D and +90D lenses) and functional parameters (best corrected visual acuity, color vision, kinetic visual field).

Electroretinographic tests included standard ERG protocol for rod and cone function examination and the evaluation of rod ERG intensity-response function; pupils were dilated with tropicamide 0.5% and phenilephrine 10% solutions; ERGs were recorded with Henkes-type corneal contact lens electrodes placed on both eyes under dim red light after corneal anesthesia with 2 drops of 0.4% oxibuprocaine chlorhydrate solution. The bandpass used for ERG recording was between 0.1 and 400 Hz. A 0.5 Hz-frequency white flash placed outside of the faradic cage, connected to a Ganzfeld dome by a fiber optic system was used to elicit the retinal signal (31), with a time window of 512 msec. The recording procedure (24) consisted in recording 25 iterations for each luminance level, starting 6 log units below the standard flash intensity (2 cd·s/m^2). Attenuation of the flash intensity was controlled with neutral density (ND) filters inserted in a slot placed in front of the flash, progressively reducing the ND intensity by 0.3 log unit steps. Off-line analysis allowed us to split sequencial groups of the recorded samples and to average them separately. In so doing, the onset, growth and subsequent disappearance of the STR could be monitored, and thereafter rod b-wave amplitudes and latencies were measured and analyzed with a dedicated software to fit the Naka-Rushton equation.

The maximum peak amplitude (MPA), the maximum peak latency (MPL), and the peak luminance intensity (PLI) at which the maximum peak was obtained were used as major investigating parameters for the evaluation of the STR.

Patients with systemic diseases like diabetes, hypertension, renal failure or neurological disorders were excluded. Neither subjects with myopia higher than -2D, ocular hypertension, or other retinal diseases, nor patients who had undergone surgical or para-surgical ocular treatments were included in the study.

Statistical evaluations for comparison between the two groups of subjects were performed with the Stat View SE+Graphics program (Abacus Concepts, Inc., Berkeley, CA) for Apple computers (Cupertino, CA), using the Student's unpaired two tail t-Test.

RESULTS

A. STR values (see Table 1)

The MPA of the STR in xLRP carriers was significantly lower ($p = 0.001$) and the MPL was slightly slower ($p = 0.056$) than in normal females. No significative differences were found for PLI ($p = 0.58$).

B. Intensity-response function (see Table 1)

The parameters of the Naka-Rushton equation fitting the rod ERG intensity-response function showed different abnormalities. Log V_{max} was virtually superimposable between the two groups ($p = 0.16$); logK was found to be significantly higher and the slope of the function (n) significantly smaller in xLRP carriers ($p = 0.0001$ and 0.008, respectively).

We also investigated the possible existence of interocular differences for each of the above parameters, in view of the reported existence of cases of interocular asymmetry of visual functions in xLRP heterozygotes (5,6). None of the parameters showed significant differences between the two eyes. No false positive or false negative results were observed. Two possible carriers found to be genotypically normal had responses within normal limits.

DISCUSSION

There are at least 3 major amacrine cells' subsets involved in the transmission and modulation of rod-driven signals (32). Two of these, the AII and A_{17} ones, are composed of frequent cells with narrow dendritic fields that regulate the flow through the retina in deep scotopic conditions: their efficiency is critical to night vision. Their regulatory role is modulated by dopaminergic amacrine (DA) cells via direct synapses. This purely scotopic pathway also has a local negative feedback ensured by the so-called indolamine-accumulating amacrine (IAA) cells.

Sieving and co-workers (13,18,19) have provided strong evidence that the STR could result from a Müller cell response to light-evoked K^+ fluxes originating in rod-driven depolarizing amacrine cells (20). On the basis of the known properties of rod-driven amacrine cell subsets, the A_{17} population (sustained depolarizing response to dim light) has been particularly thought to underlie the STR, facilitating the integration of rod information and increasing rod pathway sensitivity at the IPL level (17).

Table 1. Summary of ERG findings

	Scotopic Threshold Response (STR)			Rod Intensity-Response Function		
	MPA (mV) ±S.D.	MPL (msec) ±S.D.	MPI (logU) ±S.D.	Log V_{max} ±S.D.	LogK ±S.D.	Slope (n) ±S.D.
xL-RP carriers (n = 10)	14.77 ± 6.29	157.43 ± 19.84	3.49 ± 0.42	2.75 ±0.17	2.11 ± 0.17	0.92 ± 0.09
controls (n = 10)	34.6 ± 10.6	146.8 ± 16.8	3.06 ± 0.32	2.79 ± 0.07	1.83 ± 0.19	1.05 ± 0.14
T-test (p value)	0.001	0.056	0.58	0.16	0.0001	0.008

STR characteristics can be altered by administrating several neurotransmitters and their inhibitors in experimental conditions (12,16,17), as well as in different human retinal affections. This has been demonstrated both in overt [diabetic retinopathy (21), xL juvenile retinoschisis (22), central retinal vein occlusion (23), CSNB (24,25), and Goldmann-Favre syndrome (26)] and subclinical conditions [diabetic patients with no evidence of diabetic retinopathy (27,28), mild myopia (29)]. All these clinical observations have confirmed the post-receptoral origin of the STR.

The STR abnormalities, the significant reduction of the slope of the intensity-response function, and the increased semisaturation constant detected in xLRP carriers observed here were found in the presence of normal maximal ERG parameters. On average, rod log V_{max} almost overlapped measures from control subjects, although two cases showed frankly subnormal final amplitudes. Also the intensity-latency function did not contribute to a clear-cut differentiation between the two groups. In fact, in spite of being slightly slower than normal on average (at least at higher intensities), rod b-wave latencies did not differ significantly from controls. These findings are not in line with many of the previous reports, in which various alterations of standard ERG procedures were found, as well as for the rod log V_{max}. We wonder whether this might simply reflect genetic heterogeneity (e.g., Latin vs. Anglosaxon pedigrees), or if other explanatory mechanism should be claimed (e.g., technical).

Data from this investigation further suggest that xLRP carriers might have a disturbance of the integration of rod responses. This alteration could be at any level of the rod intraretinal microcircuitry, i.e. affecting either the synapsis between rods and rod bipolars, rod bipolars themselves, or rod-driven amacrine cells (particularly, A II and A_{17} subsets).

An abnormality of the rod to rod-bipolar (ON-) synapsis could account for a disturbed transmission to second order neurons, and therefore to an overall impairment of post-receptoral amplifying mechanisms. An additional hypothesis could be an altered release of neurotransmitters (e.g., glycine or GABA) at the level of these second neurons. Most likely, either one of these possibilities should also determine altered logK values at the rod b-wave intensity-response function analysis, and this was actually the case.

However, carriers in this study eventually reached normal rod response amplitudes, suggesting that their defect affects rod sensitivity only at threshold and deep scotopic stimuli, but not at higher intensities. The reduced sensitivity of their rod pathways might therefore affect only the slow and sensitive π_0 channel, leaving unaffected the transmission at higher intensities along the faster π_0' channel (33,34). It can be suggested that the first one is the STR channel, passing through either A II or A_{17} amacrine cells' subsets, and the latter shunted laterally to the cone ON-bipolars. Therefore, the level where the π_0 pathway could be affected remains to be determined: the defect could well be also at the level of rod-driven amacrine cells themselves, either intracellular or at the level of their surface neurotransmitter receptors, or secondary to an overall defective transmission of the signals along the π_0 pathway preceding the rod-bipolar to amacrine-cell synapsis.

Finally, some abnormality of the Müller cell response to the light-evoked K^+ fluxes could be postulated, either due to an intrinsic defect of Müller cells or to an anomalous intraretinal environment (e.g., ionic, metabolic, or structural). However, this last hypothesis seems less likely, in that a disturbance at the level of Müller cells should also affect ERG b-wave characteristics, which was not observed in this study.

The mechanisms underlying the above described alterations of the STR and rod intensity-response function remain so far speculative. More studies on experimental models of retinal degenerations need to be performed as well as further electroretinographic data need to be collected from clinical cases to attempt a reliable explanation of such abnormalities. Whatever the meaning of such rod electroretinographic alterations, this study suggests that they might provide an additional diagnostic tool for the identification of xLRP carriers, especially those with minimal fundus changes and/or in case of normality of other functional and electroretinographic tests. The logK values of the rod ERG intensity-response function and the STR amplitude appear to be the most appropriate parameters to identify all xLRP carriers, irrespective of genetic heterogeneity, and should always be performed in those individuals at risk of being carriers with other normal or uncertain findings. It will be the object of our future investigations to see whether xL-RP2 and -RP3 carriers can be distinguished by different abnormalities of the rod-driven retinal microcircuitry, and if so, to what extent.

REFERENCES

1. Berson, E.L., Rosen, J.B., and Simonoff, E.A., 1979, Electroretinographic testing as an aid in detection of carriers of X-chromosome-linked retinitis pigmentosa. *Am. J. Ophthalmol.* 87: 460-468.
2. Arden, G.B., Carter, R.M., Hogg, C.R., Powell, D.J., Ernst, W.J., Clover, G.M., Lyness, A.L., and Quinlan, M.P., 1983, A modified ERG technique and the results obtained in X-linked retinitis pigmentosa. *Br. J. Ophthalmol.* 67: 419-430.
3. Fishman, G.A., Weinberg, A.B., and McMahon, T.T., 1986, X-linked recessive retinitis pigmentosa. Clinical characteristics of carriers. *Arch. Ophthalmol.* 104: 1329-1335.
4. Peachey, N.S., Fishman, G.A., Derlacki, D.J., and Alexander, K.R., 1988, Rod and cone dysfunction in carriers of X-linked retinitis pigmentosa. *Ophthalmology* 95: 677-685.
5. Jacobson, S.G., Yagasaki, K., Feuer, W.J., and Roman, A.J., 1989, Interocular asimmetry of visual function in heterozygotes of X-linked retinitis pigmentosa. *Exp. Eye Res.* 48: 679-691.
6. Jacobson, S.G. Roman, A.J., Cideciyan, A.V., Robey, M.G., Iwata, T., and Inana, G., 1992, X-linked retinitis pigmentosa: functional phenotype of an RP2 genotype. *Invest. Ophthalmol. Vis. Sci.* 33: 3481-3492.
7. Andréasson, S.O.L., and Ehinger, B., 1990, Electroretinographic diagnosis in families with X-linked retinitis pigmentosa. *Acta Ophthalmol.* 68: 139-144.
8. García-Sandoval, B. Del Rio, T., Ayuso, C., Serrano, J.M., and Zato, M.A., 1994, 30-Hz flicker ERG in carriers of X-linked retinitis pigmentosa. Proceedings of the 34th AER Meeting, Granada, Spain, p. 76 (Abstract).
9. Ponte, F. Anastasi, M, and Lauricella, M.R., 1990, Retinitis pigmentosa and inner retina. Functional study by means of oscillatory potentials of the electroretinogram. *Doc. Ophthalmol.* 73: 337-346.
10. Anastasi, M., Lauricella, M., and Ponte, F., 1992, Photopic ERG components in retinitis pigmentosa. *Acta Ophthalmol.* 70: 187-193.
11. Ernst, W.J., Clover, G., and Faulkner, D.J., 1981, X-linked retinitis pigmentosa: reduced rod flicker sensitivity in heterozygous females. *Invest. Ophthalmol.* 20: 812-816.
12. Wakabayashi, K., Gieser, J., and Sieving, P.A., 1988, Aspartate separation of the scotopic threshold response (STR) from the photoreceptor a-wave of the cat and monkey ERG. *Invest. Ophthalmol. Vis. Sci.* 29: 1615-1622.
13. Sieving, P.A., Frishman, L.J., and Steinberg, R.H., 1986, Scotopic threshold response of proximal retina in cat. *J. Neurophysiol.* 56: 1049-1061.

14. Zrenner, E., and Nelson, R., 1988, Spatial characteristics of scotopic threshold responses in the corneally recorded electroretinogram elicited by multispot patterns. *Clin. Vis. Sci.* 3: 29-44.
15. Frishman, L.J., Sieving, P.A., and Steinberg, R.H., 1988, Contributions to the electroretinogram of currents originating in proximal retina. *Vis. Neurosci.* 1: 307-315.
16. Naarendorp, F., and Sieving, P.A., 1990, Effects of strychnine and GABA-antagonists on the STR of the cat ERG. ARVO Abstracts. *Invest. Ophthalmol. Vis. Sci.* 31: 390.
17. Naarendorp, F., and Sieving, P.A., 1991, The scotopic threshold response of the cat ERG is suppressed selectively by GABA and glycine. *Vision Res.* 31: 1-15.
18. Sieving, P.A., Nino, C., 1988, Scotopic Threshold Response (STR) of the human electroretinogram. *Invest. Ophtahlmol. Vis. Sci.* 29: 1608-1614.
19. Sieving, P.A., 1991, Retinal Ganglion cell loss does not abolish the scotopic threshold response (STR) of the cat and human ERG. *Clin. Vis. Sci.* 6: 149-158.
20. Frishman, L.J. and Steinberg, R.H., 1989, Light-evoked increases in $[K^+]_o$ in proximal portion of the dark-adapted cat retina. *J. Neurophysiol.* 61: 1233-1243.
21. Aylward, G.W., 1989, The Scotopic Threshold Response in Diabetic Retinopathy. *Eye* 3: 626-637.
22. Murayama, K., Kuo, C.Y., and Sieving, P.A., 1991, Abnormal threshold ERG response in X-linked juvenile retinoschisis: evidence for a proximal retinal origin of the human STR. *Clin. Vis. Sci.* 6: 317-322.
23. Graham, S.L., and Vaegan, 1991, High correlation between absolute psychophysical threshold and the scotopic threshold response to the same stimulus. *Br. J. Ophthalmol.* 75: 603-607.
24. Iannaccone, A. Vingolo, E.M., Tanzilli, P., and Rispoli, E., 1994, The Scotopic Threshold Response in different forms of Congenital Stationary Night Blindness. ARVO Abstracts. *Invest. Ophthalmol. Vis. Sci.* 35: 1377.
25. Miyake, Y., Horiguchi, M., Terasaki, H., and Kondo, M., 1994, Scotopic Threshold Response in complete and incomplete types of Congenital Stationary Night Blindness.*Invest. Ophthalmol. Vis. Sci.* 35: 3770-3775.
26. Iannaccone, A., Forte, R., Tanzilli, P., De Bernardo, C., Vingolo, E.M., and Pannarale, M.R., 1994, Clinical and electroretinographic intrafamilial variability in the Goldmann-Favre vitreo-retinal dystrophy. Proceedings of the International Workshop on Inherited Paediatric Retinal Disorders, Regensburg, p. 36 (Abstract). *Ophthalmic Genetics* (submitted for publication).
27. Rispoli, E., Tommasini, P., Pannarale, L., Carloni, C., Perdicchi, A., Rispoli, M., and Pannarale, M.R., 1993, The Scotopic Threshold Response in diabetic patients without evidence of retinopathy. Proceedings of the 4th Schepens International Society Meeting, Hong Kong, p. 148 (Abstract).
28. Carloni, C., Malagola, R., Tommasini, P., and Rispoli, E., 1994, The Scotopic Threshold Response in non-insulin dependent diabetic patients. Proceedings of the 74th Congress of Italian Society of Ophthalmology (SOI), Mondovì (Cuneo), Italy: Arti Grafiche Dial (in press).
29. Rispoli, E., Tommasini, P., Pannarale, L., Carrozzoni, P., Giusti, C., Rispoli, M., and Pannarale, M.R., 1993, Clinical applications of the Scotopic Threshold Response in myopic subjects. Proceedings of the 4th Schepens International Society Meeting, Hong Kong, p. 126 (Abstract).
30. Del Porto, G., Vingolo, E.M., David, D., Steindl, K., Wedemann, H., Forte, R., Iannaccone, A., Gal, A., and Pannarale, M.R., 1993, Clinical features of autosomal dominant retinitis pigmentosa associated with the Gly-188-Arg mutation of the rhodopsin gene. In: Hollyfield, J.G., La Vail, M.M., Anderson, R.E., editors, Retinal Degeneration. Clinical and Laboratory Applications. New York: Plenum Press: 91-101.
31. Rispoli, E., Iannaccone, A., and Vingolo, E.M., 1994, Low-noise electroretinogram recording techniques in retinitis pigmentosa. *Doc. Ophthalmol.* (in press).
32. Masland, R.H., 1988, Amacrine cells. *TINS* 11: 405-410.
33. Sharpe, L.T., Stockman, A., and MacLeod, D.I.A., 1989, Rod flicker perception: scotopic duality, phase lags and destructive interference. *Vision Res.* 29, 1539-1559.
34. Sharpe, L.T., Fach, C.C., and Stockman, A., 1993, The spectral properties of the two rod pathways. *Vision Res.*, 33, 2705-2720.

42

ERG FINDINGS IN TWO PATIENTS WITH AUTOSOMAL DOMINANT CONGENITAL STATIONARY NIGHT BLINDNESS AND HIS258ASN MUTATION OF THE β-SUBUNIT OF ROD PHOTORECEPTOR CGMP-SPECIFIC PHOSPHODIESTERASE

T. Rosenberg,[1] A. Gal,[2] and S. E. Simonsen[1]

[1] National Eye Clinic for the Visually Impaired
1 Rymarksvej, DK-2900 Hellerup, Denmark
[2] Institut für Humangenetik
Universitäts-Krankenhaus Eppendorf
Butenfeld 32, D-22529 Hamburg, Germany

SUMMARY

Recently, a missense mutation (His258Asn) has been identified in the β-subunit of rod cGMP-specific phosphodiesterase in a large multigeneration Danish family, in which autosomal dominant congenital stationary night blindness (adCSNB) is segregating. The mutation was shown to cosegregate with the disease phenotype. Here we report the ERG findings obtained on night blind members of two additional Danish families carrying an identical mutation. Electrophysiological findings are consistent with the hypothetical pathomechanism of a light-independent hyperpolarization of rod photoreceptors in CSNB. Variable expression of the mutation may account for individual differences in residual scotopic rod activity. Reports on impaired electroretinographic cone activity in cases of adCSNB might indicate genetic heterogeneity in adCSNB.

INTRODUCTION

Early reports on electrophysiological findings on subjects with congenital stationary night blindness (CSNB) and normal fundi demonstrated two basically different electroretinographic (ERG) patterns (1,2). This distinction has been widely adopted (3-6) and, in addition to the pattern of inheritance, ERG findings have become the most frequently used criteria for classification of CSNB. Auerbach et al. (4) subdivided the negative or Schubert-Bornschein type ERG into three subgroups according to the presence and characteristics of the b-wave. Miyake et al. (6) suggested another subclassification of CSNB with negative ERG in complete and incomplete type.

There is only a very limited number of reports on ERG findings in patients from families with autosomal dominant congenital stationary night blindness (adCSNB). Two of the three patients reported by Riggs (2) belonged to an American family with adCSNB (7). Francois et al. (8) described a Riggs-type ERG in a member of the 'historical' Nougaret pedigree first reported in 1838 by Cunier (9). It is generally believed that there is not a tight correlation between the ERG type and the inheritance pattern (4). Indeed, ERG findings from subjects with adCSNB differ considerably with respect to amplitudes and temporal characteristics of both the rod and cone photoreceptor responses suggesting a pathophysiological heterogeneity and, therefore also a heterogeneity at the molecular genetic level.

From a clinical point of view there is a strong demand for correlating phenotype with genotype. Here we report the ERG findings obtained on two subjects with autosomal dominant CSNB associated with a missense mutation in the gene encoding the β-subunit of the rod cGMP-specific phosphodiesterase (βPDE), a key enzyme in phototransduction.

Subjects and methods

Two Danish night blind subjects from families with autosomal dominant transmission of the trait were examined.

Case 1 is a 31-year-old female. Her father, paternal grandfather and greatgrandfather all were reported being night blind without any visual problems at day time. The proband has been night blind as long as she remembers. A routine eye examination disclosed normal visual acuity, emmetropia, and unremarkable slit lamp and ophthalmoscopic findings. Genealogical studies revealed that the proband belonged to the large multigeneration Danish adCSNB family first described in 1909 by Rambusch (10) and reascertained recently (11).

Case 2 is a male, born in 1969, who first came to us for examination in 1987. He has been night blind from early infancy. The same feature was present in several other family members, including his deceased father, paternal grandfather, and the grandfather's mother. Of the eleven children of his grandfather, eight suffered from congenital night blindness. Many of the patient's cousins and their children were also affected, although none of these relatives were available for clinical examination. The patient's father was reported to have had tunnel vision. According to the medical records of a local eye hospital, a paternal uncle of the proband suffered from typical retinitis pigmentosa (RP). The remaining family members had no anamnestic signs of RP. Genealogical studies have suggested, although could not yet confirm a link between this kindred and the 'Rambusch family' originating 200 km further north (10,11).

For intensity-response measurements and time-response relationship during dark adaptation, normal ERGs with average recordings were selected from a control sample (21 subjects, aged between 15-67) of unaffected persons. Data of 42 unaffected subjects, aged 13-42, served as control for photopic single flash recording and photopic 32 Hz flicker response. Informed consent was obtained from both the night blind subjects and controls.

Ophthalmological examination included a measurement of visual acuity, slit lamp examination, and ophthalmoscopy through dilated pupils. Dark adaptation was measured with a Goldmann-Weekers equipment following the protocol of integral technique with diffuse lightning of the bowl. Photopic visual fields were measured with a Goldmann perimeter using object size of 4 mm^2 and relative intensity 1.00.

Corneal ERGs were recorded from one eye with a Burian-Allen monopolar electrode wetted with methylcellulose. Pupils were dilated with 1% cyclopentolate, and corneal anesthesia was achieved with 1% tetracaine. The indifferent electrode was a chlorided silver-disc placed on the middle of the forehead, while the ground electrode was placed on the ipsilateral ear lobe. The examination was done with the patient being in horizontal position.

Light flashes were produced by a Grass PS22C photostimulator illuminating a LKC Ganzfeld 2503 bowl equipped with a weak central red fixation diode. Flash intensities and background illumination were measured with a Mastersix system exposure meter (Gossen, Erlangen, Germany; ISCEV newsletter, February 1991). ERGs were recorded by a 4-channel averager system (Dantec Evomatic 4000) with digital X-Y plotter. The frequency range was 0.2-1000 Hz, except for recording oscillatory potentials, where the range was 100-1000 Hz (analog filter, -3 dB). The following ERG recordings were done: i) Scotopic recordings (after 30 min dark adaptation) as a function of flash intensity (white), from below the b-wave threshold and in steps of 0.5-1.0 log (Kodak neutral density filters) to the maximum available intensity of Grass step 16, equivalent to 5.0 cds/m^2. Flash intervals were 5 sec for weak flashes and increased stepwise to 20 sec for the higher intensities. 3-5 responses were averaged for every intensity step. ii) Responses to single white flashes of Grass step 4 (1.0 cds/m^2) as a function of increasing dark adaptation after turning off the background preadapting illumination, 17 cd/m^2 applied for 10 minutes. iii) Photopic cone recordings (white background illumination, 17 cd/m^2 as preadaptation for 10 min) with single white flashes (5-10 averaged responses to Grass step 4 flashes, 1.0 cds/m^2), 32 Hz white flicker (1.0 cds/m^2, 20 series were averaged after discarding the first responses), and scotopic cone recordings (after 30 minutes dark adaptation) with 32 Hz white flicker (1.0 cds/m^2, 20 series were averaged after discarding the first responses), iv) Oscillatory potentials (after 30 minutes dark adaptation) were evoked by white flashes, Grass step 4 (1.0 cds/m^2), 5 averaged responses after having discarded the first 2 responses. In order to obtain maximal amplitudes, flash intervals were 15 sec with a frequency range of 100-1000 Hz (analog filter, -3 dB).

RESULTS

Ophthalmological examinations of the night blind subjects revealed normal visual acuities (1,0 without glasses) and no signs of retinal dystrophy. Visual fields had normal outer limits, and no scotomas were found besides the normal sized blind spots. Dark adaptometry in case 1 showed a biphasic-like course with a break point after 5 minutes in the dark (Fig. 1). The first phase was delayed and moderately elevated, while the second one was severely attenuated with a final threshold elevation of approximately 2.5 log units. For case 2, dark adaptometry was performed in two instances (only one of them is shown) and demonstrated a rapid initial phase with approximately 2 log unit sensitivity-gain during the first 3 minutes in the dark. After about 8 minutes a second threshold attenuation of approximately 0.5 log unit was observed (Fig. 1).

Fig. 2 shows dark adapted ERGs of the two probands following single flash stimulations of different intensities.

The reduced rod-sensitivity of the night blind subjects is reflected by a log 2-increase in the flash intensity necessary for a minimal response. a-waves appear both in the night blind subjects and in the unaffected control at equal stimulus intensity indicating a normal a-wave threshold in the former ones, whereas a-wave amplitudes are diminished at all stimulation intensities in the CSNB subjects. a-Wave amplitudes increase with growing flash intensities which, together with the severely reduced b-waves, give rise to a 'negative' ERG configuration in the two night blind cases. a-Wave implicit times seem to be of the same magnitude in the night blind persons and the control.

Both patients exhibit b-waves with diminished amplitudes and reduced implicit times. In case 1 only one b-wave is seen, while in case 2 the b-wave is represented by two positive deflections, an early wave (b_1) with implicit time of about 34 msec, and a late positive one (b_2) with implicit time between 100 and 110 msec (arrows), reflecting probably

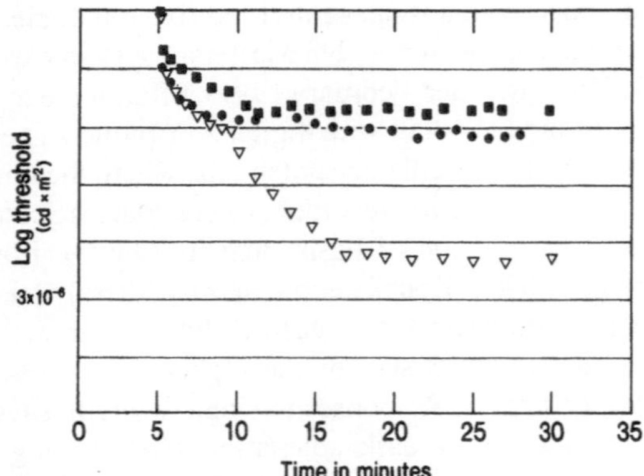

Figure 1. Course of dark adaptation (integral technique a.m. Goldmann-Weekers) in two night blind subjects. Recordings from case 1 (squares) and case 2 (circles) are compared to those from a normal subject (triangles). Final thresholds are elevated by 2-2.5 log cd/m^2 in the night blind individuals. Note the cone-rod transition with a small residual rod activity in the recording from case 2. Although both case 1 and case 2 carry the same βPDE-mutation, their clinical data differ to some extent.

residual rod activity. The early b_I -wave represents the cone component and displays marked superimposed oscillatory on-potentials.

Fig. 3 demonstrates the ERG responses to a series of single flashes (1.0 cds/m^2) during the first 30 minutes of dark adaptation.

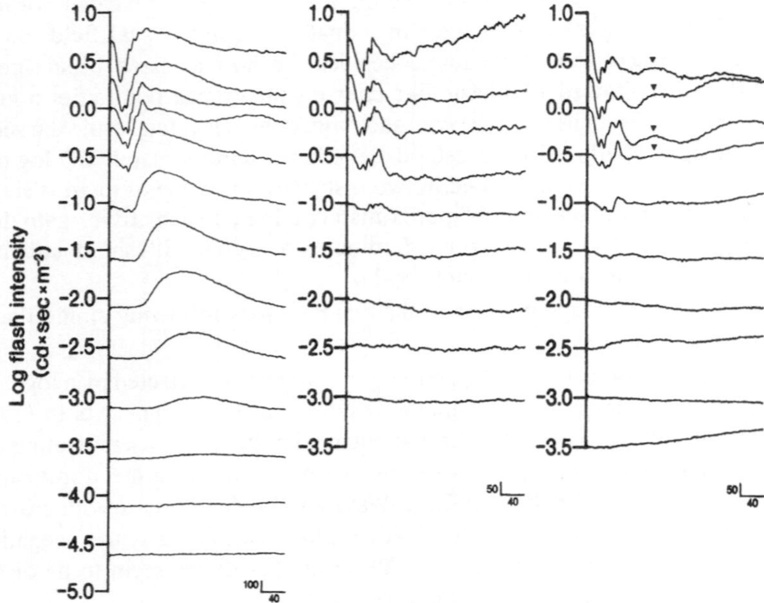

Figure 2. Scotopic ERG responses to single white flashes of increasing intensities from a normal control (left) and the two night blind subjects (case 1, middle; case 2, right). Intersection of tracings with luminance axis indicates flash intensity (log cds/m^2) and time of flash presentation. Note the different amplitude scales used for the normal subject and the night blind subjects.

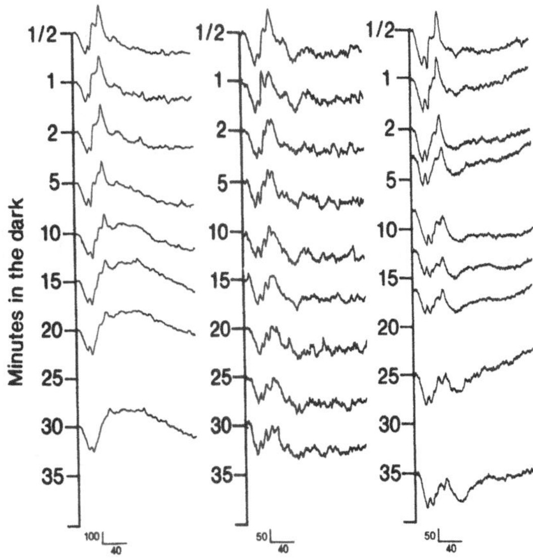

Figure 3. ERG responses to a standard flash in the dark as a function of time after the turn-off of a light adapting background. Each recording represents a single flash (1 cds/m^2) from a normal subject (left) and the two night blind subjects (case 1, middle; case 2, right). Intersection of tracings with time axis indicates the time of flash presentation. Note the different amplitude scales used in the normal subject and the night blind subjects.

Shortly after the light has been turned off, the ERG is characterized by a photopic wave-form including prominent off-responses (14) which look identical in the control and night blind subjects. Yet, the rise of a rod b_2-component, as seen in the normal person after about 10 min, is absent in case 1 and significantly reduced and remains separated from the b_1-component in case 2.

Fig. 4 presents cone responses from the two night blind subjects compared to those of an age-matched typical normal subject.

In the night blind subjects, photopic single flash elicited in case 1 and 2, respectively, b-wave amplitudes of 147 and 236 μV (normal median: 137μV with range: 62-312) with implicit times of 29.8 and 30,4 msec (normal median: 29.5 with range: 24-32.8). 32 Hz flicker responses had normal amplitudes of 100 μV and 112 μV (normal median: 107 μV, range: 56-249) and implicit times of 28.1 and 28,8 msec (normal median: 26 msec, range: 23.2-31.2) in case 1 and 2, respectively. The flicker response of case 2 was double-peaked. Responses to photopic balanced red and green single flashes and blue cone responses (not shown) were also within normal limits. Fig. 4 also shows the scotopic 32 Hz flicker responses and scotopic oscillatory potentials, compared to those of an unaffected person. It should be noted that flicker amplitudes show no dark-to-light rise in the night blind cases, whereas the dark/light ratio was normal in the unaffected subject.

DISCUSSION

Here we report extensive electrophysiological data obtained by studying two subjects with CSNB and a missense mutation in the β-subunit of rod cGMP-specific phosphodiesterase. The clinical examination excluded the presence of retinitis pigmentosa or any other progressive retinal dystrophy in both probands. Case 1 belongs to a large family, in which adCSNB is well

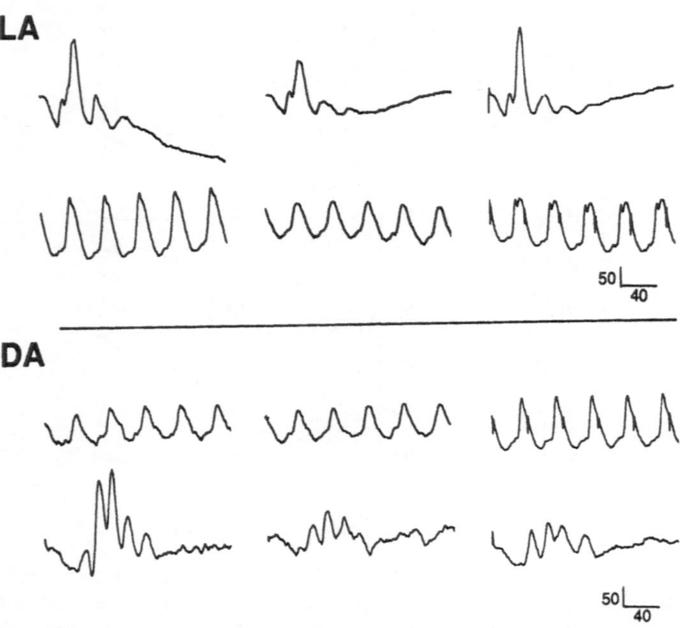

Figure 4. ERG responses from a normal subject (left) and the two night blind subjects (case 1, middle; case 2, right). The two upper lines were recorded in light adapted state (LA), while those on the bottom in dark adapted state (DA). First upper line: averaged cone responses from 5 single white standard flashes (1 cds/m^2), second upper line: averaged 32 Hz flicker responses (light adapted). First bottom line: averaged 32 Hz flicker responses (dark adapted), second bottom line: on-oscillatory potentials. Calibration scales indicate amplitudes in µV (vertical) and time in msec (horizontal).

documented. The diagnosis of CSNB in case 2 can be questioned due to the RP of his father and paternal uncle. Yet, due to a number of facts, the diagnosis of RP with largely variable expression seems to be less likely in this family. Firstly, according to anamnestic information, most night blind family members retained good visual acuity throughout life. The two individuals with reduced vision/RP were regarded clearly different from other (night blind) family members. Secondly, having observed our patient during the last 7 years, we have found no clinical signs of RP. Finally, the presence of the same βPDE-mutation as in an undisputed adCSNB family strongly supports the diagnosis of CSNB. On the other hand, we can not exclude that the occurrence of RP in this family is not incidental. As mutations in the βPDE gene have been found in patients with autosomal recessive RP (15), we have to consider the possibility of a compound heterozygosity for an RP-allele and the 'adCSNB-specific' His258Asn mutation that, together, may lead to photoreceptor degeneration. The contribution of a nonallelic mutation ('digenic RP') can be another explanation. Interestingly, RP was also ascertained in a congenitally night blind subject from the large 'Rambusch'-pedigree (11).

Based on extensive psychophysical and electrophysiologic studies on night blind subjects, it has been suggested that the pathological process, i.e. the loss of scotopic sensitivity, is within the rod photoreceptors in adCSNB possibly affecting the phototransduction enzyme cascade or the dark current of the rod outer limb membrane (5,16-18). Our finding of a mutation in the β-subunit of the rod cGMP-specific PDE in affected individuals from a family with adCSNB (12) is in keeping with the above mentioned hypothesis. In view of the predicted His258Asn amino acid substitution in the PDE β-subunit polypeptide, we proposed a hypothetical mechanism in which PDE complexes with mutant β-subunits remain permanently active due to an impaired binding of the inhibitory PDE γ-subunit, thus keeping dark adapted cytoplasmic cGMP-levels abnormally low. In consequence, cGMP-gated cation channels would remain (partially) closed in the dark, and rod photoreceptors remain permanently hyperpolarized and desensitized. This hypothesis seems to fit in well with the above mentioned deductions from clinical studies.

Dark adapted ERG responses to white flashes of varying intensities recorded in the two night blind subjects (Fig. 2) demonstrate a highly desensitized photoreceptor system with a threshold elevation of approximately 2 log units. In unaffected control persons, high intensity flashes elicit mixed cone-rod responses. In contrast, both the a- and b-wave amplitudes and implicit times seem to indicate that the corresponding responses are mediated exclusively by cones in the night blind subjects. In addition, the presumed rod-generated b_2 -waves are clearly separated from the larger and faster cone b_1 -waves in case 2. The presumed rod origin of the slow b_2-wave is supported by its absence in the dark adapted response during the first minutes in the dark following a preadaptive bleach (Fig. 3). Furthermore, the b_2-wave emerges at the same time, after about 8 min, as the cone-rod break is noticed in the psychophysical dark adaptation curve (Fig. 1).

A lack of rod contribution to dark adapted ERG in adCSNB has been reported by several authors (5,7-8,20-22). Technical imperfections of the pioneer studies might have erroneously led to this conclusion. Nevertheless, as early as in 1954, Riggs was able to demonstrate a small, though consistent increase in response during the later stage of dark adaptation and mentioned the possibility that some slight ‘scotopic’ activity is present (2). Sharp et al. (18) demonstrated also a small rod response in a patient from a Sardinian pedigree. In fact, severely depressed or absent rod activity reflects merely quantitative differences, which may be due to variable gene expression, provided that different adCSNB harbour the same mutation. The two cases in our report are good examples of such differences.

A variant type of negative ERG was reported in an adCSNB family by Noble et al. (19). We suggest that the ‘negative’ response in that case is merely the consequence of the high flash intensity used. Even in the ‘His258Asn’-type CSNB, ‘negative’ response is obtained with very bright flash stimuli (cf. Fig. 2). In contrast to the Schubert-Bornschein type ‘negative’ ERG, in which oscillatory potentials are missing and suggest a postreceptoral pathogenesis of CSNB, the presence of significant cone on-oscillations points indirectly to a photoreceptor defect in our cases.

Conflicting data exist with regard to cone photoreceptor signals in adCSNB. Photopic single flash responses as well as both photopic and scotopic flicker responses were found to be normal or slightly below normal (2,8,18-23). In contrast, other authors demonstrated severely reduced and/or delayed photopic responses (4,5,24). This may indicate genetic heterogeneity within the group of adCSNB. Alternatively, these combined rod-cone dysfunctions may be due to a retinal dystrophy of very mild course. The clinical distinction between photoreceptor dystrophies, e.g. RP, and retinal dysfunctions, e.g. CSNB, is challenged by the repeated finding that different mutations of the same gene may result either in one or the other of the above disorders (for details see ref. 12). Further molecular studies of these allelic conditions might throw light upon the process of photoreceptor cell death following impaired phototransduction in some kinds of retinitis pigmentosa.

ACKNOWLEDGMENTS

Thanks are due to Nurse Vibe Gielstrup for her competent technical assistance. Molecular genetic studies were financially supported by the Deutsche Forschungsgemeinschaft (DFG).

REFERENCES

1. Schubert, G., and Bornschein, H., 1952, Beitrag zur Analyse des menschlichen Eletroretinogramms, *Ophthalmologica* 123:396-411.
2. Riggs,L.A., 1954, Electroretinography in cases of night blindness, *Am. J. Ophthalmol.* 38:70-78.
3. Armington, J.C., and Schwab, G.J., 1954, Electroretinogram in nyctalopia, *Arch. Ophthalmol.* 52:725-733.
4. Auerbach, E., Godel,V., and Rowe, H., 1969, An electrophysiological and psychophysical study of two forms of congenital night blindness, *Invest. Ophthalmol. Vis. Sci.* 8:332-345.
5. Carr, R.E., Ripps, H., Siegel, I.M., and Weale, R.A., 1966, Visual functions in congenital night blindness, *Invest. Ophthalmol. Vis. Sci.* 5:508-514.
6. Miyake, Y., Yagasaki, K., Horiguchi, M., Kawase, Y., and Kanda, T., 1986, Congenital stationary night blindness with negative electroretinogram. A new classification, *Arch. Ophthalmol.* 104:1013-1020.
7. Carroll, F.D., and Haig,C., 1952, Congenital stationary night blindness without ophthalmoscopic or other abnormalities, *Trans. Am. Ophthalmol. Soc.* 50:193-209.
8. Francois, J., Verriest, G., de Rouck, A., and Dejean,C., 1956, Les fonctions visuelles dans l'hemeralopie essentielle nougarienne, *Ophthalmologica* 132:244-257.
9. Cunier, M.F., 1838, Histoire d'une héméralopie héréditaire depuix deux siècle dans une famille de la commune de Vendémian, près Montpellier, *Ann. Soc. Med. de Gand* 4:383-395.
10. Rambusch, S.H.A., 1909, Den medfodte Natteblindheds Arvelighedsforhold. *Oversigt over det Kgl. Danske Videnskabernes Selskabs Forhandlinger* 3:337-347.
11. Rosenberg, T., Haim, M., Piczenik, Y., and Simonsen, S.E., 1991, Autosomal dominant stationary night-blindness. A large family rediscovered, *Acta Ophthalmol. (Copenh.)* 69:694-702.
12. Gal, A., Orth, U., Baehr, W., Schwinger,E., and Rosenberg,T., 1994, Heterozygous missense mutation in the rod cGMP phosphodiesterase β-subunit gene in autosomal dominant stationary night blindness, *Nature Genet.* 7:64-68.
13. Veske, A., Orth, U., Rüther, K., Zrenner, E., Rosenberg, T., Baehr, W. and Gal, A., 1995, Mutations in the gene for the β-subunit of photoreceptor-specific cGMP phosphodiesterase in patients with retinal dystrophies and dysfunctions. This volume.
14. Nagata, M., 1963, Studies on the photopic ERG of the human retina. Jpn. J. Ophthalmol. 7:96-124.
15. McLaughlin, M.E., Sandberg, M.A., Berson, E.L., and Dryja, T.P., 1993, Recessive mutations in the gene encoding the β-subunit of rod phosphodiesterase in patients with retinitis pigmentosa, *Nature Genet.* 4:130-134.
16. Alpern, M., Holland, M.G., and Ohba, N., 1972, Rhodopsin bleaching signals in essential night blindness, *J. Physiol.* 225:457-476.
17. Ripps, H., 1982, Night blindness revisited: from man to molecules, *Invest. Ophthalmol. Vis. Sci.* 23:588-609.
18. Sharp, D.M., Arden, G.B., Kemp, C.M., Hoog, C.R., and Bird,A.C., 1990, Mechanisms and sites of loss of scotopic sensitivity: a clinical analysis of congenital stationary night blindness, *Clin. Vis. Sci.* 5:217-230.
19. Noble, K.G., Carr, R.E., and Siegel, I.M., 1990, Autosomal dominant congenital stationary night blindness and normal fundus with an electronegative electroretinogram, *Am. J. Ophthalmol.* 109:44-48.
20. Francois, J., Verriest, G., and de Rouck,A., 1963, Une nouvelle observation d'héméralopie congénitale essentielle à transmission dominante, *Bull. Soc. Belge Ophthalmol.* 135:489-497.
21. Berson, E.L., Gouras, P., and Hoff, M., 1969, Temporal aspects of the electroretinogram, *Arch. Ophthalmol.* 81:207-214.
22. Ponte, E., Lodato, G., and Lauricella, M., 1974, Electrophysiological tests in congenital stationary night-blindness with regard to the different types of inheritance, *Doc. Ophthalmol. Proc. Ser.* 10:161-167.
23. Keunen, J.E.E., van Meel, G.J., and van Norren, D., 1988, Rod densitometry in night blindness: a review and two puzzling cases, *Doc. Ophthalmol.* 68: 375-387. 24.
Krill,A.E., and Martin,D., 1971, Photopic abnormalities in congenital stationary nightblindness, *Invest. Ophthalmol. Vis. Sci.* 10:625-636.

43

DOCOSAHEXAENOIC ACID ABNORMALITIES IN RED BLOOD CELLS OF PATIENTS WITH RETINITIS PIGMENTOSA

Dennis R. Hoffman[1,2], Ricardo Uauy,[1,3] and David G. Birch[1,4]

[1] Retina Foundation of the Southwest
Dallas, Texas
[2] Department of Pediatrics, University of Texas Southwestern Medical Center at Dallas
Dallas, Texas
[4] Clinical Nutrition Unit, Institute of Nutrition and Food Technology (INTA)
University of Chile, Santiago, Chile
[3] Department of Ophthalmology, University of Texas Southwestern Medical Center at Dallas
Dallas, Texas

INTRODUCTION

As primary structural components of phospholipids, fatty acids are major building blocks of cell membranes. The ω3 long-chain polyunsaturated fatty acid, docosahexaenoic acid (DHA; 22:6ω3) is one of the most unsaturated fatty acids in the human body and is highly enriched in membrane lipids of rod photoreceptors (1). Several lines of evidence suggest that the accumulation of DHA in retinal membranes is important for normal retinal function. Abnormal electroretinographic (ERG) responses to light stimuli have been reported in DHA deficient animal models and human infants (2-6). Rat and monkey models made ω3 fatty acid deficient by feeding modifications express abnormalities in a- and b-wave ERGs (2-4). We found delayed maturation of rod ERG responses associated with reduced blood lipid levels of DHA in pre-term infants that received a commercial corn-oil-based formula compared to infants fed breast milk or experimentally-enriched DHA-containing formula (5,6). Although the physiological consequences of DHA deficiency are functionally demonstrable, the interaction of membrane-bound DHA with proteins and enzymes of the phototransduction cascade is likely to be a complex function of biochemical and biophysical parameters unique to the highly unsaturated hydrocarbon chain of DHA.

In the past decade, numerous investigators have reported abnormal levels of DHA in plasma lipids of patients with the autosomal dominant form of RP (adRP), x-linked RP (xlRP) and Usher's syndrome (7-15). Our studies have focused on red blood cell (RBC) lipids as an index of fatty acid status primarily because the fatty acids in RBCs are associated with

Degenerative Diseases of the Retina, Edited by Robert E. Anderson et al.
Plenum Press, New York, 1995

structural membrane lipids and are correlated with fatty acid profiles of both retinal and neural tissues (16,17). Furthermore, RBC lipids are less sensitive than plasma lipids to dietary fluctuations, and represent a longer-term fatty acid status in patients.

In adRP, specific mutations in the genes for proteins and enzymes vital to visual transduction presently account for RP in approximately one-third of the patients (18-20). Approximately 80 different loci in the gene for rhodopsin have been identified as mutation sites; additional mutations have been described in the peripherin gene (20,21). In some families, a combination of the unlinked alleles for two rod outer segment membrane proteins, peripherin and ROM1, appears to act in a recessive manner, resulting in digenic inheritance of the retinal degeneration (22). Gene mutations for the rod β-subunit of cGMP phosphodiesterase have been associated with photoreceptor degeneration in autosomal recessive RP (23). In xlRP, two chromosomal defects have been localized near Xp21 and Xp11.3, although candidate genes have not been identified (24). Although many of the fatty acid disproportionalities in RP are minor compared to controls, these subtle changes may have a significant impact on visual function if persistent over many years. This may be of particular importance for a patient who has visual function further compromised by a gene mutation. We suggest that a secondary factor, such as sub-normal levels of DHA in target tissues, may result in an unfavorable membrane environment and further impair the function of mutant gene products.

METHODS

Patients with known ($\geq$3 confirmed generations) or probable (2 confirmed generations) adRP (n=81, mean age $\pm$ 1SD = 42 $\pm$ 19 years) or confirmed xlRP (n=18, 22 $\pm$ 18 years) were recruited for studies. Diagnoses of xlRP was defined by the presence of at least two affected male relatives, an absence of male-to-male transmission, and either the patient's mother or daughter expressing characteristics consistent with an xlRP carrier state (25). Normally-sighted control subjects (n=62, 31 $\pm$ 18 years) were recruited only to provide blood samples. All subjects signed informed consent forms approved by the Institutional Review Board on Human Subjects of the University of Texas Southwestern Medical Center and were recruited in accordance with the tenets of the Declaration of Helsinki. Blood samples (10ml) were obtained by venipuncture from an arm vein; most patients were fasted for 12 hours. Gene mutation analysis was conducted in collaboration with Steven Daiger (University of Texas at Houston) using single-strand conformational polymorphism analysis and allele-specific oligonucleotide detection following polymerase chain reaction (21).

Fatty Acid Analysis

Lipids were extracted from red blood cells according to the procedure of Bligh and Dyer (26). Fatty acid methyl esters were prepared under N_2 using boron trifluoride/methanol heated at 100°C for 20 minutes. Thirty fatty acid methyl esters were separated and quantified using a Supelco Omegawax (0.25mm x 30m) capillary gas chromatography column. Details of the methodology have been described (15,27).

Visual Function

Full-field ERGs were obtained with a bipolar contact-lens electrode. Signals were amplified (x 10,000) and averaged (n $\geq$ 20) by computer after eliminating responses containing artifacts over twice the signal amplitude. Testing included the protocol recommended by the International Standards Committee of the International Society for the Clinical Electrophysiology of Vision (28). In addition, dark-adapted mixed cone-rod responses were obtained with 1.0 log photopic td-sec red stimuli presented at 1.0 Hz. Rod ERGs were elicited with blue flashes scotopically-matched to the red stimuli. Dark-adapted cone-only ERGs were derived by digitally subtracting the rod ERG to the blue stimulus from the mixed cone-rod response to the red stimulus. Responses were considered to be non-detectable if the peak-to-peak amplitude measured less than 2.0 μV. Details have been described (29).

RESULTS

Our studies have focused on analysis of fatty acids in RBCs primarily because the lipids in this target tissue reflect a long-term fatty acid status in patients. In contrast, plasma lipids respond rapidly to changes in dietary fat intake. In our studies of fatty acid distributions in patients with adRP, initial samples (n=35) were obtained without fasting. Fatty acid analysis of RBCs from 14 adRP individuals demonstrated no significant differences in DHA content between fasted and unfasted patient samples (3.27 ± 0.91% vs. 3.32 ± 0.89%, respectively). Nevertheless, to permit comparisons between fatty acid levels in plasma lipids with those in RBC lipids, all recent data in our laboratory have been obtained from fasted individuals.

Fatty acids were analyzed in RBC lipids obtained from 81 patients with adRP and compared to those of 62 normally sighted controls. Shown in Figure 1 is the distribution of DHA levels found in the adRP patients. Although the average level of DHA in the RBCs of adRP patients (3.13 ± 0.81% of total fatty acids) was significantly ($p < 0.05$) below the mean of the controls (3.44 ± 0.92%), there was considerable variability among the patients. Mass analysis of the DHA concentration in RBCs similarly showed a significant ($p<0.0005$) decrease in the adRP population (25.2 ± 8.2 μg DHA / ml packed RBCs) compared to control subjects (33.5 ± 10.3 μg / ml). Amplitudes and implicit times for ERG responses to standard stimuli were not significantly correlated with DHA values. However, the ratio of rod ERG amplitude to either dark-adapted cone amplitude or light-adapted cone amplitude was found to correlate significantly with the DHA content in RBCs of adRP patients ($p= 0.048$ and $p= 0.003$, respectively). Stepwise multiple regression analysis indicated that DHA but not age

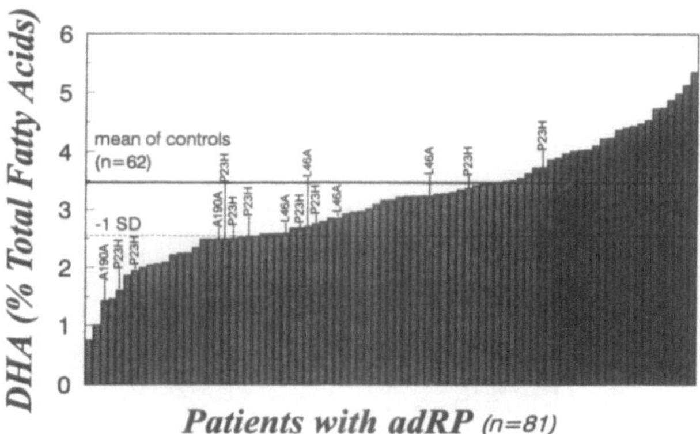

Figure 1. Distribution of docosahexaenoic acid (DHA) in red blood cells (RBCs) of patients with autosomal dominant retinitis pigmentosa (adRP). Bars represent the DHA level in RBCs of each patient. The solid line represents the mean DHA level in RBCs of normally-sighted controls ± SD (dashed line). Known rhodopsin gene mutations (Asp190Asn, Pro23His, Leu46Arg) are noted above bars.

was a determinant in the regression equations for rod-to-dark- or light-adapted cone responses. To date, rhodopsin gene mutations have been identified in fifteen of our adRP patients. As shown in Figure 1, patients with rhodopsin mutations manifest a wide range of RBC-DHA levels although all but one are below the mean of control subjects.

Environmental factors such as diet or general physical activity of the visually-impaired have been considered as possible contributing factors to alterations in blood levels of fatty acids. To address this concern, we analyzed fatty acids in RBCs of five patients having no-light-perception for least nine years. The patients had retinal detachments attributed to retinopathy of prematurity (n=3) or of undetermined origin (n=2). DHA levels in these patients were not significantly different from normal; means (±SD) were 3.67 ± 0.81% of total fatty acids and 34.4 ± 8.7 μg/ml RBC (t=0.49, p=0.6 and t=0.17, p=0.8 vs. controls, respectively). Total ω3 and ω6 long-chain polyunsaturated fatty acids in these patients were indistinguishable from control values.

Patients with xlRP are a more homogenous population with two major gene mutations. Among the two mutations, RP2 and RP3, the RP3 form is most prevalent comprising from 75% to 90% of the patients (30). The xlRP sub-type is also considered one of the most severe forms of RP due to an early onset of symptoms (age 5-10 years) often resulting in functional blindness by age 20. A cohort of 18 patients were recruited and found to have markedly reduced levels of RBC-DHA (Figure 2). As a group, the DHA level in xlRP (2.46 ± 0.55%) was significantly lower (p<0.0005) than normally sighted controls (3.44 ± 0.92%) and the levels for all patients were below the mean normal value.

By examining differences in the levels of the ω3 and ω6 fatty acids for control subjects and patients with xlRP, precursor-to-product shifts can be used to localize specific impairments in fatty acid metabolism. In Figure 3, the relative levels of each fatty acid in the ω3 pathway are presented for control subjects and xlRP patients. There was a break in ω3 fatty acid metabolism centering on eicosapentaenoic acid (20:5ω3) such that the xlRP patients had higher levels of 18-carbon precursors (18:3ω3 and 18:4ω3), no difference in 20:5ω3, and 20% to 40% lower levels of the ω3 long-chain derivatives, 22:5ω3 and 22:6ω3 than controls. In contrast, there was no obvious break in metabolism of the ω6 fatty acid series in the xlRP group compared to control subjects. Further analysis of product/precursor pairs in a sub-group of xlRP patients and controls revealed that the elongation steps (addition of 2-carbon units to the fatty acid) for saturates, monounsaturates, ω3 and ω6 fatty acids were all lower for the xlRP group than controls. By contrast, fatty acid ratios for desaturation reactions (addition of double bonds to the fatty acid) were not different between the xlRP and control groups.

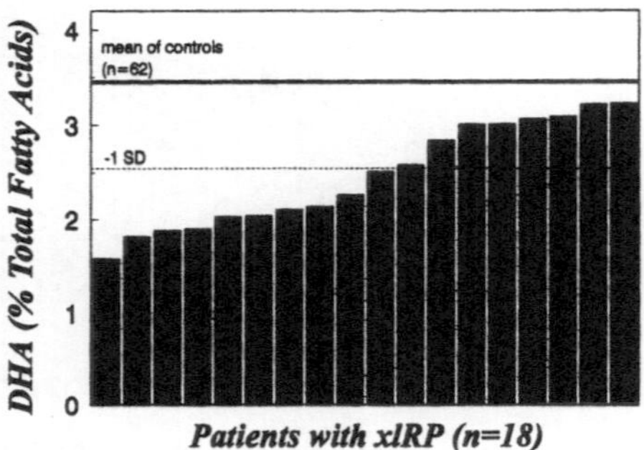

Figure 2. Distribution of docosahexaenoic acid (DHA) in red blood cells (RBCs) of patients (n=18) with x-linked retinitis pigmentosa (xlRP). The mean normal value (solid line) is 3.44±0.92% of total fatty acids (n=62).

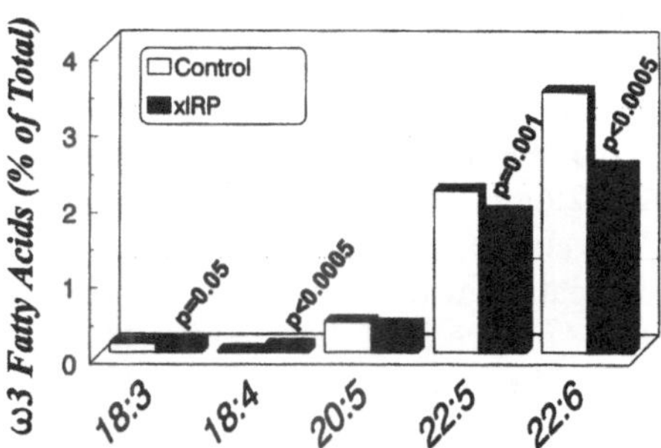

Figure 3. Profiles of ω3 fatty acids in patients with x-linked retinitis pigmentosa (xlRP, n=18) and normally-sighted control subjects (n=62). Significant differences between patients with xlRP and control subjects were determined by t-tests and indicated by p values.

To assess the physiological significance of reduced blood levels of DHA in the xlRP patients, electrophysiologic parameters of retinal function were compared to RBC-DHA content. Due to severity of the disease, fourteen of the 18 xlRP patients did not have detectable rod ERG responses preventing a comparison of rod photoreceptor status with RBC fatty acid content. All patients showed significant elevations in final visual threshold after 45 minutes of dark adaption. Visual fields showed varying degrees of constriction in general agreement with ERG amplitudes. Due to intrafamilial similarities, only one member per family was selected for statistical comparisons between retinal function and RBC-DHA levels. A stepwise multiple regression analysis revealed that RBC-DHA accounts for 55% of the cone amplitude to maximal stimulus ($p = 0.002$), 44% of cone amplitude to 30Hz flicker ($p = 0.0001$) and 38% of the cone implicit time to 30Hz flicker ($p = 0.033$).

A dietary supplementation trial utilizing fish oil enriched in the ω3 fatty acids, DHA and eicosapentaenoic acid (EPA, 20:5ω3) was conducted to evaluate the efficiency of adRP patients for gastrointestinal absorption and incorporation of these fatty acids into target tissues (i.e., RBCs). Three patients with adRP and three normally-sighted controls followed the study protocol that consisted of a 12-week ω3 fatty acid restricted diet; after the sixth week the participants were provided supplements of 3g per day of a fish oil concentrate (0.7g DHA and 1.3g EPA daily) for six weeks. After the supplementation period, the levels in RBC lipids of the major ω3 long-chain derivatives, 20:5ω3, 22:5ω3, and 22:6ω3, were elevated in both adRP patients and controls by approximately 600%, 40% and 50%, respectively,

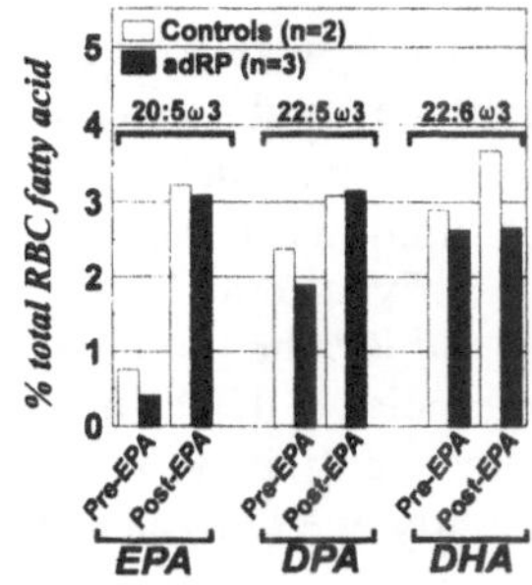

Figure 4. Incorporation into red blood cells (RBCs) of eicosapentaenoic acid (EPA, 20:5ω3) and metabolism to docosahexaenoic acid (DHA, 22:6ω3) in patients with autosomal dominant retinitis pigmentosa (adRP, n=3) and controls (n=2). Bars represent percent of total fatty acids for EPA, docosapentaenoic acid (DPA, 22:5ω3) and DHA in subjects at pre- and post-EPA supplementation time points.

compared to pre-supplementation levels. A two-way analysis of variance for each fatty acid demonstrated no significant differences in ω3 fatty acid uptake into RBCs between the adRP patient and control groups.

A subsequent supplementation trial addressed the metabolic conversion of the ω3 precursor, EPA (20:5ω3) to the intermediate docosapentaenoic acid (DPA, 22:5ω3) and to the ω3 end-product DHA (22:6ω3) in adRP patients with rhodopsin gene mutations. Participants with low DHA levels were selected for study and included two adRP patients with Pro23His mutations, one patient with a Leu46Arg mutation, and two normally-sighted controls. The protocol involved a 9-week ω3 fatty acid restricted diet; during the last three weeks the diet was supplemented with a highly purified preparation (99.4%) of EPA ethyl ester (3g per day). Shown in Figure 4 are 20:5, 22:5, and 22:6 levels in RBCs of controls and adRP patients at pre- and post-supplementation. EPA supplementation resulted in 7-fold and 4-fold increases in 20:5ω3 in RBC lipids of patients and controls, respectively. Three weeks of EPA supplementation increased the level of the intermediate, DPA (22:5ω3), relative to pre-supplementation values by 66% in adRP patients and by 30% in controls. The DHA (22:6ω3) level in RBCs of controls was elevated by 27% following the 3-week EPA enrichment period. In contrast, the DHA levels were unchanged in adRP patients during EPA supplementation. This marked difference in elevation of RBC-DHA between controls and adRP patients is consistent with a reduced rate of EPA-to-DHA conversion in some adRP patients.

Neither the brief 6-week trial with fish oil supplementation nor the short 3-week intervention with EPA ethyl ester affected retinal function over the entire trial periods (12 weeks with the fish oil supplementation, 9 weeks with the EPA supplementation) as indexed by full-field ERGs.

DISCUSSION

Any consideration of the causes of visual loss in RP must begin with genetic determinants. However, clinical correlations with gene mutation data have proven to be extremely complex because of the numerous mutations within the adRP population and there is no clear association between genotype and phenotype, even within family members. The variability may partially be due to environmental factors, including diet. Our studies are focused on defining possible extra-genetic factors that may impact on the severity of disease. Of particular significance to us is the lipid composition of the membrane environment of proteins and enzymes associated with visual function.

In our present adRP population, the levels of DHA in red blood cells showed marked heterogeneity. Comparison of rod-to-cone ERG amplitude ratios to RBC-DHA levels in our adRP patients (n=81) was statistically significant such that patients with reduced levels of DHA had correspondingly reduced rod function compared to cone function. In patients with the more severe x-linked disease, stronger correlations between cone ERG functions and RBC-DHA were found such that lower DHA values were associated with reduced cone amplitudes and longer b-wave implicit times. This association of electrophysiological responses in retinal tissue with DHA in circulating RBCs provides support for utilization of RBC lipids as an index of neural tissue fatty acid status.

Even though mutations to retinal protein-specific genes are considered a primary cause of RP, the present data together with continued reports of lipid abnormalities suggest an associated defect in processing of fatty acids, particularly ω3 long-chain derivatives. These events may, in turn, influence membrane properties relevant to visual function. Presently, we only have a preliminary comparison of RBC-DHA levels to rhodopsin gene mutations in adRP patients. Of the fifteen patients with rhodopsin mutations, there was a tendency for an association with reduced RBC-DHA levels since the majority of rhodopsin positive patients had sub-normal levels of DHA.

One possible explanation for the reduced level of DHA in blood lipids of RP patients is that the ω3 fatty acid is no longer needed by dysfunctional photoreceptors in the retina (31). Accordingly, in an otherwise healthy person with a degenerating retina due to RP, a molecular signal originating in the retina (a tissue of high DHA demand) might supersede a signal for DHA from the central nervous system (a tissue of similar high DHA demand). Patients with complete bilateral retinal detachment provide a suitable control group to evaluate this possibility since the patients have no retinal demand for DHA yet blood levels of DHA are within normal limits. This observation supports the concept of a unique metabolic defect associated with retinal degeneration in RP patients, and is not consistent with a negative feedback signal generated by the retina in RP patients to control blood DHA levels.

DHA was the most notably effected fatty acid in RBCs of patients with the X-linked form of RP. As an approach to understanding the mechanisms responsible for the reduced blood DHA levels, we have compared the product/precursor ratios of fatty acid elongation reactions. These indices for elongation of saturated fatty acids, monounsaturated fatty acids, ω6 series fatty acids and ω3 series fatty acids were all significantly reduced in the patient population compared to control subjects. An overall decrease in desaturation of membrane fatty acids was evident from the significant reduction in the unsaturation index of RBC lipids from patients with xlRP. Although the desaturation indices for ω3 and ω6 fatty acids of xlRP patients were not altered from controls, there was a significant decrease in ω3-to-ω6 ratios such that a metabolic impairment in ω3 fatty acid metabolism to DHA was evident.

Subsequent studies were designed to evaluate specific steps in metabolism of the ω3 fatty acid series. Our comparison of EPA and DHA uptake and incorporation into RBC lipids by dietary supplementation with a fish oil concentrate demonstrated equivalent handling of ω3 fatty acids by adRP patients and controls. However, when specific conversion of EPA to DHA was examined in adRP patients by oral supplementation of a highly purified EPA preparation, the final step of DPA (22:5ω3) to DHA (22:6ω3) was apparently impaired compared to controls. It is possible that the fatty acid profile in circulating RBCs reflects hepatic impairment since liver is the most likely tissue for a fatty acid biosynthetic defect to occur. Alternatively, the defective regulation of systemic levels of DHA in RP patients may be attributable to a number of metabolic factors including: dysfunctional hepatic conversion of the parent ω3 fatty acid, α-linolenic acid (18:3ω3) to long-chain derivatives (31), impaired packaging of fatty acyl groups into triglycerides or phospholipids for lipoprotein transport (7,31,32), defects in the activities of specific DHA-binding or transport proteins in either the circulation or retinal tissues (33,34), dysfunctional peroxisomal metabolism (35), or impaired fatty acid elongation mechanisms (13,27).

While specific gene mutations represent the primary cause of disease in retinitis pigmentosa, our findings raise the possibility that retinal DHA levels of patients (indexed by RBC fatty acid analysis) may contribute to the heterogeneity in severity of disease found in many families with defined gene defects. Consistent with this concept is the statistical association between ERG functional outcomes and the level of DHA in blood lipids of RP patients, particularly in the more severely affected x-linked RP population. Furthermore, we suggest that the functional activity of defective gene products having a role in transduction of visual stimuli may be potentiated by optimizing the surrounding membrane environment in rod photoreceptors. Normalization of membrane-bound DHA in target tissues such as RBCs maybe a first step to delaying ERG functional loss in patients with early detection of RP symptoms.

ACKNOWLEDGMENTS

These studies have been supported, in part, by the Foundation Fighting Blindness and National Eye Institute grant EY05235. The authors are indebted to Gary Fish, M.D., of Texas Retina Associates, Dallas, TX for ophthalmic examinations, patient reference, and medical consultation; and to Steven Daiger, Ph.D., of the University of Texas Health Science Center at Houston, TX, for gene mutation analysis. We also extend our gratitude to the patients who have participated in these investigations.

REFERENCES

1. Fliesler, S.J., and Anderson, R.E., 1983, Chemistry and metabolism of lipids in the vertebrate retina, *Prog. Lipid Res.* 22:79-131.
2. Neuringer, M., Connor, W., Lin, D., Barsted, L. and Luck, S., 1986, Biochemical and functional effects of prenatal and postnatal ω-3 fatty acid deficiency on retina and brain in rhesus monkeys, *Proc. Natl. Acad. Sci. U.S.A.* 83: 4021-4025.
3. Bourre, J.M., Francois, M., Youyou, A., Dumont, O., Piciotti, M., Pascal, G. and Durand, G., 1989, The effects of dietary α-linolenic acid on the composition of nerve membranes, enzymatic activity, amplitude of electrophysiological parameters, resistance to poisons and performance of learning tasks in rats, *J. Nutr.* 119: 1880-1892.
4. Wheeler, T.G., Benolken, R.M. and Anderson, R.E., 1975, Visual membranes: specificity of fatty acid precursors for the electrical response to illumination, *Science* 188: 1312-1314.
5. Uauy, R.D., Birch, D.G., Birch, E.E., Tyson, J.E. and Hoffman, D.R., 1990, Effect of dietary omega-3 fatty acids on retinal function of very-low-birth-weight neonates, *Pediatr. Res.*, 28: 485-492.
6. Birch, D.G., Birch, E.E., Hoffman, D.R. and Uauy, R.D., 1992, Retinal development in very-low-birth-weight infants fed diets differing in omega-3 fatty acids, *Invest. Ophthal. Vis. Sci.* 33: 2365-2376.
7. Converse, C.A., Hammer, H.M., Packard, C.J. and Shepherd, J., 1983, Plasma lipid abnormalities in retinitis pigmentosa and related conditions, *Trans. Ophthalmol. Soc. U.K.*, 103: 508-512.
8. McLachlan, T., McColl, A.J., Collins, M.F., Converse, C.A., Packard, C.J. and Shepherd, J., 1990, A longitudinal study of plasma ω-3 fatty acid levels in a family with X-linked retinitis pigmentosa. *Biochem. Soc. Trans.*, 18: 905-906.
9. Anderson, R.E., Maude, M.B., Lewis, R.A., Newsome, D.A. and Fishman, G.A., 1987, Abnormal plasma levels of polyunsaturated fatty acid in autosomal dominant retinitis pigmentosa, *Exp. Eye Res.*, 44: 155-158.
10. Holman, R.T., Bibus, D.M., Jeffery, G.H., Smethurst, P. and Crofts, J.W., 1994, Abnormal plasma lipids of patients with retinitis pigmentosa, *Lipids*, 29: 61-65.
11. Newsome, D.A., Anderson, R.E., May, J.G., McKay, T.A. and Maude, M., 1988, Clinical and serum lipid findings in a large family with autosomal dominant retinitis pigmentosa, *Ophthalmol.* 98: 1691-1695.
12. Gong, J., Rosner, B., Rees, D.G., Berson, E.L., Weigel-DiFranco, C.A. and Schaefer, E.J., 1992, Plasma docosahexaenoic acid levels in various genetic forms of retinitis pigmentosa, *Invest. Ophthal. Vis. Sci.* 33: 2596-2602.
13. Hoffman, D.R., Uauy, R. and Birch, D.G., 1993, Red blood cell fatty acid levels in patients with autosomal dominant retinitis pigmentosa, *Exp. Eye Res.* 57: 359-368.
14. Bazan, N.G., Scott, B.L., Reddy, T.S. and Pelias, M.Z., 1986, Decreased content of docosahexaenoate and arachidonate in plasma phospholipids in Usher's syndrome, *Biochem. Biophys. Res. Commun.* 141: 600-604.
15. Hoffman, D.R. and Birch, D.G., Docosahexaenoic acid in red blood cells of patients with x-linked retinitis pigmentosa, *Invest. Ophthalmol. Vis. Sci.*, in press, 1995.
16. Carlson, S.E., Carver, J.D. and House, S.G., 1986, High fat diets varying in ratios of polyunsaturated to saturated fatty acid and linoleic to linolenic acid: a comparison of rat neural and red cell membrane phospholipids, *J. Nutr.* 116: 718-726.
17. Connor, W.E., Lin, D.S. and Neuringer, M., 1993, Is the docosahexaenoic acid (DHA, 22:6n-3) content of erythrocytes a marker for the DHA content of brain phospholipids? *FASEB. J.* 7: A152.
18. Dryja, T., Hahn, L.B., Cowley, G.S., McGee, T.L. and Berson, E.L., 1991, Mutation spectrum of the rhodopsin gene among patients with autosomal dominant retinitis pigmentosa, *Proc. Natl. Acad. Sci. U.S.A.* 88: 9370-9374.

19. Sung, C-H., Davenport, C.M., Hennessey, H.C. et al., 1991, Rhodopsin mutations in autosomal dominant retinitis pigmentosa, *Proc. Natl. Acad. Sci. U.S.A.* 88: 6481-6485.
20. Humphries, P., Kenna, P. and Farrar, G.J., 1992, On the molecular genetics of retinitis pigmentosa, *Science* 256: 804-808.
21. Daiger, S., Sadler, L.A. and Rodriguez, J.A., Why do mutations in photoreceptor-specific proteins lead to retinal degenerations? in *Controversies in Neurosciences III: Signal Transduction in the Retina and Brain*. Polans, A., Series Editor, Cambridge Univ. Press, in press.
22. Kajiwara, K., Berson, E.L. and Dryja, T.P., 1994, Digenic retinitis pigmentosa due to mutations at the unlinked peripherin/RDS and ROM1 loci, *Science* 264: 1604-1608.
23. McLaughlin, M.E., Sandberg, M.A., Berson, E.L. and Dryja, T.P., 1993, Recessive mutations in the gene encoding the β-subunit of rod phosphodiesterase in patients with retinitis pigmentosa, *Nature Genetics* 4: 130-134.
24. Ott, J., Bhattacharya, S., Chen, J.D. et al.,1990, Localizing multiple X chromosome-linked retinitis pigmentosa loci using multilocus homogeneity tests, *Proc. Natl. Acad. Sci. U.S.A.* 87: 701-704.
25. Berson, E.L., Rosen, J.B. and Simonoff, E.A., 1980, Electroretinographic testing as an aid in detection of carriers of X-chromosome-linked retinitis pigmentosa, *Am. J. Ophthalmol.* 87: 460-468.
26. Bligh, E.G. and Dyer, W.J., 1959, A rapid method of total lipid extraction and purification, *Can. J. Biochem. Physiol.* 37: 911-917.
27. Hoffman, D.R., Uauy, R. and Birch, D.G., Metabolism of omega-3 fatty acids in patients with autosomal dominant retinitis pigmentosa, *Exp. Eye Res.*,in press, 1995.
28. Marmor, M.F., Arden, G.B., Nilsson, S.E.G. and Zrenner, E., 1989, Standards for clinical electroretinography, *Arch. Ophthalmol.* 107: 816-819.
29. Birch, D.G. and Fish, G.E., 1987, Rod ERGs in retinitis pigmentosa and cone-rod degeneration, *Invest. Ophthalmol. Vis Sci.* 28: 140-150.
30. Musarella, M.A., Anson-Cartwright, C.L., McDowell, C., Burghes, A.H.M., Coulson, S.E., Worton, R.G. and Rommens, J.M., 1991, Physical mapping at a potential X-linked retinitis pigmentosa locus (RP3) by pulsed-field gel electrophoresis, *Genomics* 11: 263-272.
31. Scott, B.L. and Bazan, N.G., 1989, Membrane docosahexaenoate is supplied to the developing brain and retina by the liver, *Proc. Natl. Acad. Sci. U.S.A.* 86: 2903-2907.
32. Li, J., Wetzel, M.G. and O'Brien, P.J., 1992, Transport of n-3 fatty acids from the intestine to the retina, *J. Lipid Res.* 33: 539-548.
33. Li, J., Gentleman, S., Jiao, X., Wetzel, M.G., O'Brien, P.J. and Chader, G.J., 1993, Uptake of docosahexaenoic acid (DHA) by a fatty acid-specific "receptor" in the retinal plasma membrane, *Invest. Ophthal. Vis. Sci.* 34: 1329.
34. Jiao, X., Lee, J., Rodriguez DeTurco, E.B., Bazan, N.G. and Chader, G.J., 1994, Tissue distribution of docosahexaenoic acid binding proteins in poodles with progressive rod-cone degeneration (PRCD), *Invest. Ophthal. Vis. Sci.* 35: 1611.
35. Roels, F., Fischer, S. and Kissling, W., 1993, Polyunsaturated fatty acids in peroxisomal disorders: a hypothesis and a proposal for treatment, *J. Neurol. & Psychiatr.* 56: 937.

44

LIGAND-BINDING PROPERTIES OF RECOMBINANT HUMAN IRBP

J. M. Nickerson, V. Mody, C. DeGuzman, K. A. Heron, K. Marciante, J. Boatright, J. S. Si, and Z. Y. Lin

Ophthalmology Department
Emory University
Atlanta, Georgia

INTRODUCTION

Interphotoreceptor retinoid binding protein (IRBP) is a large glycolipoprotein synthesized by photoreceptor cells of the retina and secreted into interphotoreceptor matrix (IPM) (1). It binds and transports retinoids and fatty acids in the IPM as illustrated in Figure 1.

Most studies of IRBP have employed bovine IRBP because of easy purification and accessibility of tissue. We have recently expressed the human cDNA for IRBP in insect cells using a recombinant baculovirus (7). Significant quantities of human recombinant IRBP can be obtained in this system. The goal of this study is to analyze the binding capabilities of the human protein and to begin to make and characterize mutants of IRBP. To understand the relationship between the protein structure and its functions, we need substantial quantities of recombinant human IRBP (rhIRBP). Here we report some binding properties of rhIRBP with ligands and their analogs that are likely to be physiologically important in the retina. We also report the preliminary characterization of altered rhIRBP DNA constructs that we will use to examine the relationships among the four repeats of the protein structure and putative ligand binding sites.

METHODS

RhIRBP was produced by cloning and expression in the baculovirus system as described (7). The essential steps in the cloning and expression via the baculovirus system are shown in Figure 2.

The analysis of binding properties employed spectrofluorometry. To obtain binding curves, a small aliquot, usually 0.5 to 1 microliter, of a ligand was added to a fixed amount of rhIRBP, and a measurement was made following a 100 sec equilibration. This was repeated 20 to 40 times until about a 6-8 fold excess of the ligand over the anticipated Kd was

Degenerative Diseases of the Retina, Edited by Robert E. Anderson et al.
Plenum Press, New York, 1995

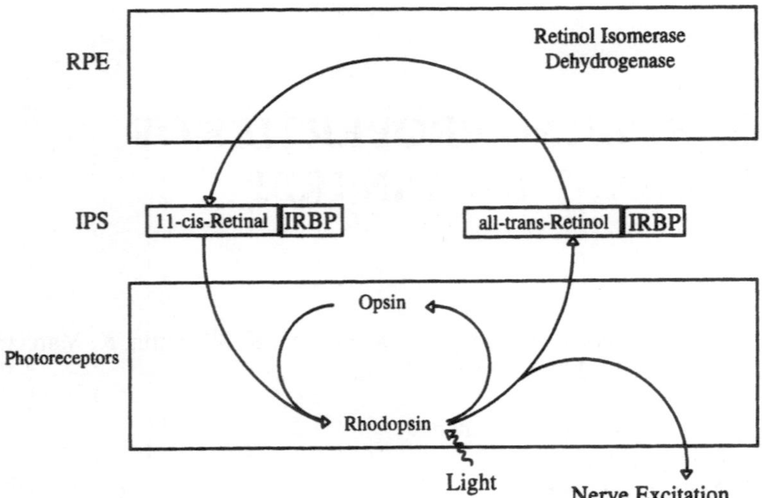

Figure 1. The Visual cycle. IRBP aids in the uptake of 11-cis-retinal into outer segments and in the uptake of all-trans-retinol into the retinal pigment epithelium (2-4). These two functions are presumed to be essential for vision in mammals as no other retinoid binding proteins are found in the IPM. Although vitamin A and its analogs are soluble in aqueous solutions, it is not clear that retinoids are sufficiently stable (5) or can pass between these two cells without the presence of IRBP. Second, IRBP possesses a buffering capacity (6), which may be needed during light bleaching or dark adaptation.

achieved. The ligand was dissolved at 40-500 micromolar concentration in ethanol. No more than 2% ethanol accumulated in the cuvette. The experiments were carried out at ambient room temperature. Plots of raw fluorescence versus total ligand concentration were converted to plots of free ligand versus bound fluorescence as described by Cogan et al. (10). To obtain the dissociation constant, we used a curvefitting algorithm, fitting the data to a simple hyperbola. Various retinoids were obtained from Sigma as indicated below. Fluorescent analogs of fatty acids were obtained from Molecular Probes Inc.

RESULTS

Previous studies established that rhIRBP is structurally and functionally equivalent to authentic human donor-eye derived IRBP. These conclusions were reached by testing several attributes of two protein preparations. The attributes included size, antigenic epitopes, charge, Stoke's radius, presence of mannose containing sugars, and amino acid sequence at the N-terminus. By all assays, the two preparations of protein are virtually equivalent (7). Thus, we believe that it is sound to use rhIRBP to study the properties of human IRBP. Only one minor difference between the two protein preparations was noticed. RhIRBP contains a five amino acid propeptide at the N-terminus as predicted from the cDNA sequence. Authentic human IRBP is a mixture of polypeptides with half containing the propeptide and the remainder lacking this N-terminal extension (11). A possible interpretation of this result is that donor-eye derived IRBP has been partially matured or degraded in the IPM, which may contain a specific protease. We do not know if this has any relation to physiologic use. Thus, it is possible that either the RPE or photoreceptor membrane may contain the specific protease as part of a speculative docking receptor for IRBP.

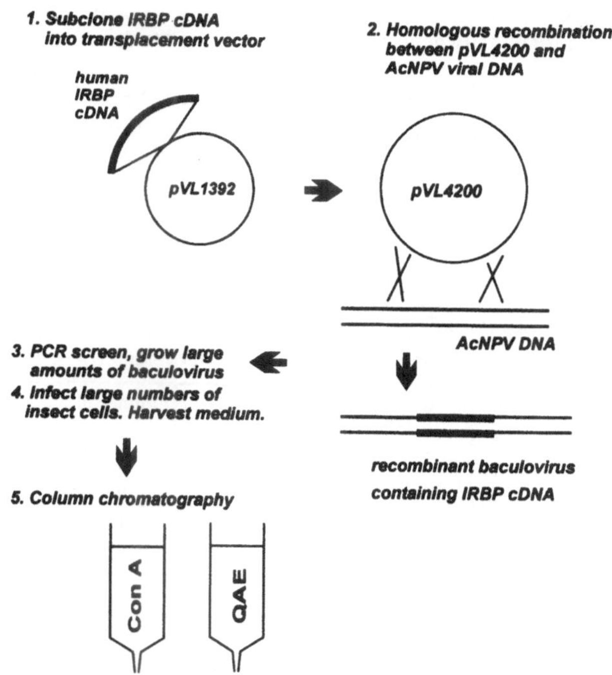

Figure 2. RhIRBP Production Scheme. 1. The human IRBP cDNA (8) was cloned into the expression transplacement vector pVL1392 (9). 2. This plasmid, pVL4200, was co-transfected into Sf9 cells with linearized AcNPV viral DNA (InVitrogen) or with linearized AcNPV missing an essential restriction fragment (Baculogold viral DNA from Pharmingen). 3. Once viral plaques formed, they were picked, small viral stocks prepared, and the resulting viral DNAs were screened by PCR analysis for the entire human IRBP cDNA. Large viral stocks were prepared from serially plaque-purified positives. 4. Large numbers of High Five cells (InVitrogen), a production cell line, were infected with the virus and medium harvested at an optimal time. The medium contained minimal amounts of fetal calf serum to simplify rhIRBP purification. 5. RhIRBP was purified from insect cell medium by Con A and ion-exchange chromatography.

LIGAND BINDING PROPERTIES

All-trans-retinol binding to rhIRBP was previously characterized by fluorometric titration. The dissociation constant is about 1 micromolar, close to that of bovine IRBP. By scanning fluorometry, the emission and excitation maxima are at 479 nm and 339 nm, respectively (7). New and more extensive analyses of ligand binding properties of rhIRBP have been carried out and the results are summarized in Table I.

Two sample fluorescence titrations are shown in Figures 3 and 4.

In Figure 3, all-trans-retinal was added to rhIRBP. About a 4-5 fold enhancement of fluorescence is observed when the ligand is bound to the protein. This is less than the roughly 10-15 fold enhancement observed when all-trans-retinol binds to rhIRBP (7). All-trans-retinal (Sigma, R-2500) is the only aldehyde with which we could detect fluorescence enhancement. We did not detect fluorescence with 9-cis-retinal (Sigma, R-5754) or 13-cis-retinal (Sigma, R-6256). The Kd of 160 nM is substantially lower than that of all-trans-retinol. Together with the relative blue shift of the emission maximum this may suggest a different binding site or more hydrophobic amino acid side chains with which the all-trans-retinal interacts.

Table I. Fluorescence spectroscopic analyses of ligands bound to rhIRBP. Several forms of retinoids and fatty acid analogs were examined to measure dissociation constants and excitation and emission maxima. These latter maxima were determined with the ligand and 1 micromolar IRBP. Also, the Kd was measured in the presence of 1 micromolar rhIRBP, which was greater than 95% pure. 12-AS refers to 12-(9-anthryloxy)-stearate, 16-AP refers to 16-(9-anthryloxy)-palmitate, 11-AU is 11-(9-anthryloxy)-undecanoate, 9-AA is 9-anthracenepropionate, 12-AD is 12-(9-anthryloxy)-dodecanoate, and 9-cis-Ral is 9-cis-retinal. "No binding" means that there is either no affinity of the ligand with rhIRBP or no fluorescence enhancement upon the binding of the ligand to the protein. "N" is the number of times the individual titration experiment was carried out. "Quinine sulfate" refers to the concentration of quinine sulfate used to define 100% on the relative fluorescence scale of the fluorometer

Ligand	Excitation max	Emission max	N	Quinine sulfate	Kd
all-trans-retinol	339 nm	479 nm	3	166 ng/ml	1.0 μM
all-trans-retinal	348 nm	460 nm	2	50	0.2 μM
13-cis-retinal	364 nm	418 nm	2	84	ND
9-cis-retinal	361 nm	445 nm	2	16	no binding
11-cis-retinal	ND	ND	0	?	ND
12-AS (stearic)	363 nm	437 nm	2	35	0.15 μM
16-AP (palmitic)	364 nm	449 nm	7	49	0.07 μM
12-AD (C12)	364 nm	447 nm	2	166	0.07 μM
11-AU (C11)	366 nm	443 nm	2	50	0.06 μM
9-AA (C2)	368 nm	415 nm	2	100000	no binding

ND = Not determined yet.

We have examined binding of some fatty acid analogs with rhIRBP also. We tested fatty acids that had been modified with a fluorescent moiety containing the anthracene ring structure. These derivatives have been used successfully previously with bovine IRBP and other fatty acid binding proteins. While it is possible that the fluorescent moiety affects ligand binding to IRBP, we first checked to see if 9-anthracenepropionic acid (9-AA) binds to rhIRBP. Were this simple model compound to bind, it would invalidate the selection of anthryloxy-series fatty acid analogs as ligands to study fatty acid binding to IRBP. As shown in Table I, 9-AA does not bind to rhIRBP, justifying the use of the anthryloxy-fatty acid series. In Figure 4, we show the fluorescence titration of the ligand 16-AP binding to rhIRBP. The Kd is about 70 nM.

Table I shows the results of the fatty acid binding experiments. Four compounds from 11 to 18 carbons in length bind to rhIRBP. The affinity of all four is high, with Kd's ranging from 60 to 150 nM. The 11, 12, and 16 carbon fatty acids have similar Kds and the stearate analog is slightly higher at 150 nM. These values are lower than those obtained by Putalina et al. (12) who found Kds for 16-AP and 9-AS of 360 nM with the bovine protein.

STUDIES IN PROGRESS TO IDENTIFY BINDING SITE LOCATIONS IN RhIRBP

Besides the normal wild type rhIRBP, we are constructing altered forms of IRBP by recombinant DNA techniques. The first series of DNA constructs includes deletions of entire repeats. With these mutated forms of IRBP we can assess the fundamental structure-function question: Given the four-repeat structure of the amino acid sequence, can each repeat bind

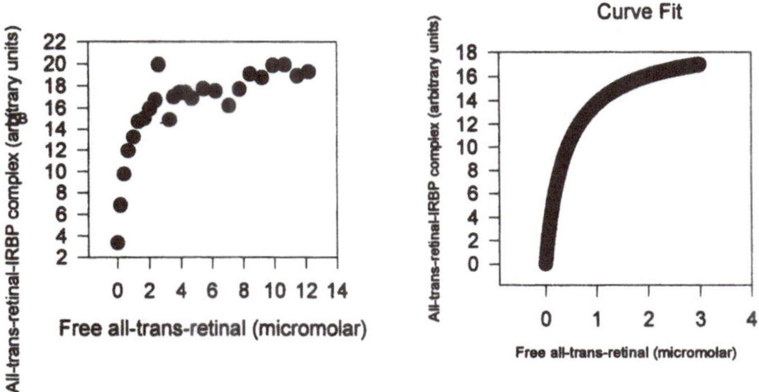

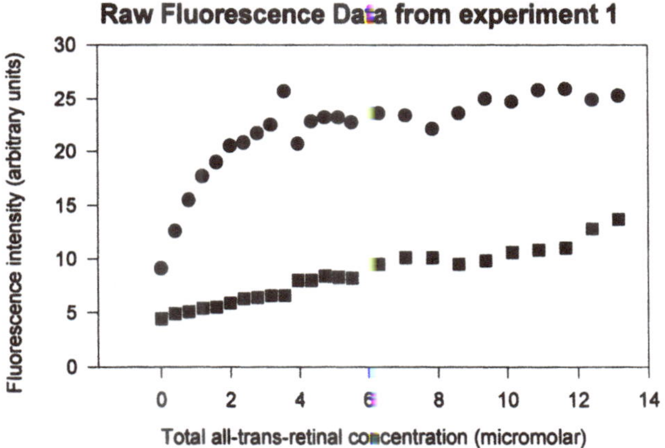

Figure 3. Binding Analysis of all-trans-retinal and rhIRBP. Raw fluorescence is plotted against the total concentration of all-trans-retinal added to the cuvettes in the lower panel. The sample cuvette containing rhIRBP is shown by the circles and without rhIRBP is shown by the squares. Using the Cogan method (10), we estimated the concentration of free retinal versus bound retinal, as shown in the upper left panel. We used the Marquardt-Levenberg algorithm (in Sigmaplot) to fit a simple hyperbola to the estimated bound versus free data. The fitted hyperbola is shown on the upper right panel.

a different ligand? Similarly, we may be able to resolve the question of the number of all-trans-retinol binding sites (13, 14). We have constructed in the pVL1392 transplacement vector several deletion alterations, which are described in Table II.

Figure 5. shows a diagrammatic representation of the various deletion proteins.

Although we have expressed the protein in several mutants, we have not yet purified enough of the intact proteins to carry out ligand binding assays. Similar approaches employing bacterial expression of bovine and Xenopus IRBP deletions are in progress (15, 16).

SUMMARY AND CONCLUSIONS

Because of the virtual identity of rhIRBP and human donor-eye derived IRBP it is valid to assess the ligand binding properties of human IRBP by using the recombinant protein

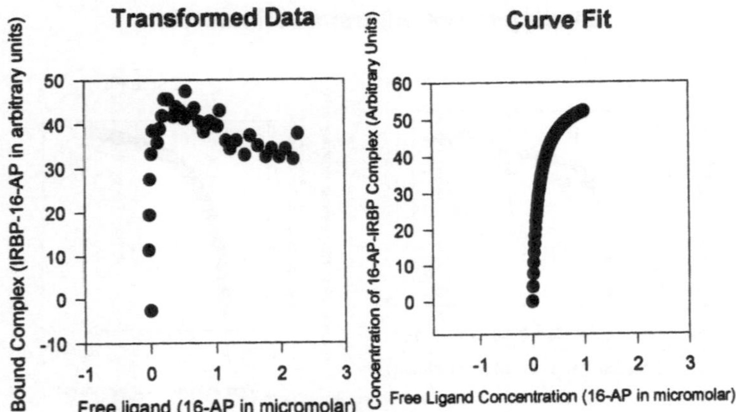

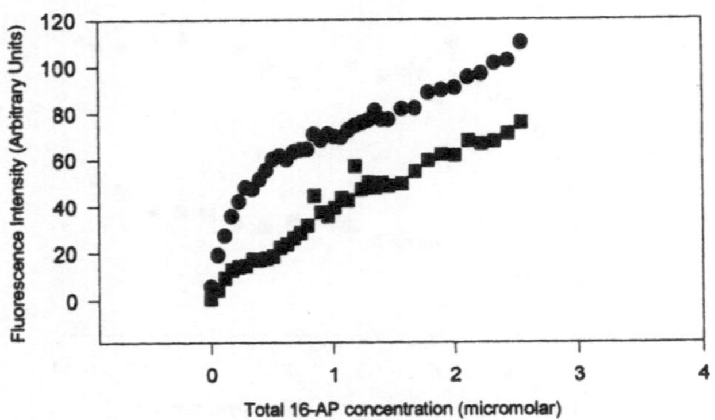

Figure 4. Fluorescence titration of a fatty acid analog: The Binding of 16-AP and rhIRBP. Conditions of this experiment are very similar to those described in Figure 3. The twofold fluorescence enhancement is not as great as with the retinoids.

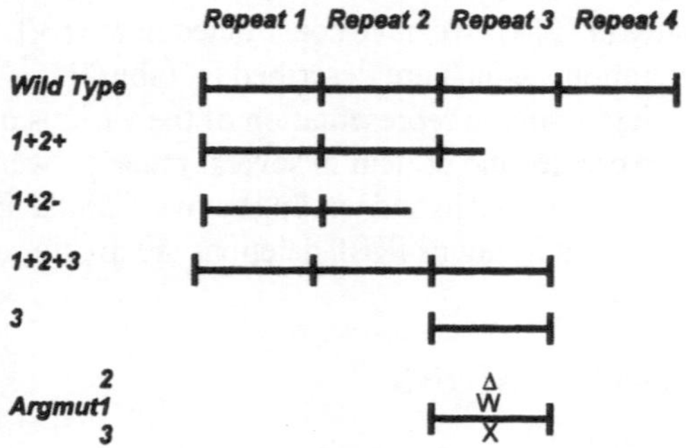

Figure 5. Mutant Forms of Human IRBP.

Table II. Status of mutant forms of human IRBP

Mutant Name*	Transfer plasmid made?	Recomb. baculovir stock made?	IRBP found? Secreted? Internal?	Protein purified? Intact? Degrad?	Used in ligand binding expts?
wild type	yes	yes	secreted	intact	yes
1+2+	yes	yes	secreted	intact + contam's	
1+2-	yes	yes	secreted	intact + contam's	
3	yes	yes	?		
1+2+3	yes	yes	secreted	intact	
Arginine mutants	yes, verifying sequence changes	no			

*Lengths of the deletion mutants and positions of the point amino acid changes, see Figure 5.

that is available in larger quantities and in purified form. The baculovirus system provides an excellent method to produce and secrete human IRBP. No renaturation steps are involved, eliminating possible complications in later analyses of the samples. The protein can be readily purified much as IRBP from eyes. Its behavior on chromatography and by binding studies suggest that the protein is very similar to the authentic protein. Small, but readily detectable differences in behavior compared to bovine material highlight the need to study the human protein.

Noy and co-workers (17) have reported about a tenfold lower Kd for bovine IRBP and all-trans-retinol. Although we report similar values for the human and bovine IRBP proteins with all-trans-retinol at 1.0 micromolar, this could be artifactually high. As many other investigators have done, we have stored our protein preparations frozen at -80 C. Noy suggests that the difference between her studies and others is that she stores her protein in 50% glycerol, never freezing the protein before ligand binding experiments.

We conclude that the ligand binding properties of rhIRBP are similar but not identical to those of bovine IRBP. We are surprised that all-trans-retinal binds more tightly than all-trans-retinol. Also, the fatty acids bind with even higher affinity. These experiments suggest different roles for IRBP in transporting these ligands, and that probably there are different structures in IRBP that bind different ligands, possibly up to four binding sites in this multiple-repeat protein. Last, alteration of the protein, especially deletion of repeats within the protein and substitutions of highly conserved amino acids, should indicate the locations of binding sites for each ligand.

Supported by:

RPB, O. K. Weiss Scholar, Fight For Sight GA92017, NIH R01 EY10553 and P30 EY 06360.

REFERENCES

1. Bunt-Milam AH and Saari JC: Immunocytochemical localization of two retinoid-binding proteins in vertebrate retina. J Cell Biol 1983;97:703-712.
2. Okajima T-IL, Pepperberg DR, Ripps H, Wiggert B and Chader GJ: Interphotoreceptor retinoid-binding protein: role in the delivery of'retinol to the pigment epithelium. Exp Eye Res 1989;49:629-644.
3. Jones GJ, Crouch RK, Wiggert B, Cornwall MC and Chader GJ: Retinoid requirements for recovery of sensitivity after visual pigment bleaching in isolated photoreceptors. Proc Nat'l Acad Sci (USA) 1989;86:9606-9610.

4. Okajima TI, Pepperberg DR, Ripps H, Wiggert B and Chader GJ: Interphotoreceptor retinoid-binding protein promotes rhodopsin regeneration in toad photoreceptors. Proc Nat'l Acad Sci (USA) 1990;87:6907-6911.
5. Crouch RK, Hazard ES, Lind T, Wiggert B, Chader G and Corson DW: Interphotoreceptor retinoid binding protein and alpha-tocopherol preserve the isomeric and oxidation state of retinol. Photochem Photobiol 1992;56:251-255.
6. Ho M-TP, Massey JB, Pownall HJ, Anderson RE and Hollyfield JG: Mechanism of vitamin A movement between rod outer segments, interphotoreceptor retinoid- binding protein, and liposomes. J Biol Chem 1989;264:928-935.
7. Lin ZY, Si JS and Nickerson JM: Biochemical and biophysical properties of recombinant human interphotoreceptor retinoid binding protein. Invest Ophthalmol Vis Sci 1994;35:3599-3612.
8. Si J-S, Borst DE, Redmond TM and Nickerson JM: Cloning of cDNAs encoding human interphotoreceptor retinoid-binding protein (IRBP) and comparison with bovine IRBP sequences. Gene 1989;80:99-108.
9. Vialard J, Lalumiere M, Vernet T, Briedis D, Alkhatib G, Henning D, Levin D and Richardson C: Synthesis of the membrane fusion and hemagglutinin proteins of measles virus, using a novel baculovirus vector containing the beta-galactosidase gene. Jour Virol 1990;64:37-50.
10. Cogan U, Kopelman M, Mokady S and Shinitzky M: Binding affinities of retinol and related compounds to retinol binding proteins. Eur J Biochem 1976;65:71-78.
11 Redmond TM, Wiggert B, Robey FA and Chader GJ: Interspecies conservation of structure of interphotoreceptor retinoid-binding protein. Similarities and differences as adjudged by peptide mapping and N-terminal sequencing [Erratum Biochem J 242 (1987) 935]. Biochem J 1986;240:19-26.
12 Putilina T, Sittenfeld D, Chader GJ and Wiggert B: Study of a fatty acid binding site of interphotoreceptor retinoid-binding protein using fluorescent fatty acids. Biochemistry 1993;32:3797-3803.
13 Adler AJ, Evans CD and Stafford WF, 3d: Molecular properties of bovine interphotoreceptor retinol-binding protein. J Biol Chem 1985;260:4850-4855.
14 Saari JC, Bunt-Milam AH, Bredberg DL and Garwin GG: Properties and immunocytochemical localization of three retinoid-bi proteins from bovine retina. Vision Res 1984;24:1595-603.
15 Redmond TM and Nickerson JM: Retinoid binding to a recombinant central two repeat segment of IRBP. Invest Ophthalmol Vis Sci 1992;33:1181.
16 Baer CA, Kittridge KL, Klinger AL, Briercheck DM, Braiman MS and Gonzalez-Fernandez F: Expression and characterization of the fourth repeat of Xenopus interphotoreceptor retinoid binding protein in E. coli. Curr Eye Res 1994;13:391-400.
17 Chen Y, Saari J and Noy N: Interactions of all-trans-retinol and long chain fatty acids with interphotoreceptor retinoid binding protein. Biochemistry 1993;32:11311-11318.

CATARACTOGENESIS IN RETINITIS PIGMENTOSA

Spectroscopic Fluorescence Analysis of Aqueous Humor Composition

Enzo M. Vingolo,[1] Andrea Bellelli,[2] Monica Santori,[1] Luigi Pannarale,[1] Renato Forte,[1] Alessandro Iannaccone,[1] and Roberto Grenga[1]

[1] Department of Electrophysiology
Center for Inherited Degenerative Retinal Disorders
[2] Institute of Biochemistry
University "La Sapienza"
V.G.Dandini 5 00154 Rome , Italy

INTRODUCTION

Retinitis Pigmentosa (RP) is a group of inherited progressive degenerative disorders of the retina genetically transmitted by recessive X-linked, recessive autosomal, dominant autosomal or digenic inheritance patterns.

These disorders are frequently associated to breadcrumb posterior subcapsular cataract. Similar lens changes are seen not only in RP, but also in other retinal diseases as in high myopia, retinal detachment, chorioretinitis, diabetic retinopathy and posterior segment inflammatory alterations (uveitis).

In experimental retinal degeneration models, such as in RCS mice in which there is a primary retinal degeneration very similar to human tapetoretinal degeneration (RP), a cataract is also present. Moreover, in 1991 Yonemura also reported in these mice corneal endothelial cell alterations, in our studies we pointed out similar corneal qualitative alterations in eyes of patients affected by Retinitis pigmentosa.

The pathogenesis of the endothelial cell changes and lens opacities, occurring in retinal degenerative disorders, are probably due to the presence of toxic products originating from the lipid metabolism of the photoreceptor membrane. In fact Simonelli and coll. (1989) also demonstrated higher concentration of malondihaldeyde (MDA), a toxic derivate of docosohexanoic acid (DHA), in RP lenses than in normal senile cataracts.

The preferential occurrence of opacities on the lens posterior layers (posterior subcapsular cataract) seems to be due to the fact that, in this region, the posterior capsule is thinner. Therefore, a toxic metabolite might derange the structure and orientation of lens fibers in the superficial layers, corresponding to the sutural connections at the level of the

Degenerative Diseases of the Retina, Edited by Robert E. Anderson et al.
Plenum Press, New York, 1995

posterior capsule, these alterations were noted in a histological analysis of cataractous R and Myopic eyes (Feher 1991). A similar pathogenesis could be claimed for the corne: endothelial cell alterations. In fact, the toxic derivates brought by the aqueous humour int the anterior chamber might damage the surface and structure of the endothelial corneal laye

Since MDA is supposed to be the involved toxic metabolite, and this molecule binc amino groups of proteins to yield a fluorescent chromophore, it is possible to measure i ocular concentration with a spectrofluorometric determination of aqueous humor sample This procedure was applied by several authors to study the composition of retinal photor ceptor membrane toxic derivatives in retinal degeneration.

In the present study, we determined the concentration of MDA chromophores in th aqueous humour of 20 cataractous RP patients, comparing the results to 20 normal age-, se and refraction-matched subjects affected with senile or pre-senile cataracts, referring to th spectroscopic studies of Goosey (1984).

MATERIALS, METHODS

All RP patients were studied according the national (F.I.A.R.P.) clinical and geneti protocol of investigation, as reported in previous studies. For the specific evaluation of ler opacities, each patient underwent clinical examination, Neitz retroillumination photograp to document the extension of cataract, and Scheimpflug photograph to evaluate the layei involved by the opacities.

Our research was performed on 40 eyes divided in two groups, age and sex matche as follows:

1. RP Group: 20 eyes, average age of 51 ± 4,38 range (40-76):
2. Control Group: 20 eyes, average age of 58 ± 7,17 range (48-79).

Surgical Procedure

ExtraCapsular Cataract Extraction (ECCE) was performed in 18 eyes of the RP grou and in all 20 of the groups with senile cataract, while in 2 eyes of RP we performed a IntraCapsular Cataract Extraction (ICCE). Aqueous samples of 300-500 μl were draw during cataract surgery before the opening of the anterior chamber. The extracted nuclei wei immediately stored in buffered Ringer Lactate solution for further evaluation.

Laboratory Procedure

The spectroscopic analysis of the aqueous samples was performed using a Jasc FP770 spectrofluorometer in a double blind designed experiment, testing 150 μl of aqueou humour diluted in 3 ml of 0.1% sodium phosphate buffered at pH 7,0. The emission spectrur was recorded with the addition of a photomultiplier as follows:

I. wavelength range: 380-480 nm (excitation wavelength: 355 nm).
II. wavelength range: 300-400 nm (excitation wavelength: 280 nm).

In both cases the emission peaks were broad and allowed us to use a slit width of 1 nm. The nuclei were homogeneized and diluted according the technique of Gossey and th MDA was extracted with Sodium Tiopenthal.

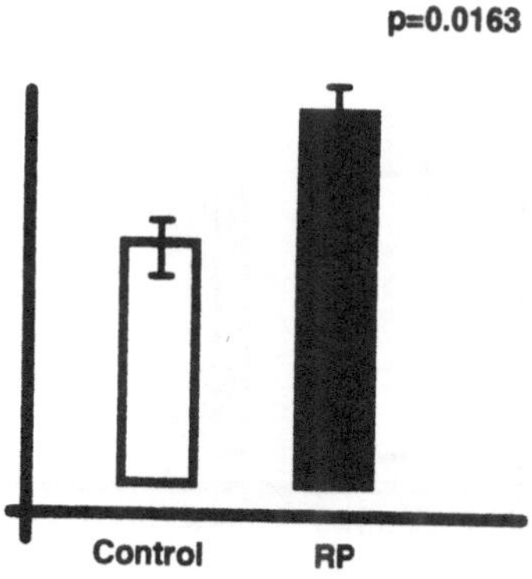

p=0.0163

Control RP

Figure 1. F310 Values. The f340 peak was 332±11.7 mu.l. in RP, while 212±11.0 mu.l. in the control group (Fig. n°2).

RESULTS

In the spectra recorded using the 355-nm excitation wavelength a marked emission peak centered at 407 nm was observed. This peak was also present in the buffer solution, and was attributed to the raman's emission of water. A broad, featureless and faint band of emission in the region of 430-460 nm was also recorded. This band was higher in samples than in the reference solution (buffer only).

The analysis of the excitation spectrum monitoring the emission at 445 nm showed two peaks, corresponding to excitation at 380 and 280 nm with relative amplitude of 1:4. These peaks have been attributed to the water's raman effect and to an aromatic chromophore, respectively.

All samples were therefore analyzed for fluorescence emission using the 280 nm excitation wavelength. In this instance, two partially superimposed peaks were detected, centered at 310 (f310) and 340 nm (f340).

These peaks were attributed to aromatic aminoacids, most probably tryptophane or tyrosine, either free or as components of proteins. The two fluorescent chromophores were detected in RP patients as well as in control group. No MDA fluorescence was observed in both groups. Concentration of f310 was in RP 266±80 mu.l., while in the control group 168±76 mu.l. (Fig. n°1).

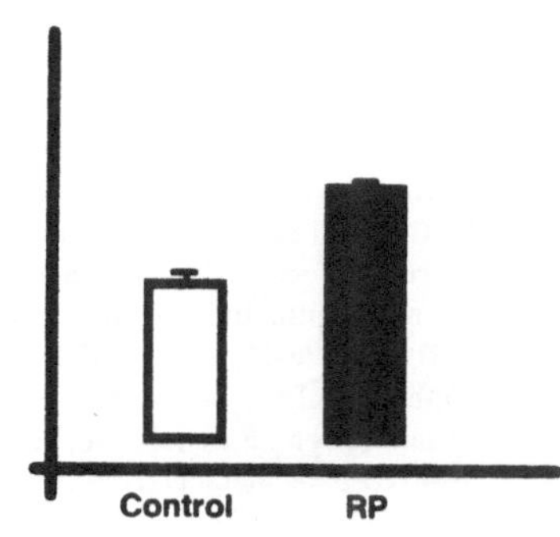

Figure 2. F340 Values. The f310/f340 ratio was 827±126 for RP eyes, and 821±92 for control subjects (Fig. n°3). Statistical analyses demonstrated significantly higher proportions of f310 (tyrosine) and f340 (tryptophane) peaks in RP eyes than in control eyes (f310: p=0.0163; f340: p=0.0181), whereas the 310/340 ratio was not significantly different between the two groups (p=0.47).

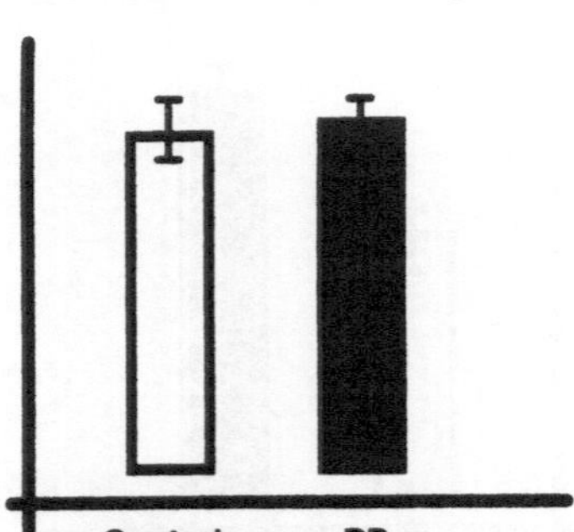

Figure 3. F310/f340 ratio values. Analyses performed on the nuclei extracted from RP patients showed no detectable levels of MDA.

DISCUSSION

Results showed that neither the aqueous humour nor nuclei from cataractous RP patients contain either MDA or its fluorescent components. However, since these substances may accumulate in the lens and in other ocular tissues, it is possible that its concentration in the aqueous humour was too small to be detected. Therefore, it cannot be excluded as a possible role of MDA in the pathogenesis of cataract and of the endothelial cell lesions.

The possible role of MDA in the pathogenesis of cataracts is supported by studies performed on animal models or cell cultures, that do not reproduce the pathophysiology of the eye in RP patients. In the few available studies on human RP patients in which MDA has been detected, its retinal origin has never been demonstrated. In conclusion, our study has been unable to demonstrate significant concentrations of MDA in the aqueous of RP patients, whereas high concentrations of aromatic aminoacids (tyrosine and tryptophane) were detected in every instance. Their higher concentration in the aqueous humour of RP patients might be due to equally probable, different causes (possibly coexisting):

1. To the leakage of the blood-retinal barrier, that is known to be significantly altered in RP patients, as demonstrated by fluoroangiographic and fluorophotometric studies, and by the identification of lymphocytes and macrophages in the vitreous body as responsible for vitreous particulation;
2. To the altered active transport of aromatic aminoacids, that could be due to an alteration in the melanin metabolism in RPE cells.
3. Increased levels of aqueous aromatic aminoacids could increase aqueous osmolarity, with damage of the lens surface and fibers. This would be more likely to happen at the level of the posterior lens capsule, because of the expected retinal origin of these substances, and particularly in its thinner portion, i.e., the axial area.
4. The absence of MDA fluorescence in the aqueous does not rule out the potential role of peroxidation derivates in RP cataractogenesis.

REFERENCES

Bhuyan K.C.: Molecular mechanism of cataractogenesis. Toxic metabolites of oxygen as initiators of lipid peroxidation and cataract. Curr. Eye Res. 3: 67-81, 1984

Bhuyan K. C. e coll.: Lipid peroxidation in cataract of the human life. Sci. 38: 1463-71, 1986.

Chen H. e coll.: Docosahexaenoic acid containing phospholipid molecular species increase in frog retinal pigment epithelial cells following photoreceptor shedding. ARVO 1991.

Cotlier E., Basking M., Kresca L.: Effects of lysophosphatidyl choline and phospholipase a on the lens. Invest Ophthalmol Vis Sci 14, 697, 1975.

Donita Garland, Zigler Samuel J. Jr., and Kinoshita Jin: Structural changes in bovine lens crystallins induced by ascorbate, metal and oxygen. Archives of Biochemistry and Biophysics Vol. 251, No. 2, December, pp. 771-776, 1986.

Donita Garland: Role of site-specific, metal-catalyzed oxidation in lens aging and cataract: A Hypothesis. Exp. Eye Res. (1990) 50, 677-682.

Duke Elder S.: System of ophthalmology. XI. 210, 1969.

Eshaghian J. e coll.: Ultrastructure of human cataract in retinitis pigmentosa. Arch. ophthalmol. vis. sci. 24 suppl.: 75, 1983.

Eshaghian J.: Human posterior subcapsular cataract. Trans. Opthalmol. Soc. UK 102: 364, 1982.

Fishman G.A., Anderson R.J. and Lourenco P.: Prevalence of posterior subcapsular lens opacities in patients with retinitis pigmentosa. Br. J. ophthalmol 69: 263-266, 1985.

Goosey J. D. e coll.: A lipid peroxidative mechanism for posterior subcapsular cataract formation in the rabbit: a possible model for cataract formation in tapetoretinal diseases. Inv. oph. 25: 608-611, 1984.

Heckenlively J.: The frequency of posterior subcapsular cataract in the hereditary retinal degenerations. Am. J. of ophtal. 93: 733-738, 1982.

Heckenlively J. R.: Retinitis Pigmentosa. JB Lippincott, Philadelphia, 1988.

Hockwin O. et Koch H.R.: Combination effects of noxious on the crystalline lens. In: Bellows J.G. edit. Cataract and abnormalities of the lens. Grune and Stratton, edit. New York, 243-254, 1975.

Hockwin O. et Ohrloff C. H.: Ageing of lens metabolism. Ophthalmic Res 11m 389-396, 1979.

Hockwin O. et Dragomirescu V.: Verlaufsbeobachtungen von Lisentrubungen mit der Scheimpflug - Photographie und densitometrischer Bildanalyse. In: Hockwin O. edit. Altern der Linse. Symposium Strasbourg. Mayr. Miesbach, 125-138, 1982.

Hoffman D.R. e coll.: w3 fatty acids in autosomal dominant retinitis pigmentosa. ARVO 1991.

Marjorie F. Lout, Dickerson Jaime E. Jr. and Rekha Garadi: The role of protein - Thiol mixed disulfides in Cataractogenesis. Exp. Eye Res. (1990) 50, 819-826.

Miglior M.: Le cataratte correlabili a malattie oculari (cataratte complicate). Presented at 2nd IACRR International Congress, Cefalu' (Palermo, Italy) October 17-19, 1986.

Newsome D.A., Michels R.G.: Detection of lymphocytes in the vitreous gel of patients with retinitis pigmentosa. Am. J. Ophthalmol. 105: 596, 1988.

Niesel P.: Spaltlampenphotographaie der Linse fur Messzwecke. Ophthalmologica 152, 387-395, 1966.

Nielsel P.: Scheimpflug - Photographic des vorderen Augenbschnittes als Methode zar Beurteilung (und Messung) der Linsentransparenz. - In: Hockwin O. edit Altern der Linse. Symposium Strasbourg, Mar Miesbach, 121-124, 1982.

Nordmann J.: Biologie du cristalin. Masson Et. 1954, pag. 355.

Pagon P.A. Retinitis Pigmentosa (Major Review). Surv. Ophthalmol. 33 (3): 137, 1988.

Pannarale M. R., Vingolo E. M., Pannarale L., Feher J.M., Arrico L., Perdicchi A. and Ricci A.: La cataratta complicata da miopia. Presentano allo IACCR 1989.

Pannarale e coll.: Cataract complicated by retinitis pigmentosa. 6_ congresso IRPA 1990.

Ramana Koppa, Nagaray H., Dasid R. Sell, Nalladi Prabhakaram, Beryl J. Artwerth and Vincent M. Monnier: High correlation between pentosidine protein crosslinks and pigmentation implicates ascorbate oxidation in human lens senescence and cataractogenesis. Proc. Natl. Acad. Sci. USA Vol. 88, pp.10257-10261, November 1991. Medical Sciences.

Simonelli F. e coll.: Lipid peroxidation and human cataractogenesis in diabetes and severe myopia. Exp. Eye. Res. 49: 181-187, 1989.

Van Boeckel M.A. and Hoenders H.J.: Glycation induced crosslinking of calf lens crystallins. Exp. Eye Res., 1991, 53, 89-94.

Van Boeckel M. A. and Hoenders H. J.: Glycation of crystallins in lenses from aging and diabetic individuals. Vol. 314, number 1, 1-4 (1992) Federation of European Biochemical Societies.

Yamaguchi K. e coll.: Corneal Endothelial Abnormalities in the Royal College of Surgeons Rat. Cornea 9 (3): 217-222, 1990.

Zigler S. e coll.: Lipid peroxidation products as possible initiators of cataract.. Invest. Ophthalmol. vis. Sci. 24 suppl.: 75, 1983.

Zigler S. e coll.: Effects of lipid peroxidation products on the rat lens in organ culture: a possible mechanism of cataract initiation in degenerative disease. Arch. biochem. biophys. 225 (1). 149-156, 1983.

INDEX

adCSNB (autosomal dominant congenital stationary night blindness)
  ERG findings in two patients with, 377–384
  mutations in gene encoding βPDE, 313–322
ADRP (autosomal dominant retinitis pigmentosa)
  see retinitis pigmentosa
Albino rat
  Fischer 87–94
  Lewis, 106
  Wistar, 88–94
Apoptosis
  in retinitis pigmentosa, 1–8
  in transgenic mice, 39–49
  inhibitory effects of cycloheximide and flunarizine on light-induced apoptosis of photoreceptor cells, 27–38
  light-induced apoptosis in the rat retina in vivo, 19–25
Argon laser photocoagulation, retinal cell responses to, 209–215
  blood-retinal-barrier breakdown, 212–213, 216
  phagocytic cells, 213–218
ARRP (qutosomal recessive retinitis pigmentosa)
  see retinitis pigmentosa

BALB/c, 168–171
Bardet-Biedl syndrome, 298
Batten disease, 79
Best disease, 327
Bruch's membrane, 87–94, 279

C3H/Hen, 163–175
Calcium, 119–129, 227–233
Campbell, 179–184
Chick, dolichol pathway in the RPE, 149–153
Chinchilla, 187
Choriocapillaris, in RP, 279
Cones
  abnormal cone receptor activity in patients with hereditary degeneration, 349–358
  in RP, 275–277
  L/M cone opsin, 195–201

Cones (*cont.*)
  macaca monkey 195–201
Cycloheximide, inhibitory effects of, 27–38

DHA (docosahexaenoic acid)
  abnormalities in RBC of patients with RP, 385–393
  x-linked RP, 385–391
DMTU (dimethylthiourea), 9
Drosophila
  abnormal Ca^{2+} mobilization and excessive photopigment phosphorylation lead to photoreceptor degeneration in, 227–234
  model for photoreceptor dystrophies and cell death, 217–226
  ninaA, 270
  ninaE, 218–219, 235–241
  norpA, 264–269
  photoreceptor proteins, 263–274
  rdg gene, 218
  role of retinal degeneration B protein in, 243–254
  role of dominant rhodopsin mutation in, 235–241
  visual transduction, a model system for human eye disease, 255–261

EAU (experimental autoimmune uveoretinitis), 105–110
EGF (epidermal growth factor), 69–76
ERG (electroretinogram)
  adCSNB, 377–384
  in drosophilia, 246–247
  cone ERG delayed, 356–357

Fischer, 87–94
Flunarizine, inhibitory effects of, 27–38

GAR1 gene, 331–338
GCAP (guanylate cyclase-activating protein), 339–347

Genetics
  ADRP, clinical features of, 303–312
  ARRP
    Bardet-Biedl syndrome, 298
    chromosome 14, 296
    in Sardinian population, 295
    mutation of the βPDE gene, 315–318
  ASN244Lys, 288
  ASN244His, 288
  GCAP, 339–347
  HPV16E7, 39–49
  molecular analysis of the human GAR1 gene, 331–338
  P53, 39–49
  peripherin/RDS, 285–292
Green light experiment, 220–223
Growth factors
  effects on nitric oxide, 62–63, 66
  EGF, 69–76

HPLC (high pH anion exchange chromatography), 150–151

IgG, 214
IRBP(interphotoreceptor retinoid binding protein), 159–160, 278–279, 395–402
Iron involvement in retinal degeneration, 177–186

Lewis, 106
Long Evans, 165

Macular Degeneration
  Arg172Trp, 290
  Arg172Gln, 290
  genetic studies of, 323–330
    dominant cystoid macular dystrophy, 327
    north carolina macular dystrophy, 327
    sorsby's fundus dystrophy, 325, 327
    TIMP3 mutation, 324–325
MDCK (Madin-Darby canine kidney), 151
MMPs (matrix metalloproteinases), 95–96
mnd (motor neuron degeneration), 79
Monkey, development of opsin and synapses of photoreceptors in, 195–202
Mouse
  C3H/Hen, 163–175
  mnd, 79
  nr, 78
  pcd, 78
  rds, 40
  rd1, 40, 78, 168
  Rd2, 78
  rd3, 80
  Rd4, 80–81
  rd5, 81
  transgenic
    IRBP-SV40 TAq, 41–42
    IRBP-E7, 42–48
  vitiligo, 155–162

Müller cells
  argon laser photocoagulation, 210, 214–215
  in RP, 280–281

Naphthalene, induced retinal degeneration in rabbit, 203–208
New Zealand, 187
NHI (nonhaem iron), 177–186
ninaA, preliminary identification and characterization, 270
ninaE
  identification and characterization of, 237–238
  retinal dystrophy models, 218–219
  rhabdomeres, 236–237
NorpA
  identification and characterization of, 265–266
  phototransduction, 264
NCMD (North Carolina macular dystrophy), 237
nr, 78

Opsin
  effects of visible light, 13–16
  L/M cone opsin, 195–201

pcd, 78
PEDF (pigment epithelium-derived factor), function, age-related expression and molecular characterization of, 51–60
Peripherin/RDS, autosomal dominant retinal degeneration, 285–292
Photoreceptor
  photoreceptors degeneration in drosophila, 227–234
  abnormal rod photoreceptor function in RP, 359–370
  drosophila as a model for, 217–226
  effect of sugars on, 129–137
  guanylate cyclase-activating protein in vertebrate photoreceptors, 339–347
  in vitiligo mouse, morphological and electrophysiological characterization of, 156–157
  morphology, 188
  normal death of, 192
Phototransduction, in vertebrate and invertebrate, 263–265

Rabbit
  chinchilla, 187
  naphthalene-induced retinal degeneration in, 203–208
  New Zealand, 187
Rat
  Campbell, 179–184
  Fischer, 87–94
  Lewis, 106
  Long Evans, 165
  RCS, 40, 119–126, 129, 171–173, 187, 191, 209–215, 257

Rat (*cont.*)
RCS-rdy$^+$, 112
rdy, 40, 187
Wistar, 88–94, 178–183
RCS, 40, 119–126, 129, 171–173, 187, 191, 209–215, 257
RCS-rdy$^+$, 112
rd1, 40, 78, 168
Rd2, 78
rd3, 80
Rd4, 80–81
rd5, 81
rds, 40
rdy, 40, 187
Retina
nitric oxide in, 61–68
response to argon laser photocoagulation, 209–216
Retinal degeneration
apoptosis of photoreceptors and lens fiber cells with cataract, 39–49
drosophila
model for photoreceptor dystrophies and cell death, 217–226
visual transduction, a model system for human eye disease, 255–261
GCAP candidate gene for human disease, 345
fractionation of interphotoreceptor MMPs, 95–103
hereditary degeneration of the retina, 177–186
inhibition of calcium mobilization, 229–231
in mouse, 77–85
in vitiligo mouse, 155–162
light-induced apoptosis in the rat retina in vivo, 19–25
molecular analysis of the human GAR1 gene, 331–338
mutations in the gene for the β-subunit of rod photoreceptor cGMP-specific PDEβ in patients with, 313–322
newly formed capillaries in experimental naphthalene-induced retinal degeneration in rabbit, 203–208
nitric oxide, 61–68
oxidative damage and responses in retinal nuclei arising from intense light exposure, 9–17
effects of visible light on the opsin gene, 13
retinal damage as assessed by rhodopsin loss, 12
role of dominant rhodopsin mutations in, 235–241
vitamin E deficiency, 111–118
Retinal development, receptor degeneration is a normal part of, 187–193
DNA fragmentation, 188–190
morphological differentiation of photoreceptors, 188
Retinal transplantation, immunological aspects of, 163–175
Retinitis Pigmentosa
abnormal cone receptor activity in patients with hereditary degeneration, 349–358
abnormal rod photoreceptor function in, 359–370
ADRP
clinical features of, 303–312
DHA, 385–393
peripherin/RDS gene, 285–292
apoptosis, 1–8
argon laser photocoagulation, 209
ARRP, genetic studies in, 293–302
cataractogenesis in, 403–407
DHA abnormalities in RBC of patients with, 385–393
retinal pathology in, 275–284
X-linked recessive RP, 371–376, 385–391
Retinoid metabolism, 157–159
Rhodopsin
clinical features of ADRP associated with Ser186Trp mutation of, 303–312
regulatory influences on the glycosylation of rhodopsin by human and bovine retinas, 139–146
role of dominant rhodopsin mutations in drosophila retinal degeneration, 235–241
Thr17met mutation, 40
Rods
abnormal rod photoreceptor function in RP
in RP, 275–277
macaca monkey, 195
RPE (retinal pigment epithelium)
characterization of PEDF by human RPE cells, 51–60
dolichol pathway in the RPE of the embryonic chick, 149–153
from RCS rats, 119–129
functional studies of, 156–157
grafts, 172–174
in RP retina, 279
in vitiligo mouse, morphological and electrophysiological characterization of, 155–156
in vitro expression of epidermal growth factor rector by human RPE cells, 69–76

SFD (Sorby's fundus dystrophy), 325–327
SMD (senile macular degeneration), 87–94
Stargardt's macular dystrophy, 327
Sugar
effect on photoreceptor outer segment assembly, 129–137
regulatory influences on the glycosylation of rhodopsin by human and bovine retinas, 139–146

Transgenic mouse, expression of S opsin, 201
Tunel, 29, 32–35, 41, 188–192,

Usher's syndrome, 385

Virus
  cytomegalovirus, 64
  human papilloma virus 16 (HPV16) E7, 39–49
Visual cycle, 157–158
Vitamin A, 159, 161
Vitamin E, deficiency, 111–118
Vitiligo mouse, 155–162

Wistar, 88–94, 178–183

X-linked RP, 371–376, 385–391
Xenopus laevis, 129–131, 134–136

Zebrafish, 200